MATERIALS FUNDAMENTALS OF GATE DIELECTRICS

MATERIALS FUNDAMENTALS OF GATE DIELECTRICS

Edited by

ALEXANDER A. DEMKOV
Freescale Semiconductor. Inc., Austin, U.S.A.

and

ALEXANDRA NAVROTSKY
University of California, Davis, CA, U.S.A.

Springer

A C.I.P. Catalogue record for this book is available from the Library of Congress.

ISBN-13 978-1-4020-3077-2 (HB) Springer Dordrecht, Berlin, Heidelberg, New York
ISBN-13 978-1-4020-3078-9 (ebook) Springer Dordrecht, Berlin, Heidelberg, New York
ISBN-10 1-4020-3077-0 (HB) Springer Dordrecht, Berlin, Heidelberg, New York
ISBN-10 1-4020-3078-9 (e-book) Springer Dordrecht, Berlin, Heidelberg, New York

---

Published by Springer
P.O. Box 17, 3300 AA Dordrecht, The Netherlands.

Printed on acid-free paper

Printed in the Netherlands.

TABLE OF CONTENTS

PREFACE

According to Bernie Meyerson, IBM's chief technology officer, the traditional scaling of semiconductor manufacturing processes died somewhere between the 130- and 90-nanometer nodes. One of the prime reasons is the low dielectric constant of SiO_2 — the choice dielectric of all modern electronics. This book presents materials fundamentals of the novel gate dielectrics that are being introduced into semiconductor manufacturing to ensure the Moore's law scaling of CMOS devices. This is a very rapidly evolving field of research and we try to focus on the basic understanding of structure, thermodynamics, and electronic properties of these materials that determine their performance in the device applications.

The volume was conceived in 2001 after a Symposium on Alternative Gate Dielectrics we had at the American Physical Society March Meeting in Seattle, upon the suggestion of the Kluwer editor Sabine Freisem. After several discussions we decided that such a book indeed would be useful as long as we could focus on the fundamental side of the problem and keep the level of the discussion accessible to graduate students and a variety of professionals from different fields. The problem of finding a replacement for SiO_2 as a gate dielectric brings together in a unique way many fundamental disciplines. At the same time this problem is truly applied and practical. It looked unlikely that the perfect new material would be found fast; rather there would be a series of evolving candidate materials and approaches. Thus we felt we could alert the next generation of scientist to an exciting problem they would have a chance to participate in solving. The book would be of interest to those actively engaged in gate dielectric research, and microelectronics in general, and to graduate students in Materials Science, Physics, Chemistry, and Electrical Engineering.

Most new gate dielectrics are transition metal or rare earth oxides. Ironically, the very d- or f-orbitals that produce the high dielectric constant also result in severe integration difficulties, thus intrinsically limiting these materials. New in the electronics industry, many of these oxides are well known in the fields of ceramics and geochemistry that offer powerful concepts and characterization techniques less familiar to the semiconductor audience. While focusing on materials fundamentals, the book tries to always keep device processing requirements in mind. The complexity of the structure-property relations in these oxides makes the use of the state of the art first-principles calculations essential. Several chapters give a detailed description of the modern theory of polarization, and heterojunction band discontinuity within the framework of the density functional theory. Experimental methods include solution and differential scanning calorimetry, Raman scattering and other optical techniques, transmission electron microscopy, and X-ray photoelectron spectroscopy. Many of the problems encountered in the world of Si CMOS are also relevant for other semiconductors such as GaAs. We conclude with a comprehensive review of recent developments in such possibilities.

SiO_2 has been the mainstay and companion of the semiconductor industry for almost 60 years. The journey of transition metal and rare earth oxides in the land of transistors is just beginning. We hope you will enjoy the unfolding story.

Alex Demkov and Alex Navrotsky
Austin and Davis
July 2004

Chapter 1

MATERIALS AND PHYSICAL PROPERTIES OF HIGH-K OXIDE FILMS

RAN LIU

Advanced Products Research & Development Laboratory, Freescale Semiconductor, Tempe, AZ 85284, USA

Rapid shrinking in device dimensions to follow Moore's law calls for replacement of SiO_2 by new gate insulators in future generations of MOSFETs. Among many desirable properties, potential candidates must have a higher dielectric constant, low leakage current, and thermal stability against reaction or diffusion to ensure sharp interfaces with both the substrate Si and the gate metal (or poly-Si). Extensive characterization of such materials in thin-film form is crucial not only for selection of the alternative gate dielectrics and processes, but also for development of appropriate metrology of the high-k films on Si. This chapter will review recent results on materials and physical properties of thin film $SrTiO_3$ and transition metal oxides (HfO_2).

1. INTRODUCTION

The continued shrinking of the CMOS device size for higher speed and lower power consumption drives the conventional SiO_2 gate oxide approaching its thickness scaling limit (1). Severe direct tunneling and reliability problems at extremely small thickness will soon set a barrier for this naturally given material. Alternative dielectric materials with a higher dielectric constant, k, and thus larger physical thickness than SiO_2 will be required to reduce the gate leakage as the gate length is scaled below 100 nm. Successful integration of high-k dielectrics into CMOS technology poses enormous challenges. Among many desirable properties, potential candidates must have a high dielectric constant, low leakage current, and good thermal stability against intermixing or diffusion to ensure sharp interfaces with both the substrate Si and the gate metal (or poly-Si). Extensive characterization of such materials in thin-film form is crucial not only for selection of the alternative gate dielectrics and processes, but also for development of appropriate metrology of the high permittivity (high-k) films on Si.

A.A. Demkov and A. Navrotsky (eds.), Materials Fundamentals of Gate Dielectrics, 1–36.

For insulating materials, there are two major contributions to the static dielectric function

$$\varepsilon_0 = 1 + 4\pi(\chi_{\text{electron}} + \chi_{\text{lattice}}), \quad (1)$$

i.e., the dielectric responses of valence electrons and lattice vibrations. The electronic dielectric constant can be estimated by

$$\chi_{\text{electron}} \sim \left(\frac{\hbar\omega_{\text{P}}}{E_{\text{PG}}}\right)^2, \quad (2)$$

where ω_{P} is the plasma frequency of the valence electrons and E_{PG} is an "average bandgap" (know as Penn gap). Since the electronic contribution is usually less than 16 and larger for insulators with smaller energy gaps, it is not wise to pursue materials with high electronic dielectric constant as high-k gate dielectrics. Therefore, the high dielectric constant should be generated from the ionic contribution

$$\chi_{\text{lattice}} \sim \frac{1}{V}\sum_i \frac{(e_i^* \cdot \xi_i)(e_i^* \cdot \xi_i)^+}{\omega_i^2}, \quad (3)$$

where e^* is the effective dynamical charge, ξ_ι the eigenvector, and ω_i the frequency of the ith phonon mode. This indicates that larger ionic polarizability leads to higher dielectric constant. Since the lattice polarization splits the longitudinal optical (LO) and transverse optical (TO) phonon degeneracy in the long wavelength limit, the total static dielectric constant can be correlated to the high-frequency electronic dielectric constant through the Lyddane–Sachs–Teller relation

$$\frac{\varepsilon_0}{\varepsilon_\infty} = \prod_i \left(\frac{\omega_{\text{LO}}^i}{\omega_{\text{TO}}^i}\right)^2. \quad (4)$$

In many high-k materials such as TiO_2 and $SrTiO_3$, some of the ratios of the frequencies of the LO and TO phonon pairs are about 2 or larger, and thus result in high dielectric constant. Since ε_0 diverges when one of the TO mode frequencies goes to zero in Eq. (4), extremely high-k can be achieved through soft phonon driven lattice instability near the paraelectric to ferroelectric phase transition. The dielectric constant in this case follows the Curie–Weiss kind of temperature dependence

$$\varepsilon_0 \propto \frac{1}{T - T_{\text{C}}},$$

where the Curie temperature T_{C} is 393 K for $BaTiO_3$ and 0 K for $SrTiO_3$.

In addition to the high dielectric constant, the other basic requirement to the physical properties is a large energy gap that gives rise to reasonable conduction and valence band offsets to ensure low leakage current. However, certain compromise needs to be effected between high-k and large bandgap since the bandgap tends to decrease with increasing dielectric constant (see Fig. 1). It can be seen again that SiO_2 and Al_2O_3 have the largest bandgaps and band offsets, but smaller dielectric constants. On the other hand, the perovskite oxides usually have very high dielectric constants,

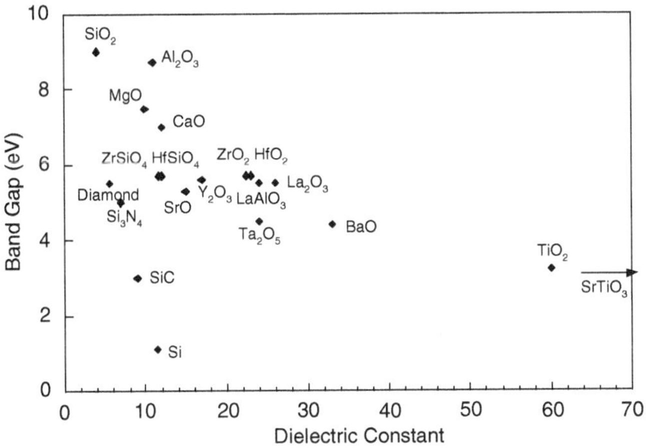

Fig. 1. Band gap vs. dielectric constant for potential candidates as gate dielectrics (from (16)).

but smaller energy gaps and band offsets, in particular, very small conduction band offsets. TiO_2 and Ta_2O_5 have similar problem. Therefore, the "medium-k" oxides with reasonably wide gaps and band offsets (2) are currently focused upon as possible replacement materials of SiO_2 as gate dielectrics.

Although there is a list of candidates that meet the near term high-k requirements in terms of the dielectric constant and band offsets, to integrate them successfully into the current CMOS process flows still posts tremendous challenges. Key issues such as thermal stability, interface chemistry and diffusion resistance need to be resolved to ensure low leakage current (<1 mA/cm^2 at ± 1 V) and electrical interface states ($<1 \times 10^{11}$ cm^{-2} eV), high electron and hole mobility in the channel as well as good reliability.

Extensive efforts have been made to develop high-k gate oxides, including transitional metal oxides and silicates (Ta_2O_5 (3), TiO_2 (4), ZrO_2 (5), HfO_2 (6), $HfSi_xO_y$ (7)), rare earth metal oxides (Gd_2O_3 (8), Pr_2O_3 (9)), and other oxides (Al_2O_3 (10)). Perovskite-type oxides such as strontium titanate, $SrTiO_3$ (STO), with much higher dielectric constant have also attracted tremendous interest as alternative gate dielectrics (11, 12). However, currently there is still no clear front runner for the next alternative gate dielectrics. This chapter will present some of the characterization results on STO films grown with molecular beam epitaxy (MBE), and HfO_2 films grown with chemical vapor deposition (CVD) and atomic layer deposition (ALD) on Si.

The structure and composition of the films were mostly characterized using TEM (transmission electron microscopy), SEM (scanning electron microscopy), SIMS (secondary ion mass spectrometry), XPS (X-ray photoelectron spectroscopy), AFM (atomic force microscopy), XRD (X-ray diffraction), RBS (Rutherford backscattering spectrometry), and AES (Auger electron spectroscopy). Since the lattice vibrations and band structures are extremely sensitive to the materials properties, the optical spectroscopy offer powerful characterization methods of structure, composition, impurity and defectivity that are crucial to the process development. It will be demonstrated that utilizing the ultra violet laser excitation with polarization configurations forbidden for the Si LO phonon can significantly suppress the Si substrate Raman signal and thus enable characterizations of ultrathin oxide films on Si. FTIR (Fourier transform infrared spectroscopy) transmission and reflection spectroscopy also provide rich information about the lattice vibration behavior. For very thin oxide layers on Si, ATR (attenuated total reflection) technique is used to enhance the surface sensitivity of the FTIR spectroscopy. FTIR ellipsometry has the advantage of measuring both parts of the complex dielectric function and offers a new method to study vibrational properties. The band structures of the oxides have been studied using spectroscopic ellipsometry. The large spectral range from near IR (0.7 eV) to deep UV (6.5 eV) and even to VUV (9 eV) make it possible to measure the band edge of most oxide materials including SiO_2. The high-k oxide materials to be discussed in this chapter are transitional metal oxides (HfO_2, ZrO_2, TiO_2) and their silicates ($Hf_{1-x}Si_xO_y$) or aluminates ($Hf_{1-x}Al_xO_y$), and perovskite oxides ($SrTiO_3$, $LaAlO_3$).

2. PEROVSKITE OXIDES: $SrTiO_3$

Perovskite forms a family of compounds of formula ABO_3 that all have the same structure. This structure is strictly cubic (see Fig. 2) only for the non-ferroelectric phase above the Curie temperature. Three of the four three-fold degenerate optical phonon modes, which contribute the most to the lattice dielectric response, are shown in Fig. 2(b)–(d). In particular, the lowest mode goes soft with decreasing temperature and thus makes the dominant contribution to the dielectric constant and also triggers the phase transition. $SrTiO_3$ has a simple cubic crystal structure and a bulk k value of $\sim$300 at room temperature. Although the lattice constant of $SrTiO_3$ ($a = 3.905$ Å) is very different from that of Si ($a = 5.431$ Å), the lattice mismatch is fairly small ($\sim$1.7%) with the STO unit cell rotated 45° around Si surface normal [001] axis to match the STO cell diagonal with the Si lattice constant.

The STO films studied in this work were deposited on up to 8 inch Si wafers by molecular beam epitaxy (MBE) with a base pressure of 5×10^{-10} mbar. Metallic Sr was used to react with the native oxide on the Si wafer at temperatures greater than 750°C and under high vacuum. Using this process, a 2×1 surface reconstruction was consistently obtained (12). The details for the film growth are described elsewhere (12). The surface structure, growth rate and stoichiometry were monitored using in-situ Reflection High-Energy Electron Diffraction (RHEED).

Figure 3(a) displays the X-ray diffraction 2θ scan from a MBE film grown on Si. Only the diffraction peaks for the cubic lattices of STO and Si are seen, indicating that the STO film is oriented with the (001) axis parallel to the (001) axis of the Si substrate. Furthermore, the electron diffraction pattern from the film (Fig. 3(b)) shows that the diffraction pattern from the STO film (larger spots) is 45° rotated around the (001) axis with respect to that of the Si substrate (smaller spots) to minimize the lattice mismatch between the two.

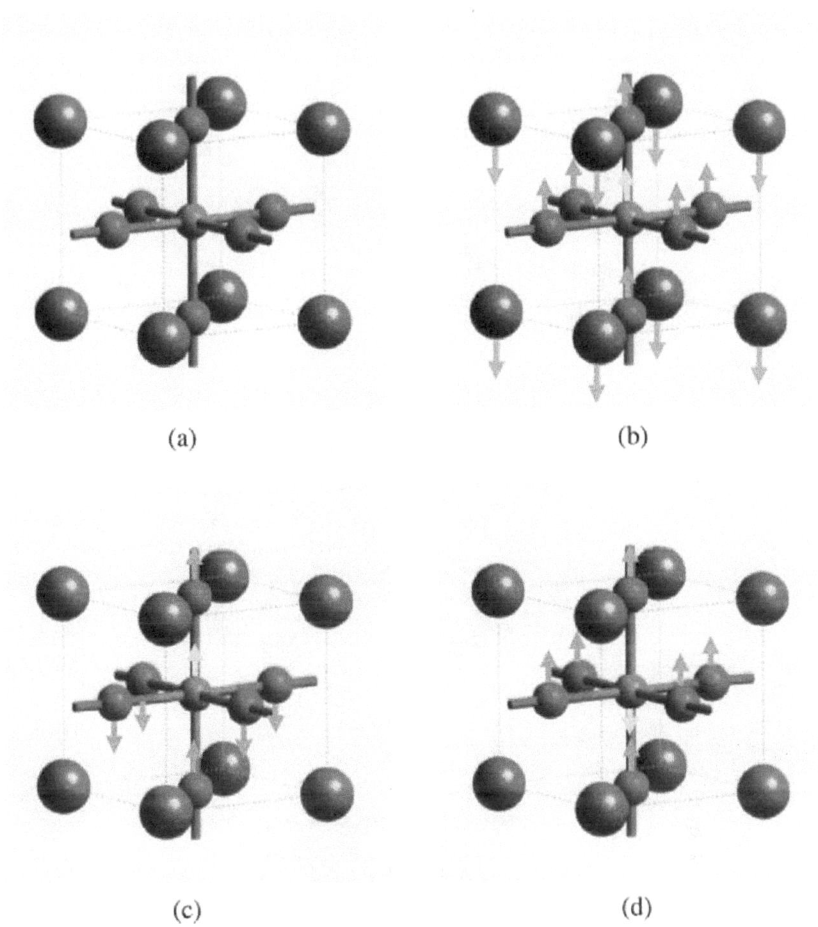

Fig. 2. (a) Unit cell, (b) the lowest, (c) the second lowest, and (d) the highest F_{1u} modes of $SrTiO_3$. Ti atoms are at the body center, O atoms at face centers, and Sr atoms at corners (from (16)).

Since both the calculation (see the Chapter by John Robertson) and XPS measurement (13, 14) showed about 0 eV conduction band offset between STO and Si, it is expected that the n-channel devices using a single STO dielectric layer may be very leaky. Therefore, one should engineer the band off set to minimize the leakage current. The first strategy is to add an interfacial layer with wider band gap, such as SrO, BaO (11) or SiO_x. Actually, although STO can be grown directly on Si (15), a thin layer of interfacial SiO_x is formed in most cases between STO and Si. This oxide layer can naturally serve as the buffer layer with much larger band offset. Figure 4 shows the high-resolution TEM (HRTEM) image of the interfacial region of a MBE STO film on Si. An amorphous interfacial layer of about 7 Å thick can be seen sandwiched between the single crystalline STO film and the Si substrate. High-resolution

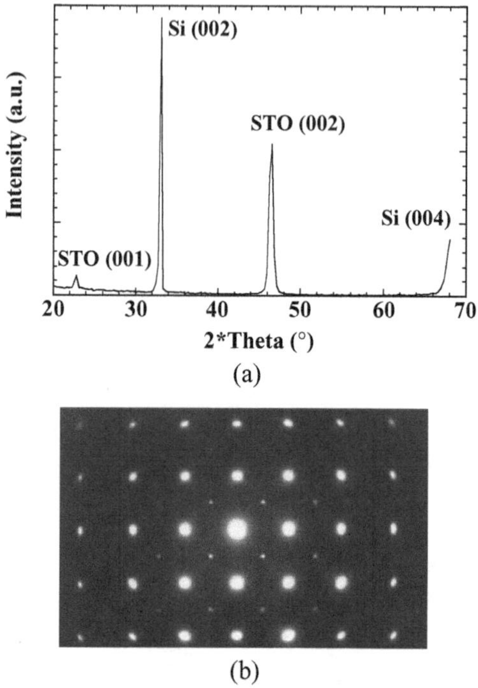

Fig. 3. X-ray diffraction 2θ scan and electron diffraction pattern from a MBE STO film on Si (from (16)).

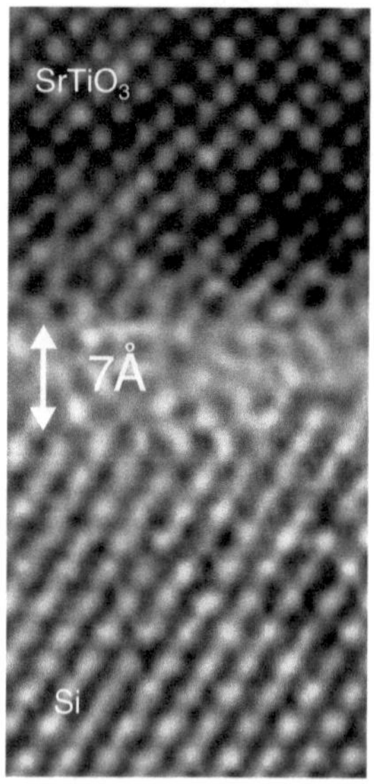

Fig. 4. High-resolution TEM image of the interfacial region between a MBE STO film and Si (from (16)).

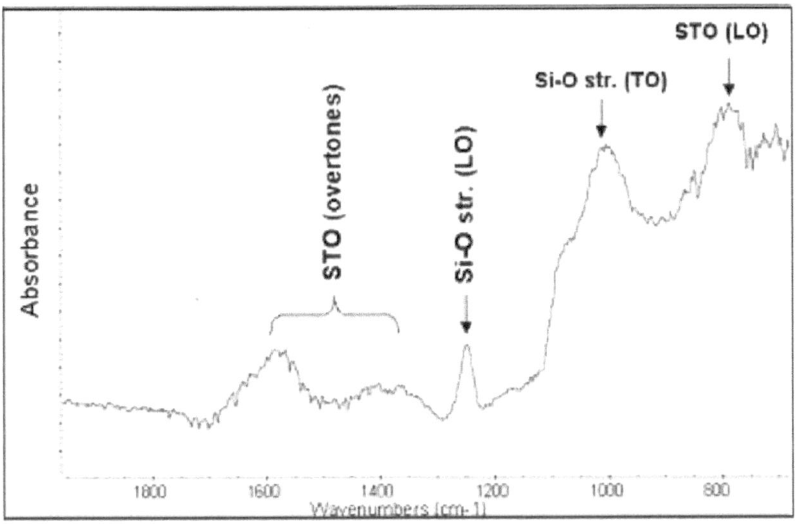

Fig. 5. Infrared absorbance spectrum from an 8 nm thick MBE STO film on Si obtained using the ATR technique. Vibration bands from both STO as well as the interfacial SiO_x were seen.

EELS mappings of similar samples using TEM indicated that the amorphous layer is silicon oxide (12). The existence of the interfacial layer has also been revealed by infrared transmission measurements. Figure 5 displays the absorbance spectrum from an 8 nm thick MBE film on Si using ATR technique. Both the phonon absorption peaks from STO as well as from interfacial SiO_2 were observed. To further reduce the leakage current, efforts have also been directed to enlarge the STO band gap by changing the composition and to neutralize or trap oxygen vacancies in STO by Al-doping. The Sr to Ti ratio has been found rather sensitive to the growth parameters and Sr-rich films can be easily formed. TEM images show that the stoichiometric films are usually very uniform both in thickness and in cross section (Fig. 6(a)), while the Sr-rich films (Sr/Ti = 1.13 measured by RBS) exhibit some non-uniform TEM contrast

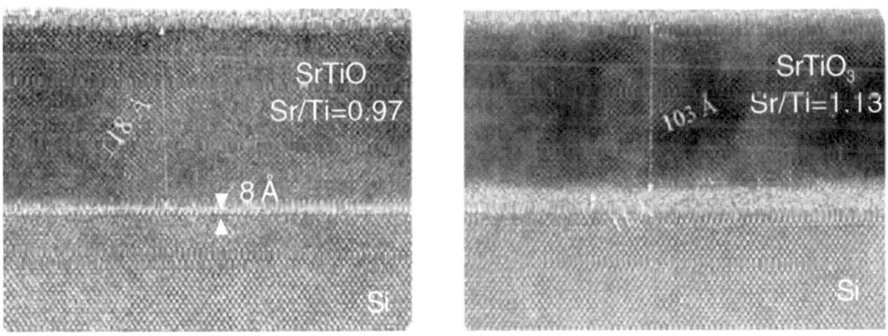

(a) (b)

Fig. 6. High-resolution TEM images of a stoichiometric STO film (a) and an Sr-rich STO film (b) on Si (from (16)).

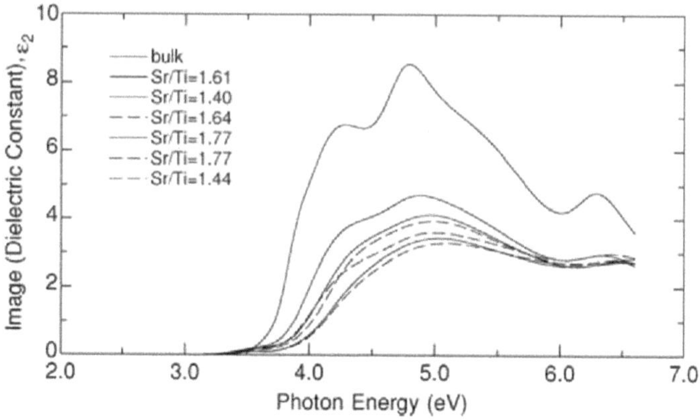

Fig. 7. Imaginary part of the dielectric function vs. photon energy derived from spectroscopic ellipsometry measurements on STO films with different Sr/Ti ratios in comparison with the bulk single crystal spectrum (from (16)).

(Fig. 6(b)). This non-uniformity can be attributed to non-uniform composition distribution or local strains. The diffraction patterns obtained from the lattice fringes of different areas of the Sr-rich film revealed noticeable lattice constant changes across the film. At some locations, tetragonal lattice distortions were also observed. It is not clear at this point whether the local strains are directly related to possible microscopic compositional changes. The energy gaps of the Sr-rich films have been investigated using spectroscopic ellipsometry. Figure 7 shows the imaginary part, ε_2, of the complex dielectric functions of Sr-rich STO films of $\sim$10 nm thickness derived from the spectroscopic ellipsometry data using a general parametric oscillator model. Since a single layer was used to simulate the film as well as the interfacial oxide layer, ε_2 for the thin films is considerably reduced by the interfacial SiO_x that has no absorption in this energy range because of its much larger band gap. Further decrease in the absorption with increasing Sr/Ti results from the decrease in the density of states of the STO conduction band, which is mainly made up of Ti 3d states. It is very interesting to notice from Fig. 7 that the STO band gap blue shifts by as much as $\sim$0.5 eV with increasing Sr/Ti. This result suggests that the Sr-rich films might give rise to lower leakage current. Incorporation of Al into STO was also found very effective in reducing the leakage current (16).

Defectivity is another important factor that potentially affects the device performance. Raman spectroscopy has been found to be rather sensitive to defects in STO, and can be used to characterize ultrathin STO films on Si when UV laser is used (17). As pointed out at the beginning of this section, STO has four three-fold degenerate optical modes: three infrared-active modes (F_{1u}) and one silent mode. There is no Raman-active mode in the ideal perovskite lattice. Defects can, however, break the local symmetry and make the Raman-forbidden modes active. Figure 8 shows the Raman spectrum from a 20 nm thick MBE STO film on Si substrate in comparison with that of a STO bulk single crystal. The single crystal spectrum exhibits only the

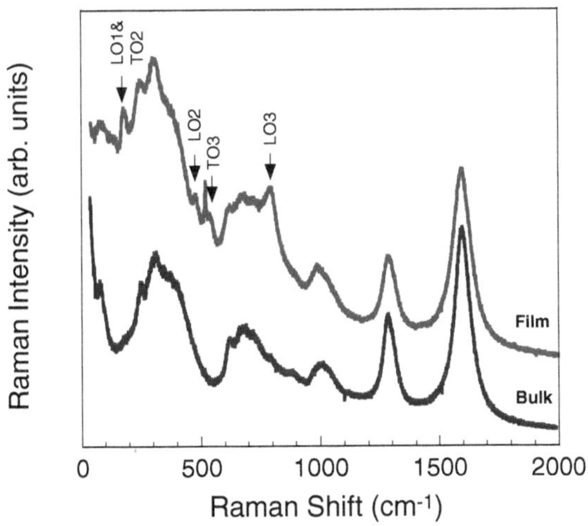

Fig. 8. UV–Raman spectra of a 20 nm thick MBE STO film on Si and of a bulk STO single crystal. The defect-induced first-order Raman features appear in the spectrum of the film (from (17)).

over-tone (multi-phonon) scattering bands that are not forbidden in Raman scattering by the symmetry, while the spectrum from the film also shows Raman features from the first-order phonon modes associated with LO_1 or TO_2 (180 cm^{-1}), LO_2 (479 cm^{-1}), TO_3 (540 cm^{-1}) and LO_3 (795 cm^{-1}). This indicates the presence of considerable amount of defects inside the film. Many kinds of defects can induce the Raman-activity. Figure 9 shows a HRTEM image of a threading dislocation with a Bergers

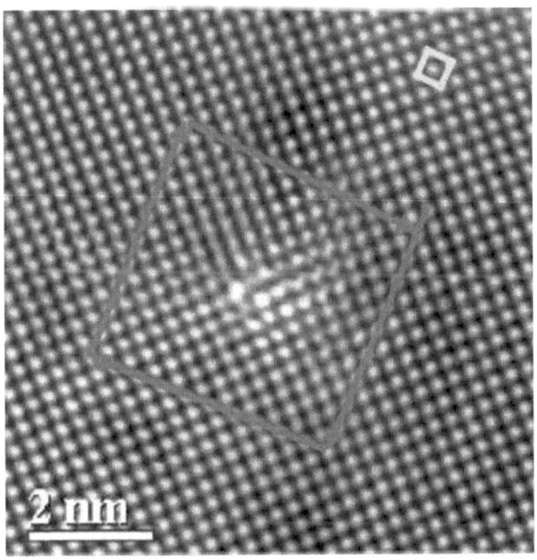

Fig. 9. Plan-view HRTEM image of a dislocation with a Burgers vector $\boldsymbol{b} = [010]$. Unit cell and axis choice are shown in upper right.

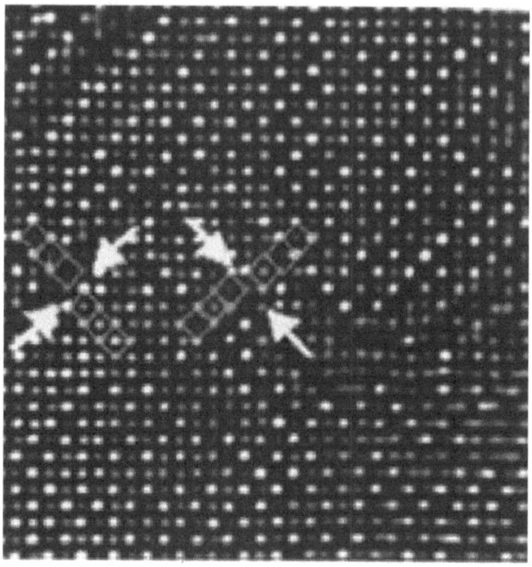

Fig. 10. HRTEM image of an Sr-rich film (The image has been Fourier-filtered to bring out the weak difference between the intensity maxima at the cell corners and cell center). The squares represent the STO unit cell. The crystal in the lower left of the image is shifted with respect to the crystal in the upper part of the image by a shift vector of [½ ½ 0] (from (17)).

vector of [010] acquired in plan view geometry. Another kind of defects is shown in the plan-view HRTEM of a lightly Sr-rich film (Fig. 10), where a shift between crystals by a vector of approximately [½ ½ 0], which is consistent with the insertion of SrO layers. The rock salt SrO layers periodically form so called Ruddlesden–Popper (RP) phase in the perovskite structure during growth to accommodate the excess Sr, which allows the epitaxial film to maintain unstrained growth.

3. TRANSITION METAL OXIDES: HfO_2

HfO_2 and its silicate or aluminate are currently the leading candidates as alternative gate dielectrics in the near term. HfO_2 is found to have high permittivity ($k > 20$) as well as excellent chemical and thermal stability in contact with silicon (18). Many deposition techniques for gate dielectrics have been explored such as physical vapor deposition (PVD), chemical vapor deposition (CVD), and atomic layer deposition (ALD). Among the methods examined, CVD and ALD pose some advantages such as good thickness control, conformality and low temperature deposition. This section reviews some of the physical properties of typical CVD and ALD films.

3.1. CVD HfO_2 Films Grown via TDEAH

Tetrakis Diethyl Amido Hafnium (TDEAH) is one of candidate precursors for the deposition of Hafnium Oxide thin films grown via Metal Organic Chemical Vapor Deposition (MOCVD). There are five classes of precursors typically used for

Chemical Vapor Deposition of Group IVB materials based upon the ligands (19), halides, alkoxides, beta-diketonates, nitrates, and alkylamides. TDEAH is an alkylamide, and one of the most promising precursors for application in a manufacturing environment. TDEAH is a low viscosity, liquid precursor that can be repeatably delivered using a direct liquid injection system. TDEAH is also completely compatible with TDMAS (tetrakis dimethyl amido silicon). The combination of the two precursors can be used to deposit hafnium silicate films, which are currently of interest because they posses higher crystallization temperatures than hafnium oxide films.

A cold walled CVD reactor was used for these depositions (20). The reaction chamber is outfitted with a direct liquid injection (DLI) delivery system. The liquid precursor is metered through the delivery lines to a vaporizing unit mounted on the side of the chamber. An Argon dilution gas (Ar-A) enters the precursor delivery lines immediately before the vaporizer unit. At the vaporizer, an argon carrier gas (Ar-B) is introduced at the point of vaporization to carry the chemical over a short distance from the vaporizer unit to the shower head. Additional process gases such as O_2, Ar, N_2, and N_2O are mixed with the vaporized precursor just before entering the shower head.

3.1.1. Microstructure analysis

TDEAH HfO_2 films were evaluated with X-ray Diffraction, Transmission Electron Microscopy, and Transmission Electron Diffraction techniques to evaluate film microstructure. Films deposited at 325°C were found to be amorphous as-deposited, but crystallized into a poly-crystalline tetragonal phase upon annealing (Fig. 11). Note the featureless diffraction data for the as-deposited samples (black) compared to the

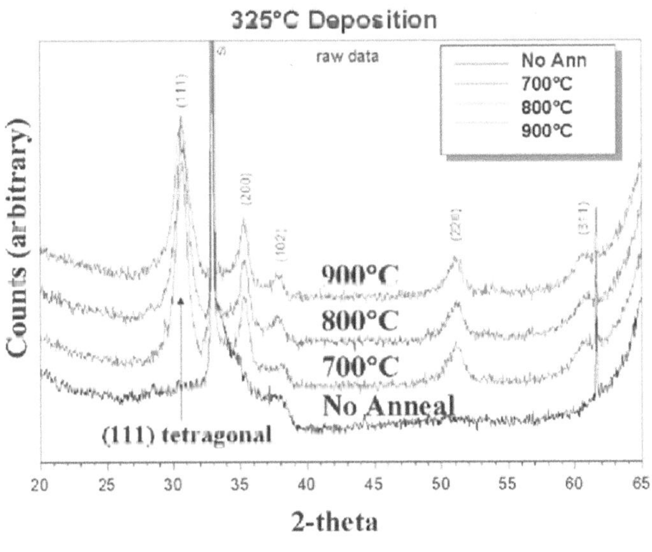

Fig. 11. Diffraction data for TDEAH HfO_2 film deposited at 325°C (from (20)).

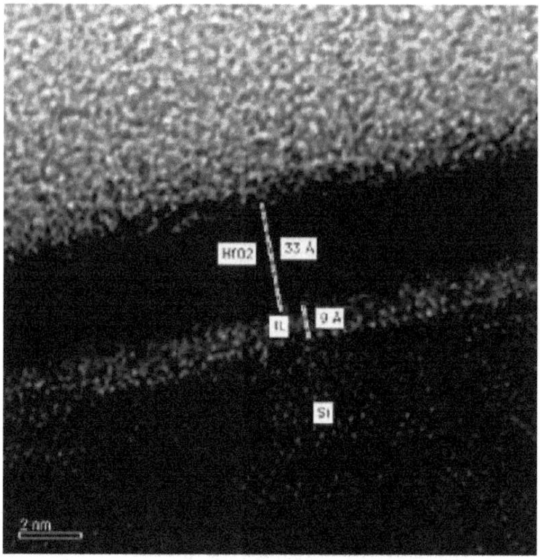

Fig. 12. TEM of a TDEAH HfO_2 film deposited at 325°C (from (20)).

annealed samples that exhibit strong reflections from a tetragonal phase of HfO_2. The amorphous HfO_2 films crystallize easily and transform into a poly-crystalline phase even at the lowest annealing temperature of 700°C. XRD data indicates that the HfO_2 grain size is approximately 100 Å for a 200 Å thick film. Transmission electron microscopy of the TDEAH HfO_2 films also clearly shows an amorphous structure in

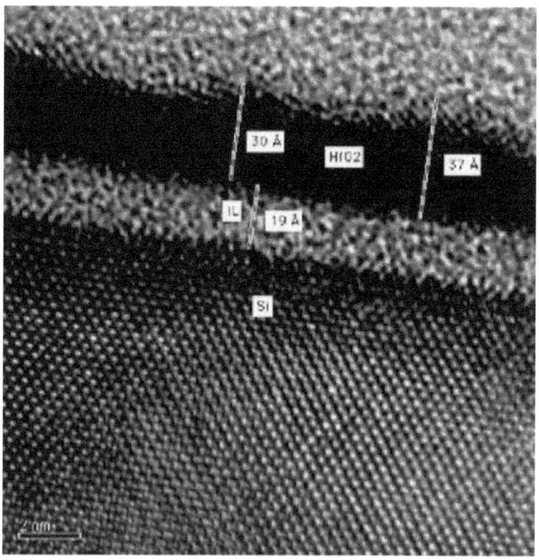

Fig. 13. TEM of a TDEAH HfO_2 film deposited at 325°C and annealed in N_2 at 900°C for 60 s (from (20)).

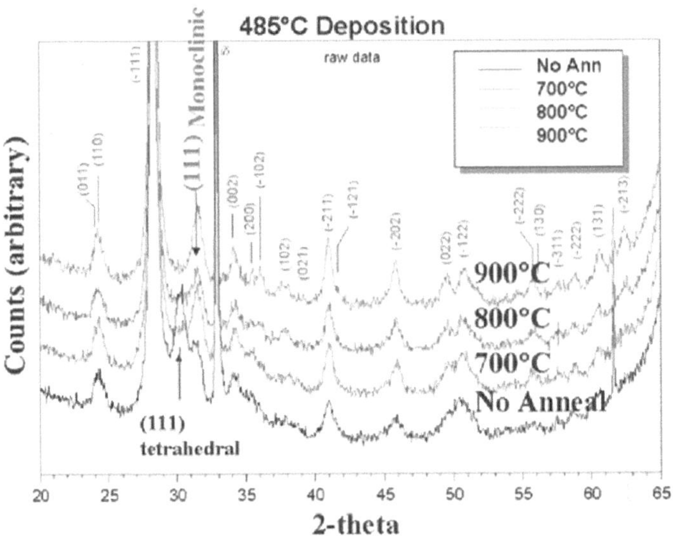

Fig. 14. Diffraction data for TDEAH HfO_2 films deposited at 485°C (from (20)).

the as-deposited film and crystallizes after annealing (Figs. 12 and 13). Although the TDEAH anneal is in a N_2 ambient, the TEM image (Fig. 13) indicates a significant growth of the interfacial layer results after annealing. Anneals on these samples were performed at atmospheric pressure, and a small partial pressure of oxygen in the RTA (rapid thermal anneal) may be responsible for the ~9 Å of interfacial layer growth. It is also speculated that the films may possess excess oxygen as-deposited, and annealing at high temperatures activates out-diffusion of oxygen, and subsequent growth of the interfacial layer. Excess oxygen in HfO_2 films deposited using $Hf(NO_3)_4$ precursor has been reported (19, 21).

The films deposited at 485°C are a mixed phase of tetragonal and monoclinic HfO_2 as-deposited. The reflection at $2\theta = 30.02°$, provides evidence for the existence of the tetragonal phase in the as-deposited samples. These films convert from the tetragonal to a mostly monoclinic, or baddeleyite, phase upon annealing (Fig. 14). XRD data indicates the grain size is approximately 150 Å for a 200 Å film, a 50 Å increase in diameter from the films deposited at 325°C. The Transmission Electron Micrographs of the TDEAH HfO_2 films deposited at 485°C show the poly-crystalline microstructure for both the as-deposited and the annealed conditions. The interfacial layer is observed to increase by about 6 Å after the 900°C-60s-N_2 anneal. For growth at 550°C, the XRD data indicates that the films are poly-crystalline with a (−111) textured monoclinic microstructure for both as-deposited and annealed films (Fig. 15). The grain size is approximately 160 Å for 200 Å films. Transmission electron micrographs of the TDEAH HfO_2 films deposited at 550°C, shows a polycrystalline HfO_2 layer for both as-deposited and annealed samples (Figs. 16 and 17). The interfacial layer increases from 15 to 21 Å after the 900°C-60s-N_2 anneal.

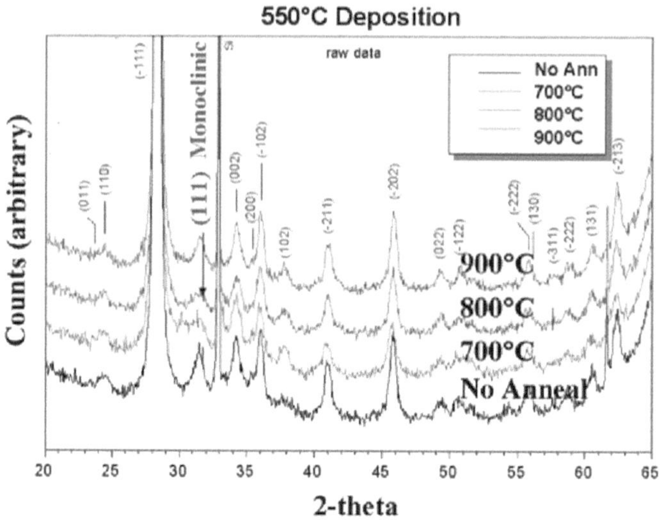

Fig. 15. Diffraction data for TDEAH HfO_2 films deposited at 550°C (from (20)).

In summary of the microstructure analysis, TDEAH films deposited at 325°C were amorphous as-deposited and crystallized into a tetragonal phase upon annealing. At 485°C, the films were a mixed tetragonal/monoclinic phase, with a transition to a purely monoclinic phase with subsequent annealing. At the highest deposition temperature of 550°C, the TDEAH HfO_2 films were monoclinic as-deposited and after

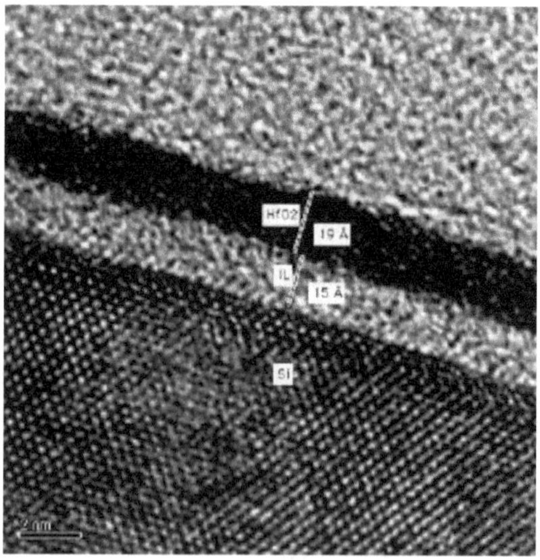

Fig. 16. TEM of a TDEAH HfO_2 film deposited at 550°C (from (20)).

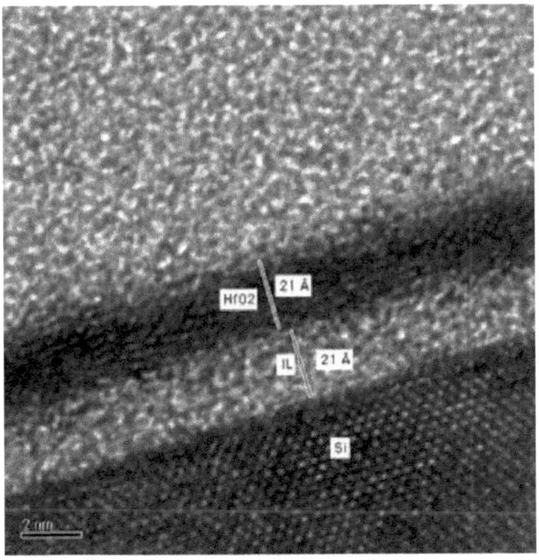

Fig. 17. TEM of a TDEAH HfO_2 film deposited at 550°C and annealed in N_2 at 900°C for 60 s (from (20)).

annealing. It has been noted that at lower deposition temperatures (or lower thermal processing conditions), higher temperature phases of HfO_2 predominate. Increasing deposition temperatures, and annealing temperatures, results in a transition from the higher temperature phase (tetragonal) to the lower temperature phase (monoclinic) of HfO_2. The formation of metastable ZrO_2 and HfO_2 allotropes at low growth temperatures, with a transition to lower temperature phases of ZrO_2 and HfO_2 at higher growth temperatures has previously been reported using both metal organic chemical vapor deposition and atomic layer deposition (22, 23). The stabilization of the higher temperature, metastable allotropes is attributed to an additional surface energy factor that is relatively large in smaller grains. Similar correlation between the grain size and the metastable phase is also seen in this study. However, other factors, such as impurities, deposition rate, or the kinetic growth regime of the films, all of which are strongly related to the deposition temperature, cannot be excluded as possible reasons for the presence of metastable HfO_2 allotropes at lower deposition temperatures.

TEM reveals a significant interfacial layer growth for all the annealed samples. The amorphous interfacial layer grows 6–9 Å after annealing. The growth of the interfacial layer during anneal may be related to the presence of an oxygen partial pressure during the atomospheric N_2 anneals, or possibly due to a super saturation of atomic oxygen in the as-deposited TDEAH HfO_2 films. The equilibrium HfO_2 phase diagram indicates that the oxygen content in HfO_2 films can deviate from stoichiometry by about 3 at.%, and possibly more for films grown under non-equilibrium conditions. HfO_2 films deposited using $Hf(NO_3)_4$ were found to have an as-deposited stoichiometry of $HfO_{2.2}$ to $HfO_{2.4}$ from the RBS analysis (19).

3.1.2. Impurity analysis

Auger electron spectroscopy (AES) and Secondary Ion Mass Spectroscopy (SIMS) were performed on all HfO_2 films to analyze the impurity concentrations at various deposition and annealing conditions. AES depth profiling shows high amounts of carbon and nitrogen impurities in HfO_2 films deposited at 325°C (Fig. 18). The alkylamide precursor, $Hf–(N–(C_2H_5)_2)_4$, is the source of both the nitrogen and carbon impurities. Post-annealing has minimal effect on eliminating the carbon impurities in the HfO_2 films. AES depth profiles for films deposited at 485 and 550°C show a dramatic reduction in the carbon and nitrogen impurity levels compared to the 325°C samples. At the 550°C deposition condition with a 900°C-60s-N_2 anneal, there appears to be some silicon diffusion into the HfO_2 film from the substrate or the poly-Si capping layer.

Secondary Ion Mass Spectroscopy (SIMS) provided additional resolution into the carbon impurity concentrations (Fig. 19) The SIMS depth profiles show a distinct reduction in the carbon impurity levels with increasing deposition temperature. SIMS depth profiles also indicate that post-annealing has minimal effect on reducing the film impurities. The samples deposited at 550°C have a non-uniform composition gradient of carbon impurities compared to samples deposited at 325 and 485°C. We currently do not have an explanation for the non-uniform carbon depth profile in the 550°C samples.

Based on the AES, SIMS, and Raman data used to analyze the impurity concentrations in the TDEAH films, the deposition temperature is the primary factor influencing the amount of precursor impurities incorporated into the film microstructure. Annealing has minimal impact on reducing the in-film impurities. Post-annealing causes an out-diffusion of silicon from the substrate into the film for the 550°C films annealed at 900°C, and influences the local bonding arrangement of carbon impurities in the as-deposited versus annealed films deposited at 325°C.

3.1.3. Optical properties

To analyze the local bonding arrangement of the carbon impurities, UV–Raman measurements were performed on the as-deposited and annealed samples at each deposition temperature (Fig. 20). The data revealed two broad Raman peaks at 1400 and 1600 cm^{-1}, corresponding to diamond-like (sp^3) and graphite-like (sp^2) bonded carbon clusters, in the 900°C-60s–N_2 annealed films deposited at lower temperatures (325 and 485°C). Contrary to the SIMS depth profiles, analysis of the 325°C films revealed no amorphous carbon in the as-deposited film with Raman Spectroscopy. This indicates that the carbon detected by SIMS might be bound to hydrogen or O and Hf, and not as an amorphous carbon cluster. Subsequent annealing of films deposited at 325°C, results in the formation of amorphous carbon clusters detected by Raman scattering. For the 485°C films, amorphous carbon is detected in both the as-deposited and annealed films. The peak intensity for the film deposited at 485°C is reduced compared to the annealed samples deposited at 325°C. Raman Spectroscopy does not detect carbon in any of the 550°C films.

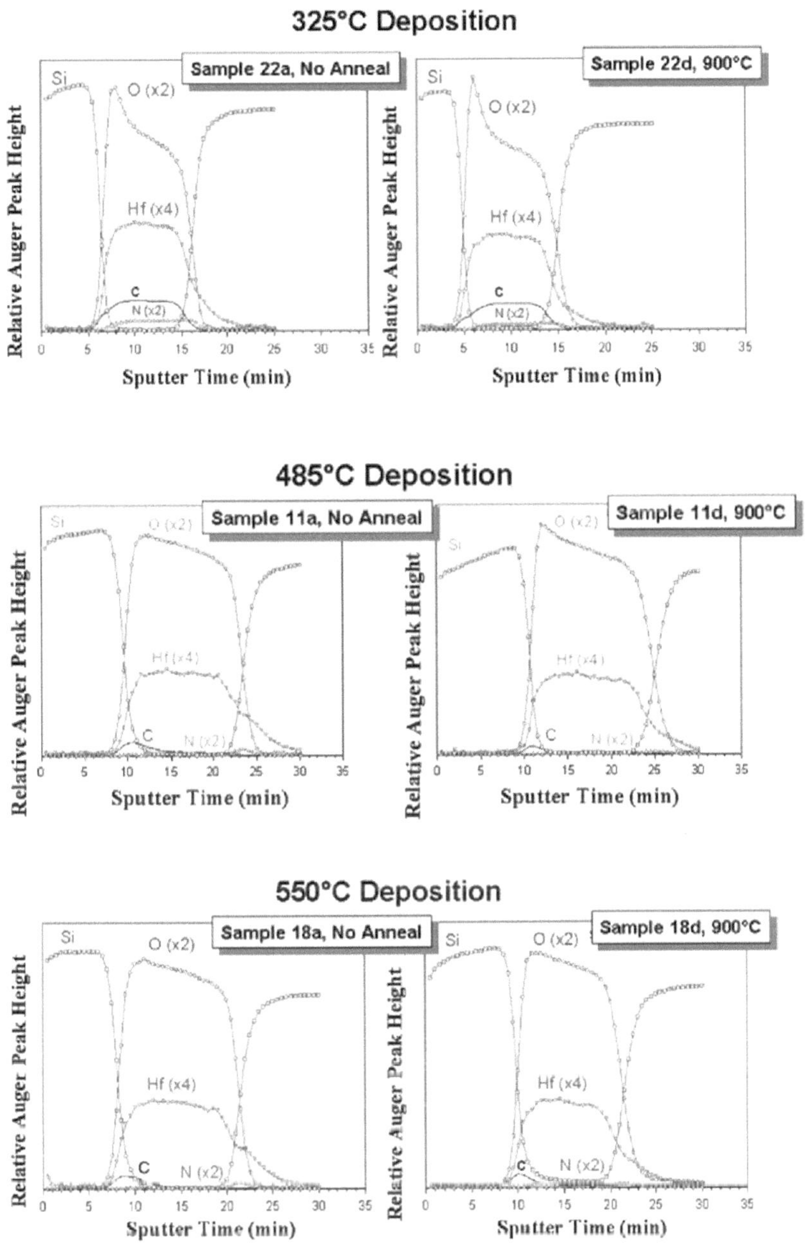

Fig. 18. Auger electron spectroscopy depth profiles of poly-Si on TDEAH HfO_2 films deposited at 325, 485, and 550°C on Si substrate. Only the as-deposited and annealed (900°C-60s-N_2) conditions are shown (from (20)).

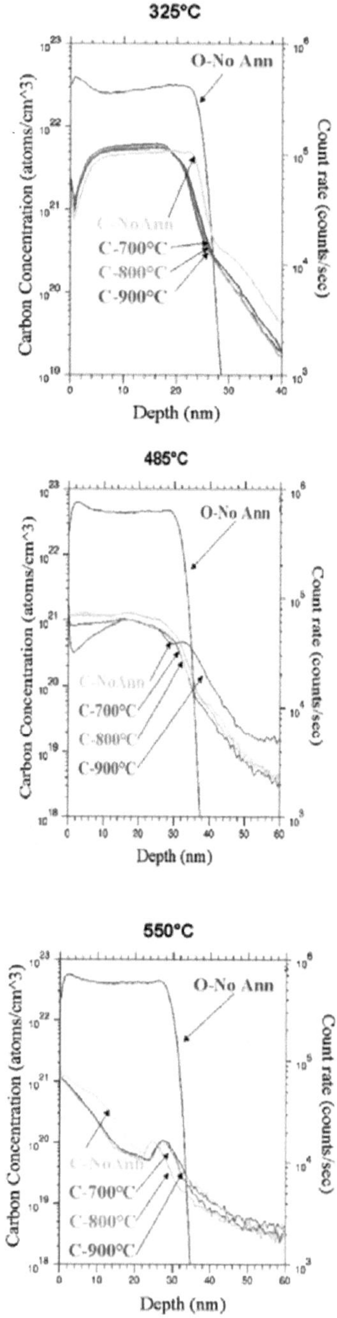

Fig. 19. SIMS depth profiles for TDEAH HfO_2 films deposited at 325, 485, and 550°C (from (20)).

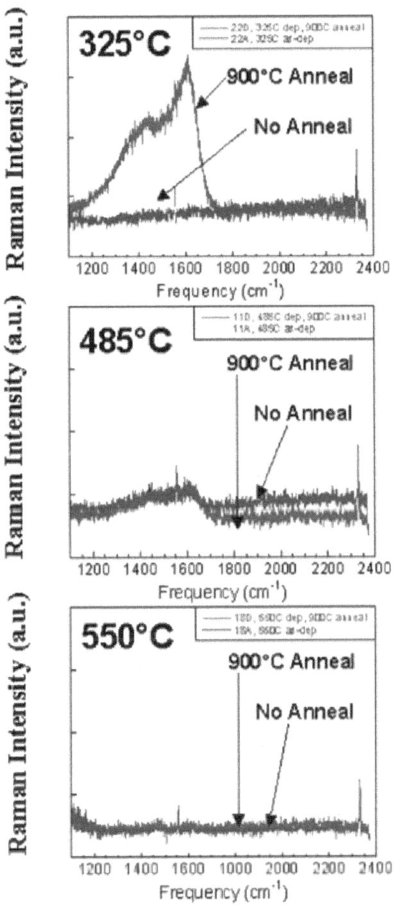

Fig. 20. Raman Spectroscopy for TDEAH HfO_2 films deposited at 325, 485, and 550°C (from (20)).

The variations of the microstructure and impurity level with growth and annealing temperature for the HfO_2 films deposited using TDEAH can also affect the band structure. To correlate the structural properties to the electronic band structure spectroscopic ellipsometry analysis was carried out on the same films. Figures 21–23 show the imaginary part of the dielectric function (ε_2) versus photon energy from 0.7 to 6.6 eV. One can see that the microstructure and impurity level manifest themselves clearly in terms of the shape of the band edge related feature near 5.8 eV and the absorption below the band edge. However, the energy position of the band edge feature and the onset of strong absorption do not appear to be related to the deposition or annealing conditions.

At 550°C, the spectra show the onset of strong absorption and a sharp band edge related feature at 5.8 eV (Fig. 21). The sharpness of the band edge related feature

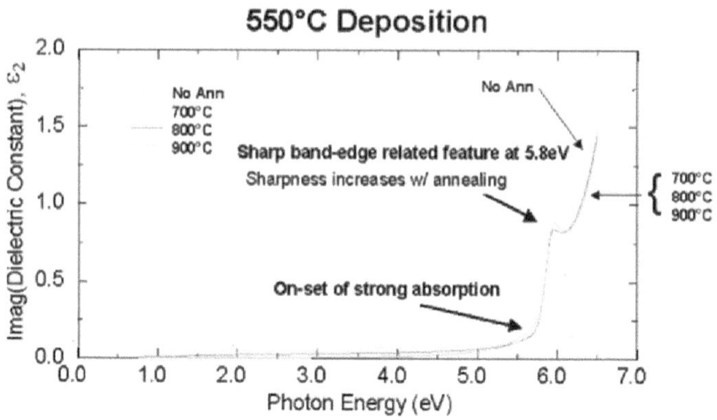

Fig. 21. Imaginary part of the dielectric function at 550°C deposition temperature and various anneal conditions.

increases with annealing temperature, and is likely due to an improvement in crystal quality and an increase in film density. At 485°C, the as-deposited film does not have the sharp band edge related feature at 5.8 eV. Notice the flat structureless character of ε_2 for the as-deposited film (see Fig. 22), which is a mixture of tetragonal and monoclinic phases based on the structural analyses. Upon annealing, the band edge related feature becomes progressively sharper. Therefore, the band edge related feature at 5.8 eV appears strongly related to the monoclinic phase of HfO_2. Finally, for the 325°C deposition condition (Fig. 23), there is no band edge related feature, due to lack of the monoclinic phase, for any of the annealed conditions. However, for the annealed tetragonal films differ from the as-deposited amorphous film by the better defined absorption edge. Substantial absorption is also seen well below the band edge for the films deposited at lower temperature and significantly

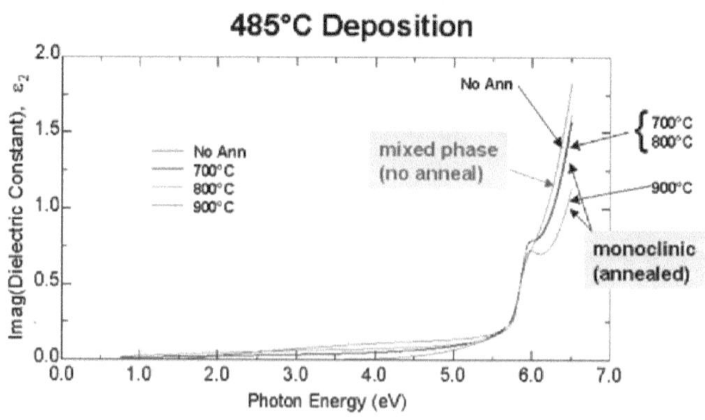

Fig. 22. Imaginary part of the dielectric function at 485°C deposition temperature and various anneal conditions.

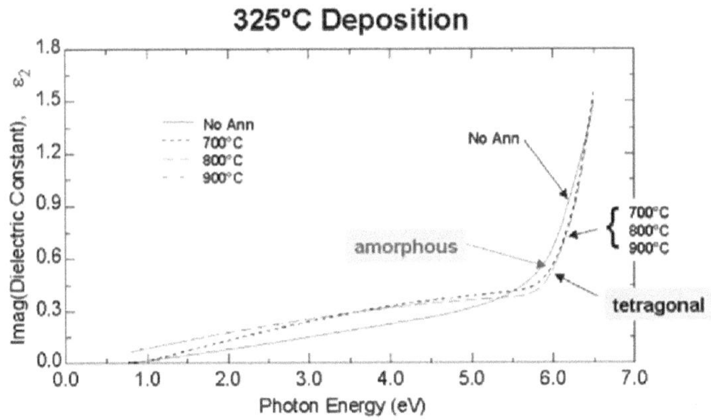

Fig. 23. Imaginary part of the dielectric function at 325°C deposition temperature and various anneal conditions.

reduced with increasing deposition and annealing temperature. This in-gap absorption might be originated from the amorphous carbon clusters observed in the UV–Raman spectra.

The sharpness of the band edge related feature also increases with annealing temperature, this is likely related to improvements in the film density and crystal quality, as shown in the X-ray Reflectometry data. An evaluation of HfO_2 films by X-ray Reflectometry (XRR) explains some of the density, and correspondingly, crystal quality, improvements that manifest in the sharpness of absorption and band-edge features in the spectroscopic ellipsometry analysis. Plots of density versus deposition temperature show an asymptotic like approach to the theoretical density of HfO_2 with increasing deposition temperature in the range of 325–550°C (Fig. 24). Subsequent post-annealing increases the film density more for the lower temperature depositions. For the films deposited at the higher deposition temperatures, the film density already approaches the theoretical density of 9.68 g/cm^3 for bulk HfO_2 as-deposited, and post-annealing has a minimal effect on additional density increases (21). Earlier we noted that the 325°C samples have a larger interfacial layer growth after anneal than samples deposited at 485 and 550°C. We believe the additional interfacial layer growth is a result of the larger density increase, and corresponding volume reduction, upon annealing HfO_2 films deposited at 325°C.

3.1.4. Correlation of electrical and physical results

Electrical characterization of the HfO_2 films has shown that increased deposition temperature dramatically improves the C–V and reduces leakage current of the films. Plots of leakage current versus HfO_2 physical thickness in Fig. 24 show dramatic slope changes based on deposition temperature. The 325°C films have a leakage behavior that is slightly dependent of HfO_2 physical thickness. The slope for the 485 and 550°C films becomes progressively steeper with increased deposition temperature.

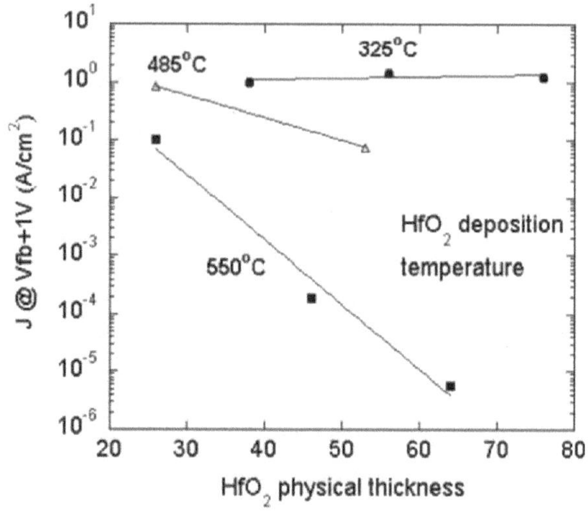

Fig. 24. Plot of leakage current vs. physical thickness of HfO_2 films deposited at different temperatures and annealed in N_2 at 800°C for 1 min (from (20)).

Based on the physical analysis results, the improvements in the electrical leakage with deposition temperature are most likely due to reduction of impurity, such as carbon, concentration, changes in microstructure or increase in the film density. More work is needed to discern the effects of the impurities, density and microstructure on the electrical characteristics.

3.2. ALD HfO_2 Films Grown via Hafnium Tetrachloride

Recently, atomic layer deposition (ALD) has emerged as the most promising technique for gate dielectric deposition, as it offers excellent film uniformity and conformality, and enables ultimate control of film thickness and composition. The HfO_2 films were formed via Atomic Layer Deposition (ALD) using hafnium tetrachloride ($HfCl_4$) and water (27). The HfO_2 film thickness is controlled by the number of cycles deposited. One deposition cycle consists of a water pulse, a nitrogen purge (to remove unreacted water), an $HfCl_4$ pulse and a nitrogen purge. HfO_2 films are grown on a chemical oxide starting surface at one of three temperatures: 200, 300 and 370°C. At each deposition temperature, films were grown with four different thicknesses, 30, 50, 70 and 200 Å, determined by spectroscopic ellipsometry. The film thickness has a linear dependence on the number of cycles and the growth rate is dependent on deposition temperature (Fig. 25). As deposition temperature is increased, growth rate decreases. This is consistent with what has been reported in the literature (25). The growth rate for 200°C deposited films is 0.8 Å/cycle whereas the growth rate for 300 and 370°C deposited films is 0.5 Å/cycle. The growth rate is higher at a lower deposition temperature because of lower water desorption rate (25). After deposition, the films were annealed in a nitrogen ambient for 60 s at 550, 800 or 900°C.

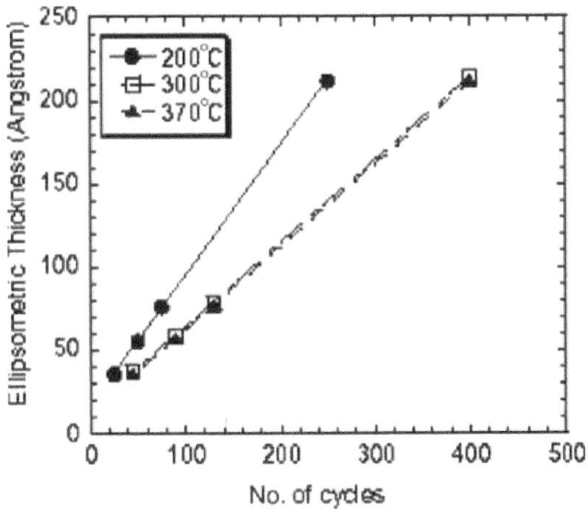

Fig. 25. Growth rate of the ALD HfO_2 films (from (27)).

XRR measurements show that film density increases as deposition temperature is increased. The film deposited at 200°C has a density of 8.8 g/cm^3 while the film deposited at a higher temperature has density of 9.6 g/cm^3 which is about the bulk density of HfO_2 (24). Annealing has significant impact in increasing the density values for the 200°C deposited films to bulk density values. A reduction in film thickness is observed following anneal indicating densification. Ellipsometry analyses show that the 200°C deposited film exhibited a higher reduction in film thickness upon annealing (12–15%) compared to films deposited at 300°C (5–7%) and 370°C (0.3–6%). Furthermore, it is observed that film density increases as film thickness increased. This trend is also observed in other physical properties examined (film composition, microstructure, impurities) and electrical properties of these ALD HfO_2 films as discussed in the rest of this chapter.

3.2.1. Film roughness

AFM analysis was performed on 50 and 200 Å films at 200, 300 and 370°C deposition temperatures and various annealing conditions. Figure 26 shows AFM images of 50 Å film deposited at 200, 300 and 370°C with and without 900°C anneal. At all deposition temperatures, the as-deposited films are smooth with root-of-mean-squared values comparable to the bare Si reference wafer. The post anneal does not seem to roughen the surface noticeably except for the 200°C deposited film after 900°C anneal. This film exhibits a patterned surface morphology with crack-like voiding. A much less pronounced surface pattern is also seen from the 300°C deposited film after annealing. Films deposited at 370°C shows fine grain like surface morphology instead of patterned structures even after 900°C anneal. However, the 370°C as-deposited film has noticeably rougher surface. AFM images of thicker film show similar trends to the thin films. In contrast to the 200°C deposited films, annealing has only minor effects on surface roughness and morphology for films

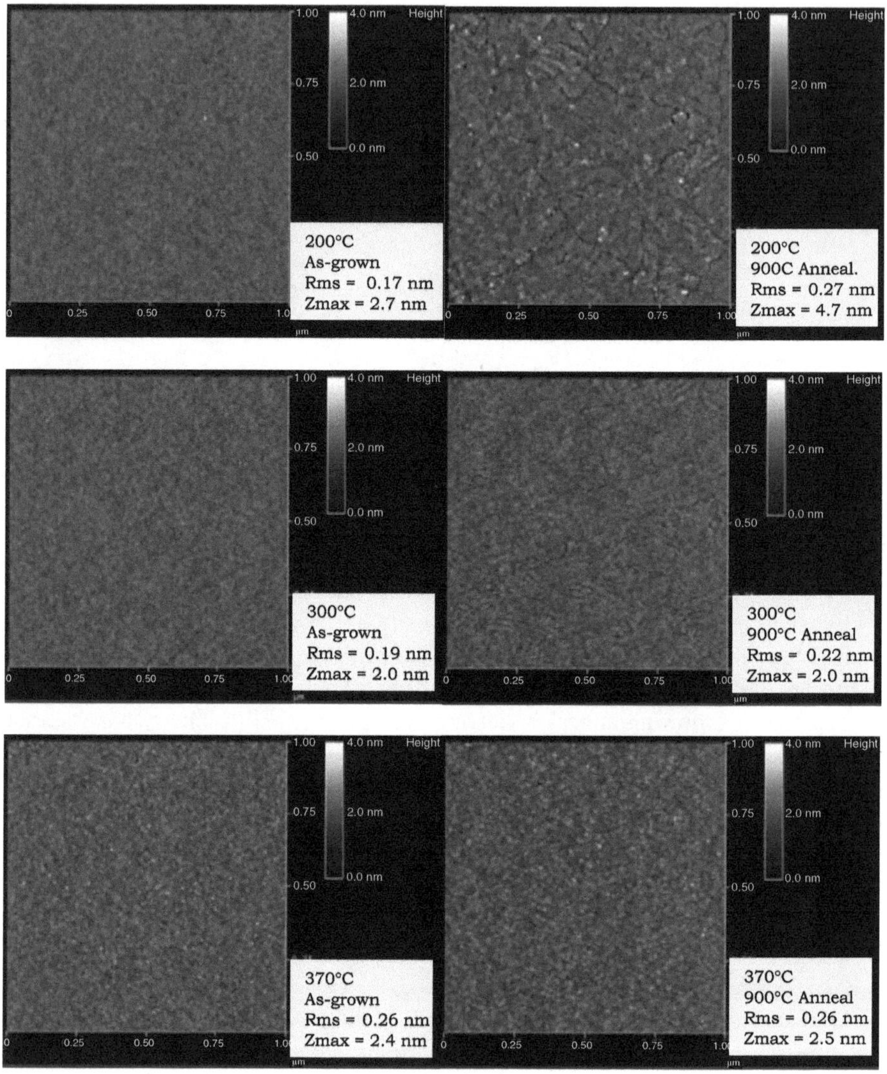

Fig. 26. AFM images of 50 Å HfO_2 films grown at 200°C (upper), 300°C (middle) and 370°C (lower) (as-deposited and after 900°C anneal) (from (27)).

grown at 300 and 370°C. Surface roughness data for HfO_2 films obtained by fitting the ellipsometry spectra also indicates slightly increased roughness increasing deposition temperature.

For both film thicknesses examined, roughness increases with increasing deposition temperature. Figure 27 plots the ratio of roughness over thickness versus annealing temperatures at all deposition temperatures. When normalized for thickness, 200 Å film is smoother than the 50 Å film. For 50 Å films deposited at 200 and 300°C,

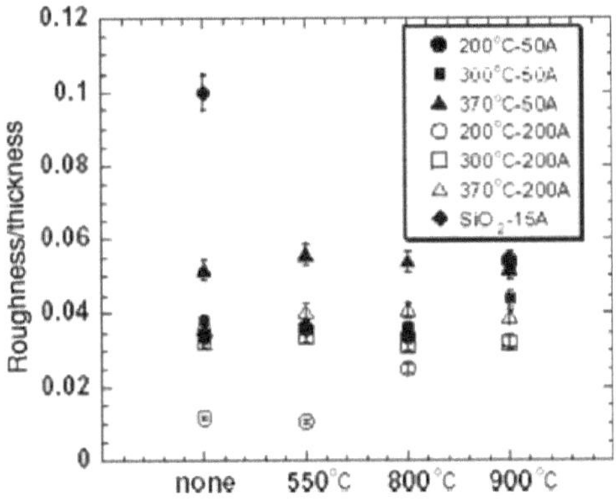

Fig. 27. AFM rms/thickness as a function of anneal temperatures for the ALD HfO_2 films (from (27)).

the roughness increases after a 900°C anneal. For 200 Å film deposited at 200°C, roughness increases after the 800 and 900°C anneals.

3.2.2. Composition and impurities

XPS depth profiles for 200, 300 and 370°C deposited films with various anneal conditions show films to be stoichiometric. Normalized high resolution spectra of the Hf 4f region and O 1s region indicate that the 370 and 300°C deposited films are indistinguishable within instrument sensitivity (Fig. 28(a) and (c)). The 200°C deposited films show a steady progression of decreases in the height of the saddle between the $4f_{7/2}$ and $4f_{5/2}$ doublet as-anneal temperature rises (Fig. 28(b)), due to the narrowing of the peaks, as the Hf bonding become more uniform. The O 1s spectra for 200°C deposited films show a steady drop in OH peak region as films are annealed in higher temperatures (Fig. 28(b)). Hf–OH peak region also changes with annealing, suggesting that increasing anneal temperature drive off water either as adsorbed water vapor or as dehydration of Hf–OH to Hf–O. Angle resolved spectra show that adsorbed water is confined to the surface of the films. Further studies of OH bonding in films are conducted using ATR-FTIR (date not shown). Consistent with XPS data, the O–H vibration bands are strongly suppressed by increasing growth temperature and are almost gone in the 370°C deposited films. All of these results may indicate that the 300 and 370°C deposited films are deposited so that no more densification or bond re-arrangement is possible whereas the 200°C deposited films have not yet reached their most stable configuration as-deposited.

Since HfO_2 films are deposited using a halide precursor, it is of interest to study the halide (chlorine) incorporation in the films at various deposition and annealing conditions. To investigate Cl impurities, low energy SIMS was performed for all 200 Å films. SIMS profiles of Cl are shown in Fig. 29. Levels of C and H on all films are low and are near the SIMS detection limit (data not shown) whereas Cl presents

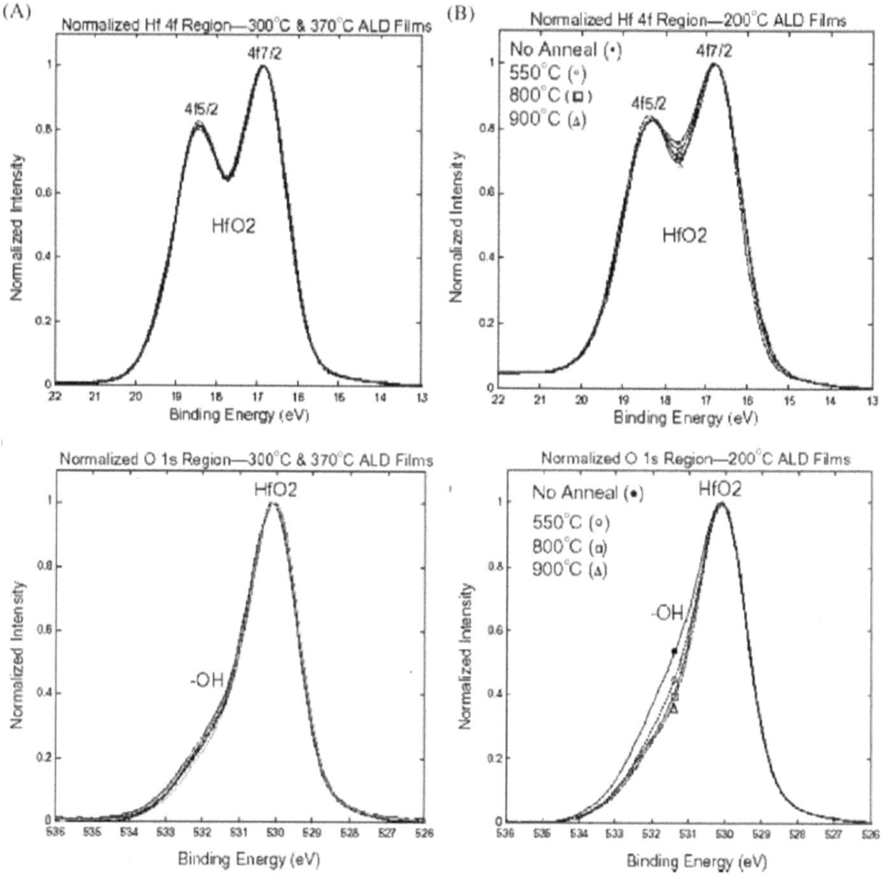

Fig. 28. Normalized XPS high resolution spectra of the Hf 4f and O 1s regions for (a and c) 300 and 370°C as-grown films, (b and d) 200°C deposited films with and without anneal (from (27)).

in detectable levels in all films. Films deposited at 200°C have the highest level of chlorine and more uniform depth profile as compared to 300 and 370°C grown films. Annealing at 550°C has little effect on reducing the Cl impurities whereas 800 and 900°C anneals lead to substantial reduction in the Cl levels. Depth profile of Cl at 300 and 370°C indicate that Cl piles up near the interface. The 370°C deposited films with a 900°C anneal has the lowest chlorine impurities of all samples examined. This finding that higher deposition temperatures lead to a more pure film is in agreement with other reports on various ALD films (25, 28). The observed trend that higher temperature deposition leads to lower impurities incorporation is consistent with what have been reported for MOCVD HfO_2 deposited using TDEAH, in which case the level of carbon incorporation decreases significantly as deposition temperature is increased from 325 to 550°C (27).

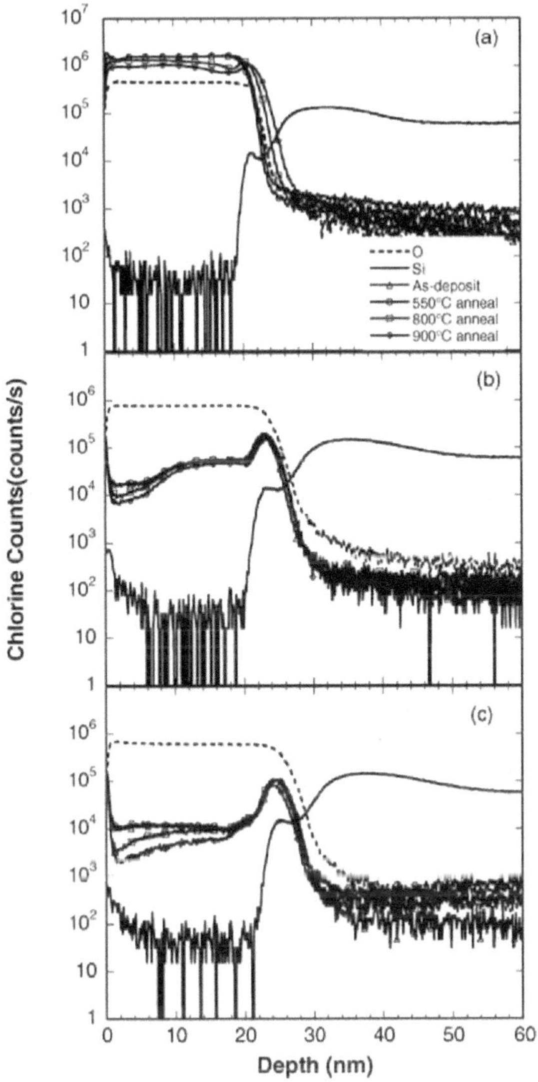

Fig. 29. SIMS profiles of chlorine on HfO_2 films deposited at (a) 200°C, (b) 300°C, and (c) 370°C (from (27)).

3.2.3. Microstructure

To study microstructure, X-ray diffraction (XRD), transmission electron microscopy (TEM), ATR-FTIR, and VUV-spectroscopy ellipsometer (VUV-SE) were performed on HfO_2 films. The 200°C deposited films are amorphous as-deposited and after the 550°C anneal as indicated by the featureless X-ray diffraction data in Fig. 30. After an 800°C anneal, peaks are observed showing film transformation into monoclinic phase. Monoclinic peak intensities increase after the 900°C anneal. TEM cross-section

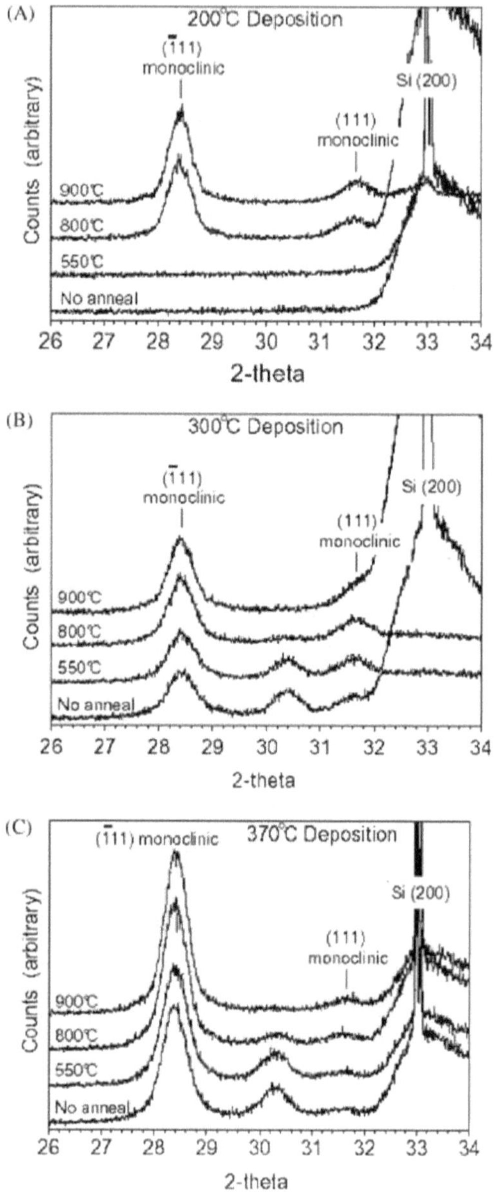

Fig. 30. XRD scans showing (−111) and (111) peaks for monoclinic HfO_2 (from (27)).

images also reveal an amorphous HfO_2 film with a distinct interfacial layer, which becomes thinner after 550°C anneal (Fig. 31). Electron energy-loss spectroscopy (EELS) data at the O K-edge (not shown here) show a spectrum characteristic of Hf-silicate at the interfacial layer. ATR-FTIR was also employed to analyze the microstructure

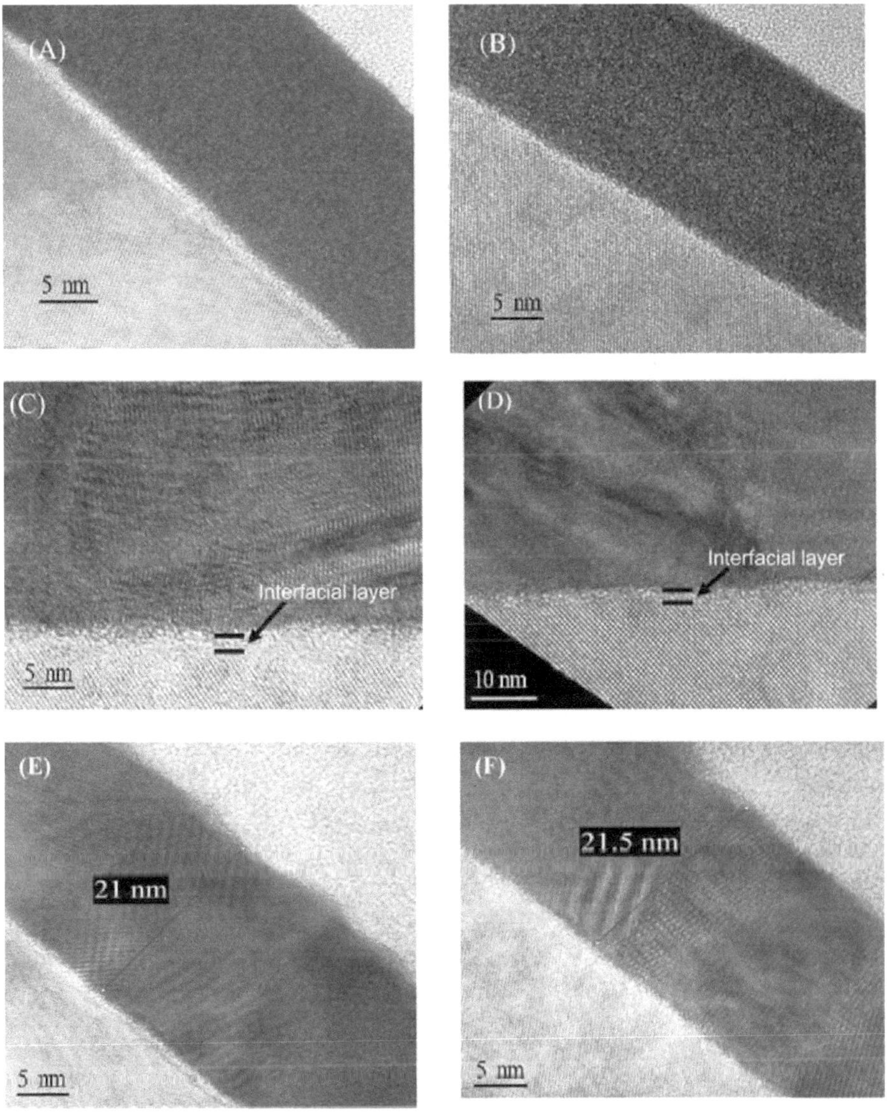

Fig. 31. HRTEM images for (a) 200°C as-deposited, (b) 200–550°C anneal, (c) 300°C as-deposited, (d) 300–550°C anneal, (e) 370°C as-deposited, and (f) 370–550°C anneal HfO_2 (from (27)).

of the 200 Å films and the results are in excellent agreement with XRD and TEM results. Figure 32(a) shows the absorbance spectra in the range of the highest phonon mode around 740 cm^{-1} in the monoclinic hafnia phase from films deposited at 200, 300 and 370°C. No sharp peak is seen for the as-grown and the 550°C-annealed

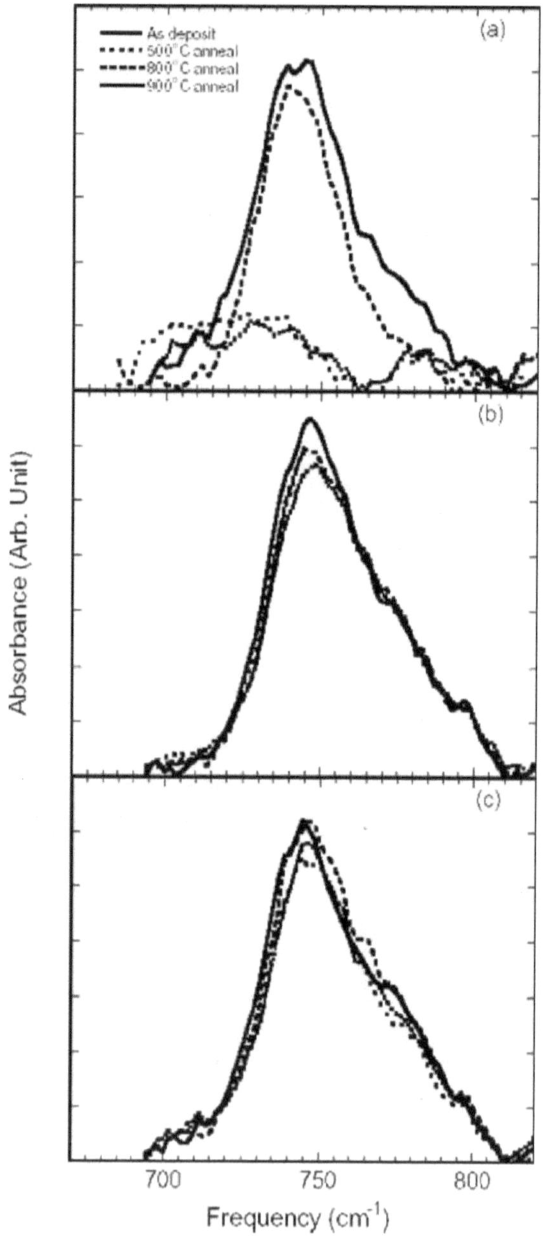

Fig. 32. ATR-FTIR spectra for HfO_2 films deposited at (a) 200°C, (b) 300°C, and (c) 370°C (from (27)).

films deposited at 200°C, indicating the films are amorphous. The phonon peak associated to the monoclinic phase emerges after the 800 and 900°C anneal. The change in microstructure will affect the electronic band structure and thus be also sensed by the VUV-SE measurements. The SE data were analyzed using a model consisting of a surface roughness layer, a hafnia layer and an interfacial layer on Si. Figure 33(a) displays the real (ε_1) and imaginary part (ε_2) of the dielectric functions of films deposited at 200°C with and without anneal. The as-deposited and the 550°C annealed films show only one broad maximum in ε_2, reflecting the amorphous nature of these films. After the 800 or the 900°C anneal, two sharp features associated to two critical points in the interband transitions appear around 6 and 7.3 eV. For both amorphous and polycrystalline films, the onset of absorption occurs at the 5.2 eV.

The 200 Å film deposited at 300°C is a mixture of tetragonal and monoclinic phase as-deposited. XRD scans in Fig. 30(b) shows that with increasing anneal temperature the film transforms from a mixture of tetragonal and monoclinic phase into a more monoclinic film. After 900°C anneal, the film becomes completely monoclinic with grains mostly (−111) textured. Kim and co-workers (26) have reported that thin HfO_2 films deposited at 300°C are amorphous as-grown, the onset of crystallization is at around 500°C and complete crystallization occurs at 700°C. TEM images of the 50 and 200 Å films indicate a thickness dependent crystal structure for the 300°C deposited films. The thinner film is amorphous (data not shown) while the thicker film is polycrystalline (Fig. 31(a) and (b)). In both cases, however, a 7–10 Å of interfacial layer is observed, which seems to become slightly thinner after the 550°C anneal. EELS spectra (not shown here) for thicker film show a drop in intensity at the interface, suggestive of a layer with slightly lower density. ATR-FTIR spectra for 300°C deposited films (Fig. 32(b)) are in consistent with the XRD and TEM results and show the monoclinic phonon peak that appears for as-deposited film and becomes stronger upon annealing. VUV-SE analysis on the film deposited at 300°C show the sharp features near 6 and 7.3 eV in ε_1 (Fig. 33(b)). Sharpening of the spectral features are observed with increasing annealing temperature presumably as the film becomes more completely single phase monoclinic.

The microstructure of the 370°C deposited film is similar to the 300°C deposited films in many respects. XRD data shown in Fig 30(c) indicate the film is a mixture of tetragonal and monoclinic phase as-deposited and transformed into fully monoclinic phase after a 900°C anneal. It is noted that the overall monoclinic peak intensity for ALD HfO_2 is three to five times weaker than that of MOCVD HfO_2 (27), suggesting that ALD film is less crystalline than the MOCVD film. This is not surprising since MOCVD films are deposited at a higher temperature (550°C) than ALD films (200–370°C). Furthermore, TEM images reveal that both thin and thick samples are polycrystalline as-deposited. The 370°C samples show no clear evidence of an interfacial layer. ATR-FTIR and VUV-SE spectra from the 370°C deposited films are similar to the 300°C deposited films (see Figs. 32(c) and 33(c)).

In summary, the microstructure of ALD HfO_2 was studied with a number of techniques and the results are in excellent agreement with one another. The 200°C

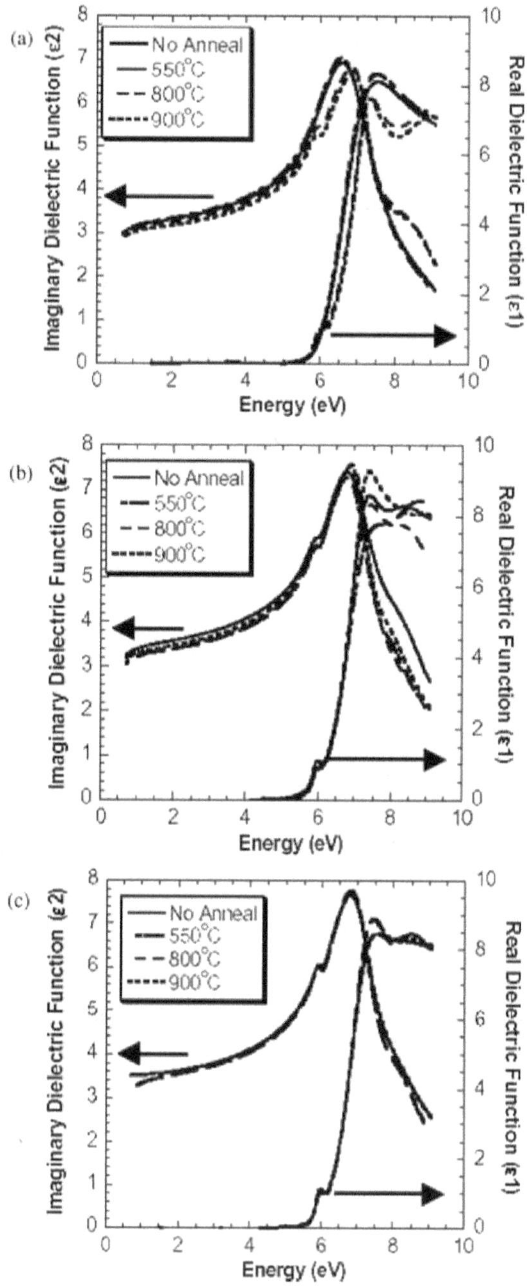

Fig. 33. VUV-SE real and imaginary part of the dielectric function of the (a) 200°C deposited films, (b) 300°C deposited films, and (c) 370°C deposited films with and without anneal (from (27)).

deposited films are amorphous and remain amorphous after a 550°C anneal. Following an 800 and 900°C anneals, the films become polycrystalline with the monoclinic structure. The 300°C deposited films exhibit a thickness dependent crystal structure. The 50 Å film is amorphous as-deposited and becomes polycrystalline after a 550°C anneal. The 200 Å film, however, is a mixture of tetragonal and monoclinic phases as-deposited. All 370°C deposited films are polycrystalline and contained a mixture of tetragonal and monoclinic phases as-deposited. With increasing annealing temperature, the 300 and 370°C deposited film become more monoclinic. At 900°C, all films are fully monoclinic. This finding that higher deposition temperatures lead to a more pure film is in agreement with other reports on various ALD films (23, 28).

3.2.4. Correlation between materials and electrical properties

Materials and electrical characterizations indicate that the deposition temperature is the main factor controlling material and electrical properties. Deposition temperature controls deposition rate, impurity concentration, and crystal structure, all of which in turn have significant impact on electrical properties.

Films deposited at 200°C have the highest chlorine and adsorbed water or Hf–OH. These films are amorphous and very smooth as-deposited but significantly roughens and show signs of crack-like voiding with high temperature anneal. Films are also less dense than those deposited at higher temperatures. Annealing has a significant impact on the microstructure of these films. It crystallizes and densifies the films, as well as drives off adsorbed water or hydrated hafnium in the films. Material characterization performed on the films helps to understand some of the trends observed in the electrical data. The C–V curves obtained using the Hg probe show kinks due to charge trapping and higher impurities in the film (Fig. 34). The 900°C anneals negatively impact device performance. This is not surprising since AFM images revealed that the 900°C anneal creates voids in the films. The 550°C anneal yields low leakage and good CV characteristics while the 800 and 900°C anneals lead to higher leakage current.

Films deposited at 300°C have similar material properties to those deposited at 370°C. Films are a mixture of tetragonal and monoclinic crystals as-deposited and become fully monoclinic with 900°C anneal. The films do have slightly more chlorine and consequently are slightly leakier than films deposited at 370°C (Fig. 34). Films deposited at 370°C have better material properties and as a consequence, better electrical characteristics such as leakage and CV characteristics (see Fig. 34) than those deposited 200°C. At 370°C deposited film denser, void free and contains less chlorine impurities. Unlike 200°C deposited films, annealing these films only change density slightly. Microstructure of 370°C deposited films is less sensitive to annealing conditions compare to the 200°C deposited films. Electrical properties of these films are also less sensitive to annealing conditions. Regardless of annealing temperature, films have good CV characteristics and reasonable leakage current. Flatband voltage also does not vary significantly with anneal temperatures.

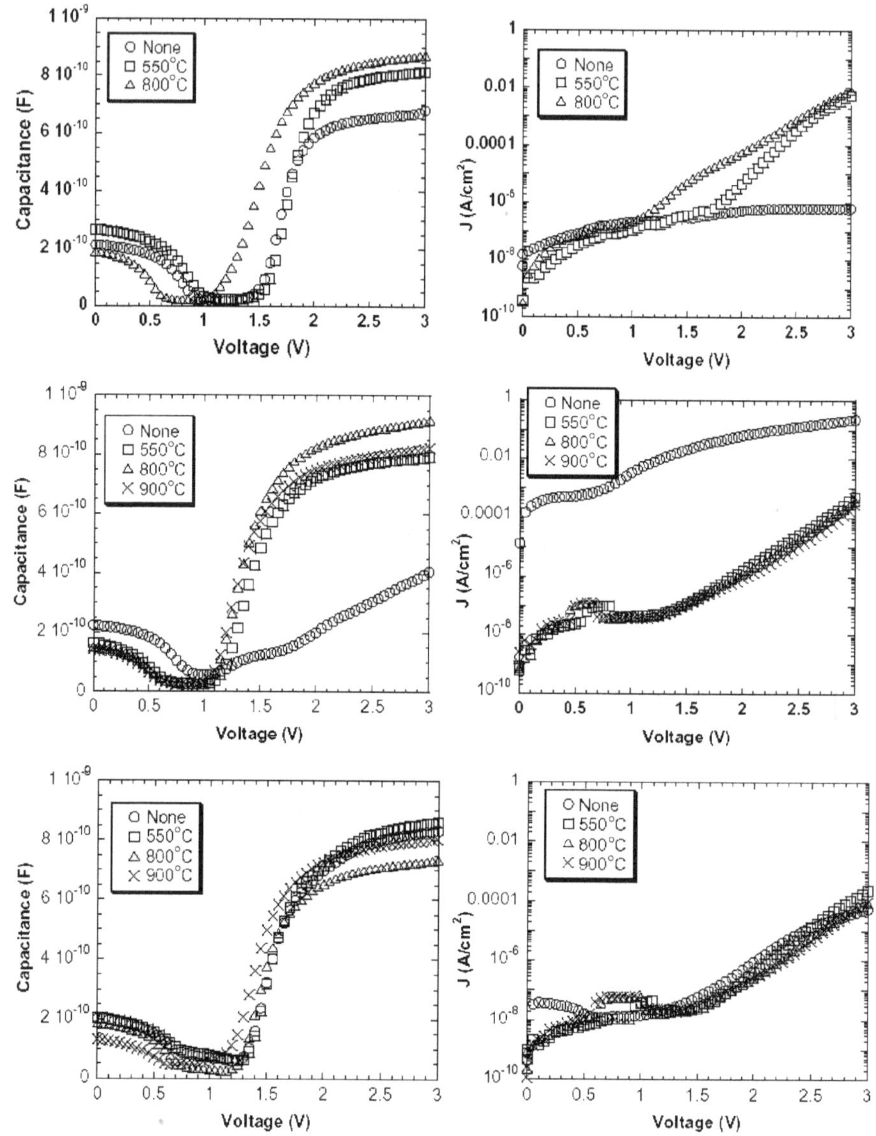

Fig. 34. CV and IV characteristics for HfO_2 films deposited at 200°C (upper), 300°C (middle), and 370°C (lower) (from (27)).

4. SUMMARY

In this chapter we overviewed the materials and physical properties of $SrTiO_3$ and HfO_2 thin films grown on Si by MBE, CVD and ALD for potential high-k gate dielectric application, and discussed briefly the correlation between the materials and electrical properties. Although perovskite-type oxides such as $SrTiO_3$ has much higher dielectric constant, their smaller energy bandgaps, which leads to smaller band offsets and thus higher leakage current, post strong challenges for their application as alternative gate dielectrics for future CMOS technology. Most recent research and development for near-term solutions have focused on HfO_2 and its silicates or aluminates because of its better thermal and chemical stability. HfO_2 films tend to be polycrystalline either as-deposited or after thermal annealing and, therefore, are subjected issues due to grain-boundary enhanced diffusion and leakage. Hf silicates and aluminates, on the other hand, have higher crystallization temperature and thus more possibly remain amorphous after the CMOS gate stack processes. Other materials such as rare-earth metal oxides, which have also attracted great attention for potential application as gate dielectrics, are not reviewed in this chapter.

ACKNOWLEDGEMENTS

The author gratefully thanks D. Triyoso, J. Schaeffer, Z.Y. Yu, A. Demkov, G. Edwards, S. Zollner, J. Kulik, G. Tam, R. Gregory, X.-D. Wang, E. Duda, S.F. Lu, D. Werho, P. Fejes, D. Roan, M. Ramon, B. Hradsky, R. Nieh, R. Rao, R. Raw, C. Hobbs, R. Garcia, J. Baker, L.B. La, K. Reed, P. Tobin, B.-Y. Nguyen, B. White, R. Droopad, J. Curless, J. Finder, and K. Eisenbeiser of Motorola, as well as J. Noilien and S. Campbell of the University of Minnesota for various technical contributions.

REFERENCES

1. M. Schulz, *Nature* **399**, 729–730 (1999).
2. J. Robertson, *J. Vac. Sci. Technol. B* **18**, 1785 (2000).
3. H.F. Luan, B.Z. Wu, L.G. Kang, B.Y. Kim, R. Vrtis, D. Roberts, D.L. Kwong, *IEDM Tech. Dig* 609 (1998).
4. B. He, T. Ma, S.A. Cambell, W.L. Gladfelter, *IEDM Tech. Dig.* 1038 (1998).
5. W.J. Qi, R. Nieh, B.H. Lee, L. Kang, Y. Jeon, K. Onishi, T. Ngai, S. Banerjee, J.C. Lee, *Proceedings of International Electron Devices Meeting 1999* (IEEE, Piscataway, NJ, USA), p. 145.
6. L. Kang, Y. Jeon, K, Onishi, B.H. Lee, W.J. Qi, R. Nieh, S. Gopalan, J.C. Lee, *2000 Symposium on VLSI Technology. Tech. Dig.* 44.
7. G.D. Wilk, R.M. Wallace, *Appl. Phys. Lett.* **74**, 2854 (1999).
8. J.C. Chen, G.H. Shen, L.J. Chen, *Appl. Surf. Sci.* **142**, 120 (1999).
9. H.J. Osten, J.P. Liu, P. Gaworzewski, E. Bugiel, P. Zaumseil, *Proceedings of International Electron Devices Meeting 1999* (IEEE, Piscataway, NJ, USA), p. 653.
10. L. Manchanda, W.H. Lee, J.E. Bower, F.H. Bauman, W.L. Brown, *IEDM Tech. Dig.* 605 (1998).
11. R.A. McKee, F.J. Walker, M.F. Chisholm, *Phys. Rev. Lett.* **81**, 3014 (1998).
12. Z. Yu, J. Ramdani, J.A. Curless, C.D. Overgaard, J.M. Finder, R. Droopad, K.W. Eisenbeiser, J.A. Hallmark, W.J. Ooms, V. Kaushik, *J. Vac. Sci. Technol. B* **18**, 2139 (2000).
13. S.A. Chambers, Y. Liang, Z. Yu, R. Droopad, J. Ramdani, K. Eisenbeiser, *Appl. Phys. Lett.* **77**, 1662 (2000).

14. X. Zhang, A.A. Demkov, H. Li, X. Hu, Y. Wei, J. Kulik, *Phys. Rev.* **B68**, 125323 (2003).
15. H. Li, X. Hu, Y. Wei, Z. Yu, X. Zhang, R. Droopad, A.A. Demkov, J. Edwards, K. Moore, W. Ooms, J. Kulik, P. Fejes, *J. Appl. Phys.* **93**, 4521 (2003).
16. R. Liu, S. Zollner, P. Fejes, R. Gregory, S. Lu, K. Reid, D. Gilmer, B. Nguyen, Z. Yu, R. Droopad, J. Curless, A. Demkov, J. Finder, K. Eisenbeiser, Gate stack and silicide issues in silicon processing II. *Mat. Res. Soc. Symp. Proc.* **670**, K1.1 (2001).
17. L. Hilt Tisinger, R. Liu, J. Kulik, X. Zhang, J. Ramdani, A.A. Demkov, *J. Vac. Sci. Technol.* **B21**, 53 (2003).
18. A.I. Kingon, J.P. Maria, S.K. Streiffer, S.K., *Nature (London)*, **406**, 1032 (2000).
19. R.C. Smith, T. Ma, N. Hoilien, L.Y. Tsung, M.J. Bevan, L. Colombo, Ro, *Adv. Mater. Opt. Electron.* **10**, 105–114 (2000).
20. J. Schaeffer, N.V. Edwards, R. Liu, D. Roan, B. Hradsky, R. Gregory, J. Kulik, E. Duda, L. Contreras, J. Christiansen, S. Zollner, P. Tobin, B.-Y. Nguyen, R. Nieh, M. Ramon, R. Rao, R. Hegde, R. Rai, J. Baker, S. Voight, *J. Electrochem. Soc.* **150**, F67–74 (2003).
21. D.G. Colombo, D.C. Gilmer, V.G. Young, S.A. Campbell, W.L. Gladfelter, *Chem. Vapor Deposit. Germany* **4**, 6 (1998).
22. C.J. Taylor, D. Gilmer, W.L. Gladfelter, S. Campbell, J.T. Roberts, *Electrochem. Soc. Proc.* **98**, 23 (1999).
23. K. Kukli, K. Forsgren, J. Aarik, T. Uustare, A. Aidla, A. Niskanen, M. Ritala, M. Leskela, A. Harsta, *J. Cryst. Growth* **231**, 262–272 (2001).
24. D.R. Lide, *Handbook of Chemistry and Physics*, 78th Edition (CRC Press, Boca Raton, New York, 1997).
25. M. Ritala, M. Leskela, *Handbook of Thin Films Materials*, ed. H.S. Nalwa (Academic Press, San Diego, 2001), p. 103.
26. H. Kim, P.C. McIntyre, K.C. Saraswat, *Appl. Phys. Lett.* **82**, 106 (2003).
27. D. Triyoso, R. Liu, D. Roan, M. Ramon, N.V. Edwards, R. Gregory, D. Werho, J. Kulik, G. Tam, E. Irwin, X-D. Wang, L.B. La, C. Hobbs, R. Garcia, J. Baker, B.E. White Jr., P. Tobin, *J. Electrochem. Soc.* H 151, F220 (2004).
28. K. Kukli, M. Ritala, J. Aarik, T. Uustare, M. Leskela, *J. Appl. Phys.* **92** (2002).

Chapter 2

DEVICE PRINCIPLES OF HIGH-K DIELECTRICS

KURT EISENBEISER

Motorola Inc., Tempe, AZ 85284, USA

1. INTRODUCTION

The growth of the semiconductor industry in the last few decades has largely been driven by the growth of integrated circuits (IC) based on complementary metal–oxide–semiconductor (CMOS) technology. CMOS technology uses n-type and p-type field effect transistors (FETs) to produce digital logic elements that are superior to other available logic technologies for many applications. The dominance of CMOS over other logic technologies is based on its low power consumption as well as the ability to scale CMOS and achieve simultaneous improvements in power consumption, speed and cost. One of the key aspects of CMOS technology is the metal–oxide–semiconductor (MOS) capacitor that acts as the control element in a CMOS device. While there are other semiconductors that have either been used before silicon or may have better material properties than silicon, silicon has become by far the dominant material in the semiconductor industry mainly due to its native oxide, silicon dioxide (SiO_2), and the MOS capacitor that can be easily manufactured on silicon using this oxide. The properties of silicon dioxide and its interface to silicon are far superior to the native oxides on other commonly available semiconductors and enable the implementation of high performance CMOS.

2. CMOS DEVICE OPERATION

The basic device technology used in CMOS circuits is a metal–oxide–semiconductor field effect transistor (MOSFET). In this device a gate electrode serves as one electrode on a MOS capacitor. An energy band diagram for this structure is shown in Fig. 1. When a bias is applied between the metal and the semiconductor, band bending in the semiconductor is modified and charges are moved. Charges induced on the gate from this applied bias are balanced by charges in the semiconductor under the insulating oxide layer. These charges can be used to accumulate, deplete or invert the surface of the semiconductor closest to the insulator. In accumulation the applied bias causes an

A.A. Demkov and A. Navrotsky (eds.), Materials Fundamentals of Gate Dielectrics, 37–55.

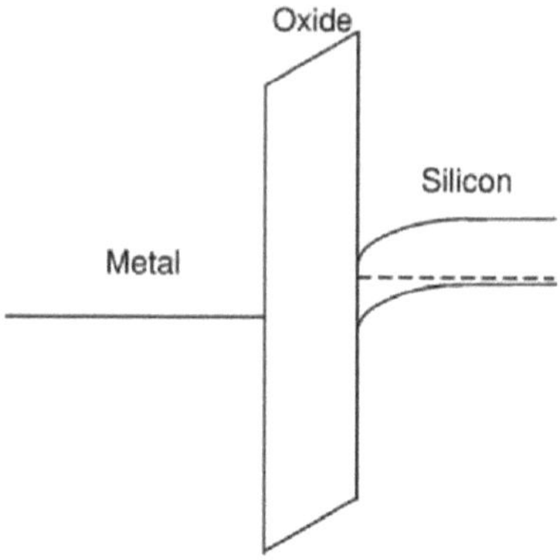

Fig. 1. Energy band diagram of MOS capacitor.

increase in the carrier concentration compared to the zero bias state. In depletion the applied bias causes a decrease in the carrier concentration and in inversion the bias causes the majority carrier concentration to change from one type of carrier, such as an electron, to the other type of carrier, a hole. The other electrodes in the MOSFET, the source and drain, are formed from heavily doped regions in the semiconductor that are of opposite type (n-type/p-type) from the channel region (p-type/n-type) which forms the semiconductor portion of the MOS capacitor as shown in Fig. 2.

Since the source and drain are of opposite type to the channel, junctions are formed around both the source and drain with depletion regions that prevent current flow from the source to the drain in the absence of a gate bias. This low current state is the OFF state in CMOS logic. When the gate is biased with enough potential, the threshold voltage, to invert the surface of the semiconductor in the MOS capacitor, a

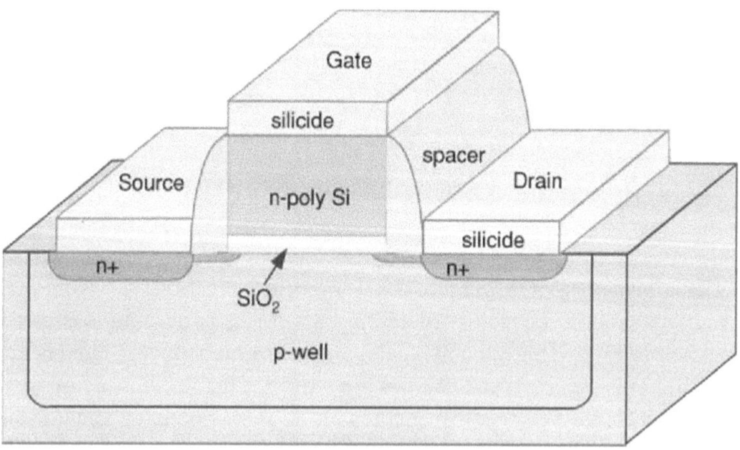

Fig. 2. Schematic of typical nMOS silicon transistor.

surface channel is formed between the source and the drain with no junction barriers to current flow. Once these junction barriers are eliminated, current will flow between the source and drain in response to an applied bias between these electrodes. This high current state is the logical ON state of the device. The current flow in the ON state between the source and the drain is directly related to the amount of charge in the channel. Since this charge is induced in response to an applied gate bias, changes in the gate bias will dramatically change the current flow between the source and the drain. This capacitive control of the drain current gives rise to gain in the device. The gain of the MOSFET allows a single device to drive many other devices, called fan out, and enables a wide variety of logic function implementations.

In the ideal case, current flow is entirely from the source to the drain when the gate is biased above threshold and there is no current flow when the gate is biased below the threshold voltage. In real devices, however, there is always some leakage current either from the gate to the channel or from the source to drain when the device is in the OFF state. The insulator in the MOS capacitor is essential to minimize this leakage current since its properties determine how much charge leaks into the channel from the gate as well as how much charge is accumulated in the channel in response to the applied gate voltage. Since digital logic circuits now contain tens of millions of transistors on each chip, even small amounts of leakage current can create significant issues such as heating and power consumption. The dominance of silicon CMOS devices in these digital logic applications is largely the result of the low power consumption in these devices which is in turn mainly the result of the outstanding performance of the silicon dioxide based capacitor used in silicon CMOS.

3. SILICON DIOXIDE GATE DIELECTRIC

The properties of silicon dioxide that make it ideal for MOSFETs are numerous. SiO_2 has a wide bandgap, 9 eV, which makes it a very good insulator between the gate of the device and the channel. This insulating property is very important to prevent current leakage from the gate into the channel. The insulating properties of SiO_2 are maintained even at high fields (SiO_2 has a breakdown field of about 15 MV/cm) allowing very thin films to be used. SiO_2 also has a low bulk fixed charge density, $<5 \times 10^{10}$ charges/cm^2, and when properly passivated, a low interface state density, $<5 \times 10^{10}$ states/eV-cm^2, with silicon. A low bulk fixed charge density is important since charges in the bulk will screen some of the charges on the gate electrode and shift the threshold voltage. These fixed charges can also interact with carriers in the channel through coulombic forces and reduce carrier mobility. For these reasons charges in the bulk of the insulator must be minimized. The interface states are even more important since they are near the channel. These states can trap or expel charge during device operation. These changes in channel charge will affect the threshold voltage and reliability of the device. Also since the charge trapping process is fairly slow, interface states will lead to frequency dispersion in the MOSFET characteristics. Another important feature of SiO_2 is that it has good stability in the bulk of the film and at the interface to silicon over a wide range of thermal and electrical stressing conditions. This means that as a device is operated in a changing environment its properties will remain constant and hence its reliability will be enhanced. A final important aspect of silicon dioxide in relation to MOSFETs is its manufacturability. The thermal oxidation of silicon can be controlled in thickness down do a few atomic layers with acceptable uniformity across a large area wafer and with repeatable results over many runs (1). This manufacturability is one of the key features that have allowed integrated circuits with millions of transistors to be made with acceptable yield.

Silicon dioxide has been a nearly ideal gate dielectric for several decades, however, scaling of CMOS devices over this time period has exposed a problem with silicon dioxide. As mentioned previously one of the reasons silicon CMOS technology has endured is because these devices can be scaled. Scaling means that the dimensions of the transistor as well as the electrical bias conditions on the transistor are reduced by a common factor. For example in a strict scaling case the oxide thickness, channel length, channel width and bias voltage are all decreased by the same factor, while the channel doping is increased by the same factor. This scaling leads to an increase in the device density, an increase in the device operating frequency and a decrease in the power dissipation. These advantages of scaling have led to higher performance at a lower cost and have driven the growth in semiconductor productivity. This growth in productivity is captured in the famous Moore's Law that states that the number of transistors per integrated circuit will double every 18 months (2). This scaling trend has held for over 30 years, however, there are some possible roadblocks in the future of this process with the continued scaling of the gate oxide thickness being one of the most critical.

There is a physical limit to how thin you can make SiO_2 and have it still behave like bulk SiO_2. This limit was explored by evaluating the chemical composition and electronic structure across thin oxide/Si interfaces (3). It was found that the silicon conduction band wave functions penetrate into the oxide and limit the oxide thickness to >0.7 nm to perform as an insulator. Theoretical studies agree with this estimate and suggest very low SiO_2/Si band offsets below this thickness (4). While this fundamental limit is many generations from being reached, other limits are already being experienced. As the SiO_2 is thinned, it still acts as a barrier to classical transport; however, quantum mechanical tunneling current becomes more significant. As the thickness of the gate dielectric becomes comparable to the distribution in the electron wave function, a significant portion of the electron population can tunnel through the energy barrier presented by the gate dielectric instead of needing to go over it, see Fig. 3. This tunneling current is dependent upon both the thickness of the dielectric as well as its bandgap and how this bandgap is aligned to the gate and the silicon as well as several other factors such as effective mass in the dielectric. Since the current from this kind of transport is exponentially dependent on the dielectric thickness, leakage currents through the gate oxide have increased dramatically as the gate oxide thickness has scaled. As the thickness of silicon dioxide in a CMOS device approaches 10–15 Å, tunneling current becomes significant (5–8). In addition to the power consumption from this leakage current it also creates defects such as

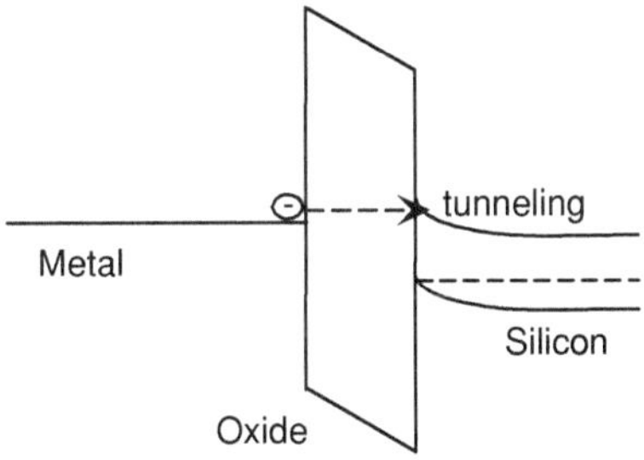

Fig. 3. Energy band diagram of MOS capacitor showing tunneling mechanism.

traps, interface states and charged states in the insulator. These defects build up over time and eventually the oxide destructively breaks down. These trends of increasing reliability issues with thinning gate dielectrics have been reported and at some point will limit the scalability of silicon dioxide (9, 10).

The effects of the increased gate leakage can be seen in the subthreshold power levels of scaled devices. The subthreshold power in a device is the power used by the device when it is in the OFF state or being operated below its threshold. This subthreshold power has two major components: leakage through the gate insulator and leakage through the substrate. As Fig. 4 shows, the subthreshold power in integrated circuits has been increasing with scaling generations as the operating power has remained largely constant (11). If these trends continue, at some point the

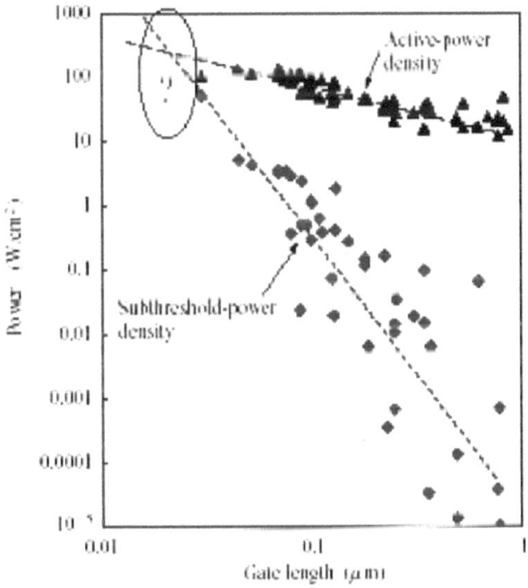

Fig. 4. Trends in active power density and subthreshold power density with IC scaling. Figure from (11). Reproduced with permission from *IBM Journal of Research and Development*.

subthreshold power will be comparable to the operating power and conventional logic circuits will fail to function properly. An even more immediate consequence of this increase in subthreshold power is an increase in the power density on the chip that increases the heat generation on the chip. This heat must be dissipated and leads to packaging problems. The heat also degrades device performance further exacerbating the problem. The large power dissipation also creates battery lifetime issues in portable applications. Besides these practical limits, at some point this power can also impose a physical limit on further scaling (12).

4. HIGH-K GATE DIELECTRICS

One possible solution to the leakage current and reliability issues created by scaling silicon dioxide is the use of a higher permittivity insulator in place of silicon dioxide. The saturated drive current, ON current, in a MOSFET can be approximated as

$$I \approx \frac{W}{2L}\mu C\left(V_{\mathrm{g}} - V_{\mathrm{t}}\right)^{2}, \tag{1}$$

where W is the transistor width, L is the channel length, μ is the carrier mobility, C is the capacitance density of the gate insulator, V_{g} is the gate voltage and V_{t} is the threshold voltage. The gate insulator capacitance density can be approximated as

$$C \approx \frac{\kappa\varepsilon_0}{t}, \tag{2}$$

where κ is the relative permittivity of the insulator, ε_0 is the permittivity of free space and t is the thickness of the insulator. When a device is scaled, the saturated drive current should remain constant. This means that as the gate voltage and channel length are reduced the capacitance must increase. From Eq. (2) we can see that this increase in capacitance requires that either κ be increased or t be decreased. The common approach has been a decrease in insulator thickness as discussed above. The alternative is to increase the relative permittivity of the insulator, go to a so-called high-k material, and maintain the thickness of the insulator. These high-k materials can be thicker than SiO_2 and still have the same capacitance density. In this way the saturated drive current can be maintained in the scaled device without the leakage and reliability issues that a thinner SiO_2 layer may have.

High permittivity in a dielectric is the result of high polarization of charge in the film. When an electric field is applied to an insulator, the charge within the insulator becomes asymmetrically distributed creating dipoles within the material. The net effect of these dipoles is that a negative charge is accumulated on one end of the dielectric and a positive charge is accumulated on the other end. The degree to which this polarization occurs for a given electric field is the permittivity of the dielectric. The polarization of charge can occur from either a displacement of electrons, a displacement of ions or from a combination of these. The electronic polarization occurs in all materials and over a wide frequency range. It is, however, inversely related to the bandgap of the material and limited to $\sim$20 for reasonable insulating materials.

The ionic component of charge polarization is present in only some materials and is affected by their atomic and crystalline structure. Materials with ionic polarization can exhibit extremely high dielectric constants, >6000 (13), however these parameters are much more sensitive to variations in temperature, material properties and operating conditions (14, 15). Typically the ionic polarization is limited to lower frequencies than the electronic polarization and materials with both ionic and electronic polarization exhibit frequency dependent dielectric constants (16).

While dielectric constants higher than silicon dioxide are required in a high-k material, dielectric constants that are too high can degrade device performance. In scaled CMOS devices effects called short channel effects can occur. When the gate couples directly to the source or drain instead of to the channel, it loses some control over the charge in the channel and short channel effects occur. These effects can be manifested in several forms such as output conductance where the drain voltage changes drain current in the saturation region, drain induced barrier lowering where the drain voltage affects the threshold voltage of the device or threshold voltage roll-off where the threshold voltage becomes a function of the gate length. These short channel effects lead to degraded device performance such as higher leakage current from a threshold voltage roll-off or reduced threshold voltage margins from high output conductance and impose scaling limits on a technology.

One cause of short channel effects is a gate dielectric that is too thick. Fringing-induced barrier lowering (FIBL) from improperly designed high-k structures leads to short channel effects (17, 18). For a constant gate capacitance that is determined by the scaling factor, as the dielectric constant of the gate dielectric increases the thickness of the dielectric increases. If the dielectric constant is too high, the dielectric thickness to achieve a given capacitance will also be too large and short channel effects will occur. A dielectric stack with a low dielectric constant interface layer can reduce the short channel effects (19). In general the physical thickness of the gate dielectric should be less than 1/10th the length of the gate (20).

While an increase in dielectric constant is the reason high-k materials are under investigation for gate dielectric applications, care must be taken to address the real reason to use these materials: a reduction in gate leakage. The gate leakage that dominates in thin silicon dioxide gate dielectrics is quantum mechanical tunneling, and by replacing the silicon dioxide with a thicker high-k layer, this tunnel current can be dramatically decreased. In these high-k materials, however, other forms of gate leakage current may dominate. Thermionic emission current occurs when carriers gain enough energy to go over the energy barrier created by the wide bandgap insulator. There is also a leakage mechanism called thermionic field emission. As a bias is applied across the insulator, the shape of the energy barrier of the insulator becomes triangular. In thermionic field emission hot carriers tunnel through a partial thickness of the barrier and have enough energy to go over the rest of the barrier. Figure 5 shows a schematic representation of these two processes. The key feature in determining the thermionic and thermionic field current through an insulator is its band gap and the conduction and valance band edge energy discontinuities. Since the bandgap of silicon dioxide is large, 9 eV, and the band offsets between silicon dioxide and

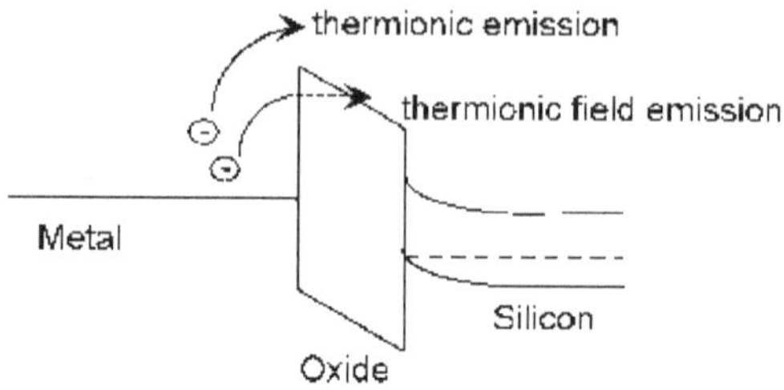

Fig. 5. Energy band diagram of MOS capacitor showing thermionic emission and thermionic field emission leakage mechanisms.

silicon are large, the thermionic emission current is small in this system. The bandgap of many high-k materials, however, is smaller. In fact empirical data suggests an inverse relationship between bandgap and dielectric constant in many of the metal oxides under investigation (21). Besides the bandgap, the band alignment is also important since the band alignment together with the difference in bandgap between the insulator and the silicon determine the energy barriers to electrons and holes. These energy barriers or band offsets in the conduction band and valance band are what determine the thermionic emission current for electrons and holes, respectively. They also are important for tunneling current as well as for hot carrier injection, where carriers are injected from the channel into the insulator during operation of the device in the ON state. As a general rule the conduction and valance band energy barriers need to both exceed 1 eV to assure low thermionic leakage current as well as well-behaved transport properties.

Reduced conduction or valance band energy barriers can also lead to enhanced Fowler–Nordheim (FN) tunneling current. In this type of leakage the carrier tunnels through a triangular barrier into the conduction band or valance band of the insulator, see Fig. 6. Since FN current is exponentially related to the energy barrier height, small barriers in either the conduction band or valance band lead to large leakage. This type of leakage has been found to play an important role in leakage current in high-k insulators under high field conditions (22, 23).

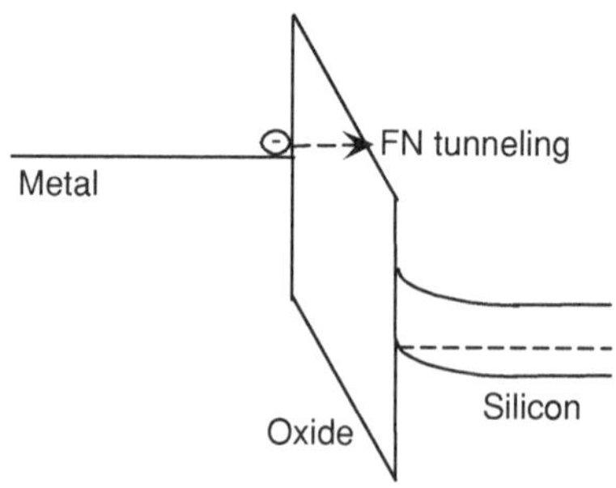

Fig. 6. Energy band diagram of MOS capacitor showing Fowler–Nordheim leakage mechanism.

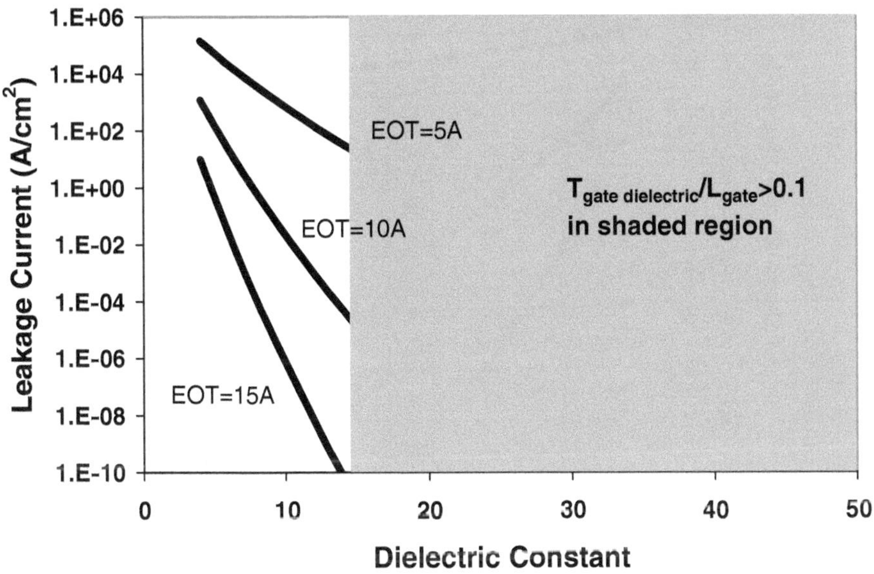

Fig. 7. Leakage current as a function of dielectric constant for different EOTs for an ideal high-k gate dielectric.

The observed empirical relationship between bandgap and dielectric constant can be used with these various leakage mechanisms to estimate the leakage current through an ideal dielectric as a function of the dielectric constant of the material. Figure 7 shows the results of these calculations. Equivalent oxide thickness (EOT) is a term used to relate the capacitance of a given thickness of a dielectric to the equivalent capacitance in silicon dioxide and is expressed as

$$\mathrm{EOT} = \frac{\kappa_{\mathrm{SiO_2}}}{\kappa_{\text{high-k}}} t_{\text{high-k}}, \tag{3}$$

where $\kappa_{\mathrm{SiO_2}}$ is the relative permittivity of silicon dioxide, $\kappa_{\text{high-k}}$ is the relative permittivity of the high-k material and $t_{\text{high-k}}$ is the thickness of the high-k material. From Fig. 7, for a given EOT, the leakage current decreases as the dielectric constant increases. If, however, the dielectric constant is too high, short channel effects become significant as denoted by the shaded region. For a gate dielectric stack with only a single material, dielectric constants in the range of 12–17 produce the best performance. Figure 8 shows a similar plot for the case where the high-k material is separated from the channel by a thin layer of silicon dioxide. Many high-k gate stacks have some sort of low-k interface layer such as this to improve process manufacturability or device performance. In this case higher dielectric constant materials are needed.

In addition to the leakage in an 'ideal' dielectric, trap states can also play a role in the leakage through the gate dielectric. In Poole–Frankel emission carriers transport through the insulator by trapping and detrapping processes. This is shown

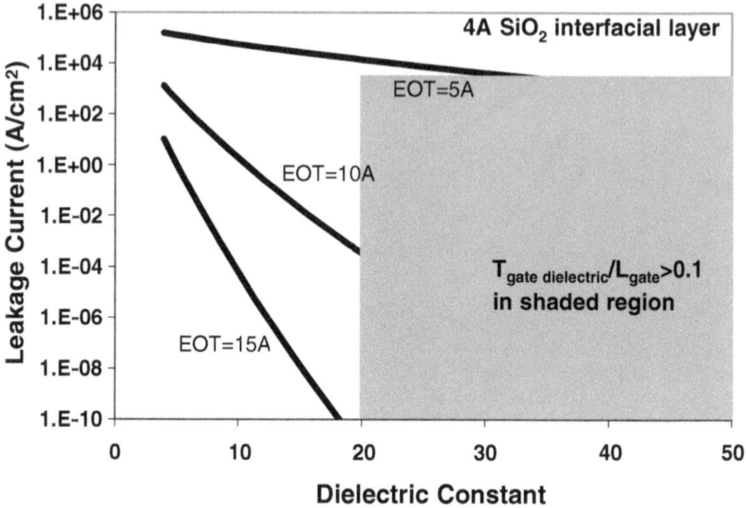

Fig. 8. Leakage current as a function of dielectric constant for different EOTs. The dielectric stack used in this calculation includes a 4A SiO_2 layer between the ideal high-k dielectric and the channel.

schematically in Fig. 9 and is one of the more significant leakage mechanisms for many high-k gate dielectrics under low field conditions (24). The traps can also play a role in other leakage mechanisms such as FN tunneling by trapping charges and modifying the band bending and barrier heights of the system (25).

The crystallinity of the dielectric can also affect its leakage properties. A polycrystalline film has grain boundaries. These can serve as low resistance leakage paths for ions or electrons. Since these grain boundaries are not present in either amorphous or single crystal materials, leakage current is usually lower for a given material in either an amorphous film or a single crystal film than in a polycrystalline film. In addition to increased leakage current, the grain boundaries can also increase the diffusion rates of impurities through the film and lead to nonuniformities in the film properties (21, 25, 26). These nonuniformities can cause device nonuniformity and circuit yield issues. For these reasons polycrystalline gate dielectrics are highly undesirable.

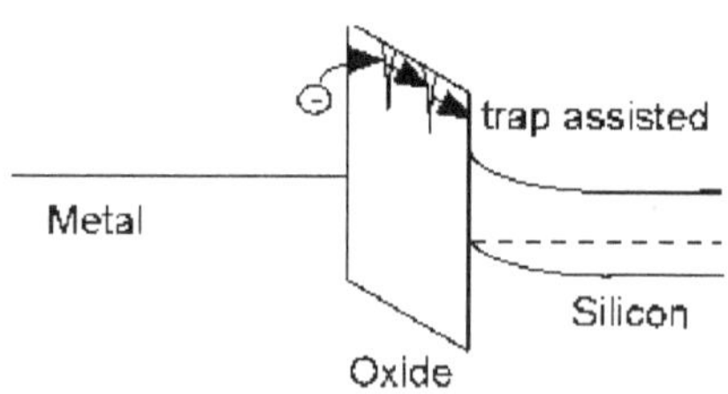

Fig. 9. Energy band diagram of MOS capacitor showing Poole–Frankel leakage mechanism.

The leakage current through these thin gate oxides also complicates characterization. Capacitance–voltage (CV) measurements are commonly used to determine the inversion and accumulation capacitance of the insulator, the EOT, and the threshold voltage of the device. This measurement can also be used to determine bulk and interface trap density, mobility and several other device characteristics. However, as the leakage current through the oxide increases, low frequency CV characterization becomes more difficult. Capacitance rolloff may be seen and may be device size dependent (27). The high leakage current and series resistance effects from the leaky gate dielectric mean that the two-element lumped circuit model commonly used to extract capacitance from impedance measurements may not be adequate, so three-element or more complicated circuit models are used requiring additional measurements (28–30). Once the true capacitance has been determined, this data can be matched to models to correct for quantum mechanical effects, poly depletion, interface states and many other features of the system (31, 32). Other measurements such as tunneling current measurements can be used to verify these results (33, 34). These techniques have been largely developed to characterize thin SiO_2/Si capacitors and produce good results in this well behaved system. High-k systems may not be as well behaved and may have different physical processes occurring that affect leakage current, interface states and a host of other parameters. These differences from the SiO_2/Si system raise questions about the validity of using the standard extraction techniques for thin high-k dielectric characterization and are driving the development of techniques tailored specifically to high-k gate dielectrics (35).

While gate leakage and dielectric constant are two of the initial considerations in selecting a gate dielectric, many other characteristics are needed for a useful material. One of the most important is low interface state density. In many material systems when two dissimilar materials are brought in contact with each other, electronic states caused by dangling bonds or other imperfections occur at the interface. In the SiO_2/Si system these states can be passivated with hydrogen and the resulting density of interface states is very low. If these states are not passivated, they can act as traps for charged carriers. During device operation the charging and discharging of these states with changing electric fields can cause undesirable device characteristics such as hysterisis and threshold shifts. The interface states can also be charged, and due to their proximity to the carriers in the inversion channel of the device in the ON state can reduce effective channel mobility.

In many high-k dielectric systems the interface must be passivated to achieve low interface state density. This passivation must have an acceptable thermal budget and be done with elements that do not degrade other device characteristics. This passivation process must be optimized for each change in materials used in the gate stack. As an alternative to passivation, the interface properties can be improved by inserting an intermediate layer between the high-k and the channel, so the poor interface is improved and is moved further from the channel (26, 36). This interface layer can have a significant effect on the channel mobility in the device (37). Silicon dioxide is often used as this interface layer at the cost of reduced capacitance (38). While the electronic properties of the gate dielectric-channel interface are critical, the mechanical characteristics of the interface are also important. Interface roughness can dominate the channel mobility since the inversion layer carriers are very close to this interface (39,40). Interfaces that are nearly atomically abrupt are needed to maintain the mobility seen in current SiO_2/Si MOSFETs.

A high density of electronic states or charges in the bulk of the dielectric is also undesirable. These states and charges can cause problems in the device similar to interface states. Since they are further from the channel than interface states and separated from the channel by an energy barrier, carrier injection into the states is less of a problem. The charges, however, can cause a shift in the threshold voltage of the device and degrade circuit performance, and an excess of these charges or states must be avoided in the dielectric (41,42).

The gate dielectric must also be able to withstand high fields without breaking down. In an ideal scaling case as the dimensions of the device are reduced, the operating voltage is also reduced and fields remain constant. In real cases two dimensional effects cause high fields in certain regions of the device and these effects increase as the device dimensions shrink. Also the scaling of the operating voltage much below 1 V is in question. To maintain high ON current and low OFF current in a device, a large difference between the operating voltage and the threshold voltage is needed, see Eq. (1). For this reason the threshold voltage must be scaled down as operating voltages drop below 1 V. This lower threshold voltage, however, leads to more OFF state leakage current and lower noise margins (7). The slower scaling of the operating voltage has already started (43) and will get worse at lower voltages.

For these reasons operating voltages especially in sub-45 nm devices will not scale as quickly as the device geometry, and fields in the device may increase more quickly than two-dimensional effects alone would cause. This means that a new gate dielectric must be able to withstand these high fields. Further complicating this situation is the fact that an inverse relationship between the dielectric constant of a material and its breakdown strength has been shown (44). From this work the breakdown field follows a $(\kappa)^{-1/2}$ dependence, which suggests that maintaining sufficient breakdown strength at high κ values will be challenging.

Besides the initial properties of the dielectric and its interfaces, the ability to maintain these properties during a full CMOS process is critical. Thermal budgets in a CMOS process can exceed 1000°C. During this high temperature processing the gate dielectric must maintain its desirable properties and also resist intermixing with either the gate above it or the channel below it. In addition the gate dielectric must serve as a diffusion barrier to prevent dopants or other elements from the gate from diffusing through the dielectric into the channel. One significant problem seen with many gate dielectrics is that boron doping from a polysilicon gate diffuses through the gate dielectric into the channel during processing and causes threshold shifts in the device (36,45,46), so if polysilicon gates are used, the gate dielectric stack must act as an effective barrier to boron diffusion.

Another process related demand on the gate dielectric is that it have good etch properties. In general the gate conductor is etched using reactive ion etching. This etch is designed with high selectivity so that it can etch through the relatively thick gate conductor layer and stop on the relatively thin gate dielectric. To minimize process changes the gate dielectric must have a chemistry compatible with this type of selective etching. At a later step at least a portion of the gate dielectric must be removed from the silicon surface. This is needed to reduce short channel effects as well as to make contact to the silicon. For this the chemistry of the gate dielectric must be compatible with an etch that will attack the gate dielectric but be selective to the underlying silicon.

Etching or other processing of the high-k gate dielectric can also introduce elements from this dielectric into the fab. Since many of the materials under consideration for high-k gate dielectrics contain elements not commonly found in a semiconductor fab, care must be taken to evaluate the level of these elements produced in the device fabrication as well as the effect that this level of contamination will have on device and circuit performance (47).

Plasma induced charging effects are another issue that high-k gate dielectrics must withstand. These effects occur when plasma-assisted processing is used in the manufacture of CMOS circuits. Charges generated during this processing can accumulate on the gate electrode and create high fields and leakage current through the gate dielectric. This can lead to current generated defects and other reliability issues (48). Since high-k materials have greater physical thickness for the same capacitance as silicon dioxide, there will be more volume for trap generation from the plasma charging effects and greater chance of damage from these effects (49).

As mentioned above, a hydrogen anneal is typically used to passivate the interface between SiO_2 and silicon. This hydrogen anneal is a standard part of the CMOS process and is typically done at the end of the front end processing. This hydrogen anneal, however, can be detrimental to the bulk properties of some high-k gate dielectrics and to their interfaces. Since many of the applications envisioned for high-k gate dielectrics also include devices with silicon dioxide gate dielectrics, the high-k gate dielectric may be exposed to a hydrogen anneal. For this reason, it would be advantageous that the high-k dielectric material be able to withstand this anneal.

Another consideration for the gate dielectric is its manufacturability. Since the gate dielectric properties and thickness have a dramatic impact on the performance of the device, these properties must be very well controlled to manufacture large integrated circuits with a high yield. The dielectric must have repeatable, well-controlled properties across a large area wafer and also from run-to-run. The deposition of this dielectric also needs to be a cost effective process with a high throughput.

5. ALTERNATIVE CMOS STRUCTURES

Most of the discussion above has focused on the integration of high-k gate dielectrics with conventional CMOS devices. While the gate dielectric changes are some of the most radical deviations from standard CMOS, other changes to CMOS materials and structures are also under consideration for future CMOS devices. These changes can impact the requirements for the gate dielectric.

For example, depletion of the polysilicon gate in conventional CMOS devices can decrease the capacitance in the device. One solution is to use a metal as the gate electrode. For conventional bulk CMOS devices, the polysilicon can be doped either n-type or p-type for nMOS or pMOS devices to achieve the correct threshold voltage. Since the work function of metals cannot be easily changed, most metal gate replacements for bulk CMOS devices are actually dual metal gate systems.

Since the metal may react differently from polysilicon in contact with the gate dielectric, the total gate stack must be considered when selecting a gate dielectric material. The choice of gate metal can affect what interface layers are needed between the gate metal and the dielectric as well as the thermal and chemical stability of the gate stack. The interaction between the gate metal and the dielectric can also affect the threshold voltage of the device (50) and cases of Fermi level pinning have been observed with polysilicon/high-k gate stacks (51). Not only must the dielectric perform well with one gate metal but in these dual gate systems, must perform equally well with two different metals.

Besides these changes in gate stack material other changes in the basic structure of CMOS devices are under consideration. In fully depleted semiconductor on insulator, FDSOI (52, 53), devices an insulating layer is inserted under the channel as shown in Fig. 10. This insulating layer reduces the junction capacitance and also the junction leakage leading to better performance. In addition FDSOI devices use very low doping in the channel which improves the carrier mobility and reduces the vertical field in the device. These changes can have a significant impact on the gate dielectric. Since

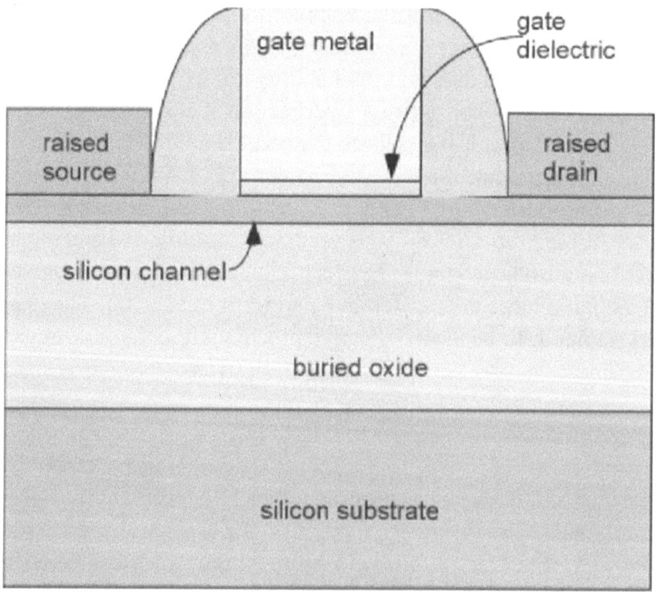

Fig. 10. Schematic drawing of fully depleted semiconductor on insulator device structure.

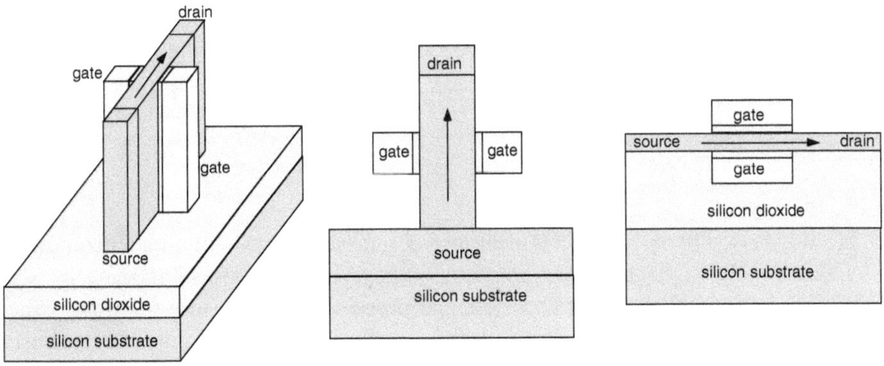

Fig. 11. Schematic representations of three multiple gate MOSFET configurations. These are a finFET, a vertical FET and a planar dual gate FET.

the vertical field is reduced, the channel carriers on average are further from the dielectric interface; so effects such as interface scattering, phonon scattering (54) or coulombic scattering from charge in the dielectric are reduced and choices for the high-k dielectric as well as the interface layer may be different from bulk devices. Also the processing of the FDSOI may affect dielectric choice. Since the channel layer in a FDSOI device is very thin, 5–10 nm, etch selectivity between the dielectric and the channel must be very good. Also the removal of the dielectric must be sufficiently complete to allow growth of high quality raised source/drain regions which are used to reduce the access resistance to the device.

Another new structure under consideration is a dual or other multiple gate structure. In this structure the single gate on top of the channel is replaced by multiple gates on two or more sides of the channel, see Fig. 11. This device reduces the junction capacitance and leakage as in the FDSOI device, but also can increase the transconductance per unit area in the device. Currently the most popular form of multiple gate device under investigation is the finFET and its derivatives (55, 56). In this device a silicon fin is formed from a SOI substrate and gates are patterned on the sidewalls of this fin. The interface to the gate dielectric then is on a patterned, vertical surface. Residual damage to this surface, misorientation of this surface due to patterning and differences in dielectric deposition on a vertical versus a horizontal surface may all affect the performance of the gate dielectric in this device.

Besides these structural changes, further material changes are also under consideration. As mentioned in the introduction the initial dominance of silicon in the digital semiconductor industry was largely predicated on its native oxide. If a high-k gate dielectric is needed anyway, several alternative materials may also work with a high-k gate dielectric and have superior properties compared to silicon. Due to this shift renewed interest has been seen in SiGe or Ge channels which have higher electron and hole mobilities as well as some interest in III–V channel devices with high-k gate dielectrics (57–61). The gate dielectric and interface will need to be tailored to the new channel materials, however, there is no fundamental reason that dielectric performance on a Ge or SiGe channel cannot meet or exceed the performance of a non-SiO_2 gate dielectric on Si.

6. CONCLUSION

From a device perspective, the ideal high-k gate dielectric would be silicon dioxide with a $4\times$–$5\times$ higher dielectric constant that real silicon dioxide. This material would have the physical, chemical and electrical properties, except permittivity, of silicon dioxide. Such a material would be ideal for minimizing the changes in device design, modeling, circuit design and manufacturing that happen with each new generation of CMOS. Since this ideal material does not exist, tradeoffs will have to be made not only with the dielectric material development but also with the device design, circuit design and manufacturing. This process will be costly and difficult but also necessary if scaling of CMOS technology is to continue for an extended time.

REFERENCES

1. D.A. Buchanan, S.-H. Lo, Growth, characterization and the limits of ultra-thin SiO_2-based dielectrics for future CMOS applications, in: *The Physics and Chemistry of SiO_2 and the Si–SiO_2 interface-3*, eds. H.Z. Massoud, E.H. Poindexter, C.R. Helms, (The Electrochemical Society, Pennington, NJ, 1996), pp. 3–14.
2. G.E. Moore, Lithography and the future of Moore's Law, *SPIE* **2438**, 2–17 (1995).
3. D.A. Muller, T. Sorsch, S. Moccio, F.H. Baumann, K. Evans-Lutterodt, G. Timp, The electronic structure at the atomic scale of ultrathin gate oxides, *Nature*, **399**, 758–761 (1999).
4. S. Tang, R.M. Wallace, A. Seabaugh, and D. King-Smith, Evaluating the minimum thickness of gate oxide on silicon using first-principles method, *Appl. Surf. Sci.* **135**, 137–142 (1998).
5. S.-H. Lo, D.A. Buchanan, Y. Taur, Modeling and characterization of quantization, polysilicon depletion, and direct tunneling effects in MOSFETs with ultrathin oxides, *IBM J. Res. Develop.* **43**(3), 327–337 (1999).
6. C.-H. Choi, K.-Y. Nam, Z. Yu, and R.W. Dutton, Impact of gate direct tunneling current on circuit performance: a simulation study, *IEEE Trans. Electron Devices*, **48**(12), 2823–2829 (2001).
7. Y. Taur, D.A. Buchanan, W. Chen, D.J. Frank, K.E. Ismail, S.-H. Lo, G.A. Sai-Halasz, R.B. Viswanathan, H.-J. C. Wann, S. J. Wind and H.-S. Wong, CMOS scaling into the nanometer regime, *Proc. IEEE*, **85**(4), 486–504 (1997).
8. D. Frank, R.H. Dennard, E. Nowak, P.M. Solomon, Y. Taur and H.-S. P. Wong, Device scaling limits of Si MOSFETs and their application dependencies, *Proc. IEEE*, **89**(3), 259–288 (2001).
9. A. Toriumi, Reliability perspective of high-k gate dielectrics—What is different from SiO2? in: *2002 7th international symposium on plasma and process induced damage*, 4–9 (2002).
10. J.H. Stathis and D.J. DiMaria, Reliability projection for ultra-thin oxides at low voltage, *Int. Electron Device Meeting*, 167–170 (1998).

11. E.J. Nowak, Maintaining the benefits of CMOS scaling when scaling bogs down, *IBM J. Res. Develop.* **46**(2/3), 169–180 (2002).
12. D.J. Frank, Power-constrained device and technology design for the end of scaling, in: *International Electron Device Meeting*, 643–646 (2002).
13. C.M. Carlson, T.V. Rivkin, P.A. Parilla, J.D. Perkins D.S. Ginley, A.B. Kozyrev, V.N. Oshadchy, and A.S. Pavlov, Large dielectric constant ($\varepsilon/\varepsilon_0 > 6000$) $Ba_{0.4}Sr_{0.6}TiO_3$ thin films for high-performance microwave phase shifters, *Appl. Phys. Lett.* **76**(14), 1920–1922 (2000).
14. C.B. Parker, J.-P. Maria, and A.I. Kingon, Temperature and thickness dependent permittivity of (Ba,Sr)TiO_3 thin films, *Appl. Phys. Lett.* **81**(2), 340–342 (2002).
15. R.J. Cava, W.F. Peck, Jr., J.J. Krajewski, G.L. Roberts, B.P. Barber, H.M. O'Bryan, and P.L. Gammel, Improvement of the dielectric properties of Ta_2O_5 through substitution with Al_2O_3, *Appl. Phys. Lett.* **70**(11), 1396–1398 (1997).
16. Z.G. Zhang, D.P. Chu, B.M. McGregor, P. Migliorato, K. Ohashi, K. Hasegawa, and T. Shimoda, Frequency dependence of the dielectric properties of La-doped $Pb(Zr_{0.35}Ti_{0.65})O_3$ thin films, *Appl. Phys. Lett.* **83**(14), 2892–2894 (2003).
17. N.R. Mohapatra, M.P. Desai, S.G. Narendra, and V.R. Rao, The effect of high-K gate dielectrics on deep submicrometer CMOS device and circuit performance, *IEEE Trans. Electron Devices*, **49**(5), 826–831 (2002).
18. G.C.-F. Yeap, S. Krishnan, and M.-R. Lin, Fringing-induced barrier lowering (FIBL) in sub-100nm MOSFETs with high-K gate dielectrics, *Electron. Lett.* **34**(11), 1150–1152 (1998).
19. C.-H. Lai, L.-C. Hu, H.-M. Lee, L.-J. Do, Y.-C. King, New stack gate insulator structure reduce FIBLE effect obviously, in: *2001 International Symposium on VLSI Technology, Systems, and Applications, Proceedings of Technical Papers*, 216–219 (2001).
20. B. Cheng, M. Cao, R. Rao, A. Inani, P. VandeVoorde, W.M. Greene, J. M.C. Stork, Z. Yu, P.M. Zeitzoff and J.C.S. Woo, The impact of high-K gate dielectrics and metal gate electrodes on sub-100nm MOSFET's, *IEEE Trans. Electron Devices*, **46**(7), 1537–1544 (1999).
21. L. Manchanda, B. Busch, M.L. Green, M. Morris, R.B. vna Dover, R. Kwo, and S. Aravamudhan, High K gate dielectrics for the silicon industry, in: *IWGI 2001*, 56–60 (2001).
22. G. W. Lee, J.-H. Lee, H.-W. Lee, M.-K. Park, D.-G. Kang, and H.-K. Youn, Trap evaluations of metal/oxide/silicon field-effect transistors with high-K gate dielectric using charge pumping method, *Appl. Phys. Lett.* **81**(11), 2050–2052 (2002).
23. B. Mereu, B. Vellianitis, B. Apostolopoulos, A. Dimoulas, M. Alexe, Fowler–Nordheim tunneling in epitaxial yttrium oxide on silicon for high-K gate applications, in: *Proceedings of CAS 2002 Semiconductor Conference*, Vol. 2, 309–312 (2002).
24. C. Chaneliere, S. Four, J.L. Autran, R.A.B. Devine, and N.P. Sandler, Properties of amorphous and crystalline Ta_2O_5 thin films deposited on Si from a $Ta(OC_2H_5)_5$ precursor, *J. Appl. Phys.* **83**(9), 4823–4829 (1998).
25. A. Kumar, T.H. Ning, M.V. Fischetti and E. Gusev, Hot-carrier charge trapping and reliability in high-K dielectrics, in: *2002 Symposium on VLSI Technology Digest of Technical Papers*, 152–153 (2002).
26. T.P. Ma, High-k gate dielectrics for scaled CMOS technology, in: *Proceedings of 6th International Conference on Solid-State and Integrated-Circuit Technology*, Vol. 1, 297–302 (2001).
27. K. Ahmed, E. Ibok, B.C.-F. Yeap, Q. Xiang, B. Ogle, J.J. Wortman and J.R. Hauser, Impact of tunnel currents and channel resistance on the characterization of channel inversion layer charge and polysilicon-gate depletion of sub-20-A gate oxide MOSFETs, *IEEE Trans. Electron Devices*, **46**, 1650–1655 (1999).
28. W.K. Henson, K.Z. Ahmed, E.M. Vogel, J.R. Hauser, J.J. Wortman, R.D. Venables, M. Xu and D. Venables, Estimating oxide thickness of tunnel oxides down to 1.4 nm using

conventional capacitance–voltage measurements on MOS capacitors, *IEEE Electron Device Lett.* **20**, 179–181 (1999).
29. K.J. Yang and C. Hu, MOS capacitance measurements for high-leakage thin dielectrics, *IEEE Trans. Electron Devices*, **46**, 1500–1501 (1999).
30. A. Nara, N. Yasuda, H. Satake and A. Toriumi, Applicability limits of the two-frequency capacitance measurement technique for the thickness extraction of ultrathin gate oxide, *IEEE Trans. Semicond. Manufact.* **15**, 209–213 (2002).
31. C.A. Richter, A.R. Hefner and E.M. Vogel, A comparison of quantum–mechanical capacitance–voltage simulators, *IEEE Electron Device Lett.* **22**, 35–37 (2001).
32. C.-H. Choi, J.-S. Goo, T.-Y. Oh, Z. Yu, R.W. Dutton, A. Bayoumi, M. Cao, P. VandeVoorde, D. Vook and C.H. Diaz, MOS C-V characterization of ultrathin gate oxide thickness (1.3–1.8 nm), *IEEE Electron Device Lett.* **20**, 292–293 (1999).
33. A. Shanware, J.P. Shiely, H.Z. Massoud, E. Vogel, K. Henson, A. Srivastava, C. Osburn, J.R. Hauser and J.J. Wortman, Extraction of the gate oxide thickness of n- and p-channel MOSFETs below 20A from the substrate current resulting from valence-band electron tunneling, in: *Proceedings of the International Electron Device Meeting*, 815–818 (1999).
34. N. Yang, W.K. Henson, J.R. Hauser and J.J. Wortman, Modeling study of ultrathin gate oxides using direct tunneling current and capacitance–voltage measurements in MOS devices, *IEEE Trans. Electron Devices*, **46**, 1464–1471 (1999).
35. S. Kar, Extraction of the capacitance of ultrathin high-K gate dielectrics, *IEEE Trans. Electron Devices*, **50**, 2112–2119 (2003).
36. E.P. Gusev, D.A. Buchanan, E. Cartier, A. Kumar, D. DiMaria, S. Guha, A. Callegari, S. Zafar, P.C. Jamison, D.A. Neumayer, M. Copel, M.A. Gribelyuk, H. Okorn-Schmidt, C. D'Emic, P. Kozlowski, K. Chan, N. Bojarczuk, L.-A. Ragnarsson, P. Ronsheim, K. Rim, R.J. Fleming, A. Mocuta, and A. Ajmera, Ultrathin high-K gate stacks for advanced CMOS devices, *Int. Electron Device Meeting*, 451–454 (2001).
37. I. Polishchuk and C. Hu, Electron wavefunction penetration into gate dielectric and interface scattering—an alternative to surface roughness scattering model, in: *2001 Symposium on VLSI Technology Digest of Technical Papers*, 51–52 (2001).
38. Y.-Y. Fan, Q. Xiang, J. An, L.R. Register, and S.K. Banerjee, Impact of interfacial layer and transition region on gate current performance for high-K gate dielectric stack: Its tradeoff with gate capacitance, *IEEE Trans. Electron Devices*, **50**(2), 433–439 (2003).
39. D. Vasilecka and D.K. Ferry, Scaled silicon MOSFETs: Universal mobility behavior, *IEEE Trans. Electron Devices*, **44**(4), 577–583 (1997).
40. A. Pirovano, A.L. Lacaita, G. Zandler, and R. Oberhuber, Explaining the dependences of electron and hole mobilities in Si MOSFET's inversion layer, *Int. Electron Device Meeting*, 527–530 (1999).
41. S. Zufar, A. Callegari, E. Gusev, M.V. Fischetti, Charge trapping in high K gate dielectric stacks, *Int. Electron Device Meeting*, 517–520 (2002).
42. I.-C. Chen, S.E. Holland, and C. Hu, Electrical breakdown in thin gate and tunneling oxides, *IEEE J. Solid-State Circuits*, **Sc-20**(1), 333–342 (1985).
43. H. Iwai and S. Ohmi, CMOS downsizing and high-K gate insulator technology, in: *4th IEEE International Caracas Conference on Device, Circuits and Systems*, 1–8 (2002).
44. J. McPherson, J. Kim, A. Shanware, H. Mogul, and J. Rodriguez, Proposed universal relationship between dielectric breakdown and dielectric constant, *Int. Electron Device Meeting*, 633–636 (2002).
45. J.R. Pfiester, F.K. Baker, T.C. Mele, H.-H. Tseng, P.J. Tobin, J.D. Hayden, J.W. Miller, C.D. Gunderson, and L.C. Parrillo, The effects of boron penetration on p+ polysilicon gated PMOS devices, *IEEE Trans. Electron Devices*, **37**(8), 1842–1851 (1990).
46. K. Onishi, L. Kang, R. Choi, E. Dharmarajan, S. Gopalan, Y. Jeon, C.S. Kang, B.H. Lee, R. Nieh, and J.C. Lee, Dopant penetration effects on polysilicon gate HfO_2 MOSFET's, in: *2001 Symposium on VLSI Technology Digest of Technical Papers*, 131–132 (2001).

47. B. Vermeire, K. Delbridge, V. Pandit, H.G. Parks, S. Raghavan, K. Ramkumar, S. Geha, and J. Jeon, The effect of hafnium or zirconium contamination on MOS processes, in: *2002 IEEE/SEMI Advanced Semiconductor Manufacturing Conference*, 299–303 (2002).
48. C.Y. Chang, T.S. Chao, H.C. Lin, and C.H. Chien, Process-related reliability issues toward sub-100 nm device regime, in: *Proceedings of 23rd International Conference on Microelectronics*, Vol. 1, pp. 133–140 (2002).
49. P.-J. Tzeng, Y.-Y. I. Chang, C.-C. Yeh, C.-C. Chen, C.-H. Liu, M.-Y. Liu, B.-F. Wu, and K.-S. Chang-Liao, Plasma-charging effects on submicron MOS devices, *IEEE Trans. Electron Devices*, **49**(7), 1151–1157 (2002).
50. Y.-C. Yeo, P. Ranade, Q. Lu, R. Lin, T.-J. King, and C. Hu, Effects of high-K dielectrics on the workfunctions of metal and silicon gates, in: *2001 Symposium on VLSI Technology Digest of Technical Papers*, 49–50 (2001).
51. C. Hobbs, L. Fonseca, v. Dhandapani, S. Samavedam, B. Taylor, J. Grant, L. Dip, D. Triyoso, R. Hegde, D. Gilmer, R. Garcia, D. Roan, L. Lovejoy, R. Rai, L. Hebert, H. Tseng, B. White, and P. Tobin, Fermi level pinning at the polySi/metal oxide interface, in: *2003 Symposium on VLSI Technology Digest of Technical Papers*, 9–10 (2003).
52. Z. Krivokapic, W. Maszara, F. Arasnia, E. Paton, Y. Kim, L. Washington, E. Zhao, J. Chan, J. Zhang, A. Marathe, M-R. Lin, High performance 25 nm FDSOI devices with extremely thin silicon channel, in: *2003 Symposium on VLSI Technology Digest of Technical Papers*, 131–132 (2003).
53. S. Bagchi, J.M. Grant, J. Chen, S. Samavedam, F. Huang, Pl Tobin, J. Conner, L. Prabhu, and M. Tiner, Fully depleted SOI devices with TiN gate and elevated source-drain structures, in: *2000 IEEE International SOI Conference*, 56–57 (2000).
54. M.V. Fischetti, D.A. Neumayer, and E.A. Cartier, Effective electron mobility in Si inversion layers in metal–oxide–semiconductor systems with a high-K insulator: The role of remote phonon scattering, *J. Appl. Phys.* **90**(9), 4587–4608 (2001).
55. B. Yu, L. Chang, S. Ahmed, H. Wang, S. Bell C.-Y. Yang, C. Tabery, C. Ho, Q. Xiang, T.-J. King, J. Bokor, C. Hu, M.-R. Lin, and D. Kyser, FinFET scaling to 10 nm gate length, *Int. Electron Device Meeting*, 251–254 (2002).
56. J. Kedzierski, E. Nowak, T. Kanarsky, Y. Zhang, D. Boyd, R. Carruthers, C. Cabral, R. Amos, C. Lavoie, R. Roy, J. Newbury, E. Sullivan, J. Benedict, P. Saunders, K. Wong, D. Canaperi, M. Krishnan, K.-L. Lee, B.A. Rainey, D. Fried, P. Cottrell, H.-S.P. Wong, M Ieong, and W. Haensch, Metal-gate FinFET and fully-depleted SOI devices using total gate silicidation, *Int. Electron Device Meeting*, 247–250 (2002).
57. T. Tezuka, N. Sugiyama, T. Mizuno, and S. Takagi, Novel fully-depleted SiGe-on-insulator pMOSFETs with high-mobility SiGe surface channels, *Int. Electron Device Meeting*, 946–949 (2001).
58. Z. Shi, D. Onsongo, K. Onishi, J.C. Lee, and S.K. Banerjee, Mobility enhancement in surface channel SiGe pMOSFETs with HfO_2 gate dielectrics, *IEEE Electron Device Lett.* **24**(1), 34–36 (2003).
59. C.O. Chui, H. Kim, D. Chi, B.B. Triplett, P.C. McIntyre, and K.C. Saraswat, A sub-400°C Germanium MOSFET technology with high-K dielectric and Metal gate, *Int. Electron Device Meeting*, 437–500 (2002).
60. H. Shang, H.Okorn-Schindt, J. Ott, P. Kozlowski, S. Steen, E.C. Jones, H.-S.P. Wong, and W. Hanesch, Electrical characterization of germanium p-channel MOSFETs, *IEEE Electron Device Lett.* **24**(4), 242–244 (2003).
61. P.D. Ye, G.D. Wilk, J. Kwo, B. Yang, H.-J.L. Gossmann, M. Frei, S.N.G. Chu, J.P. Mannaerts, M. Sergent, M. Hong, K.K. Ng, and J. Bude, GaAs MOSFET with oxided gate dielectric grown by atomic layer deposition, *IEEE Electron Device Lett.* **24**(4), 209–211 (2003).

Chapter 3

THERMODYNAMICS OF OXIDE SYSTEMS RELEVANT TO ALTERNATIVE GATE DIELECTRICS

ALEXANDRA NAVROTSKY AND SERGEY V. USHAKOV

Thermochemistry Facility and NEAT ORU, University of California at Davis, Davis, CA 95616, USA

1. INTRODUCTION

The search for gate dielectric materials superior to amorphous silica requires the input of thermodynamic data to assess materials compatibility with silicon and stability against crystallization or unwanted phase transformation during processing and/or subsequent use. Because superior dielectric properties are generally associated with materials containing heavy ions of large size and high charge, emphasis has been on trivalent and tetravalent oxides containing Ti, Zr, Hf, and the rare earths. Furthermore, the materials must be insulating (disqualifying ions of variable valence and making TiO_2 somewhat questionable) and the oxide must be less reducible than SiO_2 so that it does not oxidize silicon. Within these constraints, an amorphous film is considered superior to a polycrystalline one with grain boundaries, so it is important to understand the persistence, controlled by both thermodynamic and kinetic factors, of amorphous and glassy materials. The purpose of this paper is to summarize relevant thermodynamic data, to provide a thermodynamic and structural framework for considering new compositions and their likely properties, and to present some new calorimetric data on bulk and thin film systems based on ZrO_2 and HfO_2.

2. COMPATIBILITY WITH SILICON

One of the central issues in thermodynamics of alternative gate dielectric material is compatibility with silicon at processing conditions. Below are examples of possible reactions to be considered (1–3):

(a) Oxidizing silicon

$$Si + A_xO_y \rightarrow SiO_2 + xA \tag{1}$$

A.A. Demkov and A. Navrotsky (eds.), Materials Fundamentals of Gate Dielectrics, 57–108.

(b) Silicide or/and silicate formation

$$Si + A_xO_y \rightarrow A_xSi + SiO_2 \tag{2}$$
$$Si + A_xO_y \rightarrow A + A_xSiO_y \tag{3}$$
$$Si + A_xO_y \rightarrow A_xSi_y + A_xSiO_y \tag{4}$$

(c) A silicon oxide layer may be left on the silicon surface to aid dielectric film deposition or it may form by oxygen diffusion through the dielectric upon film growth or annealing in an oxidizing environment through the reaction:

$$Si + A_xO_y + O_2\,(g) \rightarrow SiO_2 + A_xO_y \tag{5}$$

This enables formation of crystalline or amorphous silicates through reaction:

$$SiO_2 + A_xO_y \rightarrow A_xSiO_y \tag{6}$$

(d) On processing at low oxygen pressures, oxygen deficient dielectric oxides may form which will have different thermodynamic properties, also SiO gas may form and diffuse through the oxide thin film:

$$Si + A_xO_y \rightarrow A_xSi_y + SiO\,(g) \tag{7}$$

Hubbard and Scholm (1, 2) systematically assessed binary oxides for the feasibility of reactions (1)–(4) at temperatures to 1300°C. It follows from their work that there are insufficient thermodynamic data to complete calculations for many prospective high-k candidates including Hf, Al and RE (including Y and Sc) oxides, even if interfacial energies are neglected. In thin films, however, energetics of interfaces often defines crystalline or amorphous phases formed in the above reactions and may affect usefulness of the proposed dielectrics. Unfortunately, such data for oxide systems relevant to alternative gate dielectrics are scarce and much is left to be done in this field. Thus, the second part of this review is devoted to the methods of measuring interfacial energies and to evaluation of available data and trends.

Obviously, for thermodynamic evaluation of the feasibility of the above reactions, thermodynamic data, both on surface and bulk energies, are needed both for oxides and silicides. Thermodynamic of silicides is another, largely unmapped, territory, though recently new thermodynamic data for some of bulk silicides have been obtained by high temperature direct synthesis calorimetry (4–6). However, silicides are beyond the scope of this review.

3. SOME GENERAL CONCEPTS FROM CRYSTAL CHEMISTRY AND THERMODYNAMICS

The purpose of this section is to summarize some concepts of structure and stability that may be less familiar to the semiconductor community than to the ceramics community and that are potentially useful in evaluating possible dielectric coatings.

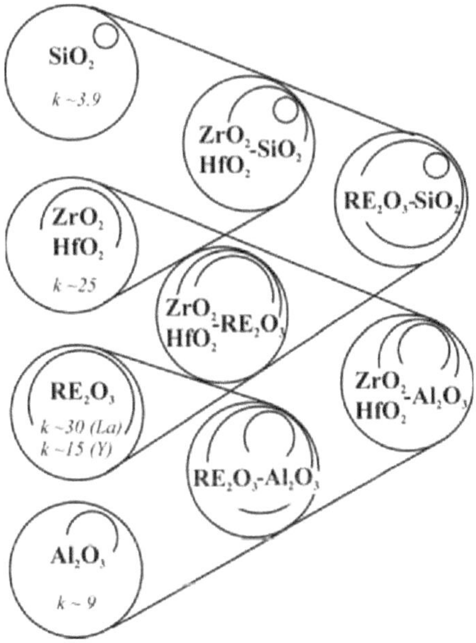

Fig. 1. Oxides and binary systems relevant to alternative gate dielectrics considered in this paper. Relative sizes of the circles on diagram reflect differences in ionic radii of cations with respect to oxygen. The dielectric constants for binary oxides are given after Wilk et al. (99).

Vitreous and amorphous silica possess structures based on a three dimensional network of linked SiO_4 tetrahedra, with these tetrahedra defining rings of different sizes. The local environment of silicon and its four oxygen neighbors is quite similar in crystalline and amorphous silica, but the latter lacks long range order. The tetrahedral Si–O bond length is about 0.16 nm, which is short compared to bond lengths for Ti–O, Zr–O, Hf–O or rare earth oxides (see Fig. 1). The large ions generally require higher coordination numbers as well. Thus if one attempts to "alloy" SiO_2 with other oxides, even those having 4+ cations, it is unlikely that these will enter the tetrahedral framework; rather they will disrupt it. Similarly Si^{4+} cannot readily substitute into the structures of these other oxides, but ternary silicate compounds, such as $ZrSiO_4$ (zircon), with totally different structures, can form. Thus ZrO_2 and HfO_2 are considered network modifiers, rather than network formers, in glass science. The rare earth oxides and the alkaline earth oxides are likewise network modifiers, and, for charge balance in the structure, their addition must break Si–O–Si bonds and disrupt the network. They also form silicate compounds, and the stability of these compounds increases with increasing difference in charge and size between Si and the other ion, that is, with increasing difference in basicity of the oxide. Here basicity is defined as the capability of an oxide to donate an oxide ion to silicon by the reaction:

$$\mathrm{Si{-}O{-}Si} + \mathrm{M{-}O} \rightarrow 2\mathrm{Si{-}O} + \mathrm{M} \quad (8)$$

Greater basicity means more stable silicate compounds.

Many binary metal oxide–silica systems show stable liquid state or metastable subliquidus immiscibility, which carries over to their quenched glasses (e.g., ZrO_2–SiO_2 (Fig. 8), RE_2O_3–SiO_2 (Fig. 12)). Typically, this two phase region extends from nearly pure silica to about 20 mol% of the oxide additive, so the solubility of an oxide in silica is limited in liquid or glass. However, systems prepared by low temperature deposition may be more continuous in composition, but one must be aware that they may be unstable, thermodynamically, to phase separation (into two amorphous phases, and the more silica-poor one may crystallize readily). The addition of alumina usually reduces the tendency toward phase separation because the tetrahedral network can be maintained by a charge-balanced substitution

$$\text{Si–O–Si} + \text{M–O} + 2\text{Al–O} \rightarrow 2\text{Si–O–Al} + \text{M} \tag{9}$$

or

$$\text{Si}^{4+}\ \text{(tetrahedral) framework} \rightarrow (1/n)\,\text{M}^{n+} + \text{Al}^{3+}\ \text{(tetrahedral framework)} \tag{10}$$

For amorphous silicates containing rare earths, zirconium, and hafnium, there is evidence that even when the amorphous phase or quenched glass may be homogeneous on the micron scale (and appear as an optically clear glass) the large ions are locally clustered, and their thermodynamics is dominated by this clustering (7). Again, the addition of alumina decreases the tendency toward clustering. Pre-existing clusters may give an easy pathway for phase separation or crystallization.

Dielectric coatings are prepared by nonequilibrium deposition techniques near room temperature. Films that do not exhibit X-ray diffraction peaks are termed “amorphous”. Their detailed structure may be complex, with small areas of crystallinity visible by electron microscopy, heterogeniety on the nanoscale, and/or short-range order that gives individual cations (Zr, Hf, La, Si, etc.) coordination environments (of oxygen and in some cases nitrogen) similar to those in macroscopic crystalline materials. These amorphous films may be similar to other amorphous materials (melt quenched glasses (8–14), radiation-damaged solids (15–18), sol–gel prepared amorphous powders (19–24), but there may be important differences in structure and properties as well. This is an area of active research where there are more questions than answers at present.

Thin films and nanophase powders have large surface areas, often up to several hundred m^2/g. Thus surface energies and interfacial energies may modify their thermodynamics. It is becoming increasingly recognized that, in systems like alumina, titania, and zirconia, which have several polymorphs, there may be a crossover in stability at the nanoscale because of the interplay of energetics of phase transitions and differing surface energies (28–36). Analogous thermodynamic effects may arise from differing interfacial energies and strain.

Bearing these issues in mind, we have organized this review as follows.

Figure 1 shows the major single component, binary, and multicomponent oxide systems which may be useful coatings. In each case, we discuss their phase diagrams, thermodynamic properties, and thermal stability (in both kinetic and thermodynamic contexts). Figures 2 and 3 show selected crystal structures in the given systems.

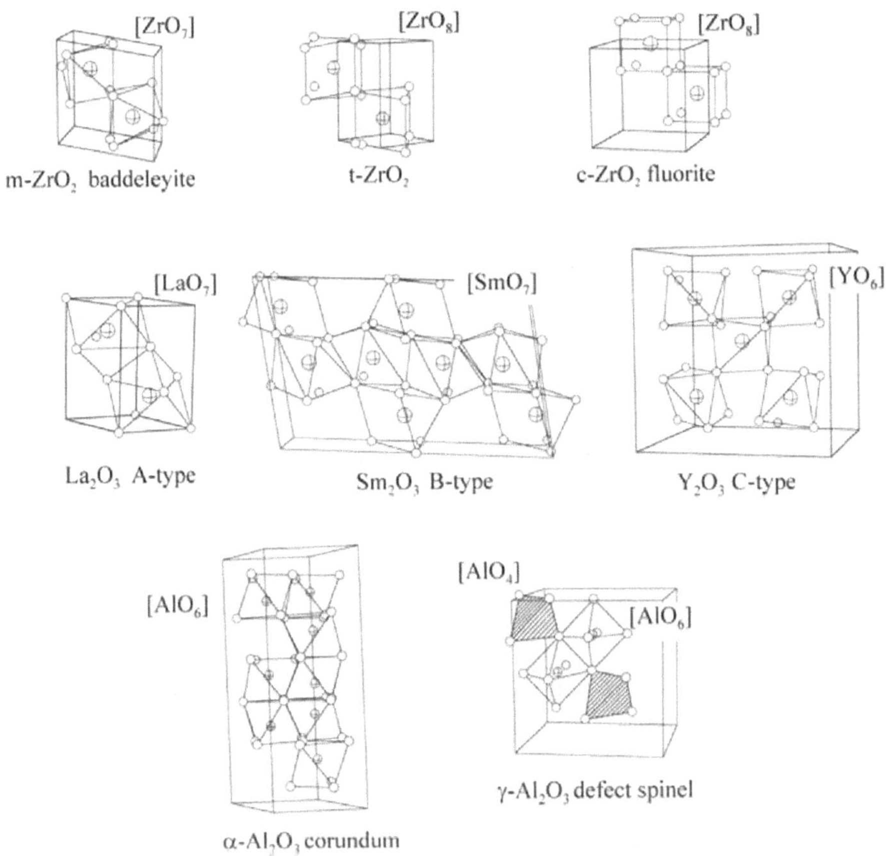

Fig. 2. Some structure types of binary oxides of zirconium (hafnium), aluminum and rare earths.

We start with the equilibrium phases and then proceed to metastable crystalline and amorphous materials. In the latter cases we include some new data recently obtained in our research group. Thermochemistry of crystalline and amorphous silica is reviewed elsewhere (37, 39).

PART I. REVIEW OF RELEVANT SYSTEMS

I.1. GENERAL REMARKS

We include equilibrium phase diagrams for the binary systems considered in this review. The diagrams are all drawn for the same temperature-composition range to provide straightforward comparison between systems with hafnia and zirconia or yttria and lanthana. The phase notation is changed from those used in original studies

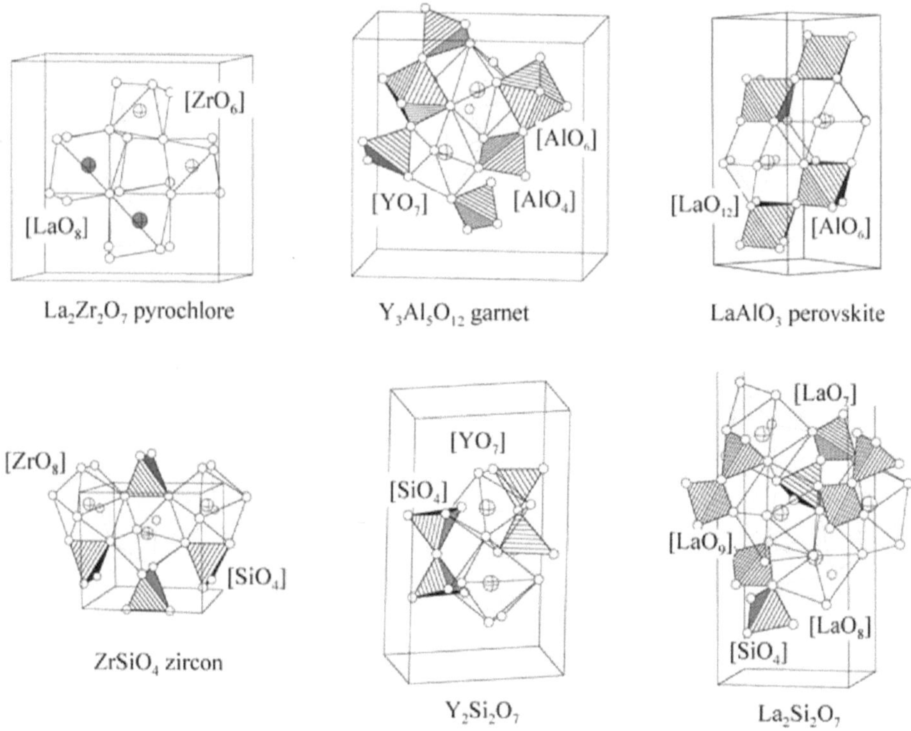

Fig. 3. Some structure types of ternary oxides of zirconium (hafnium), aluminum, rare earths and silicon.

to be consistent throughout the chapter. Only solid solution (alloy) and compound fields are labeled and experimental data points are not shown. For some systems the diagrams are complemented with newly refined transition temperatures or additional data. Some differences in the melting points of the same end member compounds can be noted between diagrams. No attempt has been made to unify them and they represent the differences between data and references in original research reports. The diagrams are thus semi-quantitative and mainly for illustrative purposes. The reader is referred to the original reports cited for all systems.

In general, diagrams with zirconia have been studied much more extensively than those with hafnia owing to the much more extensive use of zirconia in the ceramic industry (40, 41), due both to lower cost and lower temperature of the monoclinic-to-tetragonal phase transition. Because of similarity between Zr and Hf systems, phases found in zirconia systems (e.g., ordered compounds in ZrO_2–Y_2O_3) may actually exist in the corresponding hafnia systems and might be found in more detailed further studies or on reaching equilibrium.

The high-k dielectric SiO_2 replacement is most often thought to be amorphous and produced by low temperature nonequilibrium deposition techniques. Nevertheless, the equilibrium phase diagrams can give insight on the phases likely to occur on crystallization and phase separation. In this review we use the term "metastable" to indicate that a given material is higher in free energy than a coarse grained crystalline polymorph or other phase assemblage. Interface/surface energy terms can change thermodynamically stable phases and transition temperatures in systems constrained to nanophase dimensions, as discussed below. If surface area is considered as a variable, it gives another dimension to phase diagrams and effectively shrinks the list of "metastable" phases. The stabilization of tetragonal ZrO_2 and γ-Al_2O_3 at room temperature was suggested to be a particle size effect and recent calorimetric measurements confirmed this as a crossover in enthalpy (30, 36). To our knowledge, there are no systematic studies on surface/interface area effects on phase equilibrium in the binary systems discussed here, though some data were obtained for the Y_2O_3–ZrO_2 and ZrO_2–Al_2O_3 systems (41, 42) (due to their application as yttria-stabilized zirconia (YSZ) and zirconia-toughened alumina (ZTA) ceramics). In the absence of detailed information on effects of surface area on stabilization of amorphous phases, it is instructive to compare the crystallization behavior of glasses prepared by quenching with the crystallization of the same compositions prepared by low-temperature routes, e.g., sol–gel, decomposition of salts and hydroxides and thin film deposition techniques.

I.2. BINARY OXIDES

I.2.1 ZrO_2 and HfO_2—Stable Phases: Monoclinic, Tetragonal, Cubic (Fluorite)

ZrO_2 and HfO_2 form continuous solid solutions with increasing temperature and decreasing hysteresis of phase transitions (43–46). Continuous solid solutions are also common for Hf and Zr in ternary compounds (50)

Figure 4 demonstrates thermal expansion behavior of HfO_2 and ZrO_2. Hf^{4+} is just slightly smaller than Zr^{4+} (0.83Å vs. 0.84 Å) (51) and unit cell volume of monoclinic HfO_2 phase at room temperature is about 3 $Å^3$ smaller than that of ZrO_2. The critical volume of the monoclinic phase at which transformation to tetragonal occurs is about the same for hafnia and zirconia. However, the smaller cell parameter of HfO_2 and slightly lower coefficient of volume thermal expansion (21×10^{-6}°C^{-1} vs. 24×10^{-6}°C^{-1}) (56), reflect an increase of HfO_2 monoclinic-to-tetragonal transformation temperature (~1650°C for HfO_2 vs. 1160°C for ZrO_2) (46) The thermal expansion of the monoclinic phase is highly anisotropic (almost no expansion on *b* axis). There is a volume decrease on the monoclinic to-tetragonal transition (about 3.4% for ZrO_2 and somewhat smaller for HfO_2). The transition occurs rapidly on heating and on cooling and the tetragonal phases cannot be quenched to room temperature. The tetragonal phases expand almost isotropically (56) and transform into fluorite-type cubic structures at about 2300°C for ZrO_2 and about 2700°C for HfO_2 (46). The recent first-principles study of dielectric properties of HfO_2 polymorphs (188) suggests that dielectric constant for the tetragonal phase is much larger than for the cubic and monoclinic forms (70 vs. 29 and 16).

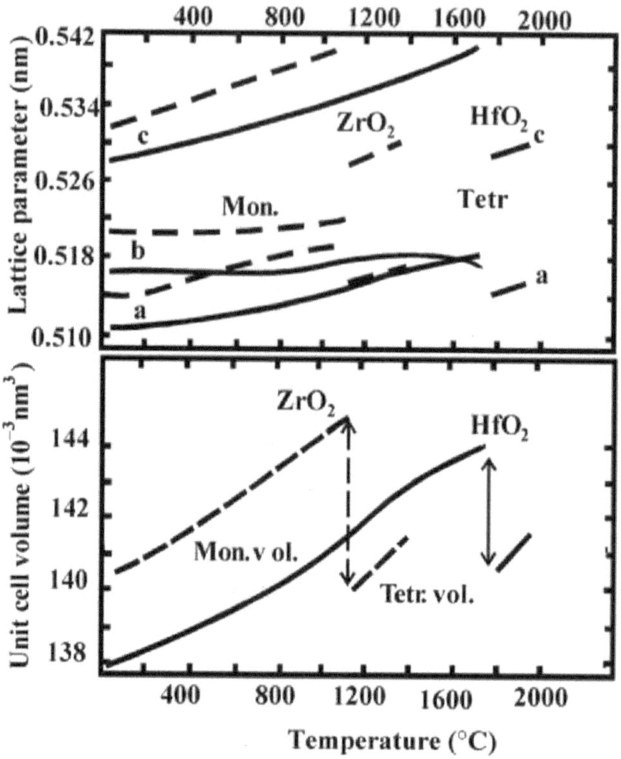

Fig. 4. The thermal expansion of HfO_2 and ZrO_2 and volume change on monoclinic-to-tetragonal phase transition (adapted from Wang et al. (46), data from Garrett et al. (49), and Patil et al. (198)).

I.2.2. ZrO_2 and HfO_2—Metastable Phases

Although the high temperature phases cannot be quenched directly, occurrence of tetragonal ZrO_2 phase as nanocrystals at room temperature has been reported for a long time (52). Stabilization of tetragonal over monoclinic in ZrO_2 samples with high surface area was interpreted to be the result of a surface energy contribution (36, 53, 54). The critical particle size at which tetragonal–monoclinic energy crossover occurs for ZrO_2 was reported to be around 30 nm, but it is strongly dependent on the stresses present at any given temperature (53, 54). This has been the basis for the wide applications of the zirconia monoclinic-to-tetragonal transition in high performance ceramics. Initially, it was thought that no such phenomena exist in the HfO_2 system (54). However, synthesis of tetragonal HfO_2 with particle size below 10 nm by decomposition of hafnium chloride and hydroxide was later reported (55). Thermal expansion of metastable tetragonal zirconia was reported (56) to be highly anisotropic below 900°C. This, however might be related to coarsening in powders and not be relevant for thin films. No such data for hafnia are reported.

Amorphous zirconia and hafnia can be synthesized in bulk by precipitation. Amorphous zirconia always crystallizes in the tetragonal modification and may partially or completely transform into monoclinic on cooling, depending on the fraction coarsened above critical size at the annealing temperature. However, in films thinner than 7 nm, where coarsening is restricted, ZrO_2 may retain the tetragonal modification after annealing (57). The exact critical sizes will depend on interfacial energies. Tetragonal hafnia is formed on crystallization of precipitates with high surface areas, while monoclinic is formed at lower surface area. Crystallization temperature in pure amorphous ZrO_2 and HfO_2 synthesized by precipitation depends significantly on particle size and in hafnia can be delayed up to 890°C at surface areas on the order of 200 m^2/g (58). However, such crystallization temperatures might not be realized even in ultra thin films, since interfacial energies normally are smaller than surface energies, and for thin film geometry, the interface area is smaller than that achievable for nanoparticles of comparable dimensions. Gusev (60) reported that crystallization temperature of 5 nm thick HfO_2 films on silicon was about 600°C, which is 170°C higher than for 40 nm films.

Tetragonal and cubic ($Fm3m$ CaF_2-type) hafnia phases were reported to occur in HfO_{2-x} 400–500 nm films synthesized by ion beam assisted deposition (IBAD) using hafnium vapor and an oxygen ion beam under conditions of oxygen starvation (61). In this case the stabilization of high temperature structures is not by surface energy but by oxygen vacancies. In the work cited, hafnia films transformed into monoclinic under annealing in oxygen at 500°C but stayed cubic or tetragonal on annealing in a vacuum at the same conditions. The tetragonal phase in substoichiometric HfO_2 was reported (61) to not be isostructural to tetragonal zirconia ($P4_2/nmc$) that occurred at high temperature or nanopowders.

Cubic zirconia was also reported to occur in films synthesized by ion-beam induced chemical vapor deposition (IBICVD) when using O_2^+ and Ar^+ ions for the decomposition of the precursor (62). The authors concluded that its occurrence is related to Ar incorporation in the structure since it was not observed when only oxygen ions were used. About 2 nm critical size for stabilization of cubic ZrO_2 at room temperature was suggested from first principles calculations and TEM observations (63). However structural identification of the cubic phase in such small particles is ambiguous.

Since there is decrease in volume on high temperature monoclinic-to-tetragonal phase transition, its temperature decreases under high pressure conditions. Two high pressure orthorhombic phases for hafnia were identified stable above 5 and 15 GPa (64). Analogous high pressure ZrO_2 polymorphs are known (65). Their formation at atmospheric and low pressures is unlikely. Though formation of orthorombic hafnia in thin films (66–68) was reported, it apparently was misidentified and is the tetragonal HfO_2 which is thought to be isostructural to tetragonal ZrO_2 (46, 55). However, to date, there are no entries for tetragonal HfO_2 in commonly used crystallographic databases (47, 48). Tetragonal HfO_2 was synthesized in nanophase powders only in a mixture with the monoclinic phase (55), and, apart from high-pressure modifications, it cannot be quenched from high temperature. Structural refinements on tetragonal HfO_2 phase

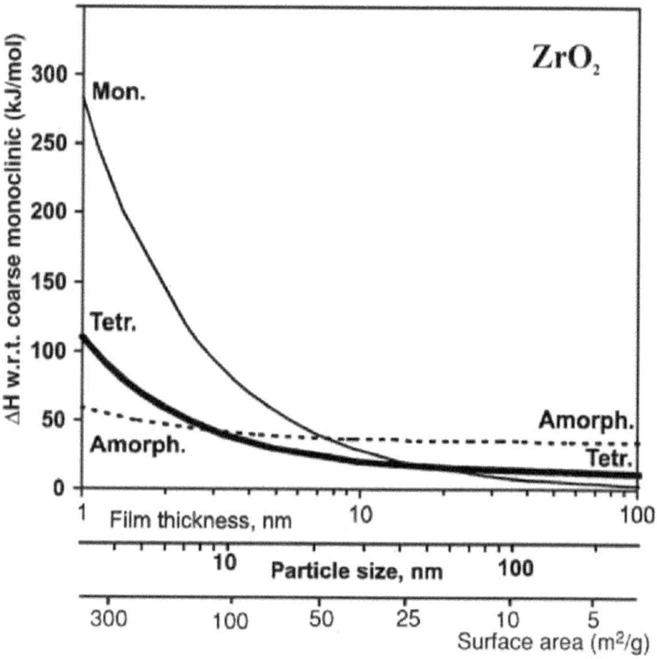

Fig. 5. Excess enthalpy of nanophase zirconia vs. particle size (logarithmic scale) as calculated from BET surface area (after Pitcher et al. (36)). The film thickness that would give similar interface areas are shown to illustrate the point that much higher surface area achievable in particles of similar dimensions. The exact critical thicknesses will depend on interfacial energies.

at high temperature is hampered by its high transition temperature (~1650°C) (46), and, to our knowledge, has not been performed.

Recently, high-temperature oxide melt solution calorimetry (see Part II) was used to measure excess enthalpies of nanocrystalline tetragonal, monoclinic and amorphous zirconia with respect to coarse monoclinic zirconia (36). Monoclinic ZrO_2 was found to have the largest surface enthalpy and amorphous zirconia the smallest (see Table 4). The surface enthalpy of amorphous zirconia was estimated to be 0.5 J/m^2. The linear fit of excess enthalpies for nanocrystalline zirconia as a function of area from nitrogen adsorption (69) gave apparent surface enthalpies of 6.4 and 2.1 J/m^2, for the monoclinic and tetragonal polymorphs respectively. Due to aggregation, the surface areas calculated from crystallite size (from X-ray diffraction) are larger than those accessible for nitrogen adsorption. The fit of enthalpy versus calculated total interface/surface area gave surface enthalpies of 4.2 J/m^2 for the monoclinic form and 0.9 J/m^2 for the tetragonal polymorph. Thus, stability crossovers with increasing surface area between monoclinic, tetragonal and amorphous zirconia (Fig. 5) were confirmed.

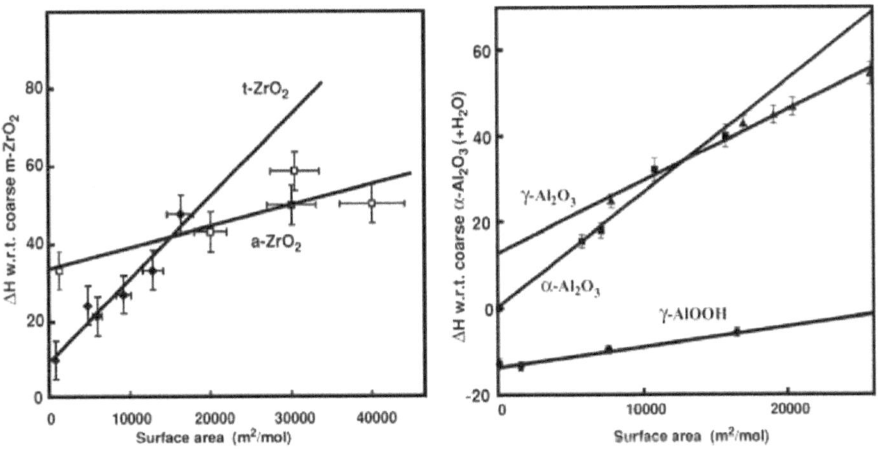

Fig. 6. *Left*: Excess enthalpies for amorphous and tetragonal zirconia samples with respect to coarse monoclinic ZrO_2 vs. surface area (after Pitcher et al. (36)). *Right*: Excess enthalpies for corundum (α-Al_2O_3), γ-Al_2O_3 (defect spinel structure) and formation enthalpy of γ-AlOOH from coarse α-Al_2O_3 and water vs. surface area (after Majzlan et al. (189)).

Using surface areas derived from XRD crystallite size, the tetragonal zirconia phase is calculated to be stabilized in particles smaller than 40 nm. Similar calculations for the tetragonal-to-amorphous crossover (Fig. 6) yield a critical particle size of 2 nm. From solution calorimetry, the amorphization enthalpy for monoclinic ZrO_2 was estimated to be 34 ± 2 kJ/mol (36). This value is close to that for HfO_2, (32.6 ± 2 kJ/mol), which can be derived from crystallization enthalpy of low-surface area amorphous precipitate (24).

Critical size for tetragonal-to-monoclinic transformation of HfO_2 crystallites in a gel with 10 mol% silica was reported as 6 ± 2 nm (24). The thickness of the film which would provide similar interface area is about 2 nm. Calorimetric study of surface/interfacial energetics for HfO_2 polymorphs is underway.

I.2.3. RE_2O_3—Stable Phases

A comprehensive comparison of binary rare earth oxides was done by Haire and Eyring (70). All rare earth elements form a sesquioxide, RE_2O_3. Below 2000°C, they are found in three forms, hexagonal, monoclinic and cubic (71–73), which were denoted as A, B, and C by Goldschmidt et al. (74) who first studied their polymorphism. For all elements but lutetium, more than one polymorph is reported. The most common form is the fluorite-related cubic C-type (*Ia*3, bixbyite type) in which most sesquioxides can be found. This structure is derived from the cubic defect fluorite structure by the ordering of oxygen vacancies. For large rare earths, hexagonal (A, $P3m1$) is common and monoclinic B form ($C2/m$) is typical in the middle of the series (Fig. 7). For all rare earth sesquioxides save Lu_2O_3, high-temperature hexagonal phase was reported, and La–Gd oxides were found to undergo yet another reversible

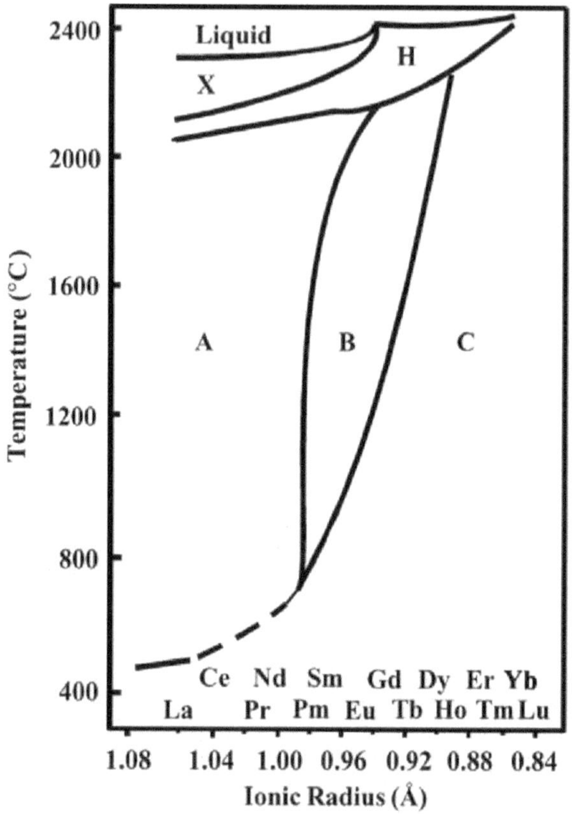

Fig. 7. Phases formed by RE oxides vs. RE radius. Adapted from Haire and Eyring (70) (data from Foex and Traverse (199), Warshaw and Roy (73), with additions for high-temperature transitions from Lopato et al. (200). C-to-A transition temperature for La_2O_3 (~550°C) after Glushkova (71).

phase transformation to a cubic phase before melting (75). The structures of the high temperature phases were refined by high temperature powder diffraction for La_2O_3 and Nd_2O_3 in space groups $P6_3/mmc$ (hexagonal, H-type) and $Im3m$ (cubic, X-type in Fig. 7) (76). There is disagreement in the literature regarding the high temperature phase of Y_2O_3. Foex and Traverse (77) reported the powder XRD pattern of Y_2O_3 at 2300°C matching to H-type phase. Swamy et al. (78, 79) reported powder pattern of Y_2O_3 at 2257°C with additional weak lines which they related to fluorite-type phase. High temperature differential thermal analysis (80) indicated that Y_2O_3 undergoes single reversible phase transition about 100°C before melting, whose enthalpy (54.8 ± 10 kJ/mol) is consistent with values derived from $YO_{1.5}$–HfO_2 (127), $YO_{1.5}$–ZrO_2 (128), and $YO_{1.5}$–CeO_2 (81) fluorite-type solid solutions.

Though all RE oxides occur as RO in gas phase at high temperature, in the solid state only the monoxide EuO and YbO can be synthesized. Dioxide is the highest established oxide of rare earths and fluorite-type dioxides have been reported only for Ce, Pr, and Tb. Cerium is the only rare earth that forms dioxide during decomposition of its compounds in air. PrO_2 is not stable in air above 390°C (82). Thermodynamic properties of rare earth oxides were reviewed by Morss (83).

I.2.4. RE_2O_3—Metastable Phases

The most frequently observed metastability in bulk RE_2O_3 phases involves A, B and C structure types. As in zirconia and hafnia, if there is a high temperature polymorph available for given rare earth oxide, it probably can be stabilized by surface energy term in small particles. For instance, it was reported that Eu_2O_3 and Y_2O_3 (cubic C-type in bulk) adopt monoclinic (B-type) structure in 13 nm particles synthesized by gas-phase condensation (84, 85).

La_2O_3 is notorious for being prone to hydration and extreme affinity for CO_2. C-type La_2O_3 can be obtained by decomposition of its nitrate in vacuum. However, on heating in air at 300°C it reacts with CO_2 and water vapor to form amorphous (by XRD) products, which decompose on further heating with formation of the A-form (71). Glushkova (71) described the C-form she synthesized as metastable with respect to the A-form. However, it is entirely possible and consistent with RE_2O_3 phase diagrams (Fig. 7) that the C form is in fact thermodynamically stable form of La_2O_3 below 550°C. The low-temperature A-to-C transformation might not be observed for kinetic reasons. Thermodynamic data are needed for C-type La_2O_3 to unambiguously answer this question.

RE_2O_3 H-type and X-type phases are not quenchable (75). However synthesis of fluorite-type (*Fm3m*) Y_2O_3 was reported (86) by quenching the C-type phase heated at 2220°C point with a laser. Gaboriaud et al. (87) reported occurence of fluorite-type Y_2O_3 in thin films deposited by ion beam sputtering.

I.2.5. Al_2O_3—Stable Phases: Corundum

Only one stable phase, corundum or α-Al_2O_3 is known. The structure is shown in Fig. 2. No high temperature phase transitions are observed. The melting point is about 2072°C (88).

I.2.6. Al_2O_3—Metastable Phases

Corundum, α-Al_2O_3, is the thermodynamically stable phase of coarsely crystalline aluminum oxide at standard temperature and pressure conditions, but syntheses of nanocrystalline Al_2O_3 usually result in γ-Al_2O_3. Based on earlier molecular dynamics simulations and their own thermochemical data, McHale et al. (29). predicted that γ-Al_2O_3 should become the energetically stable polymorph for specific surface area exceeding ~125 $m^2\ g^{-1}$ (Fig. 6). The thermodynamic stability of γ-Al_2O_3 should be even greater than implied by this energy. Due to the presence of tetrahedral and octahedral sites in its spinel-type structure, and the fairly random distribution of Al^{3+} and vacancies over these sites, γ-Al_2O_3 has a greater entropy than α-Al_2O_3. The entropy change of the α-Al_2O_3 to γ-Al_2O_3 transition, $\Delta S_{\alpha\rightarrow\gamma}$, is about +5.7 J K^{-1} mol^{-1} (29). Therefore, at room temperature, γ-Al_2O_3 could be thermodynamically stable with respect to α-Al_2O_3 at specific surface areas $> 100\ m^2 \cdot g^{-1}$, and at ~530°C (a temperature typical of oxyhydroxide decomposition) γ-Al_2O_3 might become thermodynamically stable at specific surface areas greater than only 75 $m^2\ g^{-1}$.

McHale et al. (29, 30) used high temperature oxide melt solution calorimetry (see below) to study the effect of particle size on energetics of Al_2O_3. The enthalpies of drop solution in molten lead borate of several nanosized α- and γ-alumina samples were measured. However, the surfaces of the Al_2O_3 were modified by adsorbed H_2O which could not be completely removed without severe coarsening. The surface energies of the hydrated polymorphs appeared nearly equal, indicating that the heat of chemisorption of H_2O is directly proportional to the surface energy of the anhydrous phase. Consequently, McHale et al. could not determine the anhydrous surface energies without accurate knowledge of the heats of chemisorption of H_2O. These measurements were made on two samples each of α- and γ-Al_2O_3 with a Calvet type microcalorimeter operating near room temperature (29). The differential heat of H_2O adsorption on γ-Al_2O_3 decreases logarithmically with increasing coverage (Freundlich behavior). In contrast, the differential heat of H_2O adsorption on α-Al_2O_3 does not show regular logarithmic decay, and decreases far less rapidly with increasing coverage. This indicates a greater number of high energy sites on α-Al_2O_3 per unit surface area, which are relaxed by the most strongly chemisorbed hydroxyls. This observation is strong evidence that the surface energy of α-Al_2O_3 is higher than that of γ-Al_2O_3.

A quantitative analysis of the heat of adsorption data enables the separation of hydration enthalpies and surface enthalpies for the two alumina polymorphs (29). The resulting variation of enthalpy of the anhydrous material with surface area is shown in Fig. 6. The enthalpy (and free energy) crossover postulated above is clearly demonstrated. Table 4 lists the surface and transformation energies. Calorimetric studies of water adsorption on alumina (29, 30) suggest that the higher energy surfaces have the strongest affinity for water, and that α-alumina has more strongly bonded H_2O than γ-alumina.

The addition of small amounts of SiO_2 to γ-Al_2O_3 increases the temperatures of heat treatment necessary for transformation to γ-Al_2O_3 by about 100°C (28). McHale et al. (28) reported that the spinel-type Al_2O_3–SiO_2 solid solutions with 2–10 wt.% SiO_2 are always energetically metastable by 30–35 kJ/mol (on a 4 O^{2-} per mol basis) with respect to α-Al_2O_3 and quartz.

I.3. MULTICOMPONENT SYSTEMS

In the previous section we considered the simple oxides. In the following section we look into pseudo-binary systems of hafnia, zirconia and rare earths with silica and alumina, and finally consider $HfO_2(ZrO_2)$–RE_2O_3 systems. We use lanthanum and yttrium in discussing systems with rare earths. Though yttrium does not belong to the lanthanides, being about the size of holmium, chemically it behaves very similarly to a heavy rare earth, and, due to its lower cost, systems with yttrium are studied much more extensively than those with heavy rare earths. The absence of unpaired f-electrons also makes the yttrium systems simpler from the standpoint of electronic and magnetic properties. We include the available data on crystallization in Al_2O_3–RE_2O_3–SiO_2 and RE_2O_3–ZrO_2(HfO_2)–SiO_2 with low silica content (Table 2), since this might be relevant when considering effects of formation of SiO_2-rich layers on contact with Si.

I.3.1. Systems with Silica or Alumina

The notable feature of all systems with silica or alumina under consideration is that they do not form stable solid solutions (alloys) between end members. The equilibrium solubility is limited to 3–4 mol%. All systems except HfO_2(ZrO_2)–Al_2O_3 form stable ternary compounds, however, in HfO_2(ZrO_2)–SiO_2 formation of silicate (zircon or hafnon) is kinetically hindered and it cannot be synthesized by solid state reaction below 1300°C. Synthesis of high-purity zircon was studied extensively due to its application as advanced refractory ceramics (89). Technical applications of ZrO_2-toughened Al_2O_3 (ZTA) ceramics (41) also have driven research on this system. In ZTA ceramics, crystalline zirconia is dispersed in alumina to take advantage of the stress-induced monoclinic-to-tetragonal phase transition in zirconia to dissipate the cracks. In dielectric films, alumina or silica is added to ZrO_2 (HfO_2) and La_2O_3 to retain them as amorphous to higher temperatures: the trade-off is lowering dielectric constant (90, 91). Measurement of dielectric constant for amorphous oxides and laminates of ZrO_2–Al_2O_3 (92), and HfO_2–SiO_2 and La_2O_3–SiO_2 (91) indicate that dielectric constants vary almost linearly with composition.

I.3.1.1. ZrO_2(HfO_2)–SiO_2 stable phases: zircon and hafnon

Crystalline zirconium and hafnium silicates $ZrSiO_4$ and $HfSiO_4$ (or zircon and hafnon from corresponding mineral names) are isostructural and continuous solid solution between them was established (93). The zircon structure ($I4_1/amd$, Fig. 3) contain isolated SiO_4 tetrahedra, and is common for many ABO_4 compounds (94).

According to the reported phase diagrams (Fig. 8), zircon decomposes to tetragonal ZrO_2 and SiO_2 (cristobalite) before melting sets in at 1687°C but hafnon melts incongruently at 1750 ± 15°C. However, a later report on thermal stability of zircon (95), suggests that the dissociation temperature of zircon is higher than 1700°C.

Enthalpy of formation of zircon from oxides (SiO_2 quartz and m-ZrO_2 baddeleyite) at 702°C was measured by oxide melt solution calorimetry (96) as -27.9 ± 1.4 kJ/mol. Recently, enthalpies of drop solution (ΔH_{ds}) were measured on flux-grown crystals of zircon and hafnon using lead borate solvent at 800°C ($\Delta H_{ds}(ZrSiO_4) = 156.8 \pm 2.4(8)$ kJ/mol and $\Delta H_{ds}(HfSiO_4) = 132.0 \pm 4.4(16)$) (58). Difference between sum of drop solution enthalpies of corresponding oxides measured in the same conditions ($\Delta H_{ds}(SiO_2)_{qtz} = 47.92 \pm 0.56$; ΔH_{ds}(m-ZrO_2) $= 84.7 \pm 1.2$; ΔH_{ds}(m-HfO_2) $= 61.8 \pm 1.4$ kJ/mol) and drop solution enthalpies of compounds gives their formation enthalpies from oxides at room temperature ($\Delta H^{o}_{fr.ox}$ $(ZrSiO_4) = -24.2 \pm 2.8$; $\Delta H^{o}_{fr.ox}$ $(HfSiO_4) = -22.3 \pm 4.7$). The new results agrees with previous

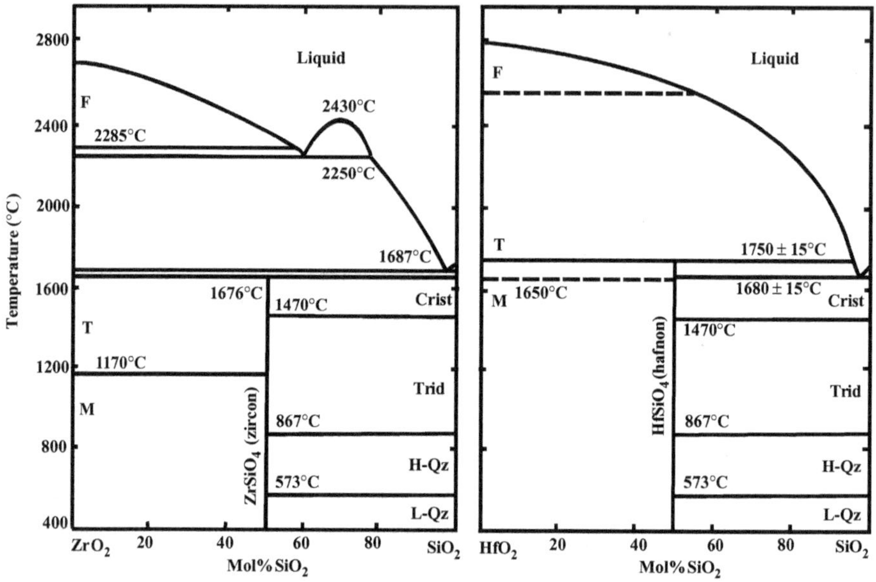

Fig. 8. Phase diagram ZrO_2–SiO_2 (after Butterman and Foster (201)), and HfO_2–SiO_2 (after Parfenenkov et al. (202)). Monoclinic-to-tetragonal and tetragonal-to-cubic transformation temperatures for HfO_2 (1650 and 2520°C) are from Ruh et al. (44).

measurements for $ZrSiO_4$ and indicate that the difference in formation enthalpy of hafnon and zircon is less than 8 kJ/mol (taking into account uncertainties). The drop solution enthalpy under the same conditions was also measured for an amorphous phase of zircon stoichiometry prepared by sol–gel (24) and dehydrated at 700°C ($\Delta H_{ds}(am.ZrO_2{\cdot}SiO_2) = 94.2 \pm 2.4$ kJ/mol). The amorphization enthalpy of zircon at 25°C can be calculated as ($\Delta H_{ds}(ZrSiO_4) - \Delta H_{ds}$ $(am.ZrO_2{\cdot}SiO_2) =$ 62.6 ± 3.4 kJ/mol). Notably, this value is close to the amorphization enthalpy determined from calorimetry on the set of natural (Zr,U,Th)SiO_4 samples with different degree of radiation-induced amorphization (59 ± 3 kJ/mol) (15). This agreement emphasizes that despite different paths of obtaining the amorphous phase, which probably result in slight structural differences, the energetics are the same within experimental uncertainty.

Though zircon and hafnon are thermodynamically stable phases with respect to oxides at 25°C (Table 1), formation of zircon by solid-state reaction usually requires temperature above 1400°C. However presence of dopants may lower the formation temperature substantially (e.g., addition of CeO_2 lower formation temperature of zircon by about 100°C) (23, 97). These effects are kinetic rather than thermodynamic.

I.3.1.2. $ZrO_2(HfO_2)$–SiO_2 metastable phases

During the last several years, amorphous zirconium and hafnium silicates have been the subject of intensive research as a high-k replacement of SiO_2 (90 references were

Table 1. Formation enthalpies from oxides and element at 25°C for some oxides and binary compounds in systems relevant to alternative gate dielectrics (ZrO_2, HfO_2, ZrO_2–SiO_2, HfO_2–SiO_2, RE_2O_3–Al_2O_3, RE_2O_3–SiO_2, RE–ZrO_2(HfO_2))

Compound/ structure[a]	ΔH_f^o ox. (kJ/mol)	ΔH_f^o el. (kJ/mol)	Compound/structure	ΔH_f^o ox. (kJ/mol)	ΔH_f^o el. (kJ/mol)
Oxides			RE aluminates[f]		
SiO_2 qz		−910.7 ± 1.0	$LaAlO_3$ pv	−67.4 ± 1.5	−1801.6 ± 1.5
ZrO_2 mon[b,c,d]		−1100.6 ± 1.7	$NdAlO_3$, pv	−52.9 ± 1.7	−1794.1 ± 1.8
HfO_2mon[e]		−1117.6 ± 1.6	$SmAlO_3$ pv	−40.6 ± 1.5	−1790.0 ± 1.6
α-Al_2O_3[b]		−1675.7 ± 1.3	$GdAlO_3$ pv	−34.1 ± 1.7	−1785.0 ± 1.8
La_2O_3(A)		−1791.6 ± 2.0	$DyAlO_3$ pv	−25.2 ± 3.1	−1794.6 ± 3.1
Y_2O_3 (C)[c,d]		−1905.3 ± 2.3	$YAlO_3$ pv	−22.8 ± 3.1	−1813.4 ± 3.1
Hf, Zr and RE silicates			$Y_4Al_2O_9$ (YAM)[g]	−6.24 ± 6.21	−5545.9 ± 5.3
$ZrSiO_4$ zrn[b]	−24.2 ± 2.8	−2035.5 ± 3.4	RE_2O_3–ZrO_2(HfO_2)		
$HfSiO_4$ hfn[h]	−22.3 ± 4.7	−2050.6 ± 5.1	$La_2Zr_2O_7$ pyr[k]	−99.5 ± 4.3	
Y_2SiO_5[i]	−52.5 ± 4.8	−2368.5 ± 5.3	$Ce_2Zr_2O_7$ pyr[k]	−94.3 ± 6.2	
$Y_2Si_2O_7$[g]	−67.1 ± 6.0	−3820.5 ± 6.7	$Nd_2Zr_2O_7$ pyr[k]	−71.6 ± 3.3	
Yb_2SiO_5[i]	−49.4 ± 8.4	−2774.8 ± 8.4	$Sm_2Zr_2O_7$ pyr[k]	−64.3 ± 3.3	
$La_{9.33}(SiO_4)_6O_2$[j]	−764 ± 23	−14560 ± 22	$Gd_2Zr_2O_7$ pyr[k]	−57 ± 3.7	
$Nd_{9.33}(SiO_4)_6O_2$[j]	−716 ± 32	−14617 ± 32	$Zr_{0.5}Y_{0.5}O_{1.75}$flrt[l]	−6.7 ± 1.3	
$Sm_{9.33}(SiO_4)_6O_2$[j]	−539 ± 23	−14562 ± 21	$Hf_{0.5}Y_{0.5}O_{1.75}$ flrt[m]	−3.3 ± 2.0	
$Gd_{9.33}(SiO_4)_6O_2$[j]	−447 ± 22	−14403 ± 28			

[a]Structure abbreviations: (zrn) and (hfn)—zircon and hafnon, (pv) perovskite, (oxy)—oxyapatite, (pyr)—pyrochlore, (flrt)—fluorite; [b]from Robie and Hemingway (177); [c]from Robie et al. (178); [d]from Glushko et al. (179); [e]from Kornilov et al. (181). There is a notable disagreement in the literature regarding the ΔH_f^o for monoclinic HfO_2. The following values were reported from combustion calorimetry in the order of appearing in the literature (converted into kJ/mol): −1136.0 (182); −1113.2 ± 1.2 (183); −1144.7 ± 1.3 (184); −1117.5 ± 2.1 (185); −1133.9 ± 6.3 (186); −1117.6 ± 1.6 (181). The latest value is the weighed mean of the results of round robin between Moscow University and Los Alamos National Laboratory and the most trustworthy. This unusual spread in the data is attributed (181) to difficulty of complete oxidation of hafnium metal, formation of the oxide $HfO_{(g)}$, and adsorption of CO_2 and H_2O by combustion products. ΔH_f^o HfO_2 values (kJ/mol) adopted in commonly used reference sources are: Barin (192) −1113.2; Glushko (179) −1117.6; Robie et al. (178) −1144.7; Robie and Hamingway (177) −1117.6. Though formation enthalpies of some RE hafnates and zirconates were also measured by combustion calorimetry (186) we are not considering these data here. [f]After Zhang et al. (10); [g]from Fabrichnaya et al. (106); [h]after Ushakov et al. (130); [i]after Liang et al. (115); [j]after Risbud et al. (110, 191); [k]after Helean et al. (180); [l]from Lee et al. (128); [m]from Lee et al. (127).

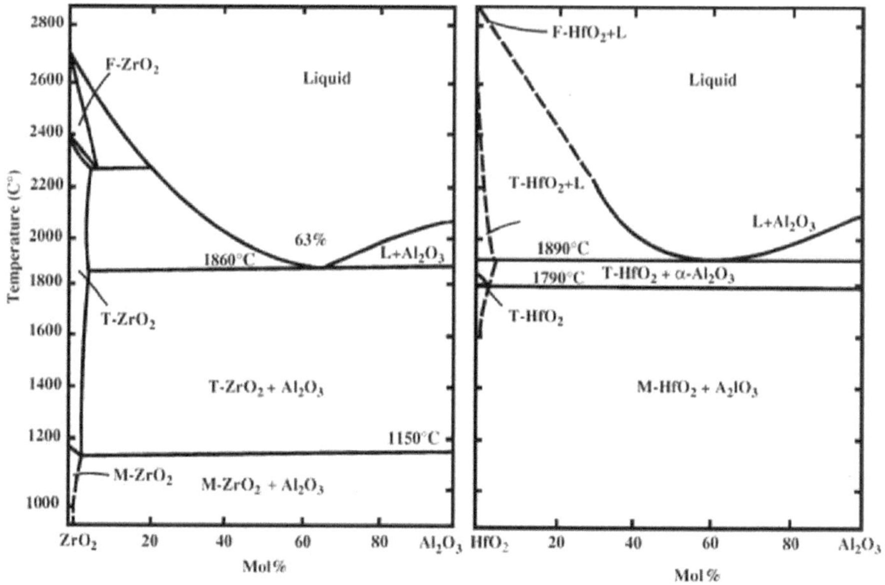

Fig. 9. Phase diagram ZrO_2–Al_2O_3 (after Lakiza (203)) and HfO_2–Al_2O_3 (after Lopato (88)).

found since year 2000 in Chemical Abstract on "zirconium silicate dielectric", e.g., (3, 90, 98, 99)). Since maximum annealing temperature at which dielectric films are studied do not normally exceed 1100°C and zircon and hafnon do not form at these conditions, the term "zirconium/hafnium silicate" and "crystallization of zirconium/hafnium silicate" in publications related to their dielectric applications refers to the amorphous oxide solid solution (alloy) and to formation of zirconia or hafnia crystallites in an amorphous silica matrix, respectively.

Solubility limits of hafnia and zirconia in silica in quenched glasses are quite low (3 and 4.6 mol%, respectively (100)). However any composition in these systems can be synthesized as an amorphous solid by sol–gel techniques or using various thin films deposition techniques. Thin films and bulk undergo amorphous phase separation prior to crystallization (3, 101). This can be interpreted based on liquid immiscibility in the ZrO_2–SiO_2 system (3).

The crystallization temperature increases with SiO_2 content. Crystallization in the bulk material prepared by sol–gel is close to that observed in thin films (Fig. 9). On crystallization, tetragonal HfO_2 and ZrO_2 crystallites form in an amorphous silica matrix. Crystallite size after crystallization increases with ZrO_2/HfO_2 content. The crystallization enthalpy per mole of zirconia decreases with decreasing crystallite size. If this phenomenon is attributed exclusively to tetragonal $ZrO_2(HfO_2)$/amorphous SiO_2 interface enthalpies, these values can be derived (24, 102) as 0.25 ± 0.08 J/m^2 for HfO_2 and 0.13 ± 0.07 J/m^2 for ZrO_2 (Fig. 10). The higher value for HfO_2 interface energy is expected from its higher density (see Part II) and agrees with observed higher crystallization temperatures.

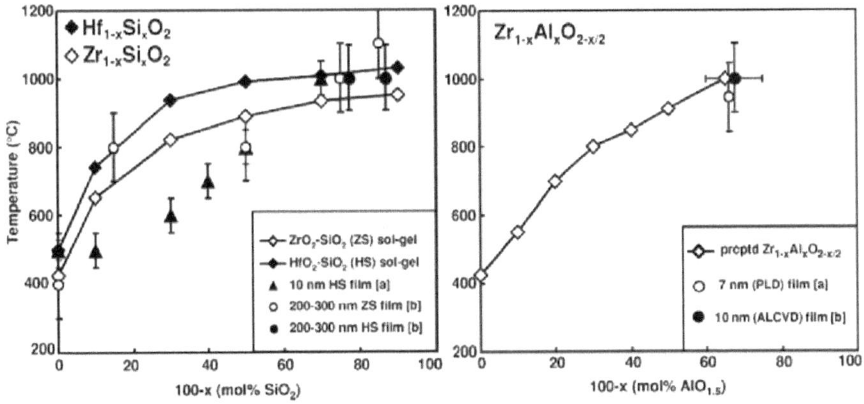

Fig. 10. *Left*: Crystallization temperatures in ZrO_2–SiO_2 and HfO_2–SiO_2 from sol–gel (24) compared with films on silicon (a) 10 nm from molecular beam epitaxy (MBE) (91); (b) 200–300 nm from chemical solution deposition (204). *Right*: Crystallization temperatures in ZrO_2–Al_2O_3 powders from precipitation (22), compared with thin films: (a) 7 nm from pulsed laser deposition (PLD) (205, 206); (b) 10 nm from atomic layer chemical vapor deposition (ALCVD) (207).

I.3.1.3. $ZrO_2(HfO_2)$–Al_2O_3 stable phases

Both zirconia–alumina and hafnia–alumina phase diagrams (Fig. 11) are of eutectic type with no compounds. The eutectic temperature in ZrO_2–Al_2O_3 (1860 ± 10°C) is lower than that for HfO_2–Al_2O_3 (1890 ± 10°C). The eutectic composition is

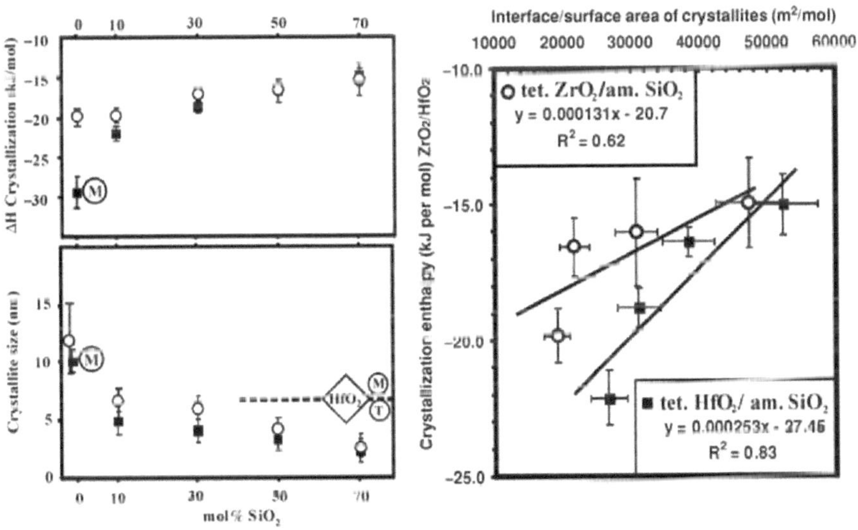

Fig. 11. *Left top*: Crystallization enthalpy per mole ZrO_2 (open circles) and HfO_2 (solid squares) in the gels with different silica content (24). *Left bottom*: Average crystallite size after crystallization. *Right*: crystallization enthalpy vs. calculated interface area (102).

67 ± 3 mol% Al_2O_3 in the system with HfO_2 and ~63 mol% in the system with ZrO_2. The HfO_2–Al_2O_3 phase diagram was studied by Lopato et al. (88) in hydrogen and helium. She did not find changes in cell parameters of monoclinic HfO_2 and corundum and concluded their negligible mutual solubility. However, based on the observation that addition of 2.5 mol% Al_2O_3 lowered the temperature of the monoclinic to tetragonal transition to 1790°C from 1830 ± 10°C for pure HfO_2, she concluded that Al_2O_3 was somewhat soluble in tetragonal HfO_2 but these solid solutions were not quenchable. The solid solubility limit of Al_2O_3 in monoclinic ZrO_2 was estimated at 0.7 ± 0.3%, while the solubility of ZrO_2 in corundum was negligible (22).

I.3.1.4. $ZrO_2(HfO_2)$–Al_2O_3 metastable phases

Stefanic and Music (22) studied crystallization in amorphous $Zr_{1-x}Al_xO_{2-x/2}$ compositions obtained by precipitation. The crystallization temperatures from DTA are reported in Fig. 10. For the compositions from 40 to 60 mol% Al_2O_3 they are close to the crystallization temperatures reported for 50–100 nm films prepared by AL-CVD (207) (Fig. 10). Stefanic (22) reported that metastable tetragonal ZrO_2–Al_2O_3 solid solution crystallizes at 550–1000°C in all cases. On further annealing of compositions with $x < 0.3$, samples decompose into tetragonal and monoclinic phases at 1000–1100°C. For $x > 0.3$, the monoclinic phase does not appear, but gamma alumina crystallizes at 1000–1100°C. For $x > 0.5$, gamma alumina crystallizes first at 800 and 1000°C a tetragonal solid solution appears. Stefanic (22) found that the cell parameter of tetragonal $Zr_{1-x}Al_xO_{2-x/2}$ solid solution is smaller than that for tetragonal zirconia, and varies insignificantly with Al content (a: 3.58–3.59 Å and c: 5.06–5.07 Å), suggesting interstitial incorporation of Al^{3+} ions. Formation of metastable tetragonal solid solutions in ZrO_2–Al_2O_3 system for compositions from 10 to 50 mol% Al_2O_3 on pyrolytic decomposition of precursors is also reported by Levi (103). ZrO_2–Al_2O_3 amorphous multilayers are found to mix on heating at temperature below crystallization temperature (57).

I.3.1.5. RE_2O_3–SiO_2 stable phases: RE_2SiO_5, $RE_2Si_2O_7$, $RE_{9.33}(SiO_4)_6O_2$

The early studies of rare earth silicates are summarized by Warshaw and Roy (72). Figure 12 shows phase diagrams for La_2O_3–SiO_2 and Y_2O_3–SiO_2 systems after Toropov et al. (104, 105). Toropov reported three compounds formed in both systems: RE_2SiO_5, $RE_2Si_2O_7$ and $RE_4Si_3O_{12}$. The third compound has not been found in Y_2O_3–SiO_2 system in the latest studies, but existence of liquid immiscibility was confirmed (see Fabrichnaya et al. (106) for recent review). RE_2SiO_5 and $RE_2Si_2O_7$ compounds with Y and La are not isostructural, four quenchable high-temperature phases were reported for $RE_2Si_2O_7$, which have different Si–O–Si angles in $Si_2O_7^{6-}$ ions (107–109). Their thermodynamic and structural characterization is incomplete.

Recent calorimetric work (110) has focused on one family of compounds, the oxyapatites, $RE_{9.33}\square_{0.67}(SiO_4)_6O_2$ (RE = La, Sm, Nd, Gd; □ = vacancy). These compounds were not identified in earlier studies (72, 104, 105) and not shown on the phase diagrams. They can, however be synthesized directly from mixture of oxides (110). Lanthanides are used as sintering aids during silicon nitride synthesis,

Table 2. Crystallization temperatures and glass transitions observed by DSC in glasses and amorphous powders in some systems (ZrO_2, HfO_2, ZrO_2–SiO_2, HfO_2–SiO_2, RE_2O_3–Al_2O_3, RE_2O_3–Al_2O_3–SiO_2)

Low temperature routes				Quenched from melt glasses				
System/ composition	T_{cr} (°C)	ΔH_{vitr} (kJ/mol)	Ph.[a]	System/composition (quenched glasses)	T_g (°C)	T_{cr} (°C)	ΔH_{vitr} (kJ/mol)	Ph.[a]
ZrO_2	426	22.4 ± 1.2[b]	T	SiO_2	1207[c]			
		(34.3 ± 2.2)[d]	M	$La_2O_3 \cdot Al_2O_3$[e]	847	868	78.5 ± 1.3	Pv
$ZrO_2 \cdot SiO_2$	970	16.0 ± 2.0[b]	T′	$Pr_2O_3 \cdot Al_2O_3$[e]	n.a.	797	66.5 ± 1.5	Pv
		(61.4 ± 3.0)[f]	(Zrn)	$Nd_2O_3 \cdot Al_2O_3$[e]	n.a.	787	69.9 ± 1.6	Pv
$0.5LaO_{1.5} \cdot 0.5ZrO_2$	818	9.7 ± 0.4[g]	F	$Sm_2O_3 \cdot Al_2O_3$[e]	823	887	63.8 ± 1.4	Pv
$0.3LaO_{1.5} \cdot 0.7ZrO_2$	803	12.8 ± 0.1[g]	F	$Gd_2O_3 \cdot Al_2O_3$[e]	842	897	63.5 ± 1.7	Pv
$0.3YO_{1.5} \cdot 0.7ZrO_2$	436	12.2 ± 1.9[g]	F					
				$Gd_2O_3 \cdot Al_2O_3 \cdot 2SiO_2$[e]	876	1019		
HfO_2	472	32.6 ± 2.0[b]	M	$Er_2O_3 \cdot Al_2O_3 \cdot 2SiO_2$[e]	890	1102		
$HfO_2 \cdot SiO_2$	1040	16.4 ± 0.5[b]	T′	$Nd_2O_3 \cdot Al_2O_3 \cdot 1.4SiO_2$[e]	855	1022		
		(73.9 ± 4.5)[f]	(Hfn)	$Y_2O_3 \cdot 1.7Al_2O_3$ (YAG)[e]	830	930	92.2 ± 1.8	G
$0.5LaO_{1.5} \cdot 0.5HfO_2$	906	10.4 ± 1.8[g]	F	$Y_2O_3 \cdot Al_2O_3 \cdot 0.2SiO_2$[h]	865	927		
$0.3LaO_{1.5} \cdot 0.7HfO_2$	865	11.6 ± 0.6[g]	F	$Y_2O_3 \cdot Al_2O_3 \cdot 7SiO_2$[h]	901	1107		
$0.3YO_{1.5} \cdot 0.7HfO_2$	522	13.2 ± 0.2[g]	F	$Y_2O_3 \cdot 1.7Al_2O_3 \cdot 1.8SiO_2$[h]	885	1087		G

[a]Structure abbreviations: (T)—tetragonal (T′)—tetragonal + amorphous silica, (M)—monoclinic, (F)—fluorite-type cubic, (Pv)—perovskite, (G)—garnet, (Zrn)—zircon, (Hfn)—hafnon; [b]from differential scanning calorimetry (DSC) measurements (24); [c]from Wang (16); [d]from high-temperature oxide melt solution calorimetry, extrapolated to zero surface area (36); [e]RE_2O_3–Al_2O_3 glasses synthesized by melting with cw-CO_2 laser and containerless quenching (14), crystallization temperatures (10) (T_{cr}) and glass transition temperatures (11) (T_g) measured by DSC at 20°C/min heating rate in Ar flow. Vitrification enthalpies are from solution calorimetry (11) per mole $REAlO_3$ relative to crystalline perovskite. [f]From high-temperature oxide melt solution calorimetry (130); [g]from DSC (58); [h]Y_2O_3–Al_2O_3–SiO_2 glasses were prepared by water-quenching from 1650°C. Glass transition and crystallization temperatures measured at 20°C/min heating rate in Ar flow (12).

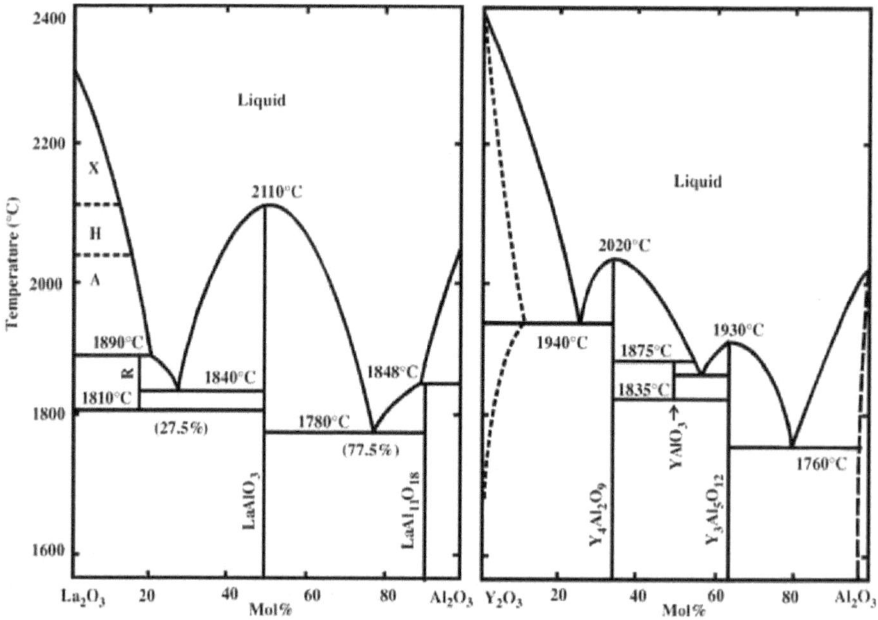

Fig. 12. Phase diagram La_2O_3–Al_2O_3 (after Mizuno et al. (25), R—orthorhombic phase with approximate composition 82.5% La_2O_3) and Y_2O_3–Al_2O_3 (after Toropov (26)), dashed lines represent solid solutions according to Noguchi (27).

resulting in RE-oxyapatite formation at grain triple junctions in silicon nitride ceramics (111). There is increasing interest in Gd-containing compounds because of their high luminescence efficiency when doped with other rare earth ions (112, 113). Many of these properties can be attributed to the unique oxyapatite structure that contains oxygen atoms located in the hexagonal tunnels parallel to the *c*-axis. These oxygen atoms are bonded to Ln cations but are not bonded to Si and are therefore isolated from Si-tetrahedra (107). RE-oxyapatites are also potentially useful for modeling the release of actinides from ceramic nuclear waste forms (114). Though oxyapatites were also reported for Y and Ho (107–109), their structures have not been refined and thus they are not in the commonly used crystallographic databases (47, 48).

The measured enthalpies of formation show that the RE oxyapatite phases studied are substantially stable with respect to their binary oxides. The stability of oxyapatites from oxides increase as the RE size increases in moving across the lanthanide series toward lanthanum. This general trend is intrinsic to changes in the RE bonding character. As follows from Fig. 19, experimentally measured (115) formation enthalpy from oxides for $Yb_2Si_2O_7$ is close to that expected from the linear trend for Sm, Nd, Gd oxyapatites (110) when values are normalized per one RE. This indicates that the phases $RE_2Si_2O_7$ and $RE_{9.33}(SiO_4)_6O_2$ are similar in energy, and agrees with synthesis patterns (107, 109).

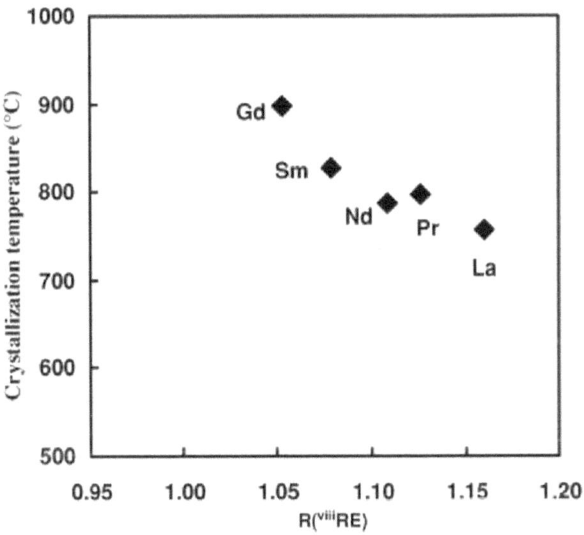

Fig. 13. Crystallization temperatures of Al_2O_3–RE_2O_3 glasses (after Zhang (10)).

However, when data for $La_{9.33}(SiO_4)_6O_2$ are considered, the stabilization effect of increasing ionic radii of the RE-site ion is not a linear function. Relatively little additional stabilization of the oxyapatite structure is gained by increasing the ionic radii of the lanthanide ion beyond Nd as shown by a flattening of the enthalpy curve (Fig. 19). In a complex, multicomponent oxide, the bonding requirements of each cation–anion polyhedron must be satisfied if the structure is to remain stable with respect to other polymorphs or phase assemblages (31). In the case of the RE-oxyapatites, substituting Sm for Gd stabilizes the structure by, presumably, better satisfying the bonding requirements of the Ln-site. This stabilization effect may continue until the RE-site ion becomes too large and begins to destabilize the structure. The eventual complete destabilization of the structure is experimentally not attained, as there is no trivalent ion with ionic radius greater than lanthanum available. The importance of this observation is that it reveals a potential pitfall with predictions of thermodynamic properties by extrapolating linearly beyond experimental data (31).

I.3.1.6. RE_2O_3–SiO_2 metastable phases

Heats of solution of La_2O_3 in a series of simple alkali and alkali earth silicate liquids were recently measured by temperature calorimetry (7). The energetics of the liquids are dominated by the exothermic reactions which form La-clusters and these phase-ordered regions do not dissociate as temperature increases to 1480°C. These calorimetric results coupled with spectroscopic measurements indicate extreme perturbation of the silicate framework by La(III), sufficient to isolate oxygen from silicon.

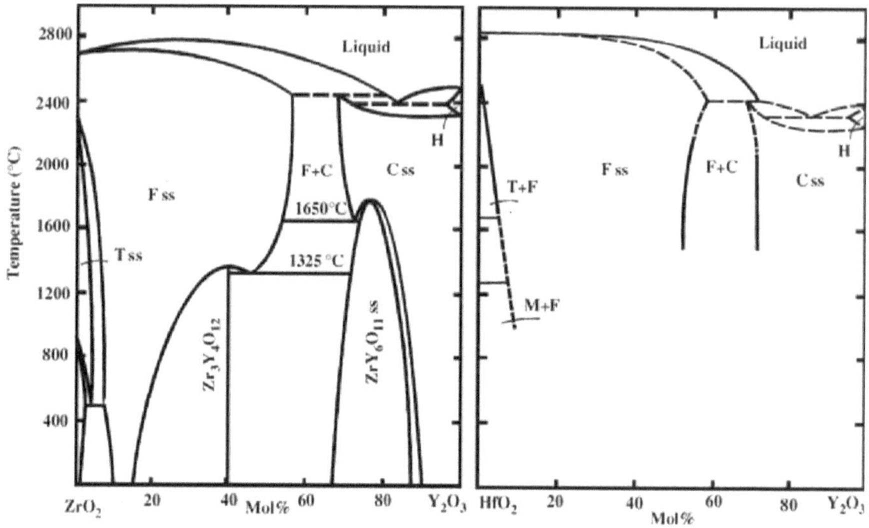

Fig. 14. Phase diagrams ZrO_2–Y_2O_3 (after Pascual et al. (208)) and HfO_2–Y_2O_3 (after Stacy et al. (209)).

This suggests the presence of phase-ordered regions rich in La(III) consistent with liquid immiscibility observed in the La_2O_3–SiO_2 system.

Crystallization of La_2O_3–SiO_2 thin films (5–20 nm) deposited on silicon by MBE technique was studied by Maria et al. (91). Likewise for ZrO_2–SiO_2, crystallization temperature increase with silica content, but silicate La_2SiO_5 was detected on crystallization.

I.3.1.7. RE_2O_3–Al_2O_3 stable phases: perovskite and garnet

The Y_2O_3–Al_2O_3 and La_2O_3–Al_2O_3 phase diagrams (Fig. 12) were studied by Mizuno et al. (25) and Toropov et al. (26). Stability of perovskite and garnet in the systems RE_2O_3–Al_2O_3 and RE_2O_3–Ge_2O_3 was studied by Kanke and Navrotsky (116). In the RE_2O_3–Al_2O_3 systems there is competition between perovskite ($REAlO_3$) and garnet ($RE_3Al_5O_{12}$ or $RE_{0.75}Al_{1.25}O_3$) phases, the former favored for larger rare earth ions (116). Figure 19 shows the enthalpy of formation from the oxides of these phases as a function of rare earth radius. Lattice match of $LaAlO_3$ perovskite with silicon (1.3% lattice mismatch) allows for epitaxial growth.

I.3.1.8. RE_2O_3–Al_2O_3 metastable phases

Figure 13 shows crystallization temperatures of Al_2O_3–RE_2O_3 glasses (10). The samples were prepared by containerless quenching methods (14). The values for crystallization temperatures are listed in Table 2. Notably, in $REAlO_3$ glasses, crystallization temperature increase with decreasing RE size, what is opposite to the

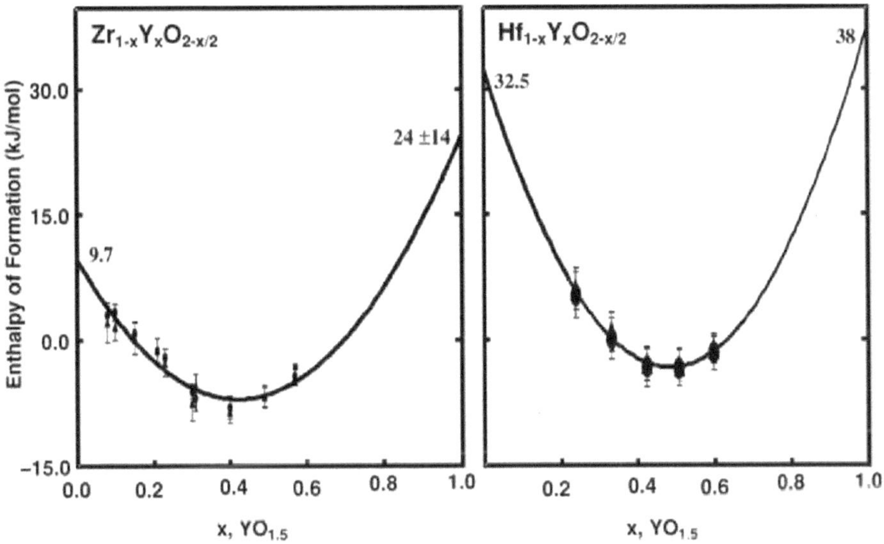

Fig. 15. Enthalpy of formation of cubic (fluorite-type) phase in ZrO_2–Y_2O_3 and HfO_2-Y_2O_3 systems from monoclinic ZrO_2(HfO_2) and C-type $YO_{1.5}$ (adapted after (127, 128)).

trend observed in amorphous $Hf_2RE_2O_7$ from precipitation (Fig. 17). No glasses with $REAlO_3$ stoichiometry with RE smaller than Gd could be produced even by containerless quenching. Adding of some silica or decreasing RE/Al ratio, can produce quenchable glasses (Table 2). On supercooling melts of composition close to $Y_3Al_5O_{12}$, separation of the amorphous phase into high- and low density phases (HDA and LDA) occurs (117–119). Since no compositional differences are observed between HDA and LDA phases, this phenomena is known as polyamorphism. Only the HDA phase is formed in low-temperature synthesis routes (120). No indication of polyamorphism was found in the La_2O_3–Al_2O_3 system.

Li et al. (187) obtained amorphous $LaAlO_3$ films by metal organic chemical vapor deposition (MOCVD) at 400–700°C, 150 nm thick films crystallized at 850–900°C, which is more than 100°C higher than the crystallization temperature in a glass of the same composition.

I.3.2. Zirconia (Hafnia) with Rare Earth Oxides

I.3.2.1. ZrO_2(HfO_2)–RE_2O_3 stable phases: fluorite and pyrochlore

Phase diagrams of ZrO_2–RE_2O_3 were reviewed by Rouanet (121), Glushkova et al. (122–124), HfO_2–RE_2O_3 phase diagrams were reviewed by Glushkova et al. (122–125), Wang et al. (46) and Kharton et al. (126). All these diagrams are characterized by narrow fields of formation of solid solutions based on monoclinic and tetragonal hafnia and zirconia. The solid solubility of rare earth oxides in these phases increases with increasing temperature and with increasing RE radius. For instance, La_2O_3 solid solubility in monoclinic HfO_2 is less than 1 mol% (50), La_2O_3 and Pr_2O_3, solid

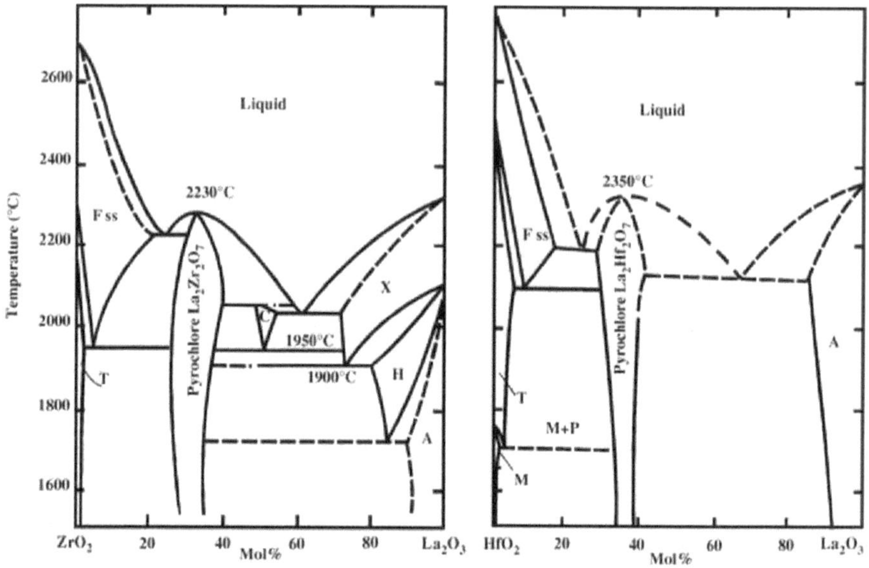

Fig. 16. Phase diagram ZrO_2–La_2O_3 after Rouanet (121) and HfO_2–La_2O_3 after Duran (210). HfO_2 rich part after Shevchenko et al. (211). Melting temperatures for $La_2Zr_2O_7$ and $La_2Hf_2O_7$ pyrochlores are from Zoz et al. (50).

solubility in tetragonal hafnia is approximately 5–6 mol% at 1500–2000°C. For other RE oxides, including neodymia, the maximum solid solubility (126) in the tetragonal hafnia does not exceed 2 mol%. Addition of RE decreases the monoclinic-to-tetragonal and tetragonal-to-cubic phase transition temperatures. Compared to zirconia-based phases, the temperatures of all phase transformations in hafnia are higher and larger additions of RE dopant are required to stabilize the cubic modification. The oxygen ionic conductivity in hafnia-based oxides is significantly lower than that of zirconia-based oxides (126).

In the zirconia and hafnia-rich parts of the systems ZrO_2–Y_2O_3 and HfO_2–Y_2O_3 (Fig. 14) fluorite-type solid solution, isostructural to high-temperature fluorite-type ZrO_2 and HfO_2, form above 1400°C in a wide range of compositions. Their energetics have recently been determined by oxide melt solution calorimetry (127, 128).

Calorimetric measurements have been made to determine the enthalpy of formation of ZrO_2–Y_2O_3 solid solutions (c-YSZ, yttria stabilized zirconia) at 25°C and at 700°C with respect to the monoclinic ZrO_2 and C-type $YO_{1.5}$ (see Fig. 15). The enthalpy of formation can be fit by a quadratic equation. The fit gives a strongly negative interaction parameter, $\Omega = -94 \pm 12$ kJ/mol, but does not imply regular solution behavior because of extensive short-range order. In this fit, the enthalpy of transition of m-ZrO_2 to c-ZrO_2, 9.7 ± 1.1 kJ/mol, is in reasonable agreement with earlier estimates and that of C-type to cubic fluorite $YO_{1.5}$, 24 ± 14 kJ/mol, is consistent with an essentially random distribution of oxide ions and anion vacancies in the high

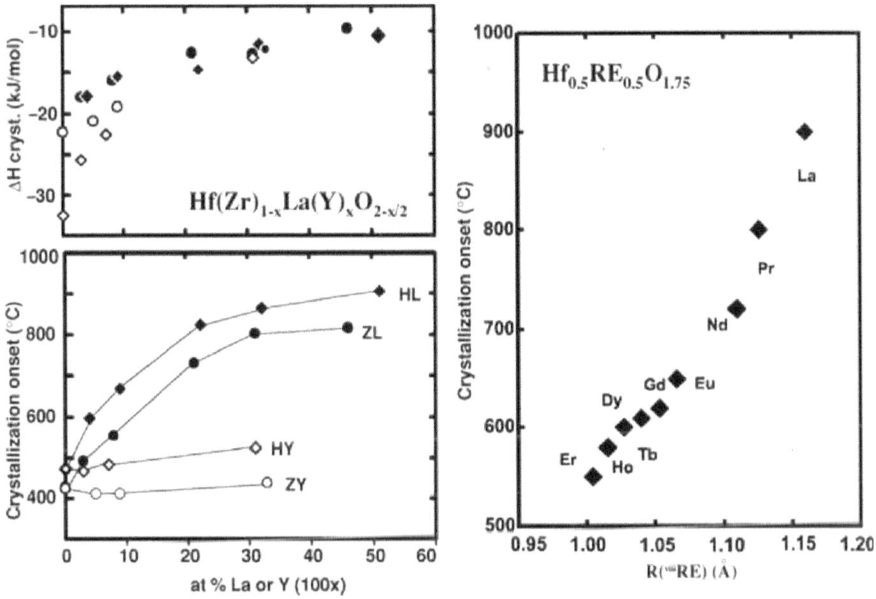

Fig. 17. *Left*: crystallization temperatures and enthalpies for precipitated pure and La- and Y-doped ZrO_2 and HfO_2. Hafnia-containing samples labeled by diamonds, zirconia by circles. La-doped samples by solid symbols, Y-doped samples by open symbols (after Ushakov et al. (58)). *Right*: crystallization temperatures of precipitated $Hf_{0.5}RE_{0.5}O_{1.75}$ (data from Glushkova et al. (59, 125)).

temperature fluorite phase. The enthalpy of transition from the disordered c-YSZ phase to the ordered δ-phase at 25°C has also been measured and is 0.42 ± 1.56 kJ/mol. No energetic difference between the disordered c-YSZ phase and the ordered δ-phase underscores the importance of short-range order in c-YSZ. Enthalpy data are

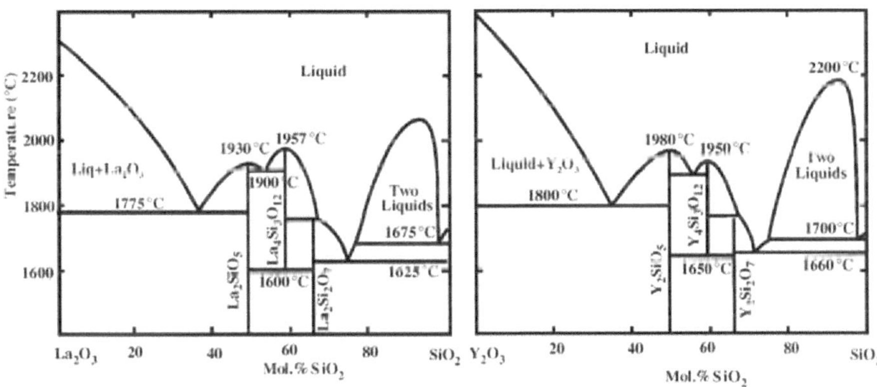

Fig. 18. Phase diagram La_2O_3–SiO_2 (after Toropov et al. (104)) and Y_2O_3–SiO_2 (after Toropov et al. (105)).

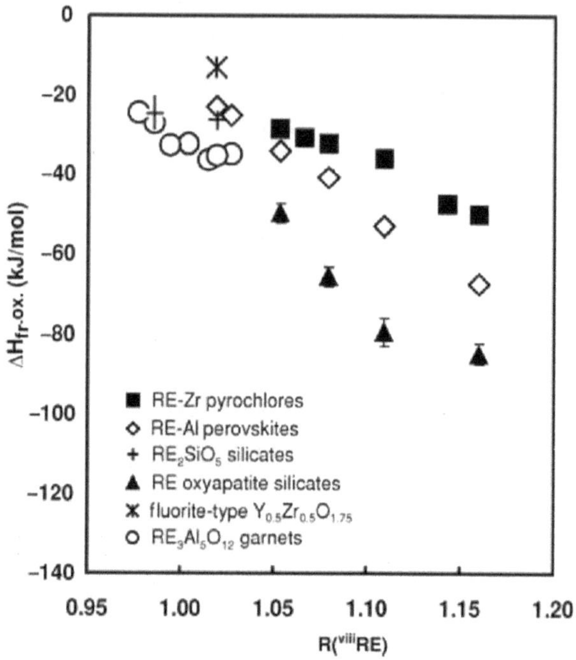

Fig. 19. Formation enthalpies from oxides of some RE compounds in systems with SiO_2, ZrO_2, and Al_2O_3. The values are normalized per one RE cation. See Table 1 for standard formation enthalpy values and references.

combined with Gibbs free energy data to calculate entropies of mixing. Using the quadratic fit, a negative excess entropy of mixing in the cubic solid solution, relative to a system with maximum randomness on cation and anion sublattices, is found and is another indication of extensive short range order in c-YSZ (128). Recent calorimetric measurements (127) for the system HfO_2–$YO_{1.5}$ show strongly negative heats of formation of the cubic phase from monoclinic hafnia and C-type yttria, similar to those in ZrO_2–$YO_{1.5}$ and also possibly indicate extensive short range order of cations and/or vacancies. The enthalpy of transformation of HfO_2 from monoclinic to cubic is about 32 kJ/mol, significantly larger than that in zirconia. The difference probably reflects the higher temperature and presumably higher enthalpy, of the monoclinic–tetragonal transition in HfO_2 compared to ZrO_2.

Zirconia and hafnia-rich parts of the systems with small rare-earths (Dy–Lu) are also characterized by a wide range of stability of fluorite-type solid solutions. However, ordered pyrochlore-type phases $A_2B_2O_7$, may form in the systems with RE larger than Dy. RE pyrochlores with Hf and Zr are reviewed by Subramanian (129). The pyrochlore structure (*Fd*3*m*) can be derived from the fluorite structure (*Fm*3*m*) by ordering on the cation sublattice and creating ordered oxygen vacancies in such a way that coordination of RE atoms remains cubic, as in fluorite, but coordination of M cations decreases to octahedral. Ordering cause the pyrochlore unit cell to double with respect to the fluorite. $RE_2Zr_2O_7$ forms for RE = La–Gd with cell parameter 10.80–10.45 Å, decreasing with RE size. In HfO_2–RE_2O_3 systems pyrochlore phase are found for RE = La–Tb with cell parameters from 10.78 Å for $La_2Hf_2O_7$ to 10.45 Å for $Tb_2Hf_2O_7$. Gd_2Hf_2O and $Gd_2Zr_2O_7$ pyrochlores reversibly transform to fluorite at high temperature (122–125). These order-disorder transitions as well as amorphization can also be induced by radiation damage (17, 18).

$La_2Hf_2O_7$ and $La_2Zr_2O_7$ pyrochlores have less than 1% lattice mismatch with silicon which allows them to be grown epitaxially by MBE (molecular beam epitaxy). However, in MBE deposited films, the fluorite-structured phase formed together with the pyrochlore phase (131, 132). Formation of fluorite-type solid solutions was also found in these systems on crystallization of amorphous powders from precipitation (58). Apparently, the fluorite-type solid solution in the Hf and Zr systems with La is metastable with respect to the pyrochlore phase (see below).

I.3.2.2. $ZrO_2(HfO_2)$–RE_2O_3 metastable phases

No glasses prepared by quenching in these systems were reported. However, amorphous solids of any composition can be prepared by precipitation. Recently, crystallization of precipitated pure and Y and La doped hafnia and zirconia (doping level from 4 to 50 at.%) was studied using thermal analysis and room- and high-temperature X-ray diffraction (HTXRD) (58). It was found that Y-doping does not significantly affect crystallization temperatures but substantial increase of crystallization temperature of amorphous hafnium and zirconium oxides could be achieved by alloying with La_2O_3 (Fig. 16). The crystallization temperature of $Hf_2La_2O_7$ composition is higher than 900°C, which makes it a candidate for advanced gate dielectrics. Measurements of the surface areas of the powders indicates that the difference in crystallization behavior between Y and La doped samples is not primarily a particle size effect (58). Nor can the difference be attributed to the effect of residual hydroxide and carbonate, because their content in the samples heated to 440°C is insignificant and similar for La- and Y-doped samples. Thus we expect that La-doped hafnia will crystallize at higher temperatures than pure and Y-doped HfO_2 in films of the same thickness.

Crystallization enthalpies of pure and doped samples reflect the changes in the phase formed. Pure ZrO_2 crystallizes as the tetragonal phase ($\Delta H_{cr} = -22.4 \pm 1.2$ kJ/mol) and pure HfO_2 as the monoclinic phase ($\Delta H_{cr} = -32.6 \pm 2.0$ kJ/mol). The largest crystallization enthalpies in doped samples were observed for hafnia samples with <8 at.% Y doping, in which the monoclinic phase was detected after crystallization. Samples containing 20–50 at.% La and Y crystallize in a fluorite-type cubic phase. The crystallization enthalpies for hafnia and zirconia compositions with larger than 20 at.% La and Y content are similar (Fig. 17). However this does not necessarily indicate that the difference in crystallization behavior is a purely kinetic phenomenon, since crystallite size after crystallization of La-doped sample is larger than for the Y-doped sample. After annealing at 1450°C for 10 hours, hafnia samples containing less than 50 at.% of La separated into monoclinic HfO_2 and $Hf_2La_2O_7$ pyrochlore phases and the sample with 50 at.% La formed $Hf_2La_2O_7$ pyrochlore.

However, the fluorite-type phase formed upon crystallization of hafnia doped with 30 at.% La, did not separate on annealing at 1100°C for several hours.

The effect La versus Y on amorphous phase stabilization is consistent with the corresponding phase diagrams. Hafnium and zirconium ions are almost the same size (0.83 and 0.84 Å, for cubic coordination (51)). The radius of Y^{3+} (1.02 Å) is much closer to the size of Hf and Zr than is La^{3+} (1.16 Å). This allows Y^{3+} to form fluorite type solid solution with hafnia over a wide range (from 10 to 50 mol% Y_2O_3). Such solid solutions are not stable in the HfO_2–La_2O_3 system below 1900°C. Instead $Hf_2La_2O_7$ pyrochlore forms with a narrow compositional range. The formation of a eutectic between HfO_2 and $Hf_2La_2O_7$ lowers the melting temperature of HfO_2 with 20 mol% La_2O_3 by about 500°C compared with a sample with the same Y-content (Fig. 16). Thus La and Y behave very differently as dopants.

Another indication of the effect of La doping on thermal stability of amorphous zirconia and hafnia comes from study of amorphization in zirconia-rare earth compounds by ion beam irradiation. While bulk ZrO_2 and $Zr_2Gd_2O_7$ cannot be amorphized even near liquid helium temperature (16), $Zr_2La_2O_7$ pyrochlore can be amorphized below 37°C by irradiation with 1.5-MeV Xe+ beam (17, 18). This suggests more reluctant recrystallization, governed by thermodynamic and/or kinetic factors, in the latter system.

Crystallization onset temperatures of precipitated amorphous solids of $Hf_2RE_2O_7$ (or $Hf_{0.5}RE_{0.5}O_{1.75}$) stoichiometry (Fig. 17) were estimated from differential thermal analysis curves published by Glushkova et al. (125). Her data indicate that crystallization temperatures increase with increasing RE radius from Er ($T_{cr} \sim 550$°C) to La ($T_{cr} \sim 900$°C). Glushkova (125) also found fluorite-type phase on crystallization. To our knowledge, there are no permittivity data for amorphous $Hf_{1-x}La_xO_{2-x/2}$. However, since La_2O_3 has a higher dielectric constant (99) than HfO_2, $Hf_{(1-x)}La_xO_{2-x/2}$ may offer increased thermal stability of the amorphous phase without sacrificing dielectric constant (58).

However, for crystalline phases isostructural to ZrO_2 and HfO_2 polymorphs, dielectric constant may depend strongly on the structure formed (188). Clearly, in view of new applications, stability of the amorphous phase in HfO_2–RE_2O_3 systems (in particularly for RE = La, Nd) begs for detailed investigation both in thin films and in bulk. The crystallization paths and relative stability of pyrochlore and fluorite phases and their electrical properties need also to be defined for these materials.

PART II. ENERGETICS OF INTERFACES

Surface related phenomena are responsible for differences in thermodynamics of thin films and bulk materials. Reduction of surface and interface energy is the driving force for grain growth with annealing at high temperatures, known as sintering in the ceramic industry. Thermodynamically, coarse material is more stable than fine grained at any temperature, but it takes thermal energy to make atoms on the surface mobile enough for coarsening to happen.

Thin film geometry implies restrictions on grain growth and coarsening, thus the interface energy term is the overriding feature, which distinguishes thin films from bulk materials. This often causes the amorphous and high temperature phases, not stable in bulk, to appear in thin films, because of their lower surface/interfacial energies. The paucity of experimental thermodynamic data on interfacial energies and theoretical difficulties in their calculations often makes it not possible to predict stability of interfaces in ultra thin films using thermodynamic data for bulk phases. Figure 5 gives a quantitative example of how surface energies may alter polymorphism in alumina and zirconia in the thin film regime, based on data on surface energies from high temperature oxide melt solution calorimetry (Table 4).

The purpose of the following sections is to review the available experimental techniques for measuring surface and interfacial energies of the solids and summarize available experimental data for relevant systems.

II.1. DEFINITIONS AND UNITS

Thermodynamic descriptions of surfaces and interfaces have a long history. It is beyond the scope of this paper to give a detailed account of surface thermodynamic phenomena and the interested reader is directed to comprehensive treatises on the subject (133–144). However, to discuss surface energies of solids and experimental techniques for their measurements, we have to define some terms and general relations.

Surface free energy (γ) is the free energy associated with unit of surface area in vacuum and measured in units of energy per units of area: usually in erg/cm^2 which is equivalent to mJ/m^2. It is often defined as the reversible work on a material to increase its surface area. Creation of the surface involves breaking bonds, which costs energy, so surface energy is always positive.

Surface free energy is often used interchangeably with terms *surface stress* and *surface tension*. The latter is measured in units of force per length: usually in dyne/cm or mN/m, which are numerically equal to erg/cm^2. The *surface tension* is defined by Gibbs as the work necessary to *stretch* the surface and *surface free energy* as the work necessary to *form* the surface. In contrast to liquids, which do not support stress, these two works may be different for solids (141), but for practical purposes we do not need to distinguish between them.

Three fundamental equations (134) in the physical chemistry of surfaces involve surface energy: (i) The Young–Laplace equation or basic equation of capillarity, which relates pressure difference across the surface (ΔP) with radius of curvature (r) (in simplified form for a sphere $\Delta P = 2\gamma/r$). (ii) The Kelvin equation which gives the effect of surface curvature on the vapor pressure of a substance on a curved surface. If P^0 is the normal vapor pressure of a substance and P is that observed on the curved surface, in simplified form for a sphere with radius (r) and ideal vapor $RT \ln(P/P^0) = 2\gamma V/r$, where V is molar volume of the substance. (iii) The Gibbs equation, which relates surface excess or deficiency of a solute on the surface (Γ_2^1) with changes of surface energy ($\Gamma_2^1 = -(a/RT)(\mathrm{d}\gamma/\mathrm{d}a)$, where a is the activity of the solute). Surface excess of solute forms if the solute lowers the surface energy ($d\gamma/da$ is negative), and vice versa.

Interfacial free energy is the free energy associated with interface area, and has the same units as surface energy. For a solid or a liquid, interfacial energy can be expressed as the surface energy minus work of adhesion (referred as Dupre's equation) (133).

$$\gamma_{12} = \gamma_1 + \gamma_2 - W_{12} \tag{11}$$

where W_{12} the work of *adhesion*.

If medium 2 is vacuum, $W_{12} = 0$; $\gamma_2 = 0$, and $\gamma_{12} = \gamma_1$. If the two media are identical, $\gamma_{11} = 2\gamma_1 - W_{11}$, and the work to separate two identical media W_{11}, called the work of *cohesion*. It is a measure of how much the interfacial energy differs from the sum of the surface energies (138). For non-polar solids, the specific surface energy of the grain boundary is less than the sum of surface energies and is a function of disorientation angle (140).

Most of the thermodynamic formalism and experimental techniques for measurements of surface energies were developed for liquid surfaces. The surface mobility is one characteristic, which distinguishes the solid surface or interface from that of the liquid. The surface of the liquids defined by thermodynamic forces acting upon it. As it is pointed by Burdon (144) experiments reported at the beginning of the last century demonstrated that, less than 0.01 second after its formation, the surface of water reached the state of constant and minimum surface energy. In contrast, the surface of solids may depend to a large extent on the path it was created, and may relax only very slowly towards its state of lowest free energy. Thus, theoretically, the surface energy for the same solid may vary with surface area. Solid surfaces/interfaces of different crystallographic orientations will have different energies. Once created, such surfaces/interfaces do not easily change orientations. Thus two crystalline solids with the same surface/interface area may have significantly different free energies.

These intrinsic difficulties led to skepticism in early treatises (133) about measuring of the surface energies of solids. Though these limitations must be always kept in mind, the development of new experimental techniques for characterization of the surfaces of the solids and measuring the surface energies, led to notable progress in the area of experimental measurements of energies of solid interfaces, as will be discussed in the following sections.

II.2. EXPERIMENTAL TECHNIQUES TO MEASURE SURFACE FREE ENERGIES OF SOLIDS

There are numerous techniques to measure surface tension of single liquids and solutions (e.g., capillary height, sessile drop, pendent drop, Wilhelmy plate, maximum pull on a cylinder, maximum pull on a cone, Du Nouy ring, drop weight or drop volume, maximum bubble pressure), and many of them are more than a hundred years old. However, the free energy of solid surfaces, even in modern textbooks, is sometimes described as "not susceptible to measurement" (137). All the usual methods applicable to liquids fail, because of non-mobility of the surface. Adam (133) in 1938 refers to solubility study and calorimetry as possible techniques but noted that they "vitiated by unavoidable and incalculable errors in experimental application". Though no new general methods for surface energy measurement have been invented since then, advances in surface characterization techniques and instrumentation have improved the situation. Below we outline principles and pros and cons of the experimental techniques (summarized in Table 3), to measure surface/interface energy of the solids.

II.2.1. Extrapolation of the Surface Energy of a Solid from the Surface Tension in the Liquid

One approach to estimate the surface energy of a solid is to measure the surface tension and its temperature dependence in the molten state using techniques developed for liquids and then extrapolate to the temperature of interest. Surface tension decreases with temperature (that implies a positive surface excess entropy (145) but not significantly, usually less than 1 erg/cm^2/°C for liquid metals (139). This approach has been used extensively for metals (146).

There are practical limitations to this technique, in particular: (i) it is troublesome to apply for refractory materials because of technical difficulties, and (ii) oxidation, reduction and decomposition reactions may invalidate observations. However, the main limitation comes from consideration of differences of solid and liquid surfaces. As discussed above, due to surface mobility of the liquid surface, energy is always minimized, so the data from liquid state may be used as a lower bound to the surface energy of a solid. For amorphous solids this estimate might be more reasonable than for crystalline.

II.2.2. Zero-Creep Method

This technique can be applied for ductile solids, such as metals or metal microlaminates (147). Strips of material of interest are exposed for prolonged time to temperature close to its melting point. Surface tension causes shrinkage of the strips to lower the surface area. Depending on the magnitude of applied weight, the strips either shrinks or elongates. The point at which force balance is reached is called the zero-creep point and surface tension can be calculated from the applied weight and geometry of the strip.

II.2.3. Crystal Cleavage

Forces related to creation of new surface by separation of a crystal along a given crystallographic plane are measured in this method. The first measurement of surface energy by cleaving was apparently performed for mica by Obreimoff (148). Mica crystals have a perfect cleavage in one plane and are easily bent in another. A procedure known as Griffith (149) approach is used to account for energy dissipation unrelated to breaking bonds in given direction (150). The technique, used extensively for silicon crystals, is known as double cantilever beam (DCB) test (150). The initial precursor crack may be introduced into the appropriate cleavage system by electrical spark

Table 3. Experimental techniques to measure surface energies of solids

Method	What is measured	Notes (references)
	Forces	
Zero-creep	Measurement of elongation or shrinkage of strips of metal as a function of applied weight. Zero-creep point corresponds to force balance.	Applicable only for ductile solids metals, at higher than room temperatures (140).
Crystal cleavage	Crystal is split along the cleavage plane and force needed to be applied to cause further development of the crack is measured.	Limited to single crystals (140, 150).
Atomic force microscopy (AFM)	Adhesion force to pull off the AFM tip from contact with the surface is calculated from measured displacement and canteliver stiffness.	May not provide absolute values for surface energies. Limited to interaction with material of AFM tip (Si_3N_4, Pt, W). Canteliver stiffness calibration is necessary (151).
	Equilibrium parameters	
Interfacial angles, equilibrium void or crystal shapes	Dihedral angles (152), thermal grooves profile (153), pore shapes.	Annealing close to melting point to provide surface mobility needed to reach equilibrium (152, 153).
Diffusion study	Difference in diffusion coefficients on grain interface (D_i) and in the lattice (D_l) $\gamma_i = 1/2(\Delta G_l - \Delta G_i) = 1/2RT\{\ln(D_i/D_l)\}$.	Mostly for metals. Limited by feasibility of self-diffusion measurements, done by tracer technique, normally at 0.3–0.5 $T_{melting}$ (145).
Nucleation study	Kinetics of nucleation.	Limited to measuring amorphous/crystalline interface energy in one component systems (139).
Solubility changes	Difference in solubility of fine grained samples and coarse grained samples. Surface energy can be calculated using Kelvin equation $RT\ln(a/a_0) = 2\gamma V/r$	Difficult to establish equilibrium (135).
	Heat	
Immersion and adsorption studies	Heat of immersion, heat of adsorption. May be used to derive differences in surface energies between solids.	May not provide absolute values for surface energies. May not provide interfacial energy estimations (156).
Aqueous solution calorimetry near room temperature	Difference in heat of solution in water or acids vs. surface area.	Not applicable for oxides with low solubility in aqueous solution (193).
High temperature oxide melt solution calorimetry	Difference in heat of drop solution in oxide melts vs. surface area.	Applicable for refractory oxides (33–36). Somewhat lower sensitivity than for calorimetry at room temperature.
Differential scanning calorimetry (DSC)	Difference in heat of crystallization vs. interface/surface area.	Limited to initially amorphous oxides (24, 102).

discharge and the tensile force needed for fracture propagation is measured. If cleaving of the crystal is accompanied by formation of significant number of dislocations, it may invalidate the results.

II.2.4. Atomic Force Microscopy (AFM)

Atomic force microscopy (AFM) can be used to measure adhesion force between material of the AFM tip and sample surface. In order to do such measurements AFM tip first brought into contact with sample surface and than pulling out. If canteliver is calibrated on known material, force of adhesion can be calculated from pull-off distance (151).

II.2.5. Equilibrium Shapes

This group of methods based on measurements of equilibrium shapes of solid surfaces, thus it requires annealing at high temperatures (usually above 1300°C) to reach high surface mobility. Hence, it is not applicable for material and phases which are not stable at these temperatures. Surface energies may be derived from measurements of dihedral angles (152), observation of equilibrium shape of pores and crystals, and from measurements of grain-boundary thermal groves (153). The latter method allows evaluating surface energy anisotropy for different crystalographic planes when combined with AFM to measure groove geometries and with electron backscattered diffraction to determine grain orientation. Using this technique Sano et al. (153). determined $SrTiO_3$ surface energies at 1400°C in air $\gamma_{100} = 0.93 \pm 0.03$ and $\gamma_{111} = 1.02 \pm 0.01$ (the units are arbitrary).

II.2.6. Diffusion

One of the reasons why gate dielectrics are preferred to be amorphous or single crystal rather than polycrystalline is that diffusion in interfaces occurs much faster than in the bulk. Interfacial energies can also be extracted from diffusion kinetics based on the assumption that the energy of the interface is the difference between the Gibbs free energies for vacancy diffusion in the lattice and the interface (145, 154). For these calculations, self-diffusion coefficients for the lattice and interface have to be obtained, preferably by tracer methods. Although most values obtained by this procedure are for metals, in a recent review Gupta (145) demonstrated that it is also applicable to compounds. He reports grain boundary energies for Au, Ag, Cu, Ni, Pb and for NiO. To our knowledge, no data for the systems which are the subject of this review were obtained by this method.

II.2.7. Nucleation

Supercooling arises because new interface has to be created, and, according to classic nucleation theory, interfacial energy defines the critical nucleus size. While there are numerous examples which use this relations to measure solid–liquid interfacial tension from nucleation data (134), to our knowledge, the only estimates of solid–solid interfacial energies from activation enthalpy of nucleation were made by Tu et al. (139) for amorphous/crystalline interfaces for Si and CoSi. The assumption is made that the macroscopic interfacial tension is valid down to critical nucleus size, and a geometrical constant depending on the shape of the nucleus has to be introduced.

II.2.8. Solubility Changes

The process of growing larger crystals at the expense of small ones, when crystal growth and dissolution rates become equal at equilibrium is known as *Ostwald ripening* (134). This process is driven by minimization of interfacial energy. As it pointed out by Ostwald in 1900 and cited by Adam (133), "as the vapor pressure of small drops of liquid is greater than that of large drops, so the solubility of small solid particles is greater than that of large".

Thus, Kelvin equation (see above), holds for solids as well, but with activities, instead of vapor pressure (134). From this equation, surface energies of solids can be derived from increase in solubility of fine grained material compared with coarse. The examples of such measurements are scarce however. Obviously, this method can give information only about hydrated surfaces if the solvent if water-based.

II.3. CALORIMETRIC TECHNIQUES

In all calorimetric techniques described below to measure surface/interface energy it is the heat transfer that is being measured and the resulting values are surface enthalpies rather than surface free energies. Contribution of surface entropy in surface free energy is small as follows from very weak temperature dependence of surface tension of the melts (see above) and, compared with other sources of errors, may be neglected for practical purposes. Recent measurements (155) by adiabatic calorimetry of C_p and S^0_{298} of bulk and nanophase CoO also confirm the excess entropy to be small.

II.3.1. Heat of Immersion and Adsorption

The heat of immersion of solid in a liquid and adsorption of a gas on a solid surface is a measure of interaction energy of the surface with the given liquid or gas molecules. These measurements cannot be used to derive absolute values for surface energies of solids, but they can indicate the differences in surface energy. Adam (133) trace the experimental reports on heat of wetting back to 1802. The heat of immersion can vary with specific surface area. Polar solids will show a large heat of immersion in a polar liquid and a smaller one in a nonpolar liquid (135). The heat of immersion measurements on zirconia performed by Holmes et al. (156) demonstrated an increase by 400 erg/cm^2 in total heat of immersion for monoclinic zirconia compared with tetragonal. It served as the first calorimetric evidence that stabilization of tetragonal zirconia in low-temperature synthesis routes is due to the small particle size (53).

Adsorption of different gas molecules is often used to characterize acid–base properties of the solid surface. Because of intrinsic inhomogeneity of solid surfaces, the different sites on the surface may adsorb different gases. For example, zirconia surfaces were found to exhibit acidic and basic properties at the same time (i.e., exhibit both CO_2 and NH_3 chemisorption) (157).

Heats of adsorption may be measured directly using calorimetry. Heats of adsorption can also be determined indirectly from kinetics of desorption using technique called temperature programmed desorption (TPD). In this method pressure of desorbing gas is recordered (e.g., using mass spectrometer) as a function of substrate temperature. For successful application of this technique, the desorption rate law and the variation of the kinetic parameters as a function of the coverage must be known (142). A semiempirical relation E_d (kJ/mol) $\sim 0.23 T_d$ (K) can be used for rough estimate of desorption energies from desorption temperature (157).

II.3.2. Aqueous Solution Calorimetry Near Room Temperature

Apparently, the first calorimetric measurements aimed to measure absolute values of surface energy of a solid were performed on NaCl by Lipsett et al. (158) in 1926–1928. The surface area of fine powders obtained by grinding and sublimation was estimated from the average size of microcrystals, as observed by optical microscopy and the differences in the heat of solution in water compare with coarse powder were measured. These experiments were not taken seriously in textbook in surface chemistry published 10 years later (133) because of two reasons: large uncertainties in surface area measurements, and probable difference in specific surface energy of small and large particles.

Introduction of BET theory by Brunauer et al. (69) in 1938 allowed for accurate measurements of surface areas by gas adsorption. Notably, Brunauer was the first to apply BET technique in conjunction with calorimetry to measure surface energies of solids. Brauner's determination of surface energy for $Ca_3Si_2O_7 \cdot 2H_2O$ demonstrated (193, 195) that heat of solution depends linearly on surface area (the scatter from linear trend corresponded less than 8% of the surface energy in the range 200–400 m^2/g). This might indicate that, at least for hydroxides, varying of surface energy with particle size may often be neglected for practical purposes.

II.3.3. High Temperature Oxide Melt Solution Calorimetry

High temperature oxide melt solution calorimetry for measurements of surface energies of solids was pioneered by Navrotsky et al. (see Navrotsky (33) for recent review)). The measurements are performed in a custom built Tian–Calvet type twin microcalorimeter using sodium molybdate or lead borate melts as calorimetric solvents. The details of the calorimeter design and experimental techniques are described elsewhere (159, 160). Enthalpy of drop solution is measured by dropping the sample from room temperature into the oxide melt at 700–800°C. As in aqueous solution calorimetry, a series of well characterized samples of the same structure but with different surface area is needed for measurement of surface enthalpy of a given compound. It is often not possible to completely prevent adsorption of water on the surfaces of the nanoparticles and corrections for evolved water must be made in the thermochemical cycles (38).

In recent years, using high temperature oxide melt solution calorimetry, surface energies of different polymorphs of Al_2O_3, TiO_2 and ZrO_2 have been measured (28–30, 34–36, 189) and phase stability reversals occurring with increasing surface area

Table 4. Some reported values for surface and interfacial energies for Si and Hf, Zr and Al oxides and techniques used for their determination

Interface or surface	Energy[a] (mJ/m^2)	Method (reference)
Si	869	From molten Si surface tension (139)
Si (111)	1140 ± 150	Cleavage (spark discharge method) (150)
Si (110)	1900 ± 200	
SiO_2 am./vac.	~300	At 1300°C extrapolated to pure silica from surface tensions of molten alkali silicates (194)
SiO_2 am./air	259 ± 3	
SiO_2 am. hydr./air	129 ± 8	Solution calorimetry at 23°C in HNO_3–HF mixture (196)
Al_2O_3 am./vac.	710	Surface tension of melt by pendant drop method (197)
Al_2O_3am./vac.	580	At m.p. by drop weight method (197)
Al_2O_3 am./vac.	905	At 1850°C (197)
a-Al_2O_3 cor./vac.	264 ± 20	Drop solution calorimetry at 700°C in $2PbO{\cdot}B_2O_3$ combined with water
$\gamma-Al_2O_3$ sp./vac.[b]	167 ± 10	adsorption study (29)
tetr. ZrO_2/ZrO_2	265	At 1850°C by equilibrium interfacial angle measurements on Ca-stabilized
tetr. ZrO_2/vac.	590	zirconia (197)
ZrO_2 am./air	500 ± 100	Drop solution calorimetry (36) at 700°C in $2PbO{\cdot}B_2O_3$
ZrO_2 tetr./air	900–2100	
ZrO_2 tetr./am.SiO_2	130 ± 70	Differential scanning calorimetry (DSC) measurements of ZrO_2 and HfO_2 crystallization from amorphous silicate (24, 102)
HfO_2 tetr./am.SiO_2	250 ± 80	

[a] $1\ mJ/m^2 = 1\ erg/cm^2 = 1\ dyne/cm = 1\ mN/m$.
[b] $\Delta H_{tr}\ \gamma\text{-}Al_2O_3 \rightarrow \alpha\text{-}Al_2O_3$ extrapolated to zero surface area is 13.4 ± 2.0 kJ/mol.

were confirmed by calorimetry (see Fig. 6, Table 4). Still, measurements of surface energies by calorimetry are by no means routine and sample-specific issues have to be resolved for each system.

II.3.4. Differential Scanning Calorimetry

Differential scanning calorimetry, or DSC, was introduced commercially in 1964 (161). Sample and reference are heated at a given rate and difference in heat flow to the sample versus reference is measured. Modern instruments are capable of quantitative measurement of enthalpies of crystallization and phase transitions for solids in milligram amounts. In some cases, DSC technique can be used to estimate interfacial energies (24, 102). Crystallization of hafnia and zirconia from amorphous silicates was studied. It was found that crystallization enthalpy per mole of hafnia and zirconia decreases with crystallite size formed after crystallization (Fig. 11). From this relation, interface energies for tetragonal $HfO_2(ZrO_2)$/amorphous silica can be calculated (Table 4). Analogously, the difference in crystallization enthalpy as a function of thin film thickness can derive interfacial energies between film and substrate. Research in this direction is underway in our laboratory.

II.4. GENERAL TRENDS IN SURFACE ENERGIES

Some relations between surface energies and bulk properties have been reported long ago, both on experimental and theoretical grounds. In 1922 empirical formulas were derived expressing surface tension as a function of difference in densities of the adjacent phases (162). Surface tension measured in liquid metals correlates relatively well with density (146). Livey and Murray (190) showed increasing surface energies with density for alkali halides. They also pointed out that for ionic compounds, since atoms on the surface are under-coordinated compared with bulk, surface energy is therefore some fraction of lattice energy, which is defined as formation energy from gaseous ions. Based on this they suggested that formation enthalpy can also be correlated to surface energies, however with caution because of difference in ionization potentials of elements.

Most metals have higher surface energies than oxides, halides and sulfides (139). Hydroxides usually have lower surface energies compared with oxides (33). Despite low surface mobility of the solids compared with liquids, the similar phenomena sometimes can be observed. For example, the wettability-formation of equilibrium shape of liquid surface on contact with solid surface is also observed in some degree for solid–solid interfaces formed in many thin films deposition techniques. Upon deposition of solid with low surface energy on surface with high surface energy, layer by layer growth is likely to occur, and island growth (ball-up) may occur in the opposite situation. For successful growing of alternating layers of different solids (superlattices or multilayers) their surface energies must be similar. This phenomenon is discussed in details in the chapters by Liang and Demkov and McKee et al. in this book.

II.4.1. Relevant Surface/Interface Energy Driven Processes

The interfacial energy can be minimized in several distinct ways:

(i) *Formation of amorphous phase on the interface.* Amorphous solids generally have lower surface and interfacial energies than crystalline surfaces. The thickness at which amorphous phase became stabilized by surface/interface energy term is called equilibrium thickness. Attempts have been made to calculate equilibrium thickness of amorphous intergranular phases for several systems. Interestingly, in the case of passivating oxide films on metals or semiconductors, amorphous film usually forms on some (e.g., Si, Ge, Ta, Nb, Al, Cr, Te) and epitaxial crystalline oxide on others (e.g., Cu, Co, Fe, Ni, Mo, Zn) (163). Jeurgenson et al. (163) demonstrated that the difference in metal–metal oxide interfacial energy for different crystallographic faces may switch the balance from amorphous to crystalline oxide film, e.g., $\{100\}$ and $\{110\}$ faces of Al substrate passivate with amorphous oxide, but $\{111\}$ with crystalline γ-Al_2O_3.

(ii) *Adsorption on the surface*. Thermodynamically, lowering interface energy by adsorption on the surface is similar to that in liquids and governed by Gibbs equation (see above). Solute segregation on interface in metals and ceramics is well documented (145). Although there are numerous qualitative examples of lowering of surface and interfacial energies in solids by impurities, examples where this effect was quantified in oxide systems are scarce. The measured surface energy for mica crystal obtained from the work of cleaving a crystal is 10 times larger in vacuum than in moist air (148). The magnitudes of these effects must be kept in mind while analyzing data on apparent surface energies from different sources. In connection with gate dielectric thin films, this implies that apparent surface energies of the same oxide deposited by different techniques (e.g., atomic layer deposition at atmospheric pressure and molecular beam epitaxy at ultra-high vacuum conditions) may be significantly different.

(iii) *Strain/epitaxy*. Minimization of interfacial energies often causes changes in the cell parameters or stabilizes normally metastable phases. The phases stabilized by strain in epitaxial thin films and nanocomposites may be different from those stabilized by lowering surface energy on solid/vacuum or solid/gas interfaces. A good example comes from study of crystallization of amorphous Al_2O_3 deposited by ALCVD on HF-last Si surface and on chemical and thermal silicon oxide surface (57). Al_2O_3 deposited on native silicon oxide remained amorphous after an annealing at 1100°C, while on the contact with crystalline silicon the cubic Al_2O_3 phase was formed epitaxially after annealing at 700–800°C. Crystallite growth by epitaxial aggregation was also observed for anatase crystallites (164).

II.4.2. Experimental Surface Energies versus Calculations

There is an increasing number of reports on calculation of surface and interfacial energies, or work of adhesion in oxides using electronic structure and force field based methods (165–176). In-depth discussion is beyond the scope of the present work. Just as experimental measurement of the surface energies for oxides are much more complicated and uncertain, than for metals, calculations of real surface and interface energies is hindered by surface inhomogeneties and anisotropy and limited by surface termination models. Even though considerable theoretical progress in this direction is achieved in the last decade, there is still a demand for experimental verifications.

The only materials for which there are enough experimental and calculated data to allow instructive comparison between measured and calculated values for surface energies are silicon and gallium arsenade (see chapter by M. Passlack on GaAs). Since large single crystals of silicon are readily available, the wealth of experimental data on silicon surface energies was obtained by cleavage method. The latest report (150) gives (111) 1.14 ± 0.15 J/m^2 and (110) 1.9 ± 0.2 J/m^2. These values are in reasonable agreement with values calculated for unrelaxed surfaces using Green's function method ((100) 1.71 J/m^2, (110) 1.69 J/m^2, (111) 1.36 J/m^2) (176).

The lower value for silicon surface energy was extrapolated from surface tension measurements on silicon melt. Surface tension of liquid silicon at melting point was determined to be 0.73 J/m^2 at 1410°C and rate of its decrease with temperature was measured as 0.001 J/m^2. Thus, it was interpolated (139) to room temperature as 0.87 J/m^2. As discussed above, it is expected that surface energies for solids estimated in this way are lower than actual energies, because of difference of the surface structure between melts and solids. Difference in the local coordination of Si in the melt from that in the crystal, make the comparison ambiguous.

Recently, the surfaces of zirconia polymorphs were studied by first-principles calculations using density functional theory and the pseudopotential formalism (168). The authors concluded that the surface energy of the most stable monoclinic (−111) and tetragonal (111) relaxed surfaces are equal within the calculational accuracy (1.246 J/m^2 vs. 1.239 J/m^2 at $T = 0$ K) and proposed that surface energy anisotropy is the key for understanding the stabilization of tetragonal zirconia in nanocrystals. These values are very different from those measured by high temperature oxide melt solution calorimetry (36) (Table 4), although the latter represent an average of an ill-controlled assemblage of faces. This discrepancy emphasizes the importance of experimentally established benchmarks in the surface energetics and of the careful comparison of theory and experiment.

4. CONCLUDING REMARKS, UNANSWERED QUESTIONS, FUTURE WORK

Phase diagrams of the systems relevant to alternative gate dielectrics, which are outlined in the first part of the chapter, are derived from experimental observations of phase equilibrium often reached at temperatures well above 1000°C. This is higher than temperatures to which gate dielectric are exposed in current manufacturing processes. Nevertheless, equilibrium phase diagrams are very useful to interpret results on phase evolution in oxide films, crystallization pathways, etc. Still, basic thermodynamic data, even for bulk materials, are often not sufficient to allow reliable calculations of phase equilibrium. Obviously, new applications should prompt more detailed investigation of thermodynamic properties of relevant bulk compounds, films and nanoparticles. Y-stabilized Zirconia (YSZ) is probably the system studied in most detail from reviewed here, due to its ceramic application. Hafnium oxide has not found wide application in ceramic industry in part because of higher cost and higher temperature of monoclinic to tetragonal phase transition. Thus, the bulk thermodynamic data for hafnium compounds are much more limited than those for zirconia. However, HfO_2 has better applicability in semiconductor industry because of larger formation enthalpy of HfO_2 than ZrO_2 and better thermal stability of the amorphous phase. The thermodynamics of bulk Hf-compounds must be explored further. In new experiments attention must be paid to characterization of sample purity and surface related effects.

As follows from our comparison, much information relevant to amorphous and nanocrystalline films can be learned from studying sol–gel and precipitation-produced solids, when particle size-related effects are considered. $HfO_2(ZrO_2)$–La_2O_3 systems have great potential since they offer both high thermal stability of amorphous phase and the pyrochlores $Hf_2La_2O_7$ and $Zr_2La_2O_7$ are lattice-matched to Si.

It is often said that if surface area is considered as a variable, it adds another dimension to equilibrium phase diagrams. Depending on experimental conditions and kinetics, surface-related effects may manifest themselves in strain and interface energy, adsorption and recombination on the surface and compositional heterogeneity. When interface energetics became the overriding feature which defines the properties of the material formed, the better analogy might be to say that we are looking not in the another dimension, but in the shadows of equilibrium phase diagram because these high-surface nanomaterials are metastable with respect to the bulk. Comparison of different experimental techniques for experimental determination of surface energies shows that they all have their own limitations, and to establish some experimental benchmarks for crystalline and amorphous oxides, it is important to converge on the estimates of surface energies obtained by different techniques in strictly controlled conditions. It is our hope that introduction of new oxides in semiconductor industry will leverage these efforts.

ACKNOWLEDGEMENTS

Some of the work reported in this review was supported by Motorola Inc. and the UC SMART program as part of the project "Thermochemistry of gate dielectric films". We acknowledge the National Center for Electron Microscopy at the Lawrence Berkeley National Laboratory for the use of its facilities. We wish to thank Dr. Alex Demkov, Dr. Bich-Yen Nguyen, Dr. Phillip Tobin (Motorola Inc.), Dr. Jean Tangeman (Containerless Research, Inc.), Dr. Martin Wilding (UCD), Prof. Susanne Stemmer and Prof. Jacob Israelachvili (UCSB) for helpful discussions.

REFERENCES

1. K.J. Hubbard, D.G. Schlom, Thermodynamic stability of binary oxides in contact with silicon, *J. Mater. Res.* **11**(11), 2757–2776 (1996).
2. D.G. Schlom, J.H. Haeni, A thermodynamic approach to selecting alternative gate dielectrics, *MRS Bull.* **27**(3), 198–204 (2002).
3. S. Stemmer, Z.Q. Chen, P.S. Lysaght, J.A. Gisby, J.R. Taylor, Investigations of the structure and stability of alternative gate dielectrics, *Proc. Electrochem. Soc. 2003-2 (Silicon Nitride and Silicon Dioxide Thin Insulating Films)* (2003) 119–130.

4. L. Topor, O.J. Kleppa, Standard enthalpies of formation of silicides M_5Si_3 (M = yttrium lutetium zirconium) and of hafnium silicide (Hf_3Si_2), *J. Less-Common Metals* **167**(1), 91–99 (1990).
5. S.V. Meschel, O.J. Kleppa, Standard enthalpies of formation of some 5d transition metal silicides by high temperature direct synthesis calorimetry, *J. Alloys Comp.* **280**(1–2), 231–239 (1998).
6. S.V. Meschel, O.J. Kleppa, Standard enthalpies of formation of some 4d transition metal silicides by high temperature direct synthesis calorimetry, *J. Alloys Comp.* **274**(1–2), 193–200 (1998).
7. M.C. Wilding, A. Navrotsky, High temperature calorimetric studies of the heat of solution of La_2O_3 in silicate liquids, *J. Non-Cryst. Solids* **265**(3), 238–251 (2000).
8. I.C. Lin, A. Navrotsky, J.K.R. Weber, P.C. Nordine, Thermodynamics of glass formation and metastable solidification of molten $Y_3Al_5O_{12}$, *J. Non-Cryst. Solids* **243**(23), 273–276 (1999).
9. H. Zhang, B. Gilbert, F. Huang, J.F. Banfield, Water-driven structure transformation in nanoparticles at room temperature, *Nature* **424**(6952), 1025–1029 (2003).
10. Y. Zhang, PhD dissertation, UC Davis (2003).
11. J.A. Tangeman, Private communication (2004).
12. Y. Zhang, A. Navrotsky, Thermochemistry of glasses in the Y_2O_3–Al_2O_3–SiO_2 system, *J. Am. Cer. Soc.* **86**(10), 1727–1732 (2003).
13. Y. Zhang, A. Navrotsky, J.A. Tangeman, J.K.R. Weber, Thermochemistry of glasses along the $2NdAlO_3$–$3SiO_2$ join, *J. Phys.: Condens. Matter* **15**(31), S2343–S2355 (2003).
14. J.K.R. Weber, J.A. Tangeman, T.S. Key, P.C. Nordine, Investigation of liquid–liquid phase transitions in molten aluminates under containerless conditions, *J. Thermophys. Heat Transfer* **17**(2), 182–185 (2003).
15. S. Ellsworth, A. Navrotsky, R.C. Ewing, Energetics of radiation damage in natural zircon ($ZrSiO_4$), *Phys. Chem. Minerals* **21**(3), 140–149 (1994).
16. L.M. Wang, S.X. Wang, W.L. Gong, R.C. Ewing, W.J. Weber, Amorphization of ceramic materials by ion beam irradiation, *Mater. Sci. Eng. A* **253**, 106–113 (1998).
17. J. Lian, L.M. Wang, J. Chen, R.C. Ewing, K.V.G. Kutty, Heavy ion irradiation of zirconate pyrochlores, *Mater. Res. Soc. Symp. Proc.* **713** (*Scientific Basis for Nuclear Waste Management XXV*), 507–512 (2002).
18. J. Lian, X.T. Zu, K.V.G. Kutty, J. Chen, L.M. Wang, R.C. Ewing, Ion-irradiation-induced amorphization of $La_2Zr_2O_7$ pyrochlore, *Phys. Rev. B* **66**(5), 054108/1-5 (2002).
19. D. Turnbull, Modes of formation of amorphous solids: an overview, *Diffusion and Defect Data* **53**(4), 9–20 (1987).
20. MacK.J.D. Kenzie, T. Kemmitt, Evolution of crystalline aluminates from hybrid gel-derived precursors studied by XRD and multinuclear solid-state MAS NMR. II. Yttrium–aluminum garnet $Y_3Al_5O_{12}$, *Thermochim. Acta* **325**(1), 13–18 (1999).
21. MacK.J.D. Kenzie, M. Schmücker, L. Mayer, Evolution of crystalline aluminates from hybrid gel-derived precursors studied by XRD and multinuclear solid state MAS NMR. III. Lanthanum hexaluminate $LaAl_{11}O_{18}$, *Thermochim. Acta* **335**(1–2), 73–78 (1999).
22. G. Stefanic, S. Music, Factors influencing the stability of low temperature tetragonal ZrO_2, *Croatica Chem. Acta* **75**(3), 727–767 (2002).
23. S.V. Ushakov, B.E. Burakov, V.M. Garbuzov, E.B. Anderson, E.E. Strykanova, M.M. Yagovkina, K.B. Helean, Y.X. Guo, R.C. Ewing, W. Lutze, Synthesis of Ce-doped zircon by a sol–gel process, *Mater. Res. Soc. Symp. Proc.* **506**, 281–288 (1998).
24. S.V. Ushakov, C.E. Brown, A. Navrotsky, A. Demkov, C. Wang, B.-Y. Nguyen, Thermal analyses of bulk amorphous oxides and silicates of zirconium and hafnium, *Mater. Res. Soc. Symp. Proc.* **745** *(Novel Materials and Processes for Advanced CMOS)*, 3–8 (2003).

25. M. Mizuno, R. Berjoan, J.P. Coutures, M. Foex, Phase diagram of the system aluminum oxide–lanthanum oxide at elevated temperatures, *Yogyo Kyokaishi* **82**(12), 631–636 (1974). *(PDFC 6438)*.
26. N.A. Toropov, I.A. Bondar, F.Ya. Galakhov, Kh.S. Nikogosyan, N.V. Vinogradova, Phase equilibriums in the yttrium oxide–alumina system, *Izv. Akad. Nauk SSSR, Seriya Khimicheskaya* **7**(7), 1158–1164 (1964) (*PDFC 2344*).
27. T. Noguchi, M. Mizuno, Liquidus curve measurements in the system yttrium oxide–aluminum oxide, *Kogyo Kagaku Zasshi* **70**(6), 834–839 (1967) (*PDFC 4370*).
28. J.M. McHale, K. Yuerekli, D.M. Dabbs, A. Navrotsky, S. Sundaresan, I.A. Aksay, Metastability of spinel-type solid solutions in the SiO_2–Al_2O_3 system, *Chem. Mater.* **9**(12), 3096–3100 (1997).
29. J.M. McHale, A. Auroux, A.J. Perrotta, A. Navrotsky, Surface energies and thermodynamic phase stability in nanocrystalline aluminas, *Science* **277**, 788–791 (1997).
30. J.M. McHale, A. Navrotsky, A.J. Perrotta, Effects of increased surface area and chemisorbed H_2O on the relative stability of nanocrystalline γ-Al_2O_3 and α-Al_2O_3, *J. Phys. Chem. B* **101**(4), 603–613 (1997).
31. A. Navrotsky, Systematic trends and prediction of enthalpies of formation of refractory lanthanide and actinide ternary oxide phases, *Ceram. Trans.* **119**, 137–146 (2001).
32. A. Navrotsky, Thermochemical studies of nitrides and oxynitrides by oxidative oxide melt calorimetry, *J. Alloys Comp.* **321**(2), 300–306 (2001).
33. A. Navrotsky, Thermochemistry of nanomaterials, *Rev. Miner. Geochem.* **44**, 73–103 (2001).
34. M.R. Ranade, S.H. Elder, A. Navrotsky, Energetics of nanoarchitectured TiO_2–ZrO_2 and TiO_2–MoO_3 composite materials, *Chem. Mater.* **14**(3), 1107–1114 (2002).
35. M.R. Ranade, A. Navrotsky, H.Z. Zhang, J.F. Banfield, S.H. Elder, A. Zaban, P.H. Borse, S.K. Kulkarni, G.S. Doran, H.J. Whitfield, Energetics of nanocrystalline TiO_2, *Proc. Natl. Acad. Sci.* **99**(suppl.2), 6476–6481 (2002).
36. M.W. Pitcher, S.V. Ushakov, A. Navrotsky, B.F. Woodfield, G. Li, J. Boerio-Goates, B.M. Tissue, Energy crossovers in nanocrystalline zirconia, *J. Am. Cer. Soc.* **88**(1), 160–167 (2005).
37. R.K. Iler, *The Chemistry of Silica* (John Wiley & Sons Inc., 1979), pp. 623–792.
38. A. Navrotsky, R.P. Rapp, E. Smelik, P. Burnley, S. Circone, L. Chai, K. Bose, H.R. Westrich, The behavior of H_2O and CO_2 in high-temperature lead borate solution calorimetry of volatile-bearing phases, *Am. Mineral.* **79**(11–12), 1099–1109 (1994).
39. A. Navrotsky, Thermochemistry of crystalline and amorphous silica, *Rev. Mineral.* **29** (SILICA), 309–329 (1994).
40. R.H.J. Hannink, P.M. Kelly, B.C. Muddle, Transformation toughening in zirconia-containing ceramics, *J. Am. Ceram. Soc.* **83**(3), 461–487 (2000).
41. M.H. Bocanegra-Bernal, S.D. de la Torre, Phase transitions in zirconium dioxide and related materials for high performance engineering ceramics, *J. Mater. Sci.* **37**(23), 4947–4971 (2002).
42. A. Suresh, M.J. Mayo, W.D. Porter, Thermodynamics of the tetragonal-to-monoclinic phase transformation in fine and nanocrystalline yttria-stabilized zirconia powders, *J. Mater. Res.* **18**(12), 2912–2921 (2003).
43. O.M. Stansfield, Thermal expansion of polycrystalline HfO_2–ZrO_2 solid solutions, *J. Am. Ceram. Soc.* **48**(8), 436–437 (1965).
44. R. Ruh, H.J. Garret, R.F. Domagala, N.M. Tallen, System zirconia–hafnia, *J. Am. Ceram. Soc.* **51**(1), 23–27 (1968).
45. A.M. Gavrish, B.Ya. Sukharevskii, P.P. Krivoruchko, E.I. Zoz, Solid solutions and polymorphism in the zirconium dioxide–hafnium dioxide system, *Izv. Akad. Nauk SSSR Neorg. Mater.* **5**(3), 547–550 (1969).
46. J. Wang, H.P. Li, R. Stevens, Hafnia and hafnia-toughened ceramics, *J. Mater. Sci.* **27**(20), 5397–5430 (1992).

47. *Powder Diffraction File (PDF): Inorganic Phases*, JCPDS-ICDD (1999).
48. Inorganic Crystal Structure Database (ICSD), The National Institute of Standards and Technology (NIST) and Fachinformationszentrum Karlsruhe (FIZ) (2004).
49. H.J. Garrett, *Am. Ceram. Soc. Bull.* **42**, 201 (1963).
50. E.I. Zoz, A.M. Gavrish, N.V. Gul'ko, Phase formation in the zirconium oxide (hafnium oxide)–lanthanum oxide system, *Izv. Akad. Nauk SSSR Neorg. Mater.* **14**(1), 109–111 (1978).
51. R.D. Shannon, Revised effective ionic radii and systematic studies of interatomic distances in halides and chalcogenides, *Acta Crystallogr.* **A32**, 751–767 (1976).
52. O. Ruff, F. Ebert, Refractory ceramics. I. The forms of zirconium dioxide, *Z. Anorg. Allgem. Chem.* **180**, 19–41 (1929).
53. R.C. Garvie, Stabilization of the tetragonal structure in zirconia microcrystals, *J. Phys. Chem.* **82**(2), 218–224 (1978).
54. R.C. Garvie, The occurrence of metastable tetragonal zirconia as a crystallite size effect, *J. Phys. Chem.* **69**(4), 1238–1243 (1965).
55. O. Hunter, R.W. Scheidecker Jr., S. Tojo, Characterization of metastable tetragonal hafnia, *Ceram. Int.* **5**(4), 137–141 (1979).
56. S.K. Filatov, *High-Temperature Crystallochemistry* (Leningrad, Nedra, 1990), p. 284.
57. C. Zhao, V. Cosnier, P.J. Chen, O. Richard, G. Roebben, J. Maes, S. Van Elshocht, H. Bender, E. Young, O. Van Der Biest, M. Caymax, W. Vandervorst, S. De Gendt, M. Heyns, Thermal stability of high k layers, *Mater. Res. Soc. Symp. Proc.* **745** (Novel Materials and Processes for Advanced CMOS), 9–14 (2003).
58. S.V. Ushakov, C.E. Brown, A. Navrotsky, Effect of La and Y on crystallization temperatures of hafnia and zirconia, *J. Mater. Res.* **19**(3), 693–696 (2004).
59. V.B. Glushkova, E.K. Koehler, L.G. Shcherbakova, Kinetics and mechanism of solid state reactions in systems of refractory oxides, in: *Science of Ceramics*, Vol. 5, *Brit. Ceram. Soc.* 219–239 (1970).
60. E. Gusev, C. Cabral, M. Copel Jr., C.D. Emic, M. Gribelyuk, Ultrathin HfO_2 films growth on silicon by atomic layer deposition for advanced gate dielectrics applications, *Microelectron. Eng.* **69**, 145–151 (2003).
61. R.R. Manory, T. Mori, I. Shimizu, S. Miyake, G. Kimmel, Growth and structure control of HfO_{2-x} films with cubic and tetragonal structures obtained by ion beam assisted deposition, *J. Vac. Sci. Technol. A* **20**(2), 549–554 (2002).
62. J.P. Holgado, F. Espinos, A. Yebero, M. Justo, J. Ocana, A.R. Benitez, Gonzales-Eipe stabilization of the cubic/tetragonal phases of ZrO_2 in thin films prepared by ion beam induced chemical vapour deposition, *Thin Solid Films* **389**, 34–42 (2001).
63. S.S. Tsunekawa, Y. Ito, J.-T. Kawazoe, Wang Critical size of the phase transition from cubic to tetragonal in pure zirconia nanoparticles, *Nano Lett.* **3**(7), 871–875 (2003).
64. O. Ohtaka, H. Fukui, T. Kunisada, T. Fujisawa, Phase relations and volume changes of hafnia under high pressure and high temperature, *J. Am. Ceram. Soc.* **84**, 1369–1373 (6) (2001).
65. O. Ohtaka, T. Yamanaka, S. Kume, E. Ito, A. Navrotsky, Stability of monoclinic and orthorhombic zirconia: studies by high-pressure phase equilibria and calorimetry, *J. Am Ceram. Soc.* **74**, 505–509 (1991).
66. S.W. Nam, S. Yoo, H. Nam, D. Choi, D. Lee, J.H. Ko, J.H. Moon, S. Ku Choi, Influence of annealing conditions on the properties of sputtered hafnium oxide, *J. Non-Cryst. Solids* **303**(1), 139–149 (2002).
67. K. Kukli, M. Ritala, T. Sajavaara, J. Keinonen, M. Leskela, Atomic layer deposition of hafnium dioxide films from hafnium tetrakis (ethylamide) and water, *Chem. Vap. Deposition* **8**, 5 (2002).
68. C. Wiemer, M. Fanciulli, B. Crivelli, G. Pavia, M. Alessandri, Evolution of crystallographic ordering in $Hf_{1-x}Al_xO_y$ high-k dielectric deposited by atomic layer deposition, *Appl. Phys. Lett.* **83**(25), 5271–5273 (2003).

69. S. Brunauer, P.H. Emmett, E. Teller, Adsorption of gases in multimolecular layers, *J. Am. Chem. Soc.* **60**, 309–319 (1938).
70. R.G. Haire, L. Eyring, Comparisons of the binary oxides, in: *Handbook on the Physics and Chemistry of Rare Earths*, Vol. 18, eds. K.A. Gschneidner Jr. L.G.R. Eyring, G.H. Choppin, Lander (North-Holland, Amsterdam, 1994), pp. 413–506.
71. V.B. Glushkova, E.K. Keler, Polymorphism of La oxide, *Doklady Akad. Nauk SSSR* **152**(3), 611–614 (1963).
72. I. Warshaw, R. Roy, Crystal chemistry of rare earth sesquioxides aluminates and silicates, in: *Progr. Sci. Technol. Rare Earths*, Vol. 1, ed. L.A. Eyring (Pergamon Press Book, The Macmilan Company, New York, 1964), pp. 203–221.
73. I. Warshaw, R. Roy, Polymorphism of the rare earth sesquioxides, *J. Phys. Chem.* **65**, 2048–2051 (1961).
74. V.M. Goldschmidt, F. Ulrich, T. Barth, Geochemische Verteilungsgesetze der Elemente. IV. *Skrifter NorskeVidenskaps-Akad. Oslo. I Mater. Naturv. Kl.* **5**, 6–24 (1925).
75. M. Foex, J.P. Traverse, Polymorphism of rare earth sesquioxides at high temperatures, *Bull. Soc. Franc. Mineral. Crist.* **89**(2), 184–205 (1966).
76. P. Aldebert, J.P. Traverse, Neutron diffraction study of the high temperature structures of lanthanum oxide and neodymium oxide, *Mater. Res. Bull.* **14**(3), 303–323 (1979). (ICSD #100213).
77. M. Foex, J.P. Traverse, Investigations about crystalline transformation in rare earths sesquioxides at high temperatures, *Rev. Int. Hautes Temp. Refract.* **3**(4), 429–453 (1966). (*H-Y_2O_3, PDF 201412, 2300 C*).
78. V. Swamy, H.J. Seifert, F. Aldinger, Thermodynamic properties of Y_2O_3 phases and the yttrium–oxygen phase diagram, *J. Alloys Comp.* **269**(1–2), 201–207 (1998).
79. V. Swamy, N.A. Dubrovinskaya, L.S. Dubrovinsky, High-temperature powder x-ray diffraction of yttria to melting point, *J. Mater. Res.* **14**(2), 456–459 (1999).
80. A. Navrotsky, in preparation.
81. W. Chen, T.A. Lee, A. Navrotsky, Enthalpy of formation of yttria-doped ceria, *J. Mat. Res.*, **20**(1), 144–150 (2005).
82. S.V. Ushakov, J. Cheng, A. Navrotsky, J.R. Wu, S.M. Haile, Formation enthalpies of tetravalent lanthanide perovskites by high temperature oxide melt solution calorimetry, *Mater. Res. Soc. Symp. Proc.* **718** (*Perovskite Materials*), 71–76 (2002).
83. L.R. Morss, Comparative thermochemical and oxidation–reduction properties of lanthanides and actinides, in: *Handbook on the Physics and Chemistry of Rare Earths*, Vol. 18, eds. K.A. Gschneidner Jr., G.R. Eyring, G.H. Choppin and Lander (North-Holland, Amsterdam, 1994), p. 239.
84. H. Eilers, B.M. Tissue, Synthesis of nanophase ZnO Eu_2O_3 and ZrO_2 by gas-phase condensation with cw-CO_2 laser heating, *Mater. Lett.* **24**(4), 261–265 (1995).
85. B.M. Tissue, H.B. Yuan, Structure particle size and annealing of gas phase-condensed Eu^{3+}:Y_2O_3 nanophosphors, *J. Solid State Chem.* **171**(1–2), 12–18 (2003).
86. S. Katagiri, N. Ishizawa, F. Marumo, A new high temperature modification of face-centered cubic yttrium oxide (Y_2O_3), *Powder Diffraction* **8**(1), 60 (1993).
87. R.J. Gaboriaud, F. Paumier, F. Pailloux, P. Guerin, Y_2O_3 thin films: internal stress and microstructure, *Mater. Sci. Eng. B* **109**(1–3), 34–38 (2004).
88. L.M. Lopato, A.V. Shevchenko, G.I. Gerasimyuk, Hafnium oxide–aluminum oxide system, *Izv. Akad. Nauk SSSR Neorg. Mater.* **12**(9), 1623–1626 (1976); *Inorg. Mater. (Eng. Transl.)* **12**(9), 1331–1334 (PDFC fig 6441).
89. S. Komarneni, R. Roy, Synthesis of zircon, in: *Zircon Science and Technology*, ed. S. Somiya (Uchidarikakuho, Japan, 1988), pp. 289–298.
90. A.I. Kingon, J.-P. Maria, S.K. Streiffer, Alternative dielectrics to silicon dioxide for memory and logic devices, *Nature* **406**, 1032–1038 (2000).

91. J.-P. Maria, D. Wickaksana, J. Parrette, A.I. Kingon, Crystallization in SiO_2–metal oxide alloys, *J. Mater. Res.* **17**(7), 1571–1579 (2002).
92. W.F.A. Besling, E. Young, T. Conard, C. Zhao, R. Carter, W. Vandervorst, M. Caymax, S. De Gendt, M. Heyns, J. Maes, M. Tuominen, S. Haukka, Characterization of ALCVD Al_2O_3–ZrO_2 nanolaminates, link between electrical and structural properties, *J. Non-Cryst. Solids* **303**(1), 123–133 (2002).
93. A.G. Karaulov, E.I. Zoz, Study of zirconia–hafnia–silica-system solid solutions and preparation of new refractory materials from them, *Ogneupory* **2**(2), 11–13 (1991).
94. S.V. Ushakov, K.B. Helean, A. Navrotsky, L.A. Boatner, Thermochemistry of rare-earth orthophosphates, *J. Mater. Res.* **16**(9), 2623–2633 (2001).
95. Y. Kanno, Thermodynamic and crystallographic discussion of the formation and dissociation of zircon, *J. Mater. Sci.* **24**, 2415–2420 (1989).
96. A.J.G. Ellison, A. Navrotsky, Enthalpy of formation of zircon, *J. Am. Ceram. Soc.* **75**(6), 1430–1433 (1992).
97. S.V. Ushakov, W. Gong, M.M. Yagovkina, K.B. Helean, W. Lutze, R. Ewing, Solid solutions of Ce U and Th in zircon, *Ceram. Trans.* **93**, 357–363 (1999).
98. G.D. Wilk, R.M. Wallace, J.M. Anthony, High-*k* gate dielectrics: Current status and materials properties considerations, *J. Appl. Phys.* **89**(10), 5243–5275 (2001).
99. G.D. Wilk, R.M. Wallace, Stable zirconium silicate gate dielectrics deposited directly on silicon, *Appl. Phys. Lett.* **76**(1), 112–114 (2000).
100. T.P. Ledneva, K.A. Bryukhova, Solubility of the refractory oxides zirconium dioxide and hafnium dioxide in silica, *Steklo* **2**, 84–87 (1973).
101. P. Lysaght, B. Forran, S. Stemmer, G. Bersuker, J. Bennett, R. Tichy, L. Larson, H.R. Huff, Thermal response of MOCVD hafnium silicate, *Microelectron. Eng.* **69**(2–4), 182–189 (2003).
102. S.V. Ushakov, A. Navrotsky, Y. Yang, S. Stemmer, K. Kukli, M. Ritala, M.A. Leskelä, P. Fejes, A. Demkov, C. Wang, B.-Y. Nguyen, D. Triyoso, P. Tobin, Crystallization in hafnia- and zirconia-based systems, *Phys. Stat. Sol. (b)*, **241**(10), 2268–78 (2004).
103. C.G. Levi, Metastability and microstructure evolution in the synthesis of inorganics from precursors, *Acta Mater.* **463**, 787–800 (1998).
104. N.A. Toropov, I.A. Bondar, F.J. Galakhov, High-temperature solid solutions of silicates of the rare earth elements, *Trans. Intern. Ceram. Congr. 8th Copenhagen* 85–103 (1962).
105. N.A. Toropov, I.A. Bondar, Silicates of the rare earth elements. III. Phase diagram for the binary system yttrium oxide–silica, *Izv. Akad. Nauk SSSR Seriya Khimicheskaya* 544–550 (1961).
106. O. Fabrichnaya, H.J. Seifert, R.L.T. Weiland, F. Aldinger, A. Navrotsky, Phase equilibria and thermodynamics in the Y_2O_3–Al_2O_3–SiO_2 system, *Zeitschrift fuer Metallkunde* **92**(9), 1083–1097 (2001).
107. A.N. Christensen, R.G. Hazell, A.W. Hewat, Synthesis crystal growth and structure investigations of rare-earth disilicates and rare-earth oxyapatites, *Acta Chem. Scand.* **51**(1), 37–43 (1997).
108. A.N. Christensen, Investigation by the use of profile refinement of neutron powder diffraction data of the geometry of the $[Si_2O_7]^{6-}$ ions in the high temperature phases of rare earth disilicates prepared from the melt in crucible-free synthesis, *Zeitschrift fuer Kristallographie* **209**(1), 7–13 (1994).
109. A.N. Christensen, A.F. Jensen, B.K. Themsen, R.G. Hazell, M. Hanfland, E. Doory-hee, Structure investigations of the high-temperature phases of $La_2Si_2O_7$, $Gd_2Si_2O_7$ and $Sm_2Si_2O_7$, *Acta Chem. Scand.* **51**(12), 1178–1185 (1997).
110. A.S. Risbud, K.B. Helean, M.C. Wilding, P. Lu, A. Navrotsky, Enthalpies of formation of lanthanide oxyapatite phases, *J. Mater. Res.* **16**(10), 2780–2783 (2001).

111. K.P. Plucknett, D.S. Wilkinson, Microstructural characterization of a microwave-sintered silicon nitride based ceramic, *J. Mater. Res.* **10**(6), 1387–1396 (1995).
112. J. Lin, Q. Su, Luminescence and energy migration in the oxyapatite $Ca_2Gd_8(SiO_4)_6O_2$ doped with several rare earth and mercury-like ions, *J. Alloys Comp.* **210**(1–2), 159–163 (1994).
113. C.R. Ronda, T. Jüstel, H. Nikol, Rare earth phosphors: fundamentals and applications, *J. Alloys Comp.* **275–277**, 669–676 (1998).
114. L.M. Wang, M. Cameron, W.J. Weber, K.D. Crowley, R.C. Ewing, *In situ* TEM observation of radiation induced amorphization of crystals with apatite structure, in: *Hydroxyapatite of Related Materials*, eds. P.W. Brown, B. Constantz (CRC, Boca Raton, 1994), pp. 243–249.
115. J.-J. Liang, A. Navrotsky, T. Ludwig, H.J. Seifert, F. Aldinger, Enthalpy of formation of rare earth silicates Y_2SiO_5 and Yb_2SiO_5 and N-containing silicate $Y_{10}(SiO_4)_6N_2$, *J. Mater. Res.* **14**, 1181–1185 (1999).
116. Y. Kanke, A. Navrotsky, A calorimetric study of the lanthanide aluminum oxides and the lanthanide gallium oxides: stability of the perovskites and the garnets, *J. Solid State Chem.* **141**(2), 424–436 (1998).
117. M.C. Wilding, P.F. McMillan, A. Navrotsky, Calorimetric study of glasses and liquids in the polyamorphic system, Y_2O_3–Al_2O_3. *Phys. Chem. Glasses* **43**(6), 306–312 (2002).
118. M.C. Wilding, P.F. McMillan, Liquid polymorphism in yttrium–aluminate liquids in: *New Kinds of Phase Transitions: Transformations in Disordered Substances*, ed. V.V. Brazhkin (2002), pp. 57–73.
119. M.C. Wilding, P.F. McMillan, A. Navrotsky, Thermodynamic and structural aspects of the polyamorphic transition in yttrium and other rare-earth aluminate liquids, in: *Physica A: Statistical Mechanics and Its Applications* (Amsterdam, Netherlands) **314**(1–4), 379–390 (2002).
120. M.C. Wilding, Private communication (2003).
121. A. Rouanet, Zirconium dioxide–lanthanide oxide systems close to the melting point, *Rev. Int. Hautes Temp. Refract.* **8**(2), 161–180 (1971) (PDFC 5232).
122. V.B. Glushkova, E.K. Keler, M. Kravchinskaya, V.V.A. Krzhizhanovskaya, A.K. Kuznetsov, P.A. Tikhonov, Zirconates of rare earth elements, in: *Rare Earth Element Compounds: Zirconates Hafnates Niobates Tantalates and Antimonates (Soedineniya Redkozemel'nykh Elementov: Tsirkonaty Gafnaty Niobaty Tantalaty Antimonaty)*, eds. V.P. Orlovskii and N.N. Chudinova (Nauka, Moscow, USSR, 1985), pp. 3–40.
123. V.B. Glushkova, M.V. Kravchinskaya, Hafnium dioxide-based refractory compounds and solid solutions. I. Phase diagrams of the systems HfO_2–M_2O_3 and HfO_2–MO, *Ceram. Int.* **11**(2), 56–65 (1985).
124. V.B. Glushkova, M.V. Kravchinskaya, A.K. Kuznetsov, P.A. Tikhonov, Hafnates of rare earth elements, in: *Rare Earth Element Compounds: Zirconates Hafnates Niobates Tantalates and Antimonates (Soedineniya Redkozemel'nykh Elementov: Tsirkonaty Gafnaty Niobaty Tantalaty Antimonaty)*, eds. V.P. Orlovskii and N.N. Chudinova (Nauka, Moscow, USSR, 1985), pp. 42–77.
125. V.B. Glushkova, V.A. Krzhizhanovskaya, Hafnium dioxide-based refractory compounds and solid solutions. 2. Kinetics and mechanism of compound formation in the systems HfO_2–M_2O_3(MO), *Ceram. Int.* **11**(3), 80–90 (1985).
126. V.V. Kharton, A.A. Yarmchenko, E.N. Naumovich, F.M.B. Marques, Research on the electrochemistry of oxygen ion conductors in the former Soviet Union. III. HfO_2-, CeO_2- and ThO_2-based oxides, *J. Solid State Electrochem.* **4**, 243–266 (2000).
127. T.A. Lee, A. Navrotsky, Enthalpy of formation of cubic yttria-stabilized hafnia (c-YSH), *J. Mater. Res.* **19**(6), 1855–1861 (2004).
128. T.A. Lee, A. Navrotsky, I. Molodetsky, Enthalpy of formation of cubic yttria-stabilized zirconia, *J. Mater. Res.* **18**(4), 908–918 (2003).

129. M.A. Subramanian, A.W. Sleight, Rare earth pyrochlores, in: *Handbook on the Physics and Chemistry of Rare Earths*, Vol. 16, eds. K.A. Gschneidner Jr., L. Eyring (Elseiver Science Publishers, North-Holland, Amsterdam, 1993), pp. 225–248.
130. S.V. Ushakov, A. Navrotsky, L.A. Boatner, in preparation
131. J.W. Seo, J. Fompeyrine, A. Guiller, G. Norga, C. Marchiori, H. Siegwart, J.-P. Locquet, Interface formation and defect structures in epitaxial $La_2Zr_2O_7$ thin films on (111) Si, *Appl. Phys. Lett.* **83**(25), 5211–5213 (2003).
132. A. Dimoulas, G. Vellianitis, G. Apostolopoulos, G. Mavrou, A. Travlos, J.C. Hooker, Z.M. Rittersma, Epitaxial and amorphous $La_2Hf_2O_7$ on silicon for high-k gates, *Abstracts Mater. Res. Soc.* 2003 Fall Meeting **146** (2003).
133. N.K. Adam, *The Physics and Chemistry of Surfaces* (Clarendon Press, Oxford, 1938).
134. A.W. Adamson, A.P. Gast, *Physical Chemistry of Surfaces* (John Wiley and Sons, New York, 1997).
135. A.W. Adamson, *Physical Chemistry of Surfaces* (John Wiley and Sons, New York, 1976).
136. R.J. Good, Surface free energy of solids and liquids: thermodynamics molecular forces and structure, *J. Colloid Interface Sci.* **59**, 3 (1977).
137. M.J. Jaycock, G.D. Parfitt, *Chemistry of Interfaces* (John Wiley and Sons, New York, 1981), p. 279.
138. J.N. Israelachvili, *Intermolecular and Surface Forces. With Applications to Colloidal and Biological Systems* (Academic Press, New York, 1985), p. 296.
139. K.N. Tu, J.W. Mayer, L.C. Feldman, *Electronic thin Film Science: For Electrical Engineers and Materials Scientists* (Macmillan Publishing Company, New York, 1992), p. 428.
140. E.D. Shchukin, A.V. Pertsov, E.A. Amelina, A.S. Zelenov, *Colloid and Surface Chemistry* (Elsevier, Amsterdam, 2001), pp. 57–58.
141. E. Orowan, Surface energy and surface tension in solids and liquids, *Proc. Roy. Soc. Lond.* **316**, 473–491 (1970).
142. E.M. McCash, *Surface Chemistry* (University Press, Oxford, 2001), p. 177.
143. F. Beschstedt, *Principles of Surface Physics. Advanced Texts in Physics* (Springer-Verlag, Berlin, Heidelberg, New York, 2003), p. 342.
144. R.S. Burdon, *Surface Tension and the Spreading of Liquids* (Cambridge University Press, Cambridge, 1940).
145. D. Gupta, Diffusion solute segregations and interfacial energies in some material: An overview, *Interface Sci.* **11**(1), 7–20 (2003).
146. P.P. Pugachevich, V.I. Yashkichev, Measurement of surface tension of liquid metals at high temperatures, in: *The Role of Surface Phenomena in Metallurgy*, ed. V.N. Eremenko (Constants Bureau, New York, 1963), p. 46.
147. A.C. Lewis, A.B. Mann, D. Van Heerden, D. Josell, T.P. Weihs, The effect of interfacial free energies on the stability of microlaminates, *Mater. Res. Soc. Symp. Proc.* **652** (Influences of Interface and Dislocation Behavior on Microstructive Evolution) Y1.3.1 6 (2001).
148. J.W. Obreimoff, The splitting strength of mica, *Proc. Roy. Soc. Lond. A* **127**, 290–297 (1930).
149. A.A. Griffith, The phenomena of rupture and flow in solids, *Philos. Trans. Roy. Soc. Lond., Ser. A* **221**, 163–198 (1921).
150. C. Messmer, J.C. Bilello, The surface energy of Si GaAs and GaP, *J. Appl. Phys.* **52**(7), 4623–4629 (1981).
151. S.P. Jarvis, Adhesion on the nanoscale, in: *Nano-Surface Chemistry*, ed. M. Rosoff (Marcel Dekker, New York, 2002), pp. 17–58.
152. W.D. Kingrey, Absolute measurement of metal–ceramic interfacial energy and the interfacial adsorption of silicon from iron–silicon alloys, in: *Study of Metal–Ceramic Interactions at Elevated Temperatures*, ed. F.H. Norton, W.D. Kingery, G. Economos, M. Humenik (U.S. Atomic Energy Commission, NYO-3144, 1953), 83 pp.

153. T. Sano, D.M. Saylor, G.S. Rohrer, Surface energy anisotropy of $SrTiO_3$ at 1400°C in air, *J. Am. Ceram. Soc.* **86**(11), 1933–1939 (2003).
154. V.T. Borisov, V.M. Golikov, G.C. Shcherbedinskii, Connection between diffusion coefficients and energies of grain boundaries, *Fizika Metallov i Metallovedenie* **17**(6), 881–885 (1964).
155. L. Wang, K. Vu, A. Navrotsky, R. Stevens, B.F. Woodfield, Boerio-J. Goates, Synthesis, surface energy and magnetic transition of CoO nanoparticles, *Chem. Mater.*, **16**(25), 5394–5400 (2004).
156. H.F. Holmes, E.L. Fuller Jr., R.B. Gammage, Heats of immersion in the zirconium oxide–water system, *J. Phys. Chem.* **76**(10), 1497–1502 (1972).
157. B. Bachiller-Baeza, I. Rodriguez-Ramos, A. Guerrero-Ruiz, Interaction of carbon dioxide with the surface of zirconia polymorphs, *Langmuir* **14**(13), 3556–3564 (1998).
158. S. Lipsett, G F.M. Johnson, G.O. Maass, The surface energy and the heat of solution of solid sodium chloride I–III, *J. Am. Chem. Soc.* **49**, 925–943 (1927) *idem. ibid.* 1940–1949 (1927); *idem. ibid.* **50**, 2701–2703 (1928).
159. A. Navrotsky, Progress and new directions in high temperature calorimetry, *Phys. Chem. Mineral* **2**, 89–104 (1977).
160. A. Navrotsky, Progress and new directions in high temperature calorimetry revisited, *Phys. Chem. Minerals.* **24**, 222–241 (1997).
161. S.C. Mraw, Differential scanning calorimetry, in: *Specific Heat of Solids*, ed. C.Y. Ho (New York, 1988), pp. 395–435.
162. A.I. Bachinskii, *Izv. Fizicheskogo Instituta pri Moskovskom Nauchnom Institute* **11**, 60 (1922). (Cited from V.K. Semenchenko Relationship between the volume and surface properties of metals and alloys, in: *The role of surface phenomena in metallurgy*, ed. V.N. Eremenko (Constants Bureau, New York, 1963), p. 9).
163. L.P.H. Jeurgens, W.G. Sloof, F.D. Tichelaar, E.J. Mittemeijer, Thermodynamic stability of amorphous oxide films on metals: Application to aluminum oxide films on aluminum substrates, *Phys. Rev. B: Condens. Matter Mater. Phys.* **62**(7), 4707–4719 (2000).
164. J. Banfield, F.H. Zhang, Nanoparticles in the environment, *Rev. Mineral. Geochem.* **44**, 1–58 (2001).
165. C.R.A. Catlow, J.D. Gale, D.H. Gay, M.A. Nygren, D.C. Sayle, Computer modelling of surfaces and interfaces, in: *Interfacial Science*, ed. M.W. Roberts (Blackwell Science, Oxford; Malden, MA, 1997), pp. 195–215.
166. R. Buczko, S.J. Pennycook, S.T. Pantelides, Bonding arrangements at the Si–SiO_2 and SiC–SiO_2 interfaces and a possible origin of their contrasting properties, *Phys. Rev. Lett.* **84**(5), 943–946 (2000).
167. A.A. Demkov, Investigating alternative gate dielectrics: a theoretical approach, *Phys. Stat. Sol. (b)* **226**(1), 57–67 (2001).
168. A. Christensen, E.A. Carter, First principles study of the surfaces of zirconia, *Phys. Rev. B* **58**(12), 8050–8064 (1998).
169. A.A. Demkov, O.F. Sankey, Growth study and theoretical investigation of the ultra-thin oxide SiO_2–Si heterojunction, *Phys. Rev. Lett.* **83**, 2038 (1999).
170. X. Zhang, A.A. Demkov, Steps on the (001) $SrTiO_3$ surface, *J. Vac. Sci. Technol. B* **20**, 1664 (2002).
171. X. Zhang, A.A. Demkov, H. Li, X. Hu, Y. Wei, J. Kulik, The atomic and electronic structure of the Si/$SrTiO_3$ interface, *Phys. Rev. B* **68**, 125–323 (2003).
172. Peacock, P.W., J. Robertson, Bonding, energies, and band offsets of Si–ZrO_2 and HfO_2 gate oxide interfaces, *Phys. Rev. Lett.* **92**(5), 057601/1-4 (2004).
173. V. Fiorentini, G. Gulleri, Theoretical evaluation of zirconia and fafnia as gate oxides for Si microelectronics, *Phys. Rev. Let.* **89**(26), 266101/1-4 (2002).
174. S. Blonski, S.H. Garofalini, Molecular dynamics simulations of α-alumina and γ-alumina surfaces, *Surface Sci.* **295**(1–2), 263–74 (1993).

175. I. Manassidis, M.J. Gillan, Structure and energetics of alumina surfaces calculated from first principles, *J. Am. Ceram. Soc.* **77**(2), 335–338 (1994).
176. M.A. Berding, S. Krishnamurthy, A. Sher, A.B. Chen, Cleavage energies in semiconductors, *J. Appl. Phys.* **67**(10), 6175–6178 (1990).
177. R.A. Robie, B.S. Hemingway, *Thermodynamic Properties of Minerals and Related Substances at 298.15 K and 1 Bar (10^5 Pascals) Pressure and at Higher Temperatures* (U.S. Geological Survey Bulletein 2131, Washington, DC, 1995).
178. R.A. Robie, B.S. Hemingway, J.R. Fisher, *Thermodynamic Properties of Minerals and Related Substances at 298.15 K and 1 Bar (10^5 Pascals) Pressure and at Higher Temperatures* (U.S. Geological Survey Bulletin 1452, Washington, DC, 1979).
179. P. Glushko, V.A. Medvedev, *Termicheskie Konstanty Veshestv* (Akademia, Nauk, Moscow, 1978).
180. K.B. Helean, A. Navrotsky, R.C. Ewing, in preparation
181. A.N. Kornilov, I.M. Ushakova, E.J. Huber Jr., C.E. Holley Jr., Enthalpy of formation of hafnium dioxide, *J. Chem. Thermodyn.* **7**(1), 21–26 (1975).
182. W. Roth, G. Becker, Atomic number and heat of formation, *Z. Phys. Chem.* **A159**, 1–26 (1932)
183. G.L. Humphrey, Heats of formation of hafnium oxide and hafnium nitride, *J. Am. Chem. Soc.* **75**, 2806 (1953).
184. E.J. Huber Jr., C.E. Holley Jr., Enthalpy of formation of hafnium dioxide, *J. Chem. Eng. Data* **13**(2), 252–253 (1968).
185. A.N. Kornilov, I.M. Ushakova, Standard heat of formation of hafnium dioxide, *Doklady Akad. Nauk SSSR [Phys. Chem.]*, **200**(6), 1382–1384 (1971).
186. Yu.N. Paputskii, V.A. Krzhizhanovskaya, V.B. Glushkova, Enthalpies of formation of rare earth hafnates and zirconates, *Izv. Akad. Nauk SSSR Neorg. Mater.* **10**(8), 1551–1552 (1974).
187. A.-D. Li, Q.-Y. Shao, H.-Q. Ling, Cheng Jin-Bo, Wu Di, Z.-G. Liu, N.-B. Ming, C. Wang, H.-W. Zhou, B.-Y. Nguyen, Characteristics of $LaAlO_3$ gate dielectrics on Si grown by metalorganic chemical vapor deposition., *Appl. Phys. Lett.* **83**(17), 3540–3542 (2003).
188. X. Zhao, D. Vanderbilt, First-principles study of structural vibrational and lattice dielectric properties of hafnium oxide, *Phys. Rev. B* **65**, 233106-1-4 (2002).
189. J. Majzlan, A. Navrotsky, W.H. Casey, Surface enthalpy of boehmite,*Clays and Clay Minerals* **48**, 699–707 (2000).
190. D.T. Livey, P. Murray, Surface energies of solid oxides and carbides, *J. Am. Ceram. Soc.* **39**(11), 363–372 (1956).
191. K.B. Helean, A. Navrotsky, Oxide melt solution calorimetry of rare earth oxides. Techniques problems cross-checks successes, *J. Thermal Anal. Calorim.* **69**(3), 751–771 (2002).
192. I. Barin, *Thermochemical Data of Pure Substances* (VCH, Weinheim, Germany, 1989).
193. S. Brunauer, Surface energy of a calcium silicate hydrate, *J. Colloid Interface Sci.* **59**(3), 433–437 (1977).
194. L. Shartsis, S. Spinner, Surface tension of molten alkali silicates *J. Research Natl. Bur. Standards* **46**, 385–390 (1951); Research Paper No. 2209 (cited from Brunauer, 1956).
195. S. Brunauer, D.L. Kantro, C.H. Weise, Surface energies of lime and calcium hydroxide, *Can. J. Chem.* **34**, 729–742 (1956).
196. S. Brunauer, D.L. Kantro, C.H. Weise, The surface energies of amorphous silica and hydrous amorphous silica, *Can. J. Chem.* **34**, 1483–1496 (1956).
197. F.H. Norton, W.D. Kingery, G. Economos, M. Humenik Jr., *Metal–Ceramic Interactions at Elevated Temperatures*, Vol. 83 (U.S. Atomic Energy Commission National Science Foundation, Washington, DC, NYO-3144, 1953).

198. R.N. Patil, E.C. Subbarao, Axial thermal expansion of zirconium oxide [zirconia] and hafnium oxide [hafnia] in the range room tempteraure to 1400°C, *J. Appl. Crystallogr.* **2**(Pt. 6), 281–288 (1969).
199. M. Foex, J.P. Traverse, Crystalline transformations induced by high temperatures in rare earth sesquioxides, *Comptes Rendus des Seances de l'Academie des Sciences Serie C: Sciences Chimiques Ser. C* **262**(8), 636–639 (1966).
200. L.M. Lopato, A.V. Shevchenko, A.E. Kushchevskii, S.G. Trevyatskii, Polymorphic transformations in rare earth oxides at high temperatures, *Izv. Akad. Nauk SSSR Neorg. Mater.* **10**(8), 1481–1487 (1974) (*PDFC 6343*).
201. W.C. Butterman, W.R. Foster, Zircon stability and the zirconium oxide–silica phase diagram, *Am. Mineral.* **52**(5–6), 880–885 (1967).
202. V.N. Parfenenkov, R.G. Grebenshikov, N.A. Toropov, Phase equilibriums in the hafnium dioxide–silicon dioxide system, *Dokl. Akad. Nauk SSSR* **185**(4), 840–842 (1969).
203. S.M. Lakiza, L.M. Lopato, Stable and metastable phase relations in the system alumina–zirconia–yttria, *J. Am. Cer. Soc.* **80**(4), 893–902 (1997).
204. D.A. Neumayer, E. Cartier, Materials characterization of ZrO_2–SiO_2 and HfO_2–SiO_2 binary oxides deposited by chemical solution deposition, *J. Appl. Phys.* **90**(4), 1801–1808 (2001).
205. J. Zhu, Z.G. Liu, Dielectric properties of YSZ high-k thin films fabricated at low temperature by pulsed laser deposition, *Mater. Lett.* **57**, 4297–4301 (2003).
206. J. Zhu, Z.G. Liu, Structure and dielectric properties of Zr–Al–O thin films prepared by pulsed laser deposition. *Microelectron. Eng.* **66**(1–4), 849–854 (2003).
207. C. Zhao, O. Richard, E. Young, H. Bender, G. Bender, G. Roebben, S. Haukka, S. De Gendt, M. Houssa, R. Carter, W. Tsai, O. Van Der Biest, M. Heyns, Thermostability of amorphous zirconium aluminate high-k layers, *J. Non-Cryst. Solids* **303**, 144–149 (2002).
208. C. Pascual, P. Duran, Subsolidus phase equilibriums and ordering in the system zirconia–yttria, *J. Am. Ceram. Soc.* **66**(1), 23–27 (1983). (PDFC 93-055).
209. D.W. Stacy, D.R. Wilder, Yttria–hafnia system, *J. Am. Ceram. Soc.* **58**(7–8), 285–288 (1975).
210. P. Duran, Phase relations in the systems hafnium dioxide–lanthanum oxide and hafnium dioxide–neodymium oxide, *Ceram. Int.* **1**(1), 10–13 (1975).
211. A.V. Shevchenko, L.M. Lopato, A.K. Ruban, Reaction studies in the hafnium dioxide–lanthanum hafnate system, *Dopovidi Akademii Nauk Ukrains'koi RSR Seriya B: Geologichni Khimichni ta Biologichni Nauki* **10**, 922–925 (1976).

Chapter 4

ELECTRONIC STRUCTURE AND CHEMICAL BONDING IN HIGH-K TRANSITION METAL AND LANTHANIDE SERIES RARE EARTH ALTERNATIVE GATE DIELECTRICS: APPLICATIONS TO DIRECT TUNNELING AND DEFECTS AT DIELECTRIC INTERFACES

GERALD LUCOVSKY

Departments of Physics, Electrical and Computer Engineering, and Materials Science and Engineering, North Carolina State University, Raleigh, NC 27695-8202, USA

1. INTRODUCTION

The primary driving force for the introduction of alternative gate dielectrics in advanced complementary metal oxide semiconductor (CMOS) devices is the dramatic increase in direct tunneling that occurs when the SiO_2 physical thickness is reduced to less than 3 nm. For example the direct tunneling current at an oxide bias of 1 V is approximately 10^{-2} A/cm^2 for an oxide physical thickness of 2.0 nm, and increases approximately one order of magnitude for each additional decrease of 0.2 nm thereby placing significant limitations on the performance and reliability of CMOS field effect transistor (FET) devices and integrated circuits (ICs). If the limitations for direct tunneling leakage current are taken to be 1–5 A/cm^2 for high performance devices, and $<10^{-2}$ A/cm^2 for mobile and low power devices, tunneling leakage limits SiO_2 physical thickness reductions to ~1–4 to 1.6 nm, and ~2.0 nm, respectively. Since there is a linear scaling relationship between lateral in-plane device dimensions, and SiO_2 physical and equivalent oxide thickness (EOT), these tunneling limitations have important implications for the ultimate decreases in device dimensions, and increases in device integration (1). One obvious solution is to replace SiO_2 by *alternative* gate dielectrics that have higher dielectric constants, k, than the nominal value of $k_o = 3.9 \pm 0.1$ for SiO_2, thereby increasing the physical thickness of the dielectric by the ratio of dielectric constants, k/k_o, without reducing the gate dielectric capacitance, C_{ox}. However, the very large decreases in tunneling currents anticipated by increases in physical thickness alone can be significantly mitigated by decreases in (i) the

A.A. Demkov and A. Navrotsky (eds.), Materials Fundamentals of Gate Dielectrics, 109–177.

effective tunneling barrier, e.g., the conduction band offset energy, E_B, between the crystalline silicon substrate, hereafter Si, and the gate dielectric, and (ii) the tunneling electron mass, m_{eff} (2), and these must be taken into account in adjusting the expectations for tunneling leakage current reductions for alternative gate dielectric materials based only on the k/k_o ratio. In this review, these mitigating effects are correlated with the electronic structure of transition metal and rare earth atom dielectrics, focusing on the lowest conduction band d* states which contribute to decreases in both E_B and m_{eff}.

The first generation of alternative gate dielectrics are silicon oxynitride pseudo-binary alloys $[(SiO_2)_{1-x}(Si_3N_4)_x]$. These alloys provide modest increases in k_a, to approximately 5.5–6.0 at optimized alloy concentrations, $x \sim 0.4$–0.6 (3–5). The small increases in k_a, are in part mitigated by reductions in E_B and m_{eff} (2), from 3.1 ± 0.1 eV and $0.55m_o$ in SiO_2, respectively, to 2.1 ± 0.1 eV and $\sim 0.25m_o$ in Si_3N_4, where m_o is the electron mass. Decreases in minimum EOT based on the tunneling leakage currents given above are modest, from ~1.4–1.6 to ~1.1–1.2 nm for high power devices, and from ~2.0 nm to 1.5–1.6 nm for mobile devices. These reductions do not provide the aggressive levels of scaling required by industry roadmaps (1, 6, 7), so that other alternative dielectrics with significantly higher ratios of k/k_o must be also considered. The optimized alloy compositions, and a mono-layer interface nitridation process have been demonstrated for devices prepared by low temperature (300°C) remote plasma processing and post deposition annealing/processing at higher temperatures (900–1000°C), have not been demonstrated in devices prepared by conventional thermal processing (4, 8, 9). However, a recent Intel Patent (10) indicates a hybrid approach combines low temperature plasma processing with conventional high temperature oxidation (11) to obtain the optimized nitrogen atom profiles identified in Refs. 4 and 9.

Transition metal and lanthanide series rare earth oxides, and their silicate and aluminate alloys have emerged as the most promising alternative gate dielectrics, and these alternative materials, generally designated as *high-k dielectrics*, are the focus of the remainder of this chapter (2). There are many formidable issues that must be addressed before their integration into manufactured devices can be contemplated and eventually achieved. The objective of this review is to identify the intrinsic limitations imposed by the electronic structure and chemical bonding of these proposed alternative high-k gate dielectrics and their interfaces within a stacked gate structure such as the one shown in Fig. 1. This gate includes (i) the Si-dielectric interface, (ii) an internal dielectric interface, and (iii) an interface between the dielectric film and a gate electrode comprised of an elemental or more complex metal. The intrinsic limitations are initially addressed by comparing the bulk properties of alternative high-k dielectrics with the corresponding properties of SiO_2, and then their interfacial properties with those included in Si/SiO_2 gate stacks. Based on applying these considerations of chemical bonding and electronic structure to direct tunneling leakage reductions and defects at internal dielectric interfaces the field of candidate high-k alternative dielectrics and be significantly narrowed.

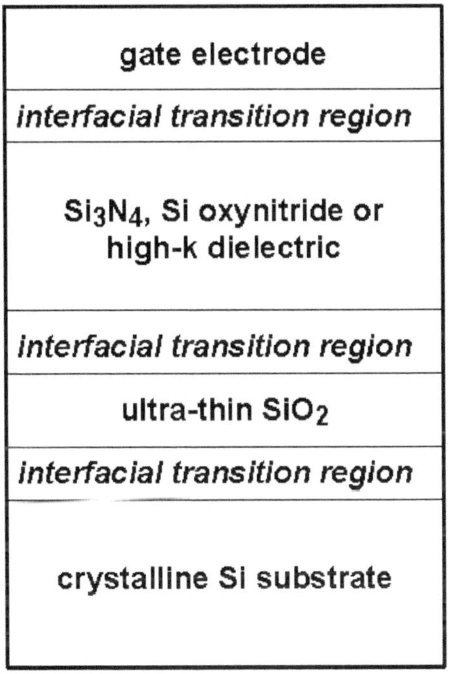

Fig. 1. Schematic representation of a stacked gate dielectric including an SiO_2 interfacial layer, a high-k dielectric and a metal gate. The interfaces between the (i) Si substrate and interfacial SiO_2 layer, (ii) SiO_2 and high-k dielectric and (iii) high-k dielectric and metal gate are indicated.

2. CHEMICAL BONDING IN HIGH-K GATE DIELECTRICS

Zallen (12) has identified three different atomic scale amorphous morphologies for non-crystalline solids: (i) continuous random networks (CRNs), as exemplified by SiO_2 with predominantly covalent bonding between the constituent atom pairs, (ii) modified continuous random networks (MCRNs), as exemplified by transition metal and rare earth atom silicate alloys in which metal atom ionic bonds *disrupt* and *modify* the covalently bonded SiO_2 CRN structure, and (iii) random close packed (RCP) non-crystalline solids comprised of negative and positive ions in a non-periodic three-dimensional packing geometry.

In an ideal CRN each atom is bonded according to its primary chemical valence. Additionally, for (i) stoichiometric compound compositions such as SiO_2, and (ii) pseudo binary alloy systems along join-lines connecting compound compositions such as $(As_2S_3)_x(As_2Se_3)_{1-x}$, chemically-ordered bonding prevails. In these examples, there are respectively, (i) Si–O bonds, but not Si–Si, or O–O bonds, and (ii) As–S and As–Se bonds, but not As–As or S–Se bonds, except at intrinsic defect levels of generally less than 1 part in 10^5. In SiO_2 the Si atoms are four-fold coordinated and the O atoms are two-fold coordinated in Si–O bonding arrangements. The randomness of the SiO_2 network is an important source of configurational entropy necessary for glass formation, and derives from at least two sources (i) a relatively large spread in bond angle at the O atom sites, $\sim 150 \pm 15°$ (13, 14), and (ii) a random distribution of dihedral angles (12).

Phillips has demonstrated that the perfection of *prototypical* CRNs such as SiO_2 is correlated with the average number of bonds/atom, N_{av}, and the average number of bonding constraints/atom, C_{av} (15, 16). It has been shown that one criterion for ideal bulk glass formation with low densities of defects is matching C_{av} to the dimensionality of the network; i.e., $C_{av} \sim 3$, for three-dimensional CRN's such as SiO_2. For CRNs in which the atoms are either two-, three- or four-fold coordinated, and the bonding geometries of the three- and four-fold coordinated atoms are non-planar, C_{av} is directly proportional to N_{av} through the following relationship (15, 16),

$$C_{av} = 2.5N_{av} - 3. \tag{1}$$

Equation (1) is derived from the symmetry-determined relationships between the bonding coordination of the constituent atoms, and the number of valence bond-stretching and bond-bending constraints that apply for a given coordination and geometry-determined local symmetry. These are discussed in Refs. 15 and 16, and additionally in seminal papers by Boolchand and his collaborators in Refs. 17 and 18.

For SiO_2, $N_{av} = 2.67$, so that $C_{av} = 3.67$ and is sufficiently high for the creation of bulk defects at levels $>10^{17}$ or $\sim 2 \times 10^{11}$ cm^{-2} (19). However, since the bending force constant at the O atom site is unusually weak, on the average one bond-bending constraint per O atom is sufficiently small to be neglected. This reduces C_{av} to a value of 3.0 (20), explaining the excellent electrical and optical properties of SiO_2 bulk glass and thin films. For thermally-relaxed SiO_2 thin films and bulk glasses, defect concentrations are $\sim 10^{16}$ cm^{-3}or equivalently, $\sim 5 \times 10^{10}$ cm^{-2}. Spectroscopic studies of defect bonding arrangement by electron spin resonance (ESR) or photoluminescence (PL) generally require significantly increasing defect levels by exposure to intense ultra-violet or X-ray irradiation, and by particle bombardment by electrons or neutrons, as for example in Ref. 21.

The low intrinsic defect, and defect precursor concentrations are associated primarily with a small density of strained network bonds that are in the tail of the bond Si–O–Si angle distribution extending to bond angles less than about 130° (22). On the other hand, the concentration of bulk and interfacial defects in other dielectrics that are on the average *over-constrained* in the context of the bond-constraint theory methodology has been shown to increase as the square of the difference between the number of bonds/atom, and the number of bonds/atom in an idealized, or low defect density ideal CRN such as SiO_2. For purposes of calculation, the ideal number of bonds/atom is set equal to 2.4, corresponding to a value of 3 for C_{av} in Eq. (2). Over-constrained bonding arrangements are the determinant factor in contributing to increased densities of defects in Si_3N_4 and silicon oxynitrides bulk films, and at their respective interfaces with Si and SiO_2 (20, 23).

The *second* amorphous morphology includes silicate and aluminate alloys, and the elemental oxides Al_2O_3, TiO_2 and Ta_2O_5, all of which have modified CRN structures that include metal atom ions (24–29). In the silicate dielectrics, the covalently bonded SiO_2 *host network* is disrupted by the introduction of ionic metals such as Na, Ca, Zr, etc. These atoms coordinate with O atoms at levels that exceed their chemical valence, and to accommodate this increase in coordination, dative bonds are formed between the metal atoms and the non-bonding pairs on the network O atoms (24). This increases the effective coordination of the O atoms, and decreases the bond-order at the metal atoms as well. The average bonding coordination of the O atoms is typically increased from two in the CRN dielectrics such as SiO_2, and SiO_2 pseudo-binary alloys with P_2O_5 and B_2O_3, to approximately three in the MCRN dielectrics. This in turn increases the number of bonding constraints per O atom in two ways: (i) by the increase in bonding coordination, and (ii) by stiffening the network on average by *reactivating* the Si O–Si bond-bending constraint that had been *broken* or *discounted* due an usually low bond bending force constant at the two-fold coordinated bonding sites (25).

The final class of non-crystalline oxides has an amorphous morphology that can be characterized as a random closed packing, RCP, of ions (12). It includes transition metal oxides, as well as Zr and Hf silicate alloys $[(Zr(Hf)O_2)_x(SiO_2)_{1-x}]$ in which the ZrO_2 and HfO_2 fractions, x, are greater than 0.5. The average oxygen coordination in Hf and Zr oxide, as well as other group IIIB transition metal oxides, Y_2O_3, La_2O_3, etc., is approximately equal to 4. The group IIIB oxides ZrO_2 and HfO_2 crystallize in distorted CaF_2 structures, and are excellent ionic conductors. High levels of ionic conductivity also occur in Y_2O_3 and La_2O_3. Finally, each of the transition metal and rare earth atom oxides mentioned above are also hydroscopic, creating potential problems in electronic device applications.

A classification scheme based on bond ionicity distinguishes between the three different classes of non-crystalline dielectrics based on bond electronegativity differences, ΔX (26). This provides a pathway for defining a bond ionicity that is correlated with the electronic structure of the constituent atoms in dielectric material bonding arrangements. A definition of bond ionicity, f_i, first introduced by Pauling, is the basis for a classification scheme that follows (26). Similar results can be obtained using alternative electronegativity scales such as the one proposed by Sanderson (27). The discussion below represents the first attempt at using bond ionicity scaling to discriminate between different *bonding morphologies* in *non-crystalline* elemental oxides and their silicate and aluminate alloys.

If $X(O)$ is the atomic electronegativity of oxygen, 3.44, and $X(Si)$ is the corresponding electronegativity of silicon, 1.90, then the electronegativity difference between these atoms, ΔX, is 1.54. Applying Pauling's empirical definition of bond ionicity, f_i, (26),

$$f_i = 1 - \exp(-0.25(\Delta X)^2), \qquad (2)$$

yields a value of f_i for Si–O bonds of ~45%. The range of ΔX values of interest in this classification scheme for gate dielectric materials ranges from about from 1.5 to 2.4, corresponding to a bond ionicity range from approximately 45 to 76%. For this range of ΔX, f_i is well approximated by a linear function of ΔX (28), so that ΔX and f_i can be used *functionally equivalent* scaling variables. The bonds in many other good glass formers including oxides such as B_2O_3, P_2O_5, GeO_2 and As_2O_3, and chalcogenides such as $As_2S(Se)_3$ and $GeS(Se)_4$ (12, 15–18) are generally less ionic in character than SiO_2. For pseudo-binary oxide and chalcogenide alloys, e.g., $(SiO_2)_x(B_2O_3)_{1-x}$, and $(GeS_2)_x(As_2S_3)_{1-x}$, respectively, compositionally-averaged values of ΔX are used to discriminate between different amorphous bonding morphologies.

The second class of non-crystalline dielectrics form MCRN's, which include ionic bonding arrangements of metal atoms that *disrupt* and *modify* the continuous random network structure (24–29). This class of dielectrics is characterized by values of ΔX between about 1.6 and 2.0, or equivalently bond ionicities between approximately 47 and 67%. The most extensively studied and characterized oxides in this group are the metal atom silicate alloys; for example, SiO_2 alloyed with Na_2O, CaO, MgO, PbO, etc., and quenched from the melt (12). This class also includes deposited thin film Al_2O_3, TiO_2 and Ta_2O_5 (29), and transition metal atom silicate alloys such as $(Zr(Hf)O_2)_x(SiO_2)_{1-x}$ in the composition range up to about $x \sim 0.3$–0.5 (24). The non-crystalline range of alloy formation in deposited thin films is significantly increased with respect to what can be obtained in bulk glasses that are quenched from the high temperature melt. The coordination of O atoms in CRNs is typically two, and increases to approximately three in the MCRNs. As examples, the average coordination of O is 2.8 ± 0.3 in thin film Ta_2O_5 in which Ta atoms are six- and eight-fold coordinated, and 3.0 in Al_2O_3 (29). The O atom coordination increases from 2 to 3 in the group IVB silicate alloys as the ZrO_2 or HfO_2 fraction is increased from doping levels to $x = 0.5$ (24).

The third class of non-crystalline oxides are those that have a random close packed ionic amorphous morphology (12). This class of oxides is defined by $\Delta X > 2$, and a Pauling bond ionicity of greater than approximately 67%. This group includes transition metal oxides deposited by low temperature techniques including plasma deposition and sputtering with post-deposition oxidation (6, 7). The coordination of the oxygen in these RCP structures is typically 4.

The coordination of O atoms in these RCP dielectrics scales monotonically with increasing bond-ionicity. This heralds a fundamental relationship between charge localization on the O atoms, and bonding coordination that has been confirmed by spectroscopic studies of Zr silicate alloys in which the coordination has been shown to vary linearly with alloy composition (24). Additionally, as shown in Fig. 2, the relative dielectric constant scales linearly with the oxygen atom coordination in non-ferroelectric, or non-anti-ferroelectric oxide dielectrics of this review (31). Crystalline ferro-electric and anti-ferroelectric oxide dielectrics have been grown expitaxially on crystalline Si (6, 7), and the properties of these dielectrics and their interfaces with Si are beyond the scope of this chapter.

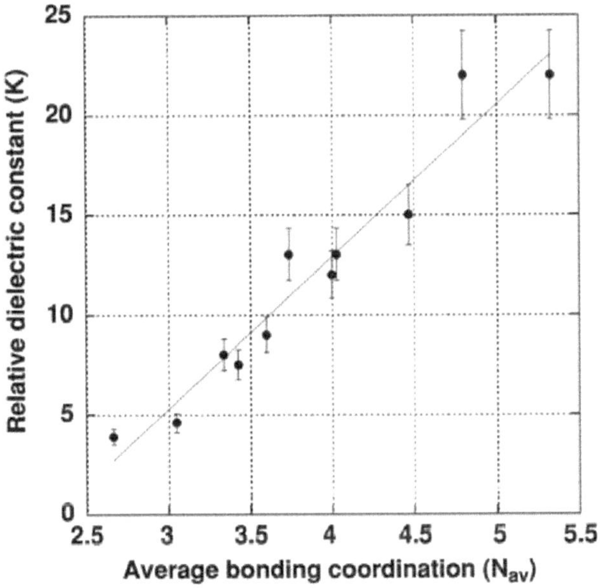

Fig. 2. Relative dielectric constant, k, as a function of the average number of bonds/atom, N_{av}, for representative dielectrics, including SiO_2, Si_3N_4, and transition metal and rare earth atom silicates, aluminates and oxides, in order of increasing N_{av}.

3. ELECTRONIC STRUCTURE CALCULATIONS

3.1. SiO_2 and Other CRN Materials

Three important aspects of the ab initio calculations of this review for the transition metal alternative dielectrics are identified and highlighted in the electric structure calculations for SiO_2. These are the (i) specification of the short range order that defines the cluster geometry, (ii) termination of the cluster, and (iii) computational approach for energy optimization (22). This approach has been applied to SiO_2 using the clusters shown in Fig. 3, and it is discussed in detail in Ref. 22. The O_3Si–O–SiO_3 bonding geometry at the center of the clusters is initially set equal to the *average* short range order (SRO) determined from radial distribution functions extracted from the X-ray diffraction studies of Ref. 32. This includes the Si–O bond length, the Si–Si second neighbor distance, or equivalently the Si–O–S bond angle at the two-fold coordinated O-atom sites, and the O–O second neighbor distance, or equivalently the O–Si–O bond angle at the four-fold, tetrahedrally-coordinated Si sites. The local cluster of Fig. 3(a) is *embedded mathematically* in a CRN structure through a one-electron potential, $V(r)$, and basis functions, S_1 and S_2, represented by Si* in Fig. 3(a). Alternatively, the clusters can be terminated by H-atoms with relatively small quantitative differences in the calculated total energy-bond angle distribution functional relationship.

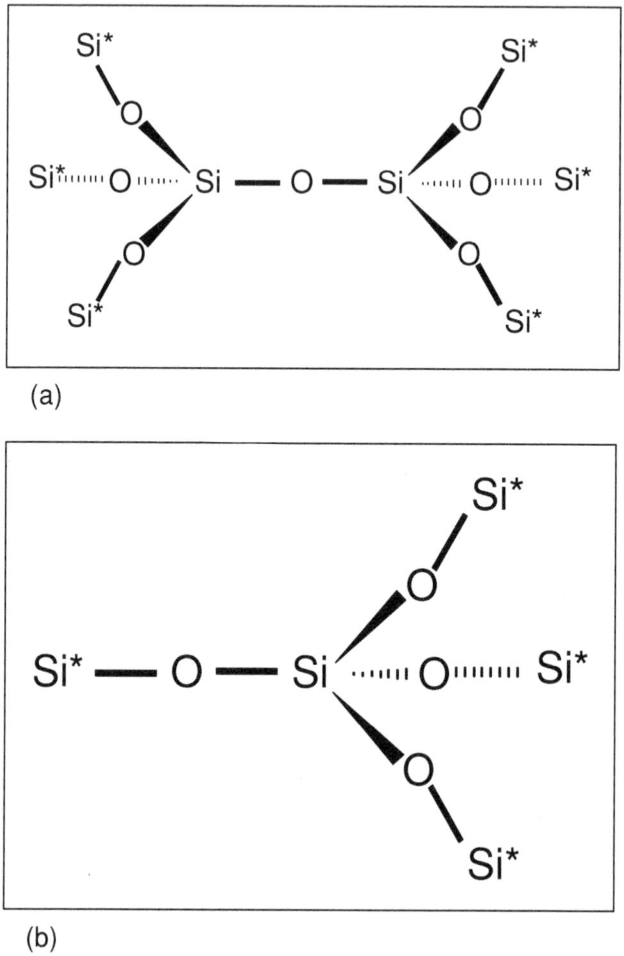

Fig. 3. (a) Schematic representation of the Si–O–Si terminated cluster used for the ab-initio calculations of this paper. The Si–O–Si bond angle, α, is 180° in this diagram, and will be varied from 120° to 150° for the calculations. The Si* represent an embedding potential that Si core eigenvalues are correct. (b) Schematic representation of a second Si–O–Si cluster that establishes the validity of the embedding potentials, Si*.

The electronic structure calculations employ variational methods in which an exact Hamiltonian is used (22). The calculations are done initially through a self-consistent field (SCF) Hartree–Fock formalism with a single determinant wave function that does not include electron correlation. Following this, there is a configuration interaction (CI) refinement of the bonding orbitals based on a multi-determinant expansion of the wave function including electron correlation. This process also includes a refinement of the local bonding parameters as well, primarily the Si–O bond length, and the Si–O–Si bond angle.

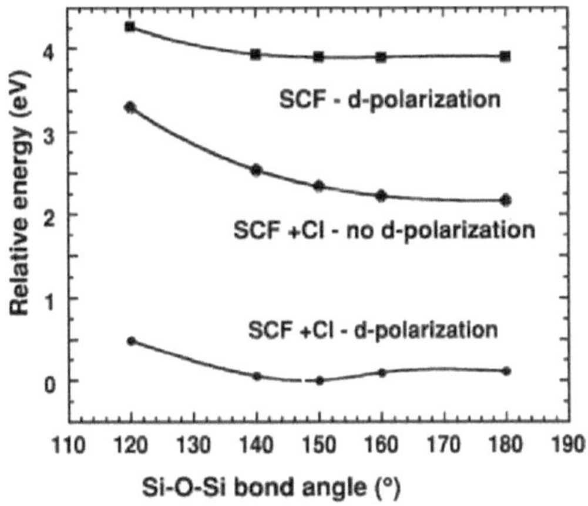

Fig. 4. Calculated energy in eV as a function of the Si–O–Si bond angle, a, for SCF with d polarization, SCF + CI (no d polarization), and SCF + CI + d polarization.

Figure 4 presents the dependence of the total energy for the cluster representation of SiO_2 as a function of the bond angle of the two-fold coordinated O-atom, and for a fixed Si–Si distance of ~0.31 nm as determined empirically in Ref. 33. Fixing this distance means that the Si–O bond length changes as the Si–O–Si angle is changed as well. This approach is consistent with empirically-determined Si and O bonding radii in the limits for (i) ionic bonding at an Si–O–Si bond angle of 180°, and (ii) covalent bonding for a bond angle of 90°. The ground state energy distribution is relatively insensitive to the dihedral angles that define the orientation of the terminating groups. More importantly, the results emphasize the importance of contributions with d-like symmetries to the Si basis set. These symmetry components of the electron distribution are equally important for Ge, S, and F, and have been included in the calculations for other CRN materials as well (22). It significant to note that the calculated minimum in total energy in our calculations occurs at a Si–O–Si bond angle of 148 ± 2°, and is different form the 144° bond angle determined in Ref. 13. However, the calculated angle is approximately equal to the average bond angle determined in the more recent studies reported in Refs. 13 and 14. The validity of this small cluster approach has been demonstrated by extension to other materials including GeO_2, GeS_2, As_2S_3 and BeF_2, where calculations have addressed local atomic structure, infrared effective charges, and ground and first excited electronic state energies as they apply to a new interpretation of photo-darkening that combines electronic structure calculations with bond-constraint theory to explain differences photo-structural changes (34, 35).

3.2. Transition Metal Oxides, and Silicate and Aluminate Alloys

The approach of Section 3.1 is now extended to transition metal oxides their respective silicate and aluminate alloys. The objective of these calculations is to provide general information about the electronic structure of the valence band, and the relative energies of the lowest anti-bonding d*-states and s*-states for transition metal and rare earth oxides, and to compare the results of these calculations to spectroscopic studies of transition metal oxides, and silicate and aluminate alloys. These calculations have been applied to two classes of X-ray and optical/UV transitions that have been studied experimentally: (i) intra-atomic transitions from deep transition metal core states such as the Zr atom 3p spin-orbit split doublet states at approximately −330 to 343 eV, to Zr atom anti-bonding 4d* and 5s* states, and (ii) inter-atom transitions in which the final states have a mixed character, as for example comprised of O 2p* and Zr atom 4d* and 5s*. This second group includes transitions originating the Zr atom 1s state at $\sim$−18 keV, the O-atom 1 s bonding state at $\sim$−530 eV, and the O 2p π non-bonding states at the top of the valence band $\sim$−6 eV, to O 2p* anti bonding*-states that are mixed with Zr atom anti-bonding 4d* and 5s* states (2). These calculations have been applied to ZrO_2, and the Zr silicate alloys, and to the corresponding intra- and inter-atom transitions in TiO_2 and HfO_2, and their respective silicate alloys. The excitation energies are then compared with experimental results in order to (i) underpin empirical models for the scaling of band gaps and conduction band offset energies (24, 36), (ii) determine the compositional dependence of conduction band offset energies in transition metal and rare earth silicate and aluminate alloys, and (iii) explore new alloy systems that provide technological advantages with respect to band offset energy limitations defined by atomic d-state energies of the constituent transmission metal and rare earth lanthanide atoms. These scaling relationships for both band gaps and conduction band offset energies are based in large part on the model calculations of John Robertson and coworkers which represent the first quantitative approach for comparing fundamental electronic structure differences among candidate high-k dielectrics (37, 38).

The ab initio calculations summarized in this review follow the same approach used for SiO_2 and other CRN oxides and sulfides in Refs. 22, 34 and 35. To reiterate, the electronic structure calculations employ variational methods in which an exact Hamiltonian is used so that the variation principle applies. The calculations are done initially through a self-consistent field (SCF) Hartree–Fock calculation with a single determinant wave function, which does not include electron correlation. Following this, there is a configuration interaction (CI) refinement of the bonding orbitals based on a multi-determinant expansion wave function, and including electron correlation effects.

This method has been applied to several relatively small 10-atom to 20-atom neutral clusters that include the bonding coordination of the transition metal atom and its immediate oxygen neighbors that are in turn terminated by H atoms (see Fig. 5). This approach is currently being extended to clusters that are centered on the O atoms as well. The first neutral cluster is comprised a four-fold coordinated Zr(Hf) atoms terminated by OH groups. The coordination of the Zr and Hf atoms in this cluster does not represent a known solid state bonding arrangement, but defines a convenient reference point for other cluster geometries. One other cluster is based on eight O atom neighbors, but does not reflect local bonding distortions in low-symmetry crystalline forms of ZrO_2 and HfO_2 in which eight neighbors are not bonded at the same distances, but instead corresponds to an idealized cubic geometry that is found in the CaF_2 structure.

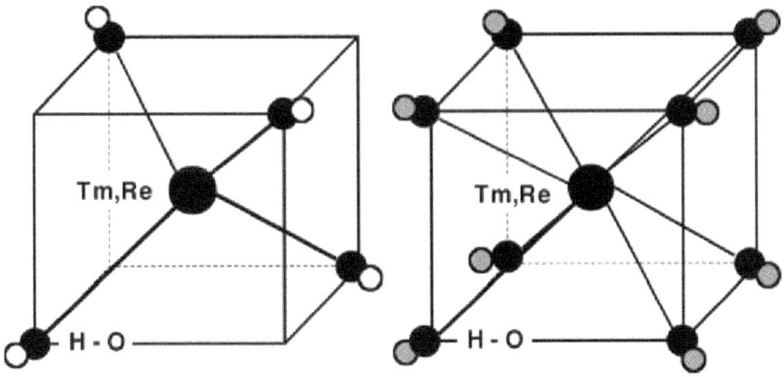

Fig. 5. Clusters for electronic structure calculations for transition metal (Tm) and rare earth (Re) oxide dielectrics. The large circles are the Tm and Re atoms and the small black circles are the O-atoms. The open circles are neutral H-atoms, and the circles with the diagonal lines are H-atoms with an excess positive charge of 0.5*e*.

This cluster is comprised of eight-fold coordinated Zr and Hf atoms. Four of the nearest neighbors are OH groups arranged in a tetrahedral geometry. The Zr and Hf bonds to these OH are predominantly ionic. The octet bonding is completed by positioning four water molecules, H–O–H, in a tetrahedral arrangement at the four remaining corners of a cube. The bonding of this group is via electrostatic donor–acceptor pair bonds in which the Zr and Hf atoms are the acceptors, and the occupied O-atom non-bond 2p π-states are the donors, and actually replicates bonding in low concentration Zr and Hf silicate alloys, e.g., $x < 0.2$.

Calculations have been made for Zr X-ray and band edge excitations ZrO_2, but also apply to Zr silicate and aluminate alloys. These calculations include (i) the ground state energy, (ii) the intra-atom Zr $M_{2,3}$ transitions, and (ii) the inter-atomic the Zr K_1, O K_1, and absorption edge transitions. The intra-atomic transitions for the Zr $M_{2,3}$ spectra are dipole allowed and localized on the Zr atoms, and therefore can be obtained from these small cluster calculations with a good degree of accuracy. The Zr K_1, and the O K_1 transitions and the absorption edge (fundamental band gap) transitions are respectively from Zr-atom K_1 core states, and O-atom K_1 core states and O 2p π non-bonding at the top of the valence band. The final states for these transitions have a mixed character: 4d* and 5s* from Zr, and O 2p* from O. The final state holes are for the Zr and O K_1 transitions are localized on the respective Zr and O atoms, whilst the final state hole for the band edge transition is delocalized on the eight O atom neighbors. Similar calculations have been applied to four and eight coordinated Hf, replicating bonding in HfO_2 and H silicate and aluminate alloys. For purposes of comparison, model calculations have been performed for four and eight fold coordinated Ti using the respectively, termination by four OH groups, and four OH and four HOH groups, each of which preserves cluster neutrality.

These calculations indicate relatively small, but significant differences for (i) the splitting of the d*-states that comprise the lowest excited or anti-bonding states, as well for (ii) the average separation between these d*-states and the s*-states that contribute to higher excitation states.

The closest correspondence between the calculated 4d*-state splitting, $\Delta(d^*_{1,2})$, and 4d*-5s*-state energy separation, $\Delta(d^*,s^*)$ and experiment, have been obtained for the intra-atom Zr $M_{2,3}$ spectrum using the eight-fold coordinated Zr cluster that has four OH groups, and four datively coupled water molecules. In contrast, the calculated splitting of the 4d*-state features in the O K_1 spectrum of ZrO_2 is well described, whilst the calculations yield a single 5s* state, rather than the band-like 5s*-doublet feature of the experimental spectrum. The energies of the 4d* and 5s* states for the eight-fold coordinated Zr cluster terminated with four OH and four HOH groups for the Zr K_1, Zr $M_{2,3}$, O K_1 and band edge transitions are currently being studied by our research group. The initial calculations indicate 4d* splittings that vary between about 1.5 and 3.5 eV, and 4d*-state-5s*-state energy difference between 9 and 12 eV.

The calculations for HfO_2 and TiO_2 yield qualitatively similar results with respect to the respective (i) d* state splittings, (ii) average d* to s* energy separations, and (iii) the quantitative differences between (a) the intra-atom transition metal N_3 and L_3 transitions, and (b) the inter-atomic transition metal K_1, O K_1 and fundamental band edge transitions (2). Much of work discussed above is still in the final stages of refinement, and will be published in the near future as identified by note-added-in proof.

3.3. Scaling of Band Gaps and Band Offset Energies with Atomic d-State Energies

Figure 6(a) compares in a schematic and qualitatively way the band edge electronic structures of TiO_2 and ZrO_2 as estimated from the calculations described above. The energies of the band gaps from taken from the model calculations of Refs. 35 and 36, which are in excellent agreement experiment (39, 40), and implemented in Fig. 6(a). The energies of the lowest excited state Ti and Zr d*-states relative to the highest O

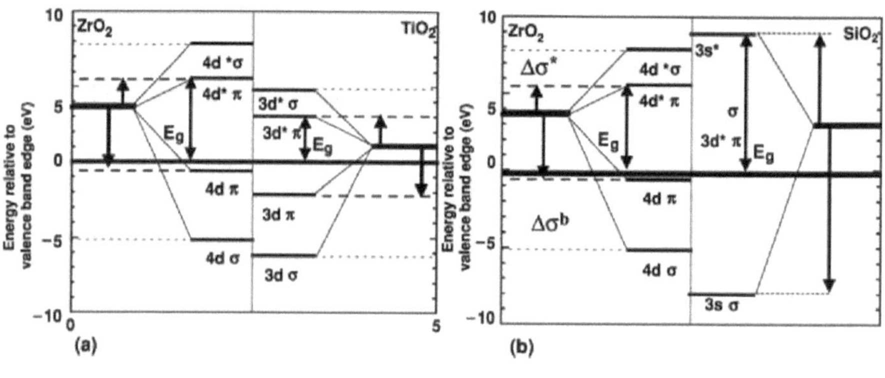

Fig. 6. Band edge energy electronic structures comparing (a) ZrO_2 and TiO_2, (b) ZrO_2 and SiO_2. The heavy lines indicate the atomic d-state energies, and the arrows indicated respectively, the band gaps, and the splittings of the Tm states with p-bonding.

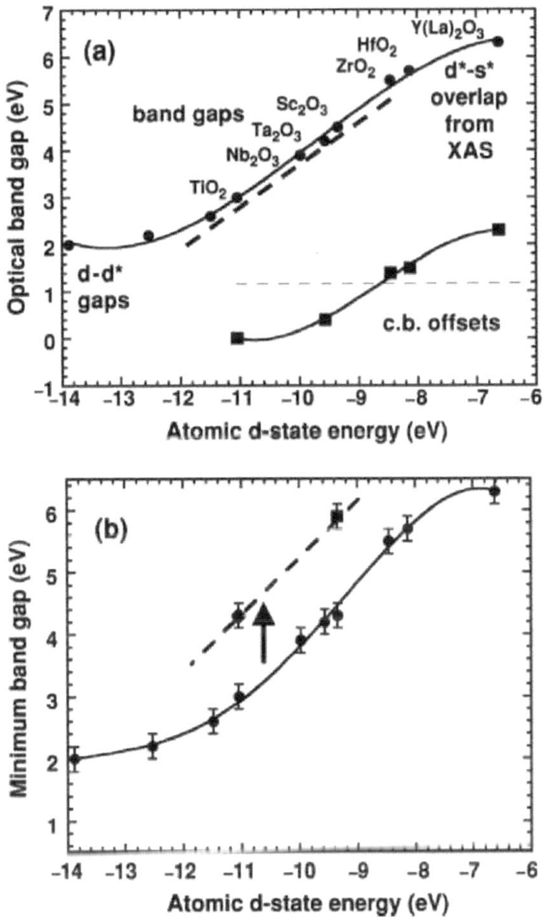

Fig. 7. Scaling of band gaps and conduction band offset energies with atomic d-state energies.

2p π non-bonding estimated from ab initio calculations presented above are typically larger than the optical band gaps by about 1 eV since they do not include solid state broadening effects; however, this does change any of the arguments below with regard to band gap scaling with atomic d-state energies. The most important as aspect of the results displayed in Fig. 6(a) is the nearly constant energy difference of 2 eV between the atomic state energies of Ti and Zr (at +1 and +3.5 eV, respectively) and the energies of the lowest d* states that define the conduction band edge. This approximately constant difference in energy is the basis for an approximately linear dependence of the optical band gap of transition metal oxides on atomic d-state energy that is shown in Fig. 7(a). The dashed line in Fig. 7(b) indicates the onset of strong optical absorption in TiO_2 and Sc_2O_3.

There are several aspects of the energy band scheme in Fig. 6(a) that are important for band gap and conduction band offset scaling in Fig. 7(a). The symmetry character of the highest valence bonding states, non-bonding O 2p π-states with an orbital energy approximately equal the energy of the atomic O 2p state, and the weak π-bonding of the transition metal atoms establishes that the lowest anti-bonding state is close in energy to the atomic n d-state of the transition metal atom.

Figure 6(b) displays a schematic representation of symmetry determined molecular orbitals based on the approach of Sections 3.1 and 3.2 that contribute to the highest occupied valence band states and the lowest conduction band states, respectively for ZrO_2 and SiO_2. The lowest Zr 4d π and Si 3s σ bonding molecular orbitals due not mix due to (i) their different symmetries, and (ii) their significant energy separation, the relative energy differences of the anti-bonding orbitals that contribute to the conduction band states of Zr silicate alloys are independent of alloy composition. Based on the results of AES and XPS measurements (36), and supported by ab initio theory, Zr silicate band gaps increase due to changes in oxygen coordination, which are incorporated into valence band offset energies (24). Similar results have been obtained for Hf silicates (36), and similar considerations apply to transition metal and rare earth aluminate alloys as well.

Figure 7(a) contains plots of the lowest optical band gap, and the conduction band offset energies, from the papers of Robertson and coworkers (37, 38), versus the absolute value of the energy of the transition metal atomic n d state in the $s^2d^{\gamma-2}$ configuration appropriate to insulators. $\gamma = 3$ for the group IIIB transition metals, Sc, Y and Lu(La), and the rare earth lanthanides, and $\gamma = 4$ for the group IVB transition metals Ti, Zr and Hf. The linearity of these plots supports the qualitative *universality* of the energy band scheme of Fig. 6(a). The band gap scaling displays a slope of approximately one between Ti and Y, indicating quantitative agreement with the energy band scheme of Fig. 6(a) which was obtained from the initial ab initio calculations discussed in Section 3.2.

The band offset energy in Fig. 7(a) is between the conduction band of Si and the empty anti-bonding or conduction band states of a high-k gate dielectric is important in metal-oxide-semiconductor, MOS, device performance and reliability. It defines the barrier for direct tunneling, and/or thermal emission of electrons from an n^+ Si substrate into a transition metal oxide. In alloys such as Al_2O_3–Ta_2O_5, or Al_2O_3–HfO_2, it also defines the energy of localized transition metal trapping states relative to the Si conduction band (41, 42).

4. EXPERIMENTAL STUDIES OF ELECTRONIC STRUCTURE

4.1. Valence Band Structure

Figure 8 includes the valence spectra for ZrO_2 and HfO_2 as determined by UPS (43). The dashed lines in the figure indicate the position of the band edge relative to the Fermi level of the spectrometer. The first dashed line at approximately 3.8 eV is at the valence band edge and is associated with O 2p π non-bonding states. On the basis of the ab initio calculations discussed above, the next two dashed lines are assigned, to Zr(Hf) 4d(5d) π states, and Zr(Hf) 4d(5d) σ states that overlap the respective

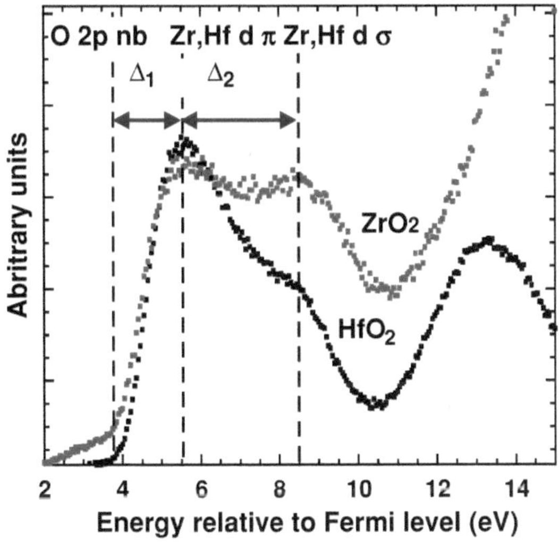

Fig. 8. UPS valence band spectra for the highest valence bands in ZrO_2 and HfO_2.

O 2p π and σ states. The energy differences of approximately 3.5 ± 0.2 eV for Δ_1, and 5.0 ± 0.2 eV for Δ_2, are in good agreement with the respective calculated differences of 3.4 ± 0.2 eV, and 4.6 ± 0.3 eV (44). The similarity between the valence band structures of ZrO_2 and HfO_2 as determined from the UPS studies is consistent with the similarity of ground state their properties, and of the ionic radii of the respective Zr and Hf atoms.

4.2. Anti-bonding Conduction Band States

Figure 9 presents a schematic representation of the XAS transitions that are addressed in this review. For ZrO_2, these include the Zr K_1 and $M_{2,3}$ edges, and the O

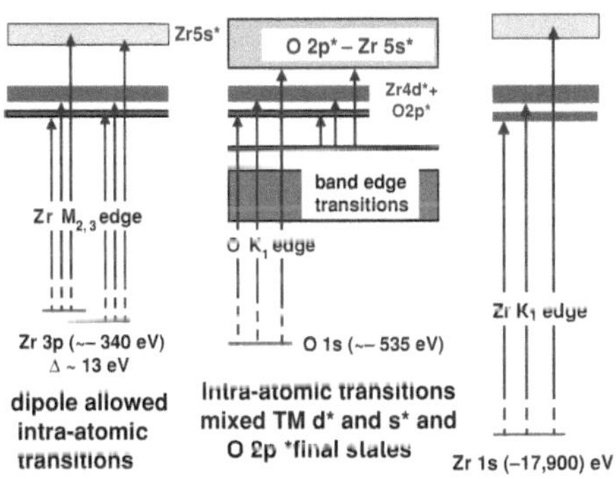

Fig. 9. Schematic representation of the intra-atomic Zr $M_{2,3}$, and inter-atomic atomic Zr K1, O K_1 and band edge transitions for ZrO_2. The ordering of the energy states is derived from ab initio molecular orbital calculations on small neutral clusters (5, 11).

K_1 edge (45). This figure also includes a schematic representation of the band gap transitions that define the optical absorption edge. The experimental studies summarized below also address the corresponding spectra for TiO_2 and HfO_2, in particular the respective Ti $L_{2,3}$ and Hf $N_{2,3}$ edges, as well as their O K_1 edges, and similar schematic representations apply to these transitions as well.

The schematic energy level diagrams of Fig. 9 include the $\Delta(d^*_{1,2})$oplittings, and the $\Delta(d^*,s^*)$ energy differences that are used to quantify the comparisons between: (i) *intra-atomic, dipole allowed transitions* in which electrons are excited from relatively deep core states of the Ti, Zr and Hf atoms into empty states that are localized on these atoms, and (ii) *inter-atomic* transitions in which electrons are excited either from TM or O atomic 1s core states, into final states have a mixed O atom–TM atom character, and therefore are not restricted by atomic dipole selection rules (46).

4.2.1. Intra-atomic, dipole allowed transitions

Figure 10(a)–(c) are the Ti $L_{2,3}$, Zr $M_{2,3}$ and Hf $N_{2,3}$ spectra for TiO_2, ZrO_2 and HfO_2, respectively (see Table 1). The features in each of these spectra are replicated for the respective spin-orbit split initial p-states, $np_{1/2}$ and $np_{3/2}$, where $n = 2$ for TiO_2, 3 for ZrO_2 and 4 for HfO_2, and are the principle quantum numbers that designate the respective L, M, and n shells (46, 47). For each of the spin-orbit split initial p-states, there are transitions to a d*-state doublet, 3d* for Ti, 4d* for Zr and 5d* for Hf, and to a 4s*, 5s* or 6s* state that is at a higher energy. Table 1 includes the positions of the spectral features for the Ti $L_{2,3}$ and Zr $M_{2,3}$ doublet components that are spectroscopically resolved, and for the energy of the single spectral Hf $N_{2,3}$

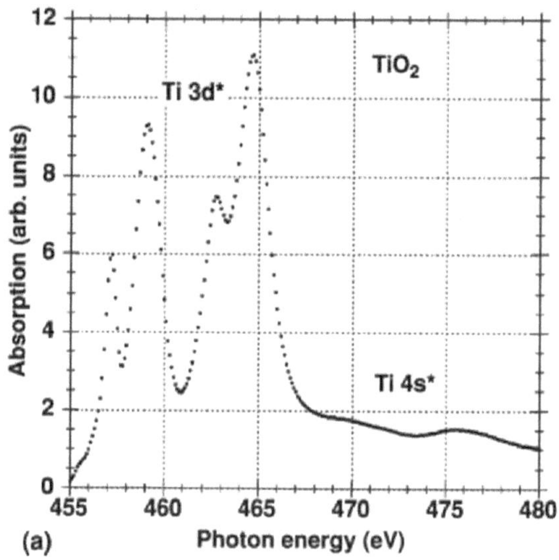

Fig. 10. (a) Ti $L_{2,3}$, (b) Zr $M_{2,3}$ and (c) Hf $N_{2,3}$ X-ray absorption spectra.

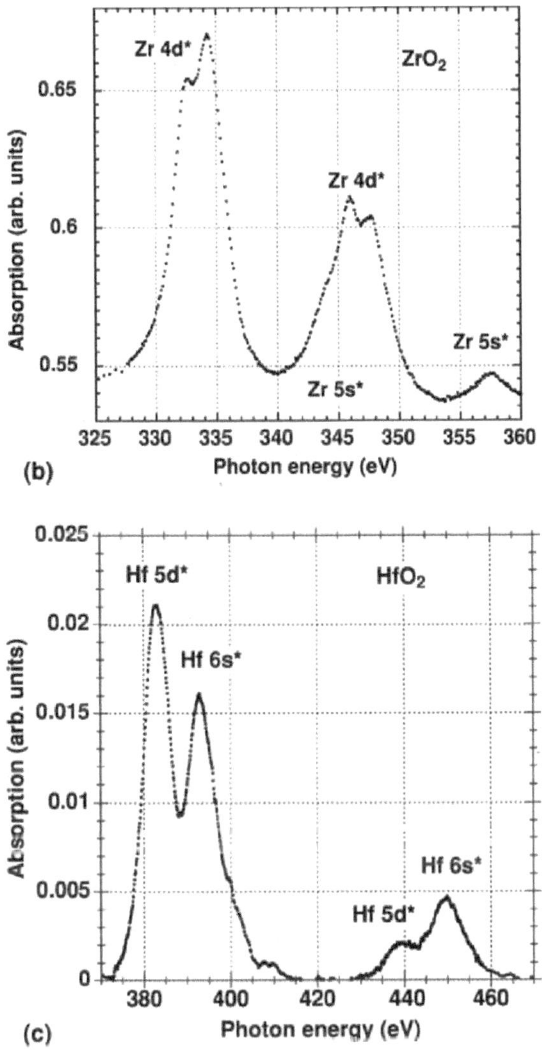

Fig. 10. (*continued*)

feature, where the doublet components are not resolved. The L_2, M_2, and N_2 features are shifted in energy with respect to the corresponding L_3, M_3, and N_3 features by the spin-orbit splittings of the respective 2p, 3p states, and 4p atomic states. These spectroscopically determined splittings are 5.6 ± 0.3 eV for Ti, 13.3 eV for Zr and for 57.6 ± 0.3 eV for Hf, and as shown in Table 2, there is very good agreement between the experimentally obtained spin-orbit splittings of this study and the *handbook values* of Ref. 48.

Table 1. Summary of experimental results for d* and s* features in XAS spectra of Figs. 2(a)–(c), 5(a)–(c), 6 and 7

	Energy (±0.2 eV)			Energy (±0.3 eV)	
Spectrum	d_1^*	d_2^*	s*	$\Delta(d_{1,2}^*)$	$\Delta(d^*,s^*)$
Ti K_1^a	4960	4962.5	4968.6, 4975.6	2.5	8.6, 15.6, 6.1, 13.1
Ti M_2	462.7	464.7	475.3	2.0	12.8
Ti M_3	457.2	459.1	469.7	1.9	12.5
O K_1(Ti)	530.1	532.8	539.5, 543.0	2.7	8.4, 12.9, 6.7, 11.2
Zr K_1	18,008[b]	18,008[b]	17,998	~3.5	~10.0[b]
Zr M_2^c	345.9	347.7	357.5	2.2	11.7
Zr M_3^d	332.6	334.8	344.5	2.2	11.9
O K_1(Zr)	532.2	535.4	542.3, 544.2	3.2	10.1, 12.0, 6.9, 8.8
Hf N_3	382.7[b]	382.7[b]	392.7	<3.5	10[b]
O K_1(Hf)	532.5	536.8	541.5, 544.2	4.3	9.0, 11.7, 4.7, 7.4

[a]Ref. 17 ; [b]not resolved; [c]Fig. 2(b); [d]deconvolved spectrum in Fig. 3.

Figure 11 is a deconvolved spectrum for the M_2 feature in the ZrO_2 XAS spectrum that displays the two 4d* and one 5s* spectral features. As indicated in Table 1, the values of $\Delta(d_{1,2}^*)$o and $\Delta(d^*,s^*)$ obtained from the positions of the M_2 spectral peaks in Fig. 10(b) are the same to within the estimated experimental uncertainty as the values obtained from the corresponding spectral peaks of the deconvolved and fitted spectrum in Fig. 11.

The relative absorption strengths for the *pairs* of d* and s* features in the spectra for TiO_2, ZrO_2 and HfO_2 are significantly different. The s* state spectral features are markedly weak in the TiO_2 L_3 and L_2 spectra, whilst the corresponding absorptions strengths increase modestly for the ZrO_2 M_3 and M_2 spectra, and then are much stronger for the HfO_2 N_3 and N_2 spectra. The $L_{2,3}$ spectrum of Sc in $GdScO_3$ an $DySco_3$, and the $M_{2,3}$ spectrum of Y in Y_2O_3, display essentially the relative absorptions as the respective group IVB neighbors, Ti and Zr. This supports the characterization of these spectra as intra-atomic. The respective s*- to d*-intensity ratios for the L_2, M_2 and N_2 spectra have been estimated to ±10% to be 3×10^{-2} for TiO_2, 8.9×10^{-2} for ZrO_2 and 8.5×10^{-1} for HfO_2.

Figure 12 is a plot of the ratios of the integrated s* state absorptions to the d* state absorptions for the TM p state spectral features, as function of calculated relative

Table 2. Spin-orbit splittings of Ti 2p, Zr 3p and Hf 4p core states

	p-state spin-orbit splitting (±0.5 eV)		
Oxide	TiO_2 2p	ZrO_2 3p	HfO_2 4p
XAS spectra	5.6	13.3	57.0
Data handbook[a]	6.4	13.7	57.5

[a]Ref. 13.

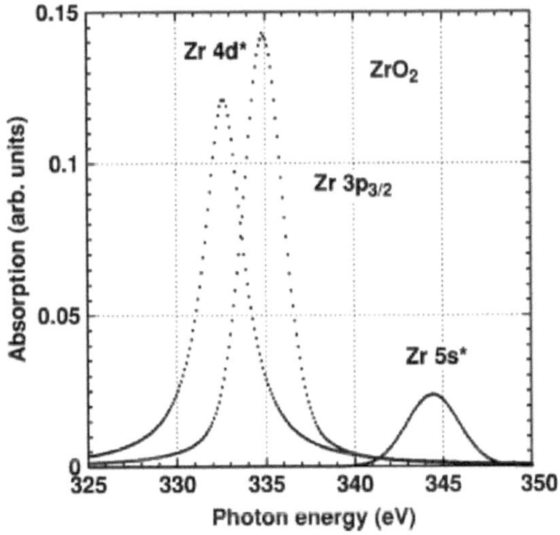

Fig. 11. Deconvolved Zr M_3 spectrum shown 4d*-state doublet, and 5s*-state singlet features.

intensities based on hydrogenic Rydberg states (46). The calculated relative intensity for HfO_2 includes a geometric correction that is based on the lanthanide contraction which makes the final 5d* and 6s* states have radial wave functions that have the same radial extent the corresponding Zr 4d* and 5s* states, but with an additional node. The x-axis value used in the plot for HfO_2 is the *geometric mean* for transitions from (i) 4p

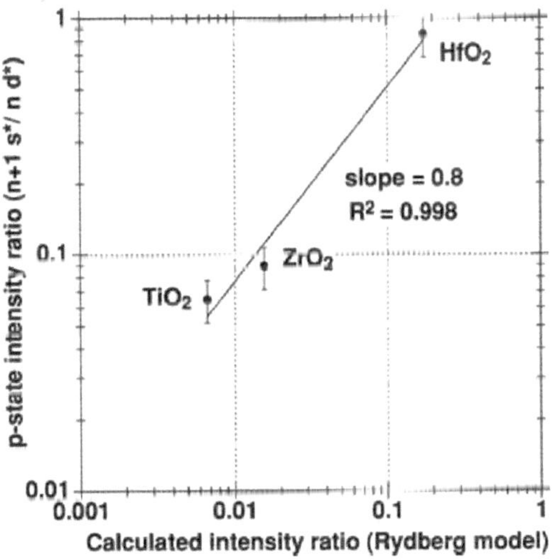

Fig. 12. Ratios of integrated s*-state absorptions to the d*-state absorptions for the Ti L_3, Zr M_3 and Hf N_3 spectra of Fig. 2(a)–(c), respectively, as a function of calculated relative intensities based on hydrogenic Rydberg states (14).

to 4d and 5s atomic states, and (ii) 4p to 5d and 6s states. The experimental intensity ratios scale linearly with the corresponding Rydberg state calculations demonstrating the atomic character of the Ti $L_{2,3}$, Zr $M_{2,3}$ and Hf $N_{2,3}$ spectra. The $L_{2,3}$ spectrum of (i) Sc in $GdScO_3$ an $DySco_3$, and the $M_{2,3}$ spectrum of (ii) Y in Y_2O_3, display essentially the relative absorptions as their respective group IVB neighbors, Ti and Zr. This supports the characterization of these corel level p state spectra as intra-atomic. In contrast, the intensity ratios for the respective K_1 spectra show a much smaller variation that is similar for d* and s* features of the band edge spectra.

4.2.2. Inter-atomic O and K_1 edge, and band edge transitions
Figure 13(a)–(c) are O K_1 edge spectra, respectively for TiO_2, ZrO_2 and HfO_2. Each of the spectra are qualitatively similar displaying a well-resolved d* doublet at the absorption threshold, and a broader s* doublet at higher energies. It is also possible that the higher energy of the assigned s* doublet has some TM p* character as well. The positions of the spectral features, and the $\Delta(d^*_{1,2})$ond respective $\Delta(s^*,d^*)$ spectroscopic splittings as determined from the peaks in Fig. 13 are included in Table 1. The $\Delta(d^*)$ splitting increase in going from TiO_2 to ZrO_2 to HfO_2, by ~0.5 and 0.8 eV, respectively. The spectral overlap between the higher energy d* state, d_2, and the threshold for the s* doublet absorption increases from TiO_2 and ZrO_2, to HfO_2. The spectral overlap is associated with (i) increases in the respective $\Delta(d^*_{1,2}, s^*)$ splittings from TiO_2 (2.7 eV) to ZrO_2 (3.2 eV) to HfO_2 (4.3 eV), (ii) decreases in $\Delta(d^*_2, s^*_1)$ energy separations of 6.7, 6.9 and 5.0 eV, respectively for TiO_2, ZrO_2 and HfO_2, and (iii) increases in the spectral half-width of the d^*_2 feature that scale with the atomic number of the TM atom.

The Zr K_1 spectrum for ZrO_2 is shown in Fig. 14. This spectrum is similar to those presented in Ref. 49. Markers in the spectra displayed in Ref. 49 indicate that

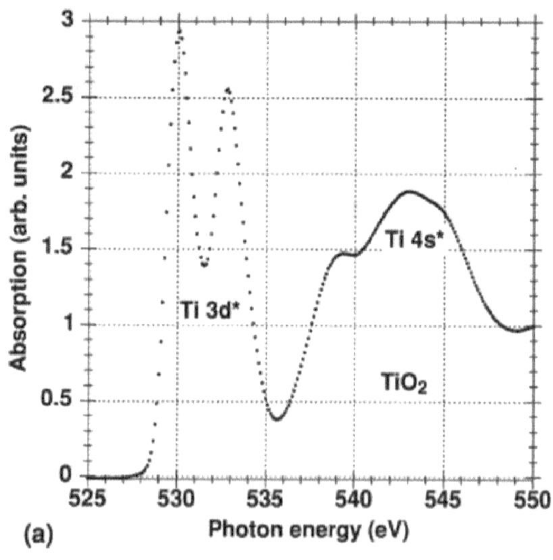

Fig. 13. O K_1 spectra for (a) TiO_2, (b) ZrO_2 and (c) HfO_2.

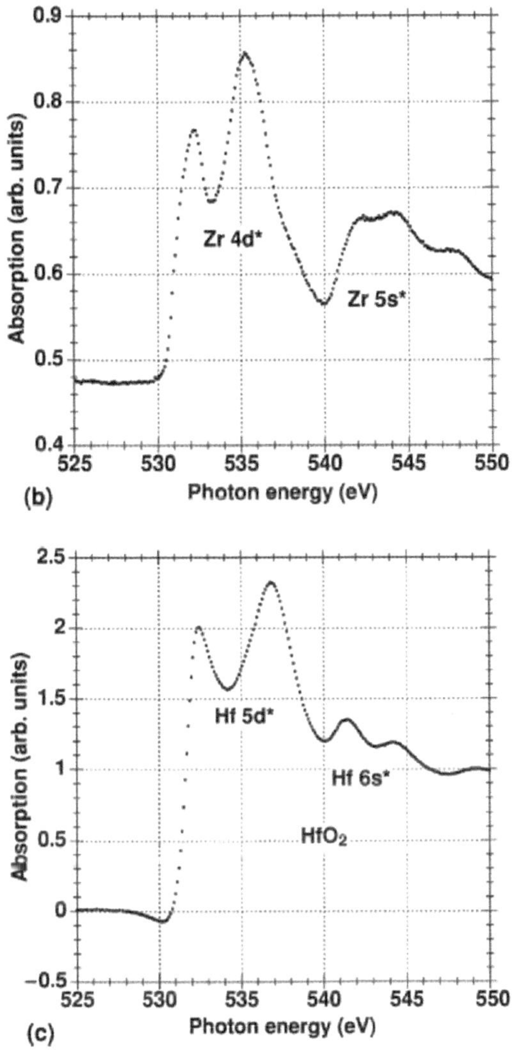

Fig. 13. (*continued*)

the positions of features that assigned to 4d* and 5s* states are essentially the same for all of the crystalline phases of ZrO_2 explored in that work. Since transitions from the Zr 1s state to Zr 4d and 5s are not dipole-allowed, the Zr K_1 edge spectrum is qualitatively similar to the O K_1 edge spectrum. Like the O K_1 spectrum, the spectral features in the Zr K_1 spectrum are consistent with mixing between Zr 4d* and 5s* states, and O 2p* states. This is also reflected in the tabulation of $\Delta(d^*_{1,2})$ and $\Delta(d^*,s^*)$ values in Table 1. The doublet 4d* features are not spectroscopically resolved in Fig. 14, or in the spectra in Ref. 49, so that these comparisons are limited by the relativistically-increased effective width of the Zr 1s state. The relative intensities of the d* and s* spectral features in the Zr K_1 in Fig. 14 and O K_1 spectra in Fig. 13(b)

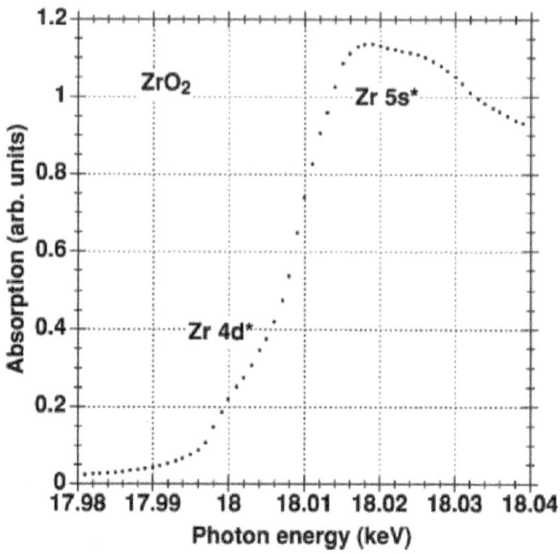

Fig. 14. Zr K_1 XAS spectrum.

are markedly different, and the d* state splitting is resolved in the O K_1 edge as well. These differences are attributed to inherent differences in these spectral transitions that are addressed in the next sub-section of this review.

Ti K_1 and Hf K_1 edge spectra have not been obtained for our thin film samples. However, there have been several published spectral studies of TiO_2 in the rutile and anatase crystal forms (50, 51). These published results for the Ti K_1 edge of TiO_2 in the rutile phase are included in Table 1. A comparison indicates a similar x-axis (energy) behavior between the O K_1 spectra for the nano-crystalline TiO_2 film of this study, and the published rutuile Ti K_1 spectrum (50, 51). In particular, the d-state splittings are the same to within experimental error, 2.5 ± 0.3 eV for the O K_1 spectrum, and 2.7 ± 0.3 eV for the Zr K_1 spectrum. The s-state energy peak separations are smaller in the O K_1 edge spectrum, 3.5 ± 0.3 eV, compared with 7.0 ± 0.3 eV in the Zr K_1 edge spectrum, but the average differences in energy between the d_1^* and d_2^* states, and the lower energy s*-state spectral peak are the same within experimental error, 8.5 ± 0.3 eV for the lower d_1^* features, and 6.4 ± 0.3 eV for the higher d_2^* feature.

Figure 15 contains a plot of the optical absorption constant, α, versus photon energy, as obtained from an analysis of vacuum ultra-violet spectroscopic ellipsometry data (52). The band edge, or threshold for optical absorption is at ~5.7 eV, essentially the same as reported from complementary measurements of the photoconductivity (53). The relative intensities of the d* state absorption between 5.7 and 6 eV, and the s* state absorption at higher photon energies is similar to the relative intensities of the same types of spectral features in the Zr K_1 edge XAS spectrum, however the energy scales are markedly different, consistent with the difference in the initial states, the Zr 1s state for K_1 spectrum, and O 2p π non-bonding valence band states for the band edge transitions.

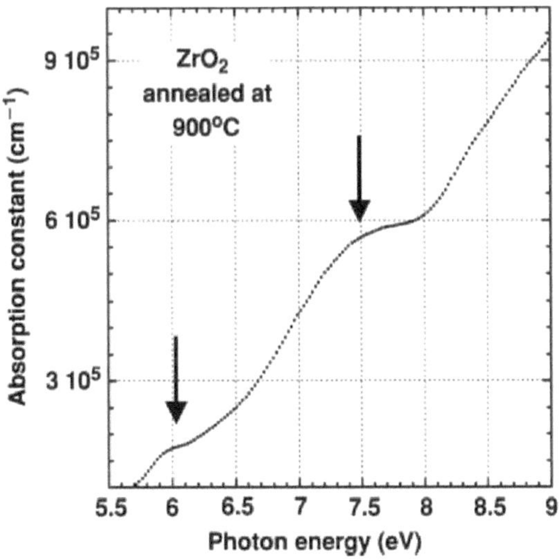

Fig. 15. Absorption edge spectrum for ZrO_2 annealed at 900°C. The arrows indicate the band edge 4d* features.

4.2.3. Comparisons between intra-atomic and inter-atomic spectra

This sub-section distinguishes between the intra-atom Ti $L_{2,3}$, Zr $M_{2,3}$ and Hf $N_{2,3}$ edge spectra of Fig. 10(a)–(c), and the inter-atomic (i) Ti O K_1, Zr O K_1 and Hf O K_1 spectra of Fig. 13(a)–(c), (ii) Zr K_1 edge spectra of Fig. 14, and (iii) fundamental band gap ZrO_2 spectra of Fig. 15.

Figure 9 indicates the intra-atomic transitions that contribute to the six-distinct features the Zr $M_{2,3}$ spectrum of Fig. 10(b) . This schematic representation applies to the group IVB transition metal, group IIIB transition metal and lanthanide rare earth atom $A_{2,3}$ spectra where A = L, M and N. For the Ti $L_{2,3}$, Zr $M_{2,3}$ and Hf $N_{2,3}$ spectra. The features in the respective $A_{2,3}$ spectra are associated with transitions from relatively deep spin-orbit split 2p, 3p and 4p-states at 300–500 eV below vacuum, to empty TM 3, 4 or 5d* and 4, 5 or 6s* states, respectively for Ti, Zr and Hf. These transitions are atomic-like in character and are not-changed by second neighbors as has been shown for the Zr $M_{2,3}$ spectra through the comparisons that include Zr homogeneous, as well as chemically phase-separated silicate alloys (47). The relative intensities of final d* states in the Ti and Zr spectra are consistent with the lower energy pair in each spectral component being the $d_{3/2}$ state with a spectral weighting of 4, and the higher energy component being the $d_{5/2}$ state with a spectral weighting of 6. Since the local symmetries of the Ti and Zr atoms are effectively six-fold coordinated in TiO_2 and eight-fold coordinated in ZrO_2, and since the ordering of the d*-states in the respective spectra are the same, this demonstrates that the d* state spitting is not directly associated with the local symmetry which would have yielded a different ordering of the d*-states for *octahedrally* coordinated TiO_2 and approximately *cubically* coordinated ZrO_2 (45).

The O K_1 edge spectra for TiO_2 in Fig. 13(a), ZrO_2 in Fig. 13(b) and HfO_2 in Fig. 13(c) are assigned to transitions from the O 1s state to final *band-like* extended states that have a mixed (i) O 2p*, and (ii) TM 3, 4 or 5d* and TM 4, 5 or 6s* state character, respectively. In contrast to the spectral features in Fig. 10(a)–(c), the d* and s* related features have relative intensities that differ by no more than a factor of 2 as contrasted with factors of more than 10 for the TiO_2 $L_{2,3}$ and ZrO_2 $M_{2,3}$ spectra of Fig. 10(a) and (b). This demonstrates that the matrix elements for absorption to these states are not determined by the Rydberg-like transition probabilities similar to those in Fig. 12 (46).

The Zr K_1 edge spectra in Fig. 14, and the Ti Zr K_1 spectra have been discussed at length in Refs. 49, 50 and 51, and are also inter-atomic spectra. The transitions between the Zr and Ti 1s states to the respective Zr 4d* and 5s* states, and Ti 3d* and 4s* states are not dipole-allowed, and the lowest energy transitions are to final states that have a mixed O 2p*, and Ti or Zr d* and s* state character. This interpretation of the Zr and Ti K_1 spectra is supported by the fact that these transitions have features at higher energy that have been used in extended X-ray absorption fine structure (EXAFS) studies to determine bond lengths and second and more distant neighbor atomic separations (54).

Even though the final states have similar atomic character, the relative absorptions of the 4d* and 5s* features in the O K_1 and Zr K_1 spectra are markedly different, and therefore reflect differences in the respective O 1s and Zr 1s ground state wave functions, and their effect on the transition probabilities for absorption.

The relative absorption strengths for the 4d* and 5s* features in the Zr K_1 spectrum in Fig. 14 bear some resemblance to features assigned to the same states in the band edge absorption spectrum in Fig. 15. For example, the threshold d* state feature has a markedly lower absorption constant than other d* and s* states features at higher energies. This is the case even though the initial states are very different; the localized Zr 1s core state for the Zr K_1 spectrum, and the more delocalized O 2p non-bonding states for the band edge absorptions.

Finally, a comparison between the energies of the first spectral peak of the respective O K_1 spectra, 530.1 eV for TiO_2, 532.4 eV for ZrO_2, and 532.6 for HfO_2, indicates that the differences between these energies are equal to within an experimental uncertainty of ± 0.3 eV to the respective differences in reported *nominal band gap energies* of 3.1 eV for rutile TiO_2 (45), 5.6 eV for ZrO_2 (53), and 5.8 eV for HfO_2 (53). This comparison carries over to complex oxide high-k dielectrics such as $GdScO_3$ and $DyScO_3$ (55).

The O K_1 edge spectra, coupled with other studies of valence band offset energies have provided additional insights into the systematic variation of conduction band offset energies with atomic d-state energies. Based on these comparisons, many of the group IVB and VB TM oxides with the highest dielectric constants, e.g., TiO_2, Nb_2O_3 and Ta_2O_3, have offset energies below 1 eV that correlate with high tunneling leakage, and or electric field assisted injection into low-lying conduction band traps associated with these atoms (41, 42). Based on scaling with atomic d states, the group IVB oxides of Zr and HF, and their respective silicate and aluminate alloys, as wells as the group IIIB, and lanthanide rare earth series oxides, and their respective silicate and aluminate alloys will have conduction band offset energies greater than about 1.2–1.3 eV and meeting roadmap targets with tunneling leakage. However, other properties of these materials, have presented challenges for implementation of high-k dielectrics into advanced CMOS devices (6, 7).

4.3. Silicate Alloys

Figure 16(a) displays the O K_1 edge for three annealed and phase-separated Zr silicate alloys (24). Based on comparisons with the O K_1 edge spectra for ZrO_2, and arguments

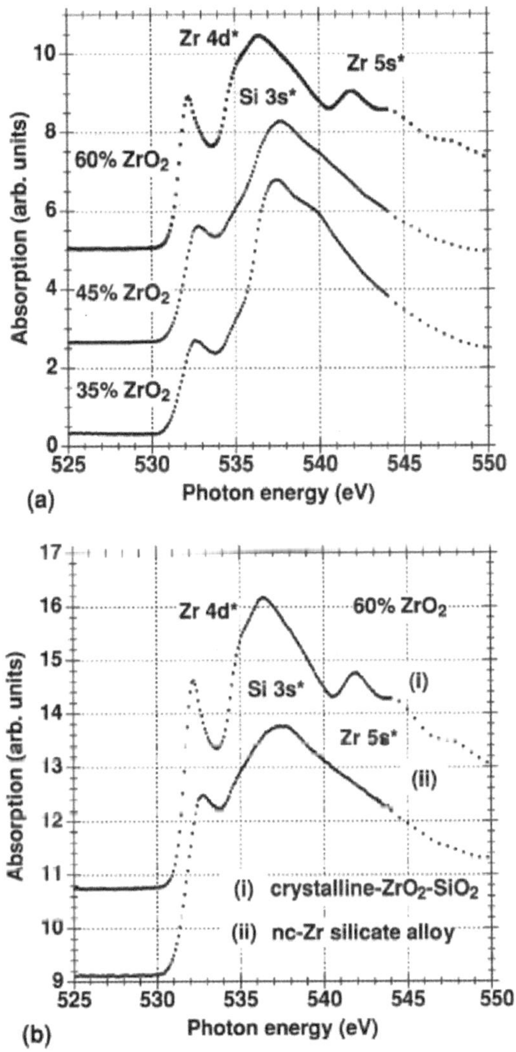

Fig. 16. (a) Zr silicate O K_1 spectra for films annealed at 900°C, and chemically phase separated into crystalline ZrO_2 and non-crystalline SiO_2. (b) Comparison between O K1 spectra for a Zr silicate alloy with 60% ZrO_2 ($x = 0.6$) as-deposited and annealed at 900°C.

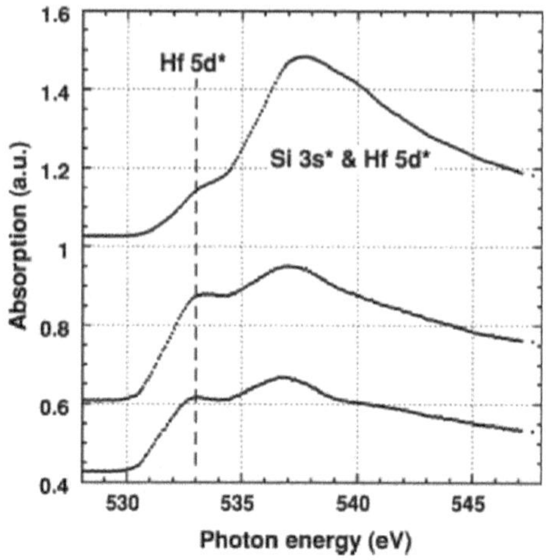

Fig. 17. Hf silicate O K_1 spectra for as-deposited non-crystalline films.

based on a molecular orbital model for Zr silicate alloys, the first spectral peak and the distinct shoulder shifted in energy by about 3 eV are associated with Zr 4d* states coupled to O 2p* states. The broad feature 536–540 eV is assigned to Si 3* states, also coupled to O 2p* states. The distinct doublet feature at about 542–544 eV for the $x \sim 0.6$ (60% ZrO_2) alloy has been assigned above to Zr 4s* states, also mixed with O 2p* states. The differences between these three spectra with respect to narrowness of the lowest energy 4d* feature, and the distinct spectral peak on the Si 3s* feature reflect different aspects of the chemical phase separation morphology. These differences are addressed in detail in Ref. 56, with the most important aspect of the separation being in the characteristic size of the crystallites or grains in Zr phase. This is ~3–5 nm for $x < 0.5$, and in excess of 10 nm for $x > 0.5$. To a good approximation the spectrum for the $x = 0.6$ sample is a linear combination of the O K_1 spectra for crystalline ZrO_2, and non-crystalline or amorphous SiO_2.

Figure 16(b) contains a comparison between the spectrum for the annealed 60% ZrO_2 alloy in trace (i) and an as-deposited alloy in trace (ii). The spectral assignments remain the same, but all of the features in trace (i) are broadened due to the non-crystalline bonding arrangements (56).

Finally, Fig. 17 displays the spectra of three non-crystalline silicate alloys with approximate concentrations of 70, 50 and 25% HfO_2, accurate to approximately ±7%. The most important aspects of these spectra are that (i) the relative intensities of the lowest energy 5d* feature scales with alloy composition, while (ii) the separation between the Hf 5d* spectral peak and center of the Si 3s* band is approximately constant. This behavior is also reflected in the XPS spectra discussed in Section 4.5.

4.4. Complex Transition Metal-Rare Earth Binary Oxides

This section introduces an additional dimension to the spectroscopic studies of binary oxides that go beyond Tm and Re silicates and aluminates as well (55). This is the coupling of d states of different Tm and/or Re through bonding to a common

O atom in *complex oxides* with compositions given by, $ReTmO_3$, and $Tm(1)Tm(2)O_4$. These bonding interactions have the potential for removing some of the restrictions on band-gap scaling that have been addressed with respect to Fig. 6(a) and (b) for *simple* Tm and Re oxide, group IIIB Tm oxides and trivalent Re oxides such as Y_2O_3 and Gd_2O_3, respectively, group IVB Tm oxides such as TiO_2, ZrO_2 and HfO_2, group VB Tm oxides such as Nb_2O_5 and Ta_2O_5.

The first complex oxides addressed are crystalline $GdScO_3$ and $DyScO_3$. These *complex oxides* have distorted perovskite structures in which the Gd and Dy atoms are nominally 12-fold coordinated, and the Sc atoms are six fold coordinated. The Gd or Dy atoms are bonded through O atoms to the Sc atoms. Before displaying the O K_1 edge XAS spectra for these two crystals, the O K_1 edges for thin film, crystalline Y_2O_3 and ZrO_2 are compared in Fig. 18(a) and (b). The most significant difference

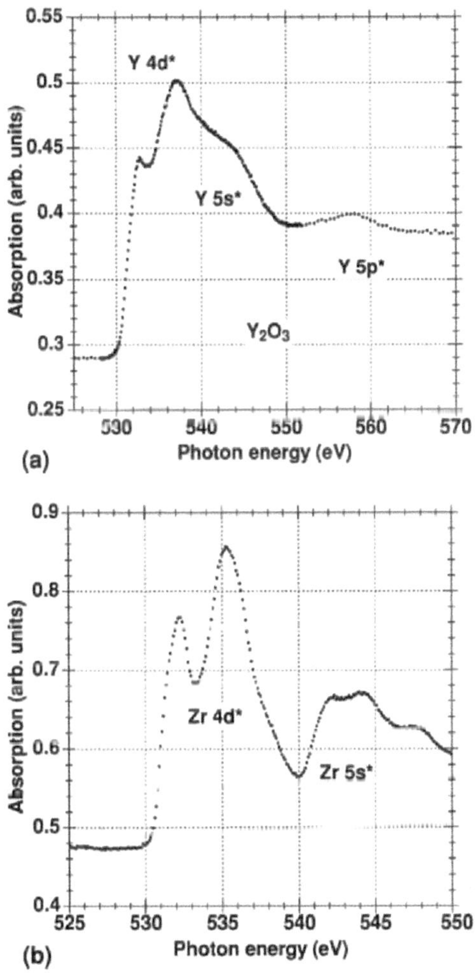

Fig. 18. O K_1 spectra for (a) $GdScO_3$ and (b) $DyScO_3$.

between these spectra is the spectral overlap between the Tm 4d* doublet and the Tm 5s* band. In Y_2O_3, there is a significant overlap, whilst in ZrO_2, the 4d* doublet and 5s* band features are spectroscopically resolved. This difference correlates with a difference in the energy separation of atomic 4d and 5s states in the $4d^{\gamma-2}5s^2$ atomic configurations, where $\gamma = 3$ for Y_2O_3 and $\gamma = 4$ for ZrO_2. These splittings are ~1.5 eV for Y_2O_3 and >3 eV for ZrO_2, and are the determinant factor in the marked differences in the spectral overlap in Fig. 18(a) and (b).

The O K_1 edge spectra in Fig. 19(a) and (b) for $GdScO_3$ and $DyScO_3$ respectively display three d* state features. Based on Fig. 18(a), these overlap the Sc 4s* and Dy 6s*

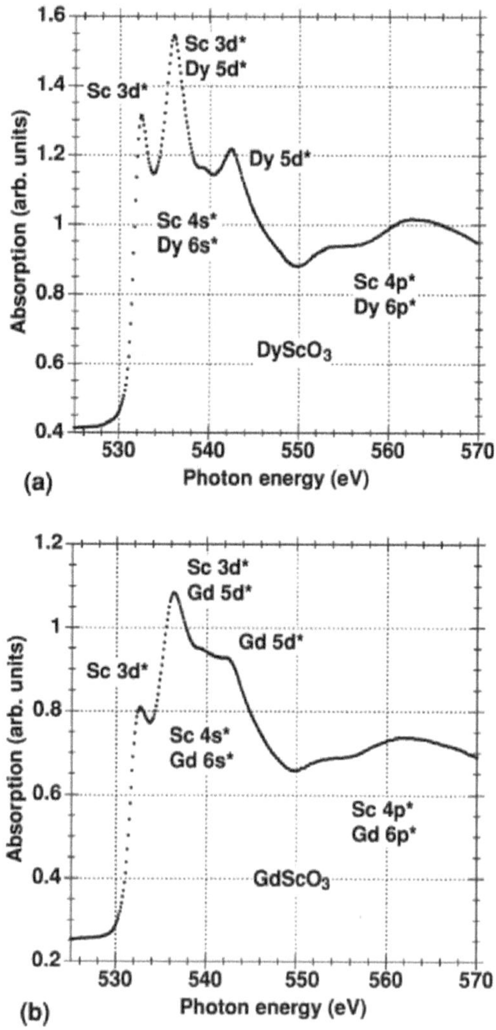

Fig. 19. O K_1 spectra for (a) Y_2O_3 and (b) ZrO_2.

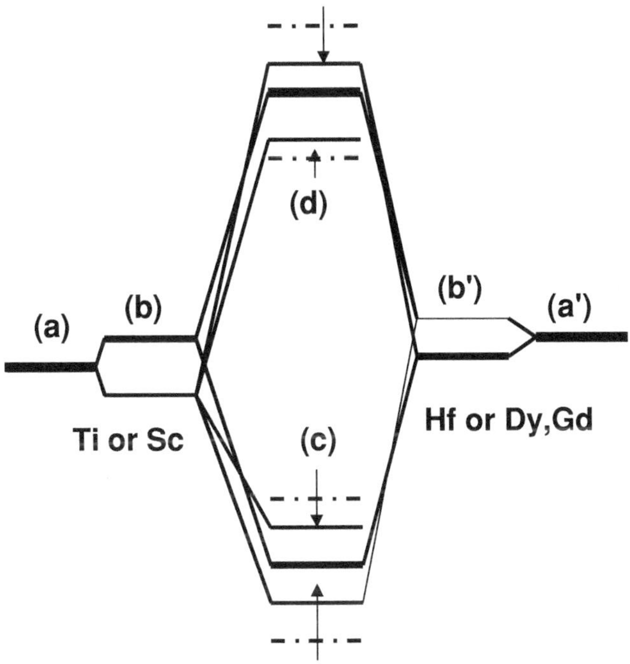

Fig. 20. Schematic representation of d state coupling in complex oxides.

features. The features at higher energy are assigned to Sc 4p* and Dy 6p* states. Since the 3d/4s atomic splitting of Sc in greater by at least 2–3 eV than the corresponding 5d/6s splittings for both Gd and Dy, this suggests that the feature labeled Sc 3d* has been shifted to higher energy through interactions between Sc 3d, and Gd or Dy 5d states bonded to a common O atom. It also suggests that the next spectral narrow d* features marked Sc 3d* and Gd 5d* in Fig. 19(a), and Sc 3d* and Dy 5d* in Fig. 19(b) have a mixed Sc–Re atom character. The may result from a near degeneracy of the 3d and 5d states as shown in the schematic bonding interaction diagram of Fig. 20.

Figure 20 presents a schematic representation of the coupling of Sc 3d*, and Dy or Gd 5d* states through bonding to a common O atom. This schematic model includes (i) the relative energies of the respective atomic d states, (ii) the symmetry splittings of these states, (iii) the coupled valence band bonding, and (iv) the anti-bonding conduction band states. The arrows in the bonding states indicate the coupling in which the overlap between 3d Sc and 2p O states is greater than the overlap between 5d Gd or Dy states, and O 2p states. This model replicates the spectral features in Fig. 19(a) and (b).

Figure 21(a) displays the optical absorption constant, α, at the band edge as a function of photon energy as obtained from the analysis of VUV SE data (55). The shoulder between about 4.8 and 5.8 eV is assigned to 3d*-state absorption associated

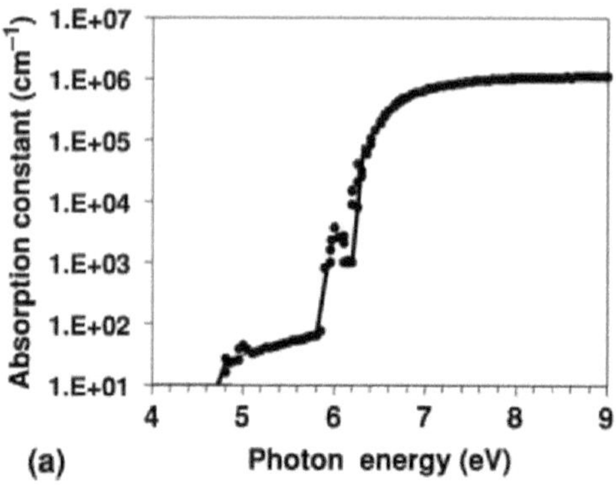

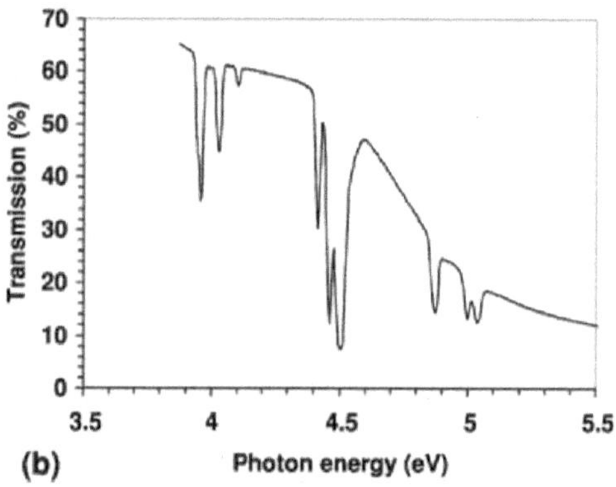

Fig. 21. (a) Absorption constant, a, as a function of photon energy for $GdScO_3$ in the spectral range form 4 to 9 eV, as obtained from the analysis of VUV spectroscopic ellipsometry data. (b) Optical transmission as a function a function of photon energy for $GdScO_3$ in the spectral range from 3.5 to 5.5 eV.

with the Sc atoms. The 3d* state band gap of Sc_2O_3 is approximately 4.6 eV, and is associated with low values of α, in the range of 100 cm^{-1} (57). This weak d state absorption is attributed to a crystalline distortion associated with the dynamic Jahn–Teller effect (58). The rapid rise of absorption at approximately 5.8 eV in Fig. 21(a) marks the onset of transitions from the top of valence band, O 2p π non-bonding states, to the lowest energy coupled d* state that has a strong component of Sc 3d* character. Since there is no distinct spectral evidence for the second d* state, the absorption above 6 eV is assigned to transitions to Sc and Gd s*-states. The relatively sharp features on the shoulder at ~4.8 and 5 eV also appear in the optical absorption spectrum of Fig. 21(b) as a singlet at 4.85 eV, and a doublet centered at 5 eV. These sharp features, along with the other two triplet bands at ~4 and 4.5 eV are characteristic 4f intra-atomic optical transitions of trivalent Gd (59).

Figure 22(a) is the O K_1 edge spectrum for a crystallized $(HfO_2)_{0.5}(TiO_2)_{0.5}$ alloy with a stoichiometric titanate composition of $HfTiO_4$. The coupled Ti 3d*

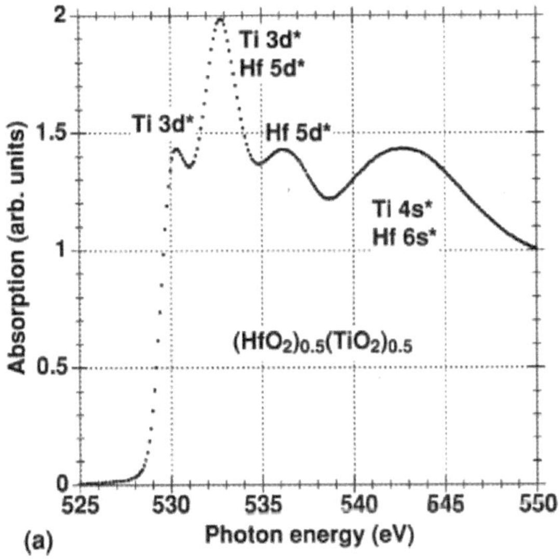

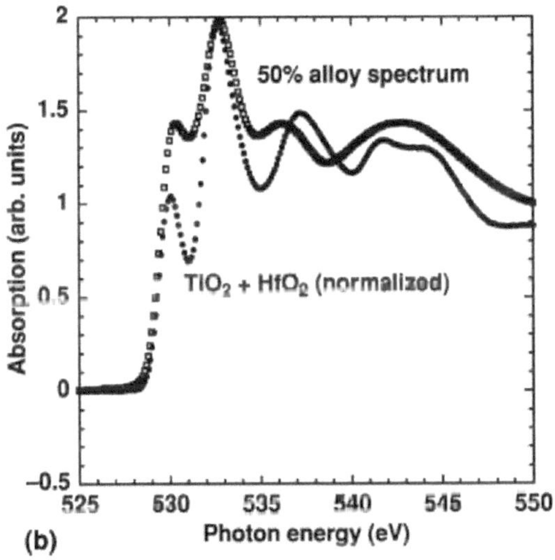

Fig. 22. (a) O K_1 spectrum for a $(HfO_2)_{0.5}(TiO_2)_{0.5}$ alloy that has been annealed at 600°C. (b) Comparison between the spectrum in (a), and a spectrum synthesized from the O K_1 spectra of HfO_2 and TiO_2.

and Hf 5d* states have been labeled as in the corresponding spectra for $GdScO_3$ and $DyScO_3$, which can also be written, respectively, in mixed oxide notations as $(Dy_2O_3)_{0.5}(Sc_2O_3)_{0.5}$ and $(Gd_2O_3)_{0.5}(Sc_2O_3)_{0.5}$. The most significant difference between the $HfTiO_4$ spectrum, and the $GdScO_3$ and $DyScO_3$ spectra is the overlap of the d* and s* states. The three d* localized states, and the s* state bands are spectroscopically distinct in $HfTiO_4$, whilst the corresponding bands overlap significantly in the $GdScO_3$ and $DyScO_3$ spectra.

Finally Fig. 22(b) compares the O K_1 edge spectrum for $HfTiO_4$ with a normalized sum of the spectra for TiO_2 and HfO_2. The 532.5 eV peak of the $HfTiO_4$ O K_1 spectrum was set normalized to sum of the 532 eV features in the O K_1 spectra of TiO_2 and HfO_2. Differences between the spectral peak energies of the features higher and lower energy features, and their respective amplitudes in the experimental and summed spectra are a clear and unambiguous indicator of the proposed d-state coupling.

4.5. XPS and AES Results for Zr Silicates

A detailed and comprehensive study of XPS and AES measurements is presented in Ref. 24 for Zr silicate alloys, $(ZrO_2)x(SiO_2)_{1-x}$. Figure 23(a)–(c) summarizes the results of XPS measurements of O 1s, Si 2p, and Zr 3d core level binding energies for the end-member oxides, SiO_2 and ZrO_2, and for 13 pseudo-binary oxide alloy compositions distributed approximately equally over the entire alloy composition range. These are for as-deposited thin films. Studies of films annealed at 500°C in Ar display essentially the same spectra, whereas films annealed at 900°C show evidence for chemical phase separation into SiO_2 and ZrO_2, independent of whether the phase separation is accompanied by crystallization (56).

Figure 16(a) indicates the compositional dependence of the O 1s binding energy. The sigmodial character of the plot is a manifestation of mixed coordination for O-atoms as anticipated by the discussion above relative to the classification scheme for oxides based on bond ionicity. The coordination of oxygen increases from 2 to 3 in the composition range from SiO_2 (coordination 2), to 3 for the 50% ZrO_2 chemically-ordered alloy that defines the stoichiometric silicate composition, $ZrSiO_4$. Derivative XPS spectra, displayed in Ref. 56 confirm that the sigmoidal dependence is due to mixed coordination. Finally, the total shift in the O 1s core level binding energy between SiO_2 and ZrO_2 is 2.45 ± 0.1 eV.

Figure 23(b) and (c) displays, respectively, similar spectra for the Si 3p and Zr $3d_{5/2}$ core levels. The Si 2p data in Fig. 23(b) shows a linear dependence consistent with a single atomic coordination of four, and a total shift of 1.85 ± 0.1 eV between the end member elemental oxides, SiO_2 and ZrO_2. Note that these core level shifts are in the same direction, with the values at the SiO_2 *end* of the alloy regime being more negative. As discussed in Ref. 24, this is consistent with partial charges calculated on the basis of electronegativity equalization (27).

The data for the compositional dependence of the Zr $3d_{5/2}$ core level show some additional structure for low values of x. The total change in binding energy across the alloy system is 1.85 ± 0.1 eV, and is essentially the same as for the 2p Si level. This means that the slopes of the plots in Fig. 23(b) and (c) in the linear regime are the same as well. The equality of these slopes is also consistent with the principle of electronegativity equalization (27). More importantly the equivalence of the slopes is also consistent with the XAS data for Zr silicate alloys. Parallel slope shifts in core level spectra are equivalent to the 4d* anti-bonding states of Zr and the 3s* band peak

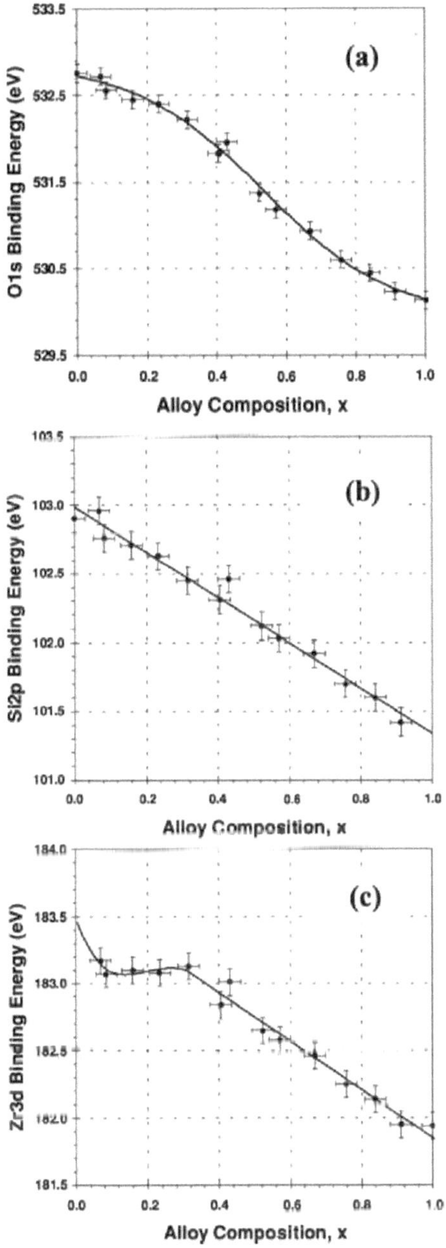

Fig. 23. XPS chemical shifts of (a) O 1s, (b) Si 2p and (b) Zr $3d_{5/2}$ core levels from as-deposited (300°C) $(ZrO_2)_x(SiO_2)_{1-x}$ alloys as a function of composition, x.

of Si maintaining a constant energy separation as a function of alloy composition. This has been demonstrated in Fig. 17 for Hf silicate alloys, and a similar situation prevails for as-deposited Zr silicate alloys as well.

Finally, the departure from linearity for $x < 0.4$ in Fig. 23(c) has been assigned to the change in the nature of the chemical bonding at the Zr site as a function of alloy composition (24). The coordination of Zr has been assumed to be eight independent of alloy composition; however, each of these eight oxygen atoms are not equivalent with respect to bonding neighbor coordination and electronic structure. The number of ionic Zr–O bonds associated with network disruption increases from four to eight with increasing x for alloys in the SiO_2 rich bonding regime. In this alloy regime, each O-atom makes at least one Zr–O bond with a bond order of one in a Si–O–Zr arrangement, and there must be at least four of these arrangements. The remainder of the eight-fold coordination is made up with donor–acceptor pair electrostatic bonds with bridging O-atoms of the non-disrupted portion of the SiO_2 continuous random network. These weaker bonds have been modeled in ab-initio calculations as components of a dipolar electrostatic field, and alternatively, and equivalently can also be described as *donor–acceptor pair or dative bonds*. The donor–acceptor bonds are replaced by Si–O–Zr ionic bonding arrangements as x increases, and the network disruption increases. At a composition of $x = 0.5$, network disruption is essentially complete, and the O-atom coordination is three, and the bond order of the Zr atoms is formally one-half with all bonds between eight-fold coordinated Zr^{4+} ions and terminal O-atoms of silicate ions, SiO_4^{4-}. Each of the terminal O atoms of a silicate ion makes bonds with two Zr^{4+} ions.

Ab-initio calculations similar to those discussed in Sections 3.1 and 3.2 have been used to identify the effects of the donor–acceptor pair bonds on the Zr core level shifts. In this model calculation, the Zr-atom has four OH-groups in a tetrahedral arrangement to emulate the ionic bonds, and four tetrahedrally-grouped water molecules with the O-atom non-bonding p-electron pair aligned in the direction of the Zr-atom to emulate the donor–acceptor pair bonding interaction. The calculations indicated that bonding is optimized at an effective inter-atomic spacing of ~0.26–0.28 nm between the Zr-atoms and the bridging O-atoms of the network. The minimum is broad and shallow opening up the possibility of a spread in inter-atomic spacing where bond-strain and configurational entropy are likely to also be contributing factors in determining a statistical distribution of these bonding arrangements in a non-crystalline solid.

The calculations indicate a positive shift in the Zr 1s bonding energy as a function of the inter-atom spacing between Zr- and bridging O-atoms. The calculations also indicate the effects of the donor–acceptor pair bond on Zr core levels are *equivalent* to a dipole field. The effect of the donor–acceptor pair bonds, or dipole fields is to reduce the binding energy of the Zr 1s core state. Since all of the core states move rigidly with respect to the Zr 1s state, this calculation explains the direction of the non-linearity of the Zr $3d_{5/2}$ core state in Fig. 23(c).

AES measurements on the as-deposited films were performed on-line immediately following film deposition. AES chemical shifts of O_{KVV} and Zr_{MVV} transitions as a

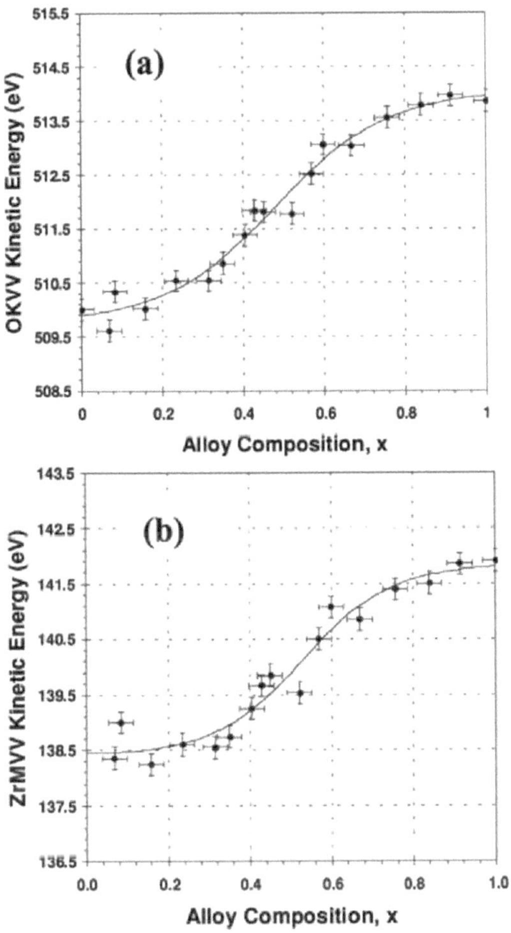

Fig. 24. AES chemical shifts of (a) O_{KVV} and (b) Zr_{MVV} kinetic energies in as deposited $(ZrO_2)_x(SiO_2)_{1-x}$ alloys as a function of composition. The plots in (a) and (b) are for the *highest energy peaks* in the respective AES derivative spectra. The solid lines are polynomial fits that are intended to emphasize the sigmoidal character of the compositional dependence.

function of composition for derivative spectra are shown respectively in Fig. 24(a) and (b). They show nearly identical non linear behaviors that are qualitatively different and therefore *complementary* to the XPS chemical shifts of the O 1s and Zr $3d_{5/2}$ core level binding energies shown in Fig. 23(a) and (c), respectively. The compositional dependence of the AES *peak kinetic energy* values display marked sigmoidal non-linear dependence. Finally, due to spectral overlap between the Zr_{MVV} and Si_{LVV} features in the AES spectra, it was not possible to track the compositional dependence of the AES Si_{LVV} feature.

The chemical shifts of the Auger electron kinetic energies for O_{KVV} and Zr_{MVV} transitions in the as-deposited films are consistent with changes in the calculated partial charges and their effects on the O and Zr core state energies, i.e., the kinetic energies of the Auger electrons increase with increasing x reflecting the *decreases in the negative* XPS binding energies, i.e., shifts to less negative values. The differences between the XPS and AES spectral features derive from differences between the XPS and AES processes. Following Ref. 60, the AES electrons of Fig. 24(a) and (b) originate in the valence band, whereas the XPS electrons of Fig. 23(a) and (b) originate in the respective core states with no valence band participation. This is addressed below where the non-linear behavior of the AES features reflect systematic shifts in valence band energy with increasing O-atom coordination.

The XPS and AES results are combined with determinations of valence band offset energies for SiO_2 and ZrO_2 (61–63) to generate an empirical model for the compositional variation of valence band offset energies with respect to Si. The O_{KVV} transition in amorphous-SiO_2 has been investigated theoretically, and it has been shown that the highest kinetic energy AES feature is associated with two electrons being released from the non-bonding O 2p π states at top of the valence band; one of these is the AES electron, and the second fills the O 1s core hole generated by electron beam excitation (49). Based on this mechanism, the XPS and AES results of this study have been integrated into a model in that provides an estimate of valence band offsets with respect to Si as a function of alloy composition. For an *ijk* AES A-atom transition, the kinetic energy of the AES electron, $E_K(A,ijk)$, is related to the XPS binding energies, $E_B(A,i)$, $E_B(A,j)$, and $E_B(A,k)$, and a term $\Omega(A)$ that includes all final state effects:

$$E_K(A,ijk) = E_B(A,i) - E_B(A,j) - E_B(A,k) - \Omega(A). \qquad (3)$$

Applied to the O_{KVV} transition, A = O, $i = k$ (O 1s) and j, k = L = O (2p π non-bonding). Equation (3) is the basis for an empirical model for the energy of the Zr silicate valence band edge with respect to vacuum, and then with respect to Si, both as functions of the alloy composition. If E_{BE}(O 1s) is the XPS binding energy, and $E_{KE}(O_{KVV})$ is the average kinetic energy of the Auger electron with respect to the top of the valence band edge, then the offset energy, $V_{OFFSET}(x)$, is given by

$$V_{OFFSET}(x) \sim - A \cdot 0.5[E_B(O\,1s) - E_K(O_{KVV})] + B, \qquad (4)$$

where A and B are determined from the experimental valence band offsets of 4.6 eV for SiO_2 and 3.1 eV for ZrO_2 (61–63). This model is presented in Fig. 25, and the sigmoidal shape is determined by the relative compositional dependencies of the XPS (O 1s) and AES (O_{KVV}) results in Figs. 23(a) and 24(a). The analysis has also been applied to the Zr_{MVV} AES and Zr $3d_{5/2}$ XPS results of Figs. 23(c) and 24(b), and gives essentially the same compositional dependence as is displayed in Fig. 25, but with different empirical constants, A′ and B′. The weakly sigmoidal dependence is a manifestation the discreteness of the O-atom coordinations as function of the alloy composition, a mixture of two-fold and three-fold for $x < 0.5$, and three-fold and four-fold for $x > 0.5$.

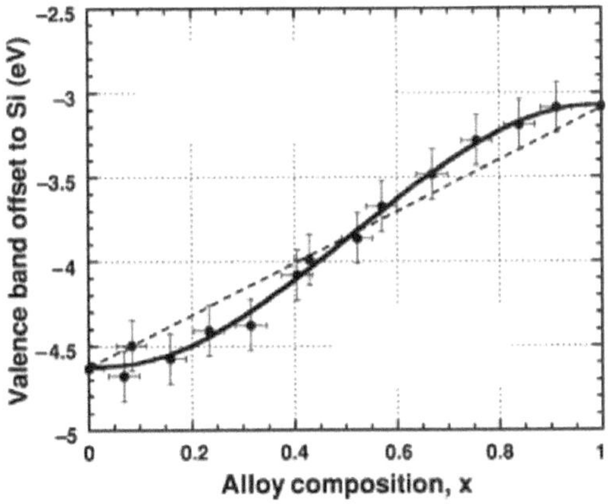

Fig. 25. Calculated values of the valence band offset energies relative the valence band of crystalline Si at ~−5.2 eV as calculated from the two parameter empirical model. The plots in are derived from O atom XPS and AES data. The signmoidal dependence results from differences between the compositional dependencies of the respective XPS and AES results used as input, and not on empirical constants.

Figure 26 contains plots of the average conduction and valence band offset energies of Zr silicate alloys as determined from the model of Eq. (7), and the experimentally determined band gaps for SiO_2, ~9 eV, and ZrO_2, ~5.6 eV. This approach demonstrates that essentially all of the band gap variation occurs in the valence band offsets, so that the offset energies of the respective Zr 4d* states and Si 3s* states are constant to $< \pm 0.2$ eV with respect to the conduction band edge of Si. The contributions of these Zr and Si states to the conduction band density of states are proportional to their alloy concentrations, with *qualitative differences* in these states playing a significant role in determining direct tunnelling currents.

Qualitatively similar results have been obtained been obtained for Hf silicate alloys using the same XPS and AES approach (36). The results presented in Fig. 27, combined with absorption spectroscopy data presented in Ref. 64, indicate the variation

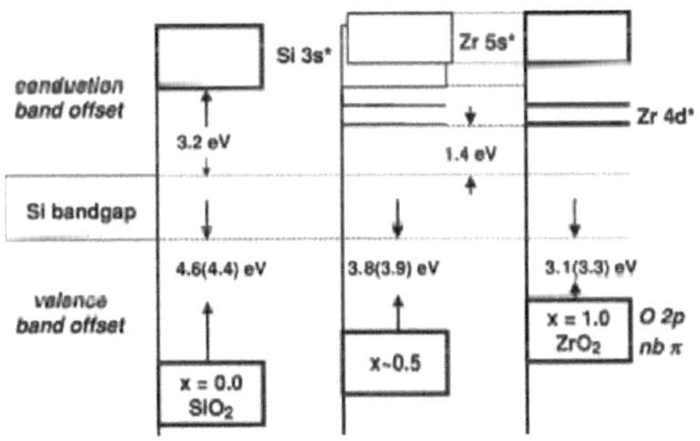

Fig. 26. Band edge electronic structure for SiO_2, an $x = 0.5$ Zr silicate alloy and ZrO_2.

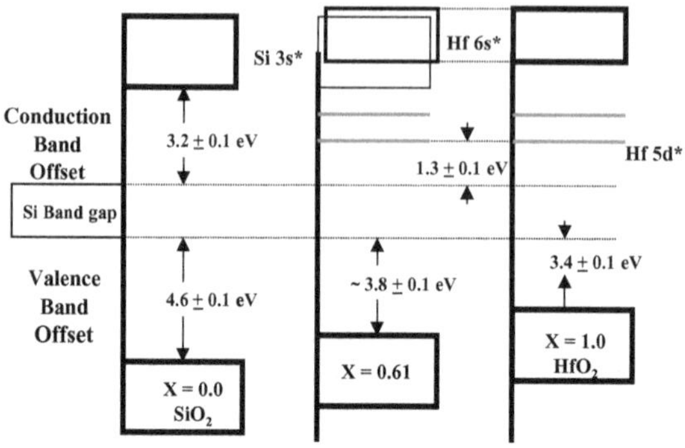

Fig. 27. Band edge electronic structure for SiO_2, an $x = 0.5$ Hf silicate alloy and HfO_2.

in the respective band gaps is reflected entirely in the valence band offset energy with respect to Si. Figure 28 displays a plot of the effective band gap for Hf silicate alloys obtained from the analysis VUV SE spectroscopic studies (65). It contains plots of the E05 band gap, defined as the photon energy corresponding to an absorption constant, $\alpha = 10^5\ \text{cm}^{-1}$, and the spectral peak of the real part of the complex dielectric constant, ε_1. The non-linearity of the plot reflects the complex nature of the band edge states, localized Hf 5d* and extended Si 3s* states, whose relative amplitudes change systematically as a function of alloy composition. In Ref. 54, absorption edge measurements were analyzed for Zr and Hf silicate alloys using a Tauc's band edge representation in which

$$\alpha = (h\nu - E_{\text{opt}})^2 \tag{5}$$

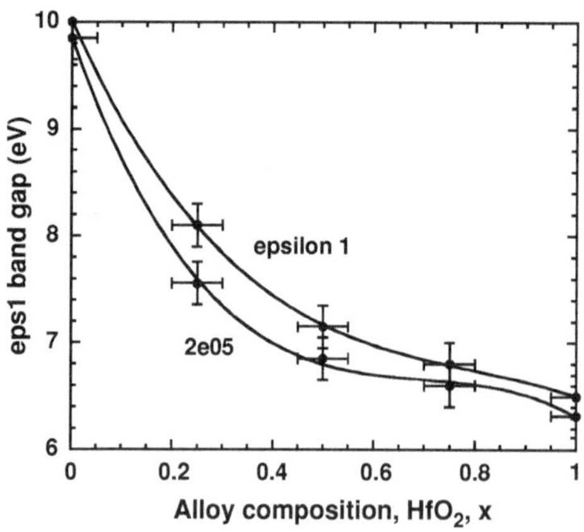

Fig. 28. Values of E05, the energy at which $a = 10^5\ \text{cm}^{-1}$, and the peak in e1 as functions of Hf silicate alloy composition.

where $h\nu$ is the photon energy, and E_{opt} is an effective band gap. The results displayed in Ref. 64 give essentially the same alloy dependence as shown in Fig. 20.

5. APPLICATION OF ELECTRONIC STRUCTURE TO DIRECT TUNNELING

In order to reduce direct tunneling in MOS devices with equivalent oxide thickness, EOT, <1.5 nm, and extending below 1 nm, there has been a search for alternative dielectrics with significantly increased dielectric constants, k, allowing increases in physical thickness proportional to k, and thereby significantly reducing direct tunneling. However, significant *increases* in k to values of 15 to 25 in transition metal and rare earth oxides are generally accompanied by *decreases* in the conduction band offset energy with respect to Si, E_B, and the effective electron tunneling mass, m_{eff}. This trade-off between increases in k, and decreases in E_B and m_{eff} is quantified by the introduction of a figure of merit, Φ_m, given by, $\Phi_m = k[E_B \cdot m_{eff}]^{0.5}$, where k, E_b and m^*_{eff} are respectively, the dielectric constant, the conduction band offset energy, and the effective electron tunnelling mass (2, 66).

The expectation was that increased values of k, which permit the use of physically thicker films for the same EOT as SiO_2, would provide significant reductions in direct tunnelling allowing scaling to continue to at least an EOT of 1.0 nm, and hopefully to values of EOT approaching 0.5 nm. The discussion presented above has demonstrated that conduction band offset energies for high-k dielectrics containing transition metal atoms are at most 1 eV less than SiO_2 for group IIIB (Y, La) and rare earth oxides, silicates and aluminates, and in group IVB (Zr, HfO) oxides, silicates and aluminates, they are reduced further to at least 1.5 eV. These limitations assume that E_B must be greater than at least 1 eV, and therefore have exclude Sc and Ti oxides, silicates and aluminates. Since the tunnelling figure of merit includes a dependence on m_{eff} as well, it is necessary to determine E_B and m_{eff}, or at least their production, and eventually to understand any inherent relationships between E_B and m_{eff} that are related specifically to electronic structure of the high-k dielectrics. Based on a new approach for experimental determination of E_B m_{eff} products as discussed in Ref. 66, E_B m_{eff} has been determined for HfO_2 to be $0.23 \pm 0.01 m_o$ eV for HfO_2. Based on the spectroscopic approach used to determine E_B for ZrO_2 (24), and the extension of this approach to other transition metal and rare earth oxides including HfO_2, a value of $E_B = 1.5 \pm 0.2$ eV has been inferred for HfO_2, and Hf silicate and aluminate alloys as well. Using this value of $E_B \sim 1.5$ eV for the Si–HfO_2 conduction band offset energy then corresponds to a value of $m_{eff} = 0.15 \pm 0.02 m_o$, in good agreement with other analyses of tunneling through HfO_2 films (67).

Next, it is important to understand the low value for tunneling mass for HfO_2, and its impact on direct tunneling in Hf silicate alloys. This mass is significantly smaller than the tunneling mass of $\sim 0.55 m_o$ for SiO_2. The microscopic origin of this differences is first addressed, then its effect on the compositional dependence of direct tunneling in Hf silicate alloys.

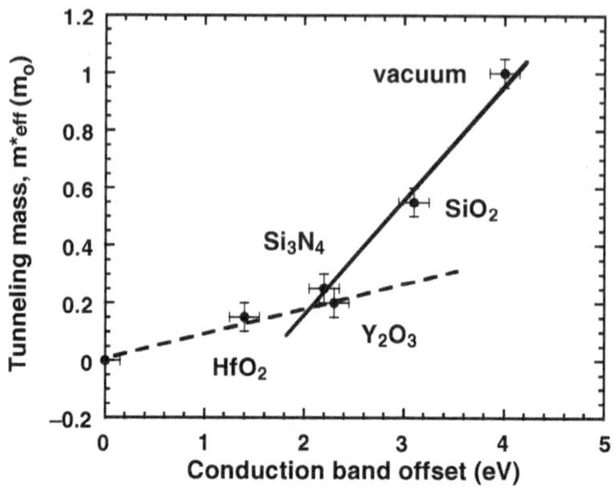

Fig. 29. Electron tunneling mass versus conduction band offset as determined from Franz two-band model. The solid line is for dielectrics with extended s* conduction bands, and the dashed line is for localized d* state band edges.

Figure 29 contains a plot of tunneling mass versus band offset energy in the context of the Franz two band model, as discussed in Refs. 68 and 69. The effective electron masses for tunneling through the vacuum, and SiO_2, Al_2O_3 and Si_3N_4 dielectrics fall on a straight line, along with the extrapolated mass for Y_2O_3; however the mass for HfO_2 does not. The Franz two-band model is an effective mass approximation that works best when the conduction and valence band states are extended and free electron like, and yields this linear dependence when symmetry is assumed between the tunneling of conduction and valence band electrons (69). This is the case for that has been assumed for to described conduction band tunneling through SiO_2 and Al_2O_3, where the lowest conduction band states are 3s* anti-bonding states, but not for transition metal oxides with d* state bands, where the effective mass approximation does not apply. However, the overlap of transition metal d* states with transition metal s* states differs for different transition metal atoms, and is proportional to the difference between the atomic nd and $(n + 1)$s states of the transition metal where n is the principal quantum number equal to 5 for Hf and 4 for Y. The point for Y_2O_3 falls on the plot for the oxides with extended free electron like conduction band states due to this overlap, and the point for HfO_2 is well removed from this fit to the data due primarily to differences in s d overlap which is greater the Y_2O_3. This difference in overlap has been illustrated in Fig. 18(a) and (b) for Y_2O_3 and ZrO_2, and based on the similarities between the O K_1 spectra of ZrO_2 and HfO_2, respectively, in Fig. 13(b) and (c), it also applies to comparisons with HfO_2. Finally, it has been shown in Ref. 66 that the low value of $m_{\text{eff}} = 0.15m_o$ coupled with an $E_B \sim 1.5$ eV gives a minimum tunneling current for a given EOT in the middle of the silicate alloy regime, whereas for Y silicates, the higher values of both $m_{\text{eff}} \sim 0.25m_o$ and $E_B \sim 2.3$ eV gives a minimum tunneling current at the Y_2O_3 composition.

Figure 30 displays this result for the compositional dependence of tunneling current density for Si oxynitride alloys, Hf silicate alloys and Y silicate alloys at an

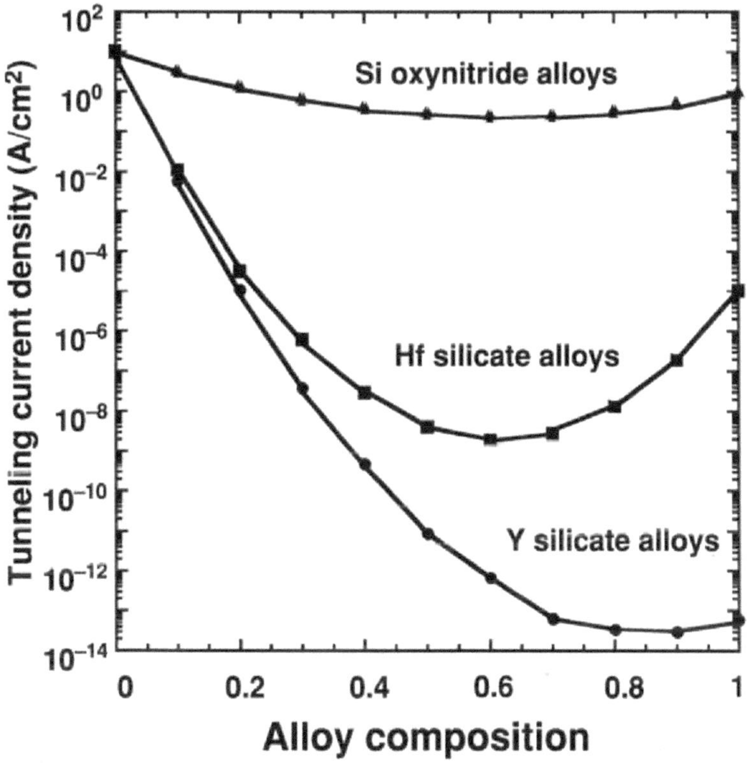

Fig. 30. Calculated tunneling at 1 V oxide bias for alloys with EOT = 1.0 nm.

oxide bias of 1 V as calculated using the WKB approximation (70, 71). The plots in Fig. 31 provide the important connection between the tunneling figure of merit, Φ_m, and the tunneling calculated tunneling currents. The plots in Fig. 31 are for the figure of merit, Φ_m, where k, E_b and m^*_{eff} have been computed for Si oxynitride alloys, Hf silicate alloys and Zr silicate alloys using compositionally averaged values of k, E_B and m_{eff}. A plot for the Si oxynitride alloy system is shown for reference. The values of k, E_B and m_{eff} for SiO_2 for this model calculation are respectively, 3.8, 3.15 eV, and $0.55m_o$ and the corresponding values for Si_3N_4 are, 7.6, 2.15 eV, and $0.25m_o$. The corresponding values for HfO_2 and Y_2O_3 have been included in Fig. 31. The plot for Si oxynitrides shows a relatively small variation across the alloy system, and accounts for (i) the relatively decreases in direct tunneling of approximately that are obtained in optimized Si oxynitride alloys with $x \sim 0.5$ (3). However, these are still significant for device scaling as discussed below. The differences between the compositional dependence for Y silicate and Hf silicate alloys are more significant with respect to narrowing the field of high-k gate dielectrics. The monotonically increasing function for Y silicates predicts that tunneling with respect to SiO_2 will be reduced over the entire alloy range, whilst the qualitatively difference behavior predicts that

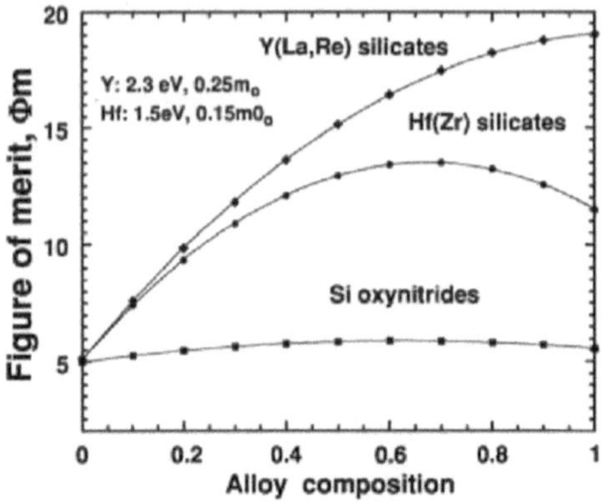

Fig. 31. Compositional dependence of tunneling figure of merit for alloys in Fig. 30.

the tunneling reduction in Hf silicate alloys will display a minimum in the middle of the alloy system.

The plots in Fig. 30 are for the direct tunneling current in n+-dielectric-N+-poly-Si at an oxide bias of in excess of 1 V above flat band for substrate accumulation. The calculation takes into account the potential drops across the poly-Si and the channel region, and there is a potential drop of 1 V across the dielectric for the gate potential used in the evaluation of the current density. The doping concentration in the substrate was 2.5×10^{17} cm^{-3}, and in the poly-Si, 9×10^{19} cm^{-3}. The values of the computed tunneling current density are independent of these values for n+ and N+ because of the corrections made for the potential drops in the poly-Si and channel regions of the dielectric stack. The differences between the calculated compositional variations of direct tunnelling in Hf and Y silicate alloys represent the importance for determining the $(E_B)(m_{eff})$ product for high-k dielectrics, which can be accomplished through the novel approach identified in Ref. 66. The correlation between the tunneling figure of merit in Fig. 31, and the calculated tunneling currents in Fig. 30 in evident in the complementary nature of the plots. In particular, the plot for Hf silicate alloys, which applies to Zr silicate as well, indicates that the tunneling current is a minimum in the middle of the alloy, paralleling a behavior for Si oxynitride alloys. These behaviors are in agreement with experimental results for both the Si oxynitride and the Hf silicate alloys (2). The behavior for Y (and other group IIIB and trivalent Re silicates) is qualitatively different with the minimum in tunneling occurring at the elemental oxide composition.

6. BOND-DEFECTS AND DEFECT PRECURSORS

6.1. The Si–SiO_2 Interface

Numerous experimental studies have demonstrated that the interface between Si and SiO_2 in field effect transistor, FET, devices which have been subjected to high temperature processing extending to at least 1000°C, are not atomically abrupt, but instead contain an interfacial transition region with Si sub-oxide, SiO_x, $x < 2$, bonding, as well as region of defective Si, Si_D, in the Si substrate (72–74). The studies in Refs. 72–74 indicate that these regions are each of the order of 0.5 nm in directions perpendicular to the plane of the Si–SiO_2 interface. This section of the chapter discusses the bonding changes in the SiO_x transition region as function of thermal annealing processes. (75–77). Bond-constraint theory is then used to identify the driving chemical bonding and physical forces for these changes, which are explained in terms of self-organization mechanism. This interfacial self-organization explains in part the unique properties of the interfacial transition regions that are important in device performance and reliability. It is therefore a necessary part of high-k stacks in which scaling metrics for performance and reliability are based on Si–SiO_2 devices (31).

The combination the single wavelength ellipsometry measurements (78) and optical second harmonic generation, SHG (75, 79) have identified two high-temperature transitions: (i) the first at ~1000°C and (ii) the second is at ~900°C. The second transition will be shown be accompanied by bonding changes in the interfacial transition between the Si substrate and the SiO_2 dielectric film. The second transition was first identified by optical second harmonic generation, SHG, a non-linear optics approach (75). Applied to the an Si–SiO_2 gate stack, no SHG signal is expected from either (i) the non-crystalline SiO_2 layer, due to lack of long range order required for development of an electric field component at two times the incident photon frequency from a coherent laser source, or (ii) the crystalline Si substrate because of the local symmetry at the Si-atom bonding sites. In contrast, SHG, is possible Si surfaces, or Si–SiO_2 interfaces, with matrix elements being different for different interface bonding symmetries.

The plot in Fig. 32 gives, the phase angel, θ as a function of annealing temperature for a sample prepared by thermal oxidation at ~850°C, and then annealed to 1050°C. The largest rate of change in θ is between 850 and 900°C, and this has been interpreted as being due to bonding relaxations on the terrace and step edges of vicinal Si(111) oxidized surfaces (75, 80, 81). There are additional smaller changes in θ, between 900 and 1050°C, and these are assigned to the relaxation of bulk growth stress as discussed above. An interface prepared by remote plasma-assisted oxidation at 300°C gives the same values of θ to ±2° as the 850°C thermal oxidation, and additionally the same value of θ to ±2° after a 900°C anneal.

6.2. Chemical Bonding Changes Associated with Interfacial Self-organization

Soft X-ray XPS, SXPS (77) and cathodo-luminescence spectroscopy, CLS (76) have provided direct evidence for interfacial bonding changes that occur after a 900°C anneal. The SXPS studies probe these changes an areal density of 10^{14}–10^{15} cm^{-2}, corresponding to the number of surface bonds per Si atom, whilst the CLS studies probe these changes at an areal density of ~10^{11}–10^{12} cm^{-2}, corresponding to the scale of interfacial bonding defects.

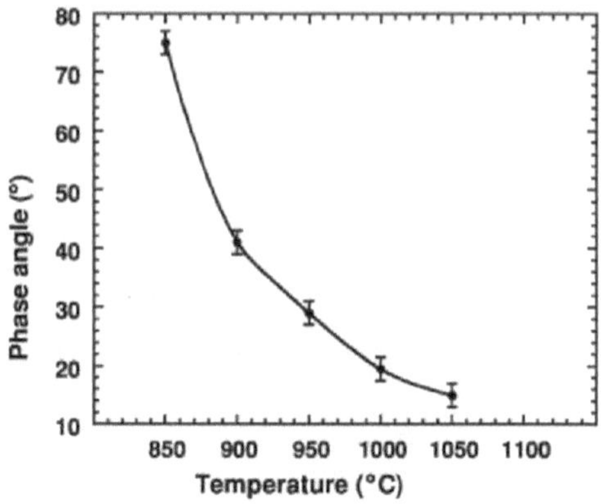

Fig. 32. Phase angle difference between terrace and step edge contributions to the optical SHG signal from vicinal Si(111) wafers off cut 5° in the 112 bar direction. The point at 1100°C does not fall on the trend line and occurs at temperature at which interface decomposition associated with the evolution of gaseous SiO is known to occur. The solid line is an interpolation that establishes the trend in the data points.

6.2.1. SXPS studies

Figure 33 summarizes the results of experiments performed on Si(111) interfaces presented in Ref. 77. The spectral features identified in this figure as I_1, I_2 and I_3, correspond respectively to bonding arrangements in which Si has one, two and three oxygen neighbors. These have been designated, respectively, as Si^+, Si^{2+}, and Si^{3+}. The figure also shows the substrate feature, labeled Si^0, and the SiO_2 feature labeled Si^{4+}. The samples for this study were prepared by remote plasma processing at 300°C, and then subjected to annealing at temperatures of 600, 800, 900 and 1000°C.

There are two aspects of this figure to note. First the most significant change in the spectra after the 900°C anneal is a reduction in the Si^{2+} component. This is the feature that is *not native*, or equivalently *foreign* to a Si(111) interface. Analysis of these spectral changes indicates that after the 900°C anneal, there is one monolayer of excess sub-oxide with an average composition of SiO. As discussed in Ref. 82, this corresponds to a physical thickness of ~0.3 nm, in good agreement with the ion scattering measurements identified above. This is in excess of the one monolayer of Si bonding expected at an ideal abrupt Si(111)-SiO_2 interface. Similar results have been obtained for Si(100)-SiO_2 interfaces, with comparable changes between the as-grown and 900°C annealed distributions of interfacial bonding arrangements (77).

The changes in bonding between as-grown, and 900°C annealed interfacial layers indicated that (i) as-grown interfaces are more homogeneous sub-oxide layers with random bonding, whilst (ii) after the 900°C anneal, they display an inhomogeneous character, consistent with a chemical phase separation. This is associated with Si-rich, Si–O–3Si and O-rich, Si–3O–Si, as revealed by the Si^{3+} and Si^+ features, both of which are significantly larger than the average Si–2SO–2Si bonding as revealed by the Si^{2+} feature prior to the anneal.

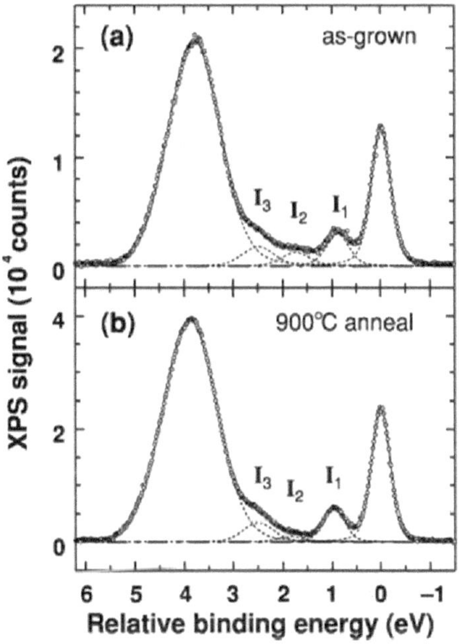

Fig. 33. Spin-orbit stripped and background subtracted Si(2p) data for Si(111). (a) shows "as-grown" sample (~1.2 nm), and (b) is for a piece of the same wafer, annealed ex situ to 900°C in Ar. The energy scale is referenced to the substrate Si feature (with four Si neighbors) at 0.0 eV, and the three features marked, I_1, I_2 and I_3, are features assigned to Si with 1, 2 and 3 oxygen neighbors. The broad feature centered at approximately 4 eV is the SiO_2 features (with four O neighbors).

6.2.2. CLS studies

These results in Fig. 34 are for an Si–SiO_2 gate stack structure prepared at 300°C, with a thickness of 5 nm (76). The CLS spectra were obtained as a function of the electron beam energy, between about 0.5 and 4.5 keV. The spectra in Fig. 34 are for an electron beam energy of 2 keV, and reveal interface and near interface features. Spectra are shown for the as-deposited sample, in which the interface remote plasma assisted oxidation (~0.6 nm) and SiO_2 remote plasma-enhanced chemical vapor deposition were performed at 300°C. Additional spectra are shown 400°C *hydrogenation anneal* in forming gas, 10% H_2/90% N_2, after a 900°C rapid thermal anneal, RTA, for approximately 1 minute in Ar, and after a 900°C RTA followed by a 400°C anneal in forming gas. The forming gas anneals passivates defects in which the bonding includes a silicon or oxygen atom dangling bond with a single unpaired electron by creation of a hydrogen atom terminal group, such as Si–H, SiO–H, etc.

As shown in Fig. 34, The combination of a rapid thermal anneal, RTA, at 900° and 400°C anneal under a hydrogen ambient significantly reduces interfacial trap cathodo-luminescence features. The energy of the electron beam, 2 keV, has

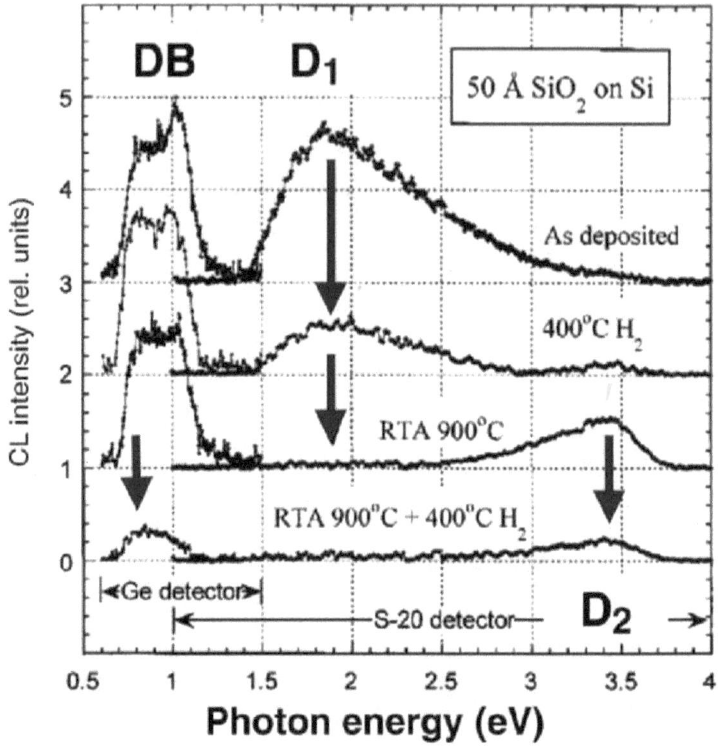

Fig. 34. Process-dependent localized states as revealed by features in the cathodo-luminescence spectra in a thin (5 nm) Si–SiO_2 structure. The combination of a rapid thermal anneal, RTA, and 400°C anneal under a hydrogen ambient reduces interfacial trap emissions almost completely. The energy of the electron beam, 2 keV, has been adjusted to give high sensitivity to luminescence features emanating from the Si–SiO_2 interface region. The defect luminescence bands are indicated by arrows. DB is an interfacial dangling bond, D_1 is a near interfacial defect in the as-grown transition region prior to the 900°C RTA, and D_2 is a substrate luminescence band.

been adjusted to give high sensitivity to luminescence features emanating from the Si–SiO_2 interface region (76). The defect luminescence bands are indicated by arrows. DB is an interfacial Si angling bond, D_1 is a near interfacial defect in the as-grown transition region prior to the 900°C RTA, and D_2 is a substrate luminescence band. The most significant change after the RTA is the reduction of D_1 below the detection limit. The DB feature establishes that this CLS technique is sensitive to defect bonding changes at the 10^{12} cm^{-2} level.

6.3. Strain Profiles in Si–SiO_2 Gate Stacks

The results presented above are consistent with the strain profile shown in Fig. 35. There is compressive elastic strain in the SiO_2 region of the hetero-structure after 900 and 1000°C anneals that derives from several different sources: (i) the molar

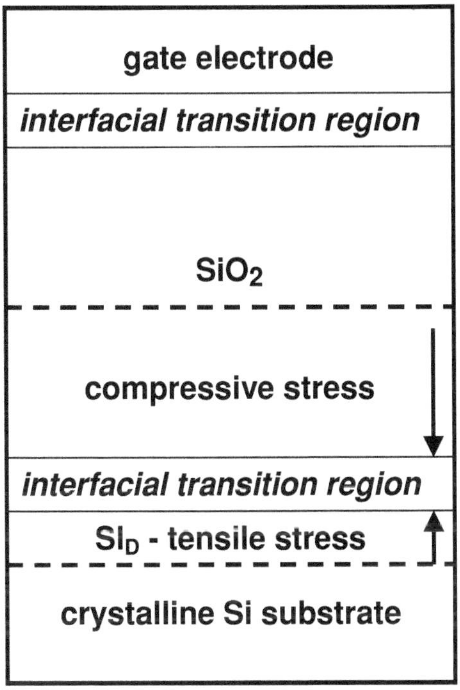

Fig. 35. Strain profile in an SiO_2 gate stack after a 900°C anneal.

volume mismatch at the Si–SiO_2, in which the Si–Si interatomic distance in the silicon substrate is 0.235 nm, and the Si–Si second neighbor distance in the SiO_2 has an average value of $\sim 0.305 \pm 0.005$ nm, with the spread in distances being in part due to a spread of $\sim \pm 20°$ in the Si–O–Si bond angle (34), and (ii) differences in the thermal expansion coefficients between Si and SiO_2, which result in net compressive elastic strain in the SiO_2 film upon cooling down after a high temperature thermal growth or annealing step. The compressive stress in the SiO_2 is compensated by tensile elastic strain in the Si substrate. The ion scattering experiments of Feldman and co-workers in Ref. 72, as well in Ref. 78, indicate strain gradients both the Si substrate and SiO_2 film, respectively. Since the strain changes sign across the metallurgical boundary between the Si substrate and the SiO_2 film, it is important to understand the nature of the strain within the interfacial transition region.

The strain profile in Fig. 35, coupled with the spectroscopic changes at the bonding and defect levels as discussed above, suggest that the 900°C relaxation can be described as a strain induced self-organization. In the next section of the review, this issue is addressed through an application of constraint theory that has been previously applied to both semiconductor dielectric interfaces (19), and to non-crystalline binary alloys such as Ge_xSe_{1-x} and Si_xS_{1-x} (17, 18).

6.4. Microscopic Model for Interfacial Self-organization

As noted Lucovsky and Phillips and coworkers applied the concepts of constraint theory developed to explain glass formation in chalcogenide and oxide glasses (15, 16), to Si-dielectric interfaces (19, 20) in which there was a discontinuity in the both the average number of bonds/atom, designated above as N_{av}, and the average number of valence bonding constraints/atom, designated above as C_{av}. This approach is now used to provide important insights to the physical mechanisms underlying (i) the formation of interfacial transition regions, and (ii) defect formation and defect relaxation at these interfaces (31). This approach builds on the studies of Boolchand and co-workers on the nature of the glass transition (17, 18), and the compositional dependence of the floppy to rigidity transitions that occur in glass forming binary alloy systems including a-$Se_{1-x}Ge_x$.

The studies of Boolchand and coworkers (17, 18) have established that (i) there are two transitions associated with the overall transition from *floppy* or under-constrained bonding in a-Se, to *rigid* or over-constrained bonding in an alloy with 33% Ge, or $GeSe_2$, and (ii) that these transitions span a *self-organized* region that is strain-free (18). The first of these transitions occurs at the onset of average *local bonding rigidity* at a composition of $GeSe_4$. This composition corresponds to an average number of bonds/atom of 2.4, and an average number of bonding constraints/atom of 3, so that alloy compositions with increasing Ge content are over-constrained on average. However, the onset of *global bonding rigidity* is delayed by self-organization into non-statistical bonding arrangements that minimize the total bond-strain. These reorganizations occur up to a composition at which local bond strain percolates throughout the entire volume of the alloy, at which point global bonding rigidity sets in, and the alloy is macroscopically over-constrained.

Lucovsky, Phillips and coworkers in Refs. 20 and 83 pointed out that Si–SiO_2 and Si–Si_3N_4 interfaces were hetero-structures in which in which the substrate Si could be considered as *rigid or over constrained* with the number of bonds/atom equaling exactly four, and the number of valence bonding constraints per atom being greater than the network dimensionality of three. These two dielectrics, SiO_2 and Si_3N_4, have different average numbers of bonds/atom, and valence bonding constraints/atom, with silicon oxynitride alloys, represented by the formula $(Si_3N_4)_x(SiO_2)_{1-x}$ spanning a composition range from (i) an ideal non-crystalline solid for SiO_2 in which the number of bonding constraints per atom is the same as the network dimensionality, three, to (ii) an over-constrained or rigid Si_3N_4 dielectric in which the number of bonding constraints per atom is substantially greater than three.

The discussions above have so far demonstrated that after a thermal anneal at 900°C, there is (i) a region of strained or defective Si in tensile stress in the Si substrate, (ii) a transition region with an average SiO composition that is approximately one molecular layer thick, ~0.3 nm, and (iii) a region in which there is a compressive stain gradient in the stoichiometric SiO_2 dielectric. It is important to note that the phase separation of bulk SiO_x films with $x \sim 1$ phase separate and that the extent of reaction is more than 90% completed at a temperature of 900°C (84). This suggests that the interfacial reaction is also limited by reaction kinetics, but that the driving force for the reaction is a reduction in bond-strain energy.

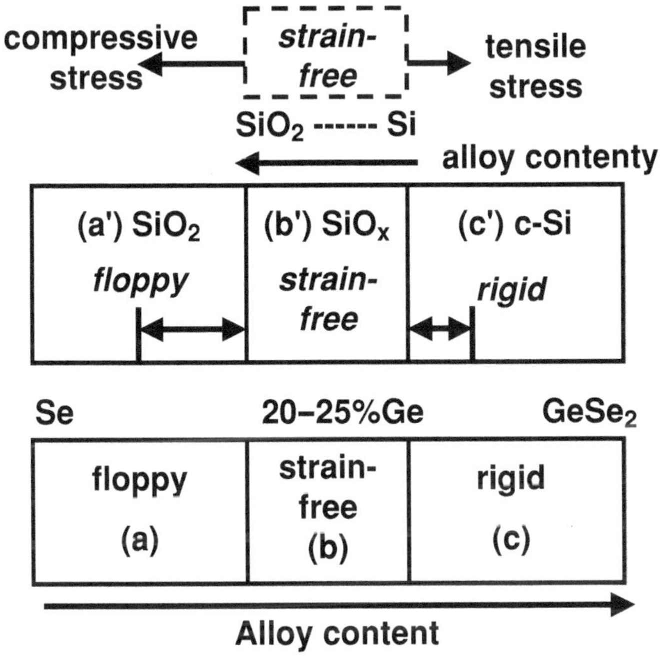

Fig. 36. Schematic representation of stress in floppy, self-organized and rigid regions of an SiO_2–SiO_x–Si structure, and the corresponding regions of the non-crystalline Ge_xSe_{1-x} alloy system in the range from Se to $GeSe_2$ ($x = 0.333$).

Building on an increased understanding of self-organization in binary glass alloys such as Ge_xSe_{1-x} (17, 18) and as discussed above, this has identified a novel approach for understanding the changes in bonding within the Si–SiO_2 interface transition region that occurs after annealing at 900°C. The interfacial transition region must provide a continuous and smooth *transition* between intrinsic and thermally induced tensile macroscopic strain in the Si substrate and compressive macroscopic strain in the SiO_2 layer. Since this transition region *bridges* a rigid silicon substrate, and a *compliant or under constrained* SiO_2 dielectric, this suggests that the interfacial transition region plays the same role as the *strain-free compositional regime* in Ge_xSe_{1-x} alloys, i.e., the bonding changes after the 900°C anneal are in effect a self-organization that prevents the percolation of rigidity in the interfacial region, and thereby provides a strain free region that connects the silicon substrate to the SiO_2 dielectric. This analogy is represented schematically in Fig. 36. If this indeed the case, then it is necessary to establish three things; first, that an interfacial transition region approximately one molecular layer thick with an average SiO bonding can indeed be strain free, second, the scale of self-organization within the transition region, and third, that the transition regions exhibits unique and special properties with respect to defect formation, and defect generation under electrical stress.

The local sub-oxide bonding configurations in the interfacial transition are not known in detail; nevertheless, important aspects of sub-oxide network stress can be inferred from general considerations of the experimental results presented above. The soft X-ray photoemission results indicate that the average sub-oxide composition in the transition region is essentially SiO and has an effective width of ~0.3–0.4 nm (77). The extent and composition of this region have also been determined from the ion scattering experiments with essentially the same result (72). Combining these results indicates that the average Si atom in this transition region forms two Si–Si bonds and two Si–O bonds. The number of bond-bending constraints associated with this average bonding can then be determined from symmetry considerations.

There is one Si–Si–Si bond angle, and one O–Si–O bond angle; each of these symmetrical angles can be constrained at the sp^3 tetrahedral angle of approximately 110°. On the other hand, it is unlikely that the asymmetric O–Si–Si bond angles will be constrained as a consequence of different local energy gaps associated with O–Si and Si–Si bonds. The difference in these gaps can be estimated from the difference of the valence band offset energies between crystalline Si and amorphous SiO_2. This difference is ~4.5 eV. Therefore, these gaps are sufficiently different so that the Pauling bonding resonance responsible for bond-bending forces is expected to be negligibly weak compared to the resonances responsible for the bond-bending forces associated with the symmetric Si–Si–Si and O–Si–O bond angles (26).

Based on this analysis, the number of constraints per SiO formula unit can now be calculated. There are three stretching constraints, two for the Si atom, and one for the O atom, one Si–Si–Si bond bending constraint, and one O–Si–O bending constraint. Finally, there is an additional constraint associated with the angle between the normals to the Si–Si–Si and O–Si–O planes. This gives six constraints per SiO, or equivalently three bonding constraints per atom, matching the atomic degrees of freedom. This leads to the unexpected conclusion that the ultra-thin, interfacial SiO_x layer is effectively strain free, and therefore provides an ideal continuous transition between the tensile and compressive stress respectively of the Si substrate and SiO_2 dielectric. This is reinforces the analogy between the bonding at this hetero-interface between Si and SiO_2 and the composition changes in stress in glass forming alloy systems such as Ge_xSe_{1-x}, as shown in Fig. 36.

The mechanism described above is supported by studies that have demonstrated a chemical phase separation of homogeneous bulk Si sub-oxide films deposited at 300°C, into crystalline Si and non-crystalline SiO_2, after isochronal annealing the range from 500 to 1000°C (84). This separation was first detected by changes and photoluminescence. Photoluminescence was detected in the as-deposited homogeneous sub-oxides, paralleling results of other studies, but was not detected after the 900°C anneal. High resolution transmission electron microscopic imaging, coupled with Fourier transformation infra-red spectroscopy, FTIR, confirmed that the phase separation products were non-crystalline SiO_2 and nano-crystalline Si. The kinetics for this transition were studied by FTIR to determined the extent of the reaction. This study and the results of Keister et al. (77), indicated that that SiO_x was indeed converted into Si and SiO_2, but with one important and significant difference. The SXPS study indicated a one molecular layer of SiO at the interface, the kinetic study of Ref. 84 had insufficient sensitivity to determine if the nano-crystallites of SI had a

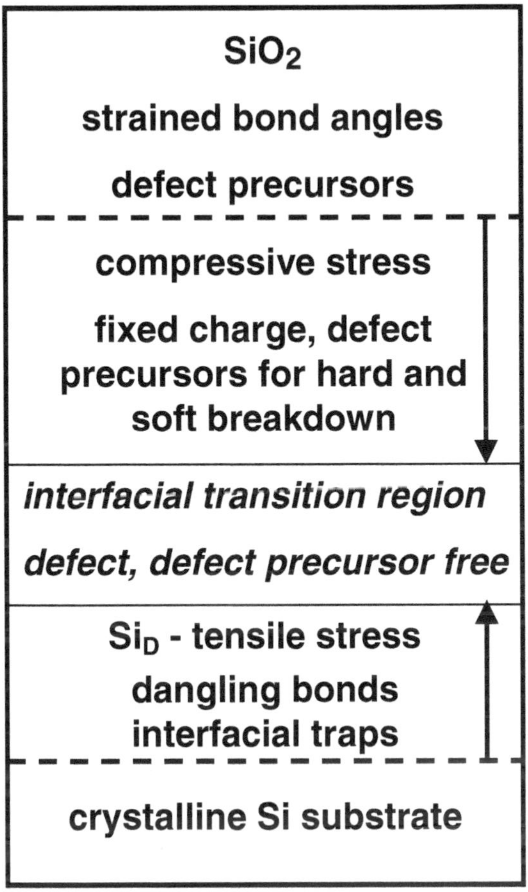

Fig. 37. Strain profile in an SiO_2 gate stack after a 900°C anneal, including the regions in which defect and defect precursors are present.

mono-molecular scale sub-oxide transition region. The absence of photoluminescence suggests a low density of defects at these interfaces, and consistent on a strain-induced self-organization as well.

6.5. Electrical Properties and Reliability Changes from Interfacial Self-organization

Figure 37 includes in a schematic representation the strain profile, defects, and defect precursors that are assumed to be present in an Si–SiO_2 heterostructure that includes a strain-free, self-organized interfacial transition region. Based that special properties with respect to aging, e.g., the absence of defect formation and/or changes in properties based on intrinsic defect precursor arrangements, that chalcogenide alloy glasses and thin films display in the compositional regime between the onset of local bonding-rigidity, and global or percolated bonding rigidity, the analogy in Fig. 28 assumes similar special properties for the interfacial transition regions between the Si substrate and SiO_2 dielectric regions. The assumption of a strain free, self-organized interfacial transition region includes implicitly a scale for in-plane self-organization. This is currently under study in our research group at North Carolina State University; however, the scale can be estimated from the results of electrical measurements on electron and hole transport in channels of their respective FET devices.

Channel mobilities have been extracted from the analysis of the current–voltage characteristics of FET devices using well-known and reliable techniques. The mobility is plotted as a function of the normal filed in the substrate region, based on an assumption that the density of charge carriers in the channel can be determined from capacitance of the SiO_2 dielectric film, i.e., from the value of EOT (85–87). This leads to the remarkable result that electron and hole channel mobilities display universal dependencies that are determined by three physical properties of the interfacial region: (i) the substrate doping that establishes the density of bulk charge defects within the channel region, (ii) a fixed charge, Q_{int}, within the SiO_2 dielectric that contributes to the scattering and determines the peak value of the mobility versus field, and (iii) an interface roughness parameter that defines the field dependence of the mobility in the high-field regime. The universality of electron and hole mobility curves derives from the empirical observation that to within an experimentally-defined uncertainty, the values of Q_{int}, and an empirical roughness factor that includes a characteristic scattering length, λ_{int}, are the same. Q_{int} is approximately 2–5 $\times$ 10^{10} cm^{-2}, whilst λ_{int} is typically 2.5–3.0 nm. The value of λ_{int} is assumed to represent the scale of self-organization within the interfacial transition region, and is the range of dimensions that are associated with strain-driven self-organization.

Prior to post metallization annealing in a hydrogen-containing ambient such as forming gas, MOS devices with Si–SiO_2 interfaces display a defect density of dangling bonds that is in the low 10^{12} cm^{-2} regime (88). These dangling bonds have been studied by electron spin resonance which reveals two interesting properties; first, the electron g-factor displays the same symmetry of the Si substrate, second, their number can reduced by more than a factor of 50 by a post metallization anneal, PMA, in a hydrogen-containing ambient, and third, that contribute to discrete interfacial defect features in the Si forbidden band gap when studied by capacitance–voltage techniques. The totality of these results is consist with these bonding defects being in the Si substrate, and in particular in the tensile strained interfacial region designed in Fig. 29 as Si_D. The density of dangling bond defects prior to the PMAs, but after thermal annealing at 900°C, is approximately constant in the low 10^{12} cm^{-2} regime, and independent of the way a device quality interfaces was formed, e.g., by thermal oxidation at a temperature in excess of 800°C, or by an optimized remote plasma-assisted oxidation. This result, coupled with other results described above is consistent with the precursor bonding arrangements for the interfacial self-organization being inherent in the empirically optimized thermal oxidation processes used throughout the semiconductor industry. The low values of D_{it} obtained, optimized remote plasma assisted oxidation processes described in Refs. 8 and 9, and by other research groups

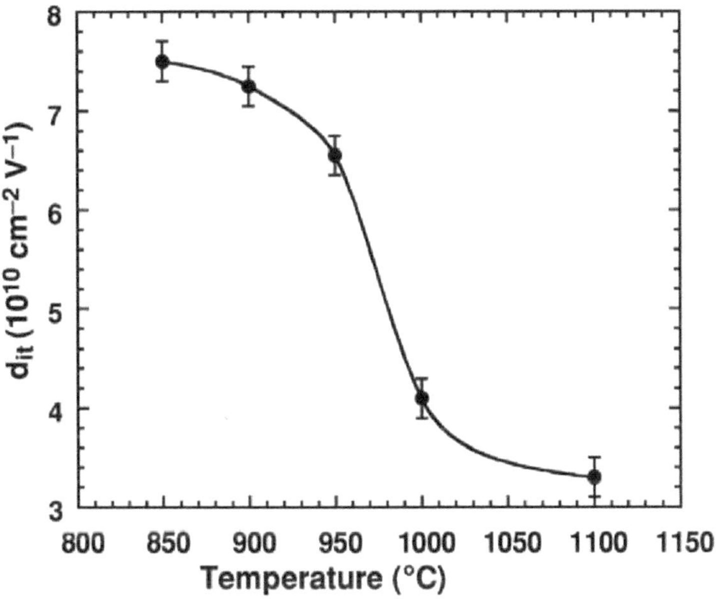

Fig. 38. Values of mid-gap densities of interface traps, D_{it}, extracted from C–V measurements on Si(111) metal-oxide-semiconductor structures as function of processing temperature. The solid line is an interpolation that establishes the trend in the data points.

world-wide, indicates the precursor state for self-organization can be obtained in many different interface preparation processes.

Figure 38 displays the density of interface traps, D_{it}, as a function of annealing temperature at a Si(111)-SiO_2 interface, prepared by thermal oxidation at 850°C, furnace annealed at temperatures up to 1100°C for 30 minutes in Ar, and then subjected to a PMA for 30 minutes at 400°C in forming gas after the initial growth and following each annealing step (75). Values of D_{it} have been determined from analysis of capacitance–voltage, C–V, traces using standard techniques. The plot in Fig. 38 demonstrates that the most significant decrease in D_{it}, defined by numerical differentiation of the trace, occurs at a temperature of ~975°C, very close to the onset of the release of growth induced stress as in Ref. 67. In marked contrast, there is only a small decrease in D_{it} after the 900°C anneal indicating that these defects are not reduced significantly by the atomic rearrangements that occur during the interfacial transition region self-organization. This is consistent with these defects being resident in the Si substrate in the immediate vicinity of the interfacial transition region. This interpretation is also consistent with the universality of the energy dependence of D_{it} respect to the Si valence and conduction band edges, and with the relatively narrow range of D_{it} reported for device-quality interfaces, independent of the processing using to create the interface. In particular, this is exemplified by the lows values of D_{it} for plasma processed and thermally-grown interfaces, subjected to annealing between 900 and

1000°C, and then to standard PMA procedures, forming gas anneals for at least 30 minutes at temperatures between 400 and 450°C.

Experiments further indicate a direct correlation between D_{it} and the duration of remote plasma-assisted oxidation process used to form the Si–SiO_2 interface prior to SiO_2 deposition by remote plasma enhanced chemical vapor deposition (8, 9 and references therein). The longer process, the thicker the plasma oxidized film (8, 9). This process is self-limiting in the sense that the thickness displays a power law dependence on time with a power law factor substantially smaller than one, ~0.2–0.25. In addition, the interface formation process determines the precursor arrangements for the interfacial layer that forms after the 900°C anneal. The equivalence of a 300°C plasma interface formation process that generates ~0.5–0.6 nm of oxide, and a thermal oxidation process at ~850–950°C, with respect to interface self-organization has been established earlier in this chapter. As an example, the effect of longer plasma interface formation processes on the defect precursors that determine device reliability is presented in Fig. 39(a)–(c).

Figure 39(a) demonstrates a six-fold increase in stress induced leakage current (SILC) after a 1000 second electrical constant-voltage stress in a device with a 0.8 nm interfacial relative to an optimized device with a 0.6 nm interfacial oxide. Figure 39(b) and (c) presented plots of time dependent dielectric breakdown, TDDB, as a function

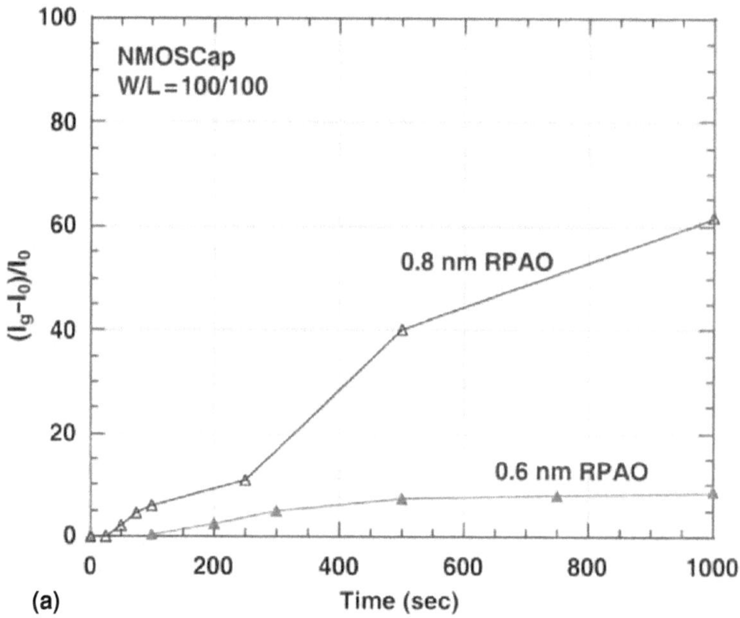

Fig. 39. (a) Stress induced leak current for Si–SiO_x–Si oxynitride, $(SiO_2)_{0.5}(Si_3N_4)_{0.5}$, gate stacks for a different RPAO thickness of ~0.6 and ~0.8 nm. (b, c) Time dependent dielectric breakdown (TDDB) a function of the reciprocal electrical stress field ($1/E$) for Si–SiO_x–Si oxynitride, $(SiO_2)_{0.5}(Si_3N_4)_{0.5}$, gate stacks for a different RPAO thickness of ~0.6 and ~0.8 nm.

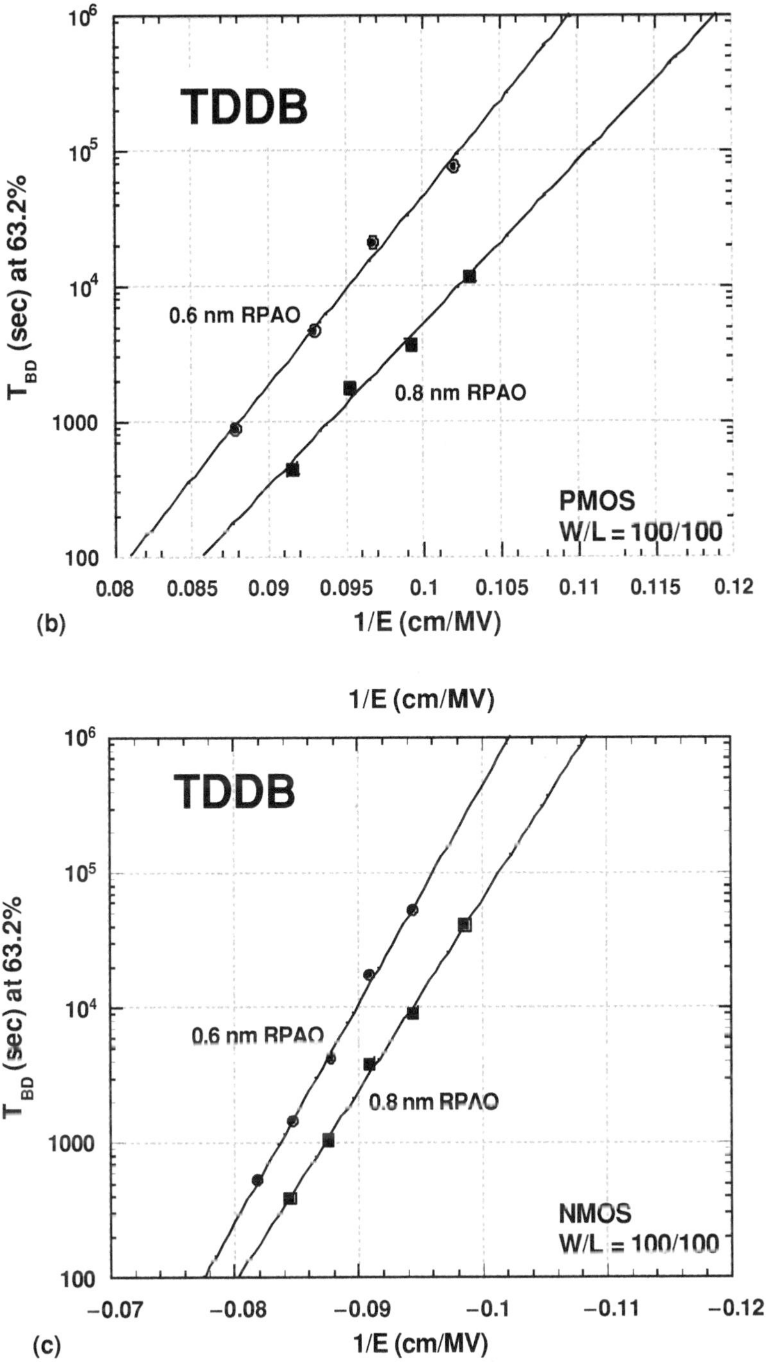

Fig. 39. (*continued*)

of oxide bias for pMOS and nMOS FETs, respectively (89). The traces marked 0.6 nm are for the optimized 300°C plasma interface formation process, and the traces marked 0.8 nm are for a longer plasma process that increases D_{it}, and reduces channel current drive as well. The effect of the longer process time, and thicker interface layer prior to SiO_2 deposition and processing including high temperature anneals and PMAs is to reduce the TDDB time scale by a factor of 10, i.e., to reduce significantly device reliability. These results suggest that interfaces formed by the longer remote plasma oxidation at 300°C create more extensive sub-oxide bonding arrangements than the optimized oxidation process, and that these defects can not be reduced after a 900°C anneal. This is of importance for device processing that requires separate and independent control of interface formation, as for example for devices with deposited high-k dielectrics (6, 7).

Finally, the defects and defect precursors associated with non-optimum interface formation, and strain in the Si substrate and SiO_2 layers have been discussed in a general way with respect to Fig. 37. Based on the results discussed in this section of the papers, it is concluded that (i) Si atom dangling bonds and interface traps, D_{it}, are located in the strained Si substrate region designated as Si_D, and the (ii) the precursor states for soft and hard dielectric breakdown are in the strained SiO_2 film. Due to the relatively large Si–O–Si bond-angle distribution of $\sim\pm 19°$ (22, 34), there is a density of defect precursors in the portions of the SiO_2 that have minimal elastic intrinsic or thermally-generated strain, but have bond angles less than about 130°. These strained bonds contribute to defects and defect precursors as well, and it is not possible to make a persuasive argument for separating defects and/or defect precursors into two groups, one associated with elastric strain, and a second associated with the Si–O–Si bond angle distribution. Si–O–Si bonds with bond angles between 130 and 150° are the least stable as based on their relative binding energy, and are therefore are most easily further strained and by the elastic compress strain in the SiO_2 layers. These sites are candidates for the bond-fission and breakdown. They are also the bonding sites that are chemically attacked by water (90), and reduced in number in the formation of SiOF low-k films (91).

7. DEFECT REDUCTION IN STACKED GATE DIELECTRICS

This section applies the approach of Section 6 to internal interfaces between pairs of dielectric films with markedly different bonding arrangements, including interfaces between SiO_2 and (i) Si_3N_4, and Si oxynitrides, $(Si_3N_4)_x(SiO_2)_{1-x}$, (ii) Al_2O_3, and (iii) transition metal and lanthanide rare earth atom silicate alloys. e.g., $(Zr(Hf)O_2)_x(SiO_2)_{1-x}$. These interfaces are important for device scaling as thermally-grown Si–SiO_2 gate stack constituents are replaced by stacks that include ultra-thin SiO_2 interfacial layers and deposited high-k alternative gate dielectrics (6, 7). The arguments for these replacements have been articulated in other papers dealing with (i) Si oxynitride alloys (3) that have relative dielectric constants, k, between 5 and 7, and (ii) the *so-called* high-k dielectrics where the relative dielectric constants up at

Table 3. Values of N_{av}, $\Delta(N_{av})$ at interfaces with SiO_2, scaled defect densities, $N_{def,cal}$, and experimentally-determined interfacial fixed charge, $N_{def,exp}$

Dielectric	N_{av}	$\Delta(N_{av})$	$N_{def,cal}$ (10^{12} cm^{-2}) ± 10%	$N_{def,exp}$ (10^{12} cm^{-2}) ± 10%
		CRN		
Si_3N_4	3.43	1.03	0.82	0.75
$(Si_3N_4)_{0.5}(SiO_2)_{0.5}$	3.05	0.65	0.26	0.20
		MCRN		
(Al_2O_3)	3.60	1.20	1.2	6.0
(Al_2O_3) interface				
(Al_2O_3)—six-fold Al^{3+}	4.80	2.40	7.0	6.0
$(Zr(Hf)O_2)_{0.25}(SiO_2)_{0.75}$	3.34	0.94	0.66	1.0
$(Zr(Hf)O_2)_{0.5}(SiO_2)_{0.5}$	4.0	1.60	2.5	2.0
$(Zr(Hf)O_2)_{0.25}(Al_2O_3)_{0.75}$	4.03	1.63	3.4	NA
$(Zr(Hf)O_2)_{0.5}(Al_2O_3)_{0.5}$	4.47	2.07	4.8	NA
$(Y(La)_2O_3)_{0.5}(SiO_2)_{0.5}$	3.74	1.34	1.6	NA
		RCP		
$Zr(Hf)O_2$	5.33	2.97	12	4
$Zr(Hf)O_2$ interface				
$(Zr(Hf)O_2)_{0.75}(SiO_2)_{0.25}$	4.67	2.47	7.5	4
$Y(La)_2O_3$	4.80	2.40	7.0	NA

least 20–25 (6, 7). For each of these classes of replacement dielectrics, as k is increased it will demonstrated that there are increases in the average number of bonds/atom, and in the average number of bonding constraints/atom. These increases lead to differences between the average number of bonds/atom in SiO_2 and the replacement dielectrics, resulting in significant densities of electronically-active defects at the between SiO_2 and the replacement dielectrics interfaces (20). A schematic representation of these gate stacks has been shown in Fig. 1.

7.1. Defects at Internal Dielectric Interfaces

Experiments have revealed significant densities of fixed charge at internal interfaces between SiO_2 and alternative gate dielectrics in stacked structures (6, 7, 42, 92–94). These results are summarized in Table 3. The experimentally determined densities of fixed charge, $N_{def,exp}$ in Table 3 have been obtained from room-temperature capacitance–voltage, $C–V$, measurements using conventional techniques such as plotting the flat-band voltage, V_{fb}, as a function of the EOT contribution from the alternative dielectric/high-k constituent . Additional contributions due to systematic shifts of V_{fb} as a function of EOT can arise from the filling of interface traps, D_{it}, and from charge injection, Q_{inj}, as detected by hysteresis (6, 42, 93). The values in Table 3 for $N_{def,exp}$, implicitly include D_{it} and Q_{inj}; however, analyses of $C–V$ data on p-type and n-type substrates, and as function of temperature have demonstrated that the values of D_{it} and Q_{inj} that have been included in Table 3 are about one order of magnitude smaller than the fixed charge Q_f, so that $Q_f \approx N_{def,exp}$.

7.2. Application of Constraint Theory to Internal Dielectric Interfaces

In network amorphous solids the bonding coordinations, m, of the constituent atoms are typically 2, 3 and 4, and the local bonding arrangements are non-planar (14, 15, 17). Under these conditions, the number of bond-stretching constraints/atom, $C_{av,st}$, is given by

$$C_{av,st} = \frac{m}{2}, \tag{6}$$

and the number of bond-bending constraints/atom, $C_{av,b}$, is given by

$$C_{av,b} = 2m - 3. \tag{7}$$

If the bonding geometry of one of three- or four-fold coordinated atoms is planar, then one bending constraint/planar bonded atom is broken, and the number of bending constraints is reduced accordingly, so that.

$$C_{av,b} = 2m - 3 - \frac{n}{N_{atom}}, \tag{8}$$

where n is number of atoms in a planar bonding configuration, 4 for the nitrogen atoms in Si_3N_4, and N_{atom} is the total number of atoms in the chemical formula representation of the solid, 7 for Si_3N_4.

For non-planar bonding, there is a linear relationship between the total number of valence bonding constraints/atom, C_{av}, and N_{av}, given by

$$C_{av} = 2.5N_{av} - 3. \tag{9}$$

C_{av} is also proportional to N_{av} when the bonding is planar at the three- or fold-coordinated bonding sites,

$$C_{av} = 2.5N_{av} - 3 - \frac{n}{N_{atom}}. \tag{10}$$

A different linear relationship also applies between C_{av} and N_{av} when the network includes terminally-bonded atoms such as hydrogen (17); however, none of the dielectrics of this chapter include this type of local bonding arrangement.

The condition for an ideal strain-free CRN is that C_{av} is equal to the network dimensionality of three. This condition for $C_{av} = 3$, corresponds to a value of 2.4 for N_{av}, when the bonding arrangements are non-planar, and this accounts for the excellent glass formation in chalcogenide amorphous semiconductors such as As_2S_3 and As_2Se_3, and chalcogenide alloys such as $Ge_{0.2}S_{0.8}$ and $Ge_{0.2}Se_{0.8}$ (14, 15, 17). $N_{av} = 2.67$ for SiO_2 so that substituting into Eq. (4) yields a value of $C_{av} = 3.67$, greater than the network dimensionality of 3.0. This suggests that SiO_2 should be strained, and therefore have a significant number of intrinsic bonding defects. However, as noted above, as the direct result of a broad Si–O–Si bond-angle distribution, $\sim 150 \pm 20°$ (22), the bond-bending for force constant is unusually small, and one bond-bending constraint associated with the Si–O–Si bonding is broken (17, 22). This removal of one bonding constraint per oxygen atom reduces the average number of bonding constraints/atom for SiO_2 to 3.0, thereby accounting for its outstanding glass formation properties, and its low density of electronically-active defects, $\sim 1 - 5 \times 10^{10}$ cm^{-2} or equivalently $0.4 - 1.1 \times 10^{16}$ cm^{-3}. This makes SiO_2 an excellent reference material to using in scaling relationships in which the value of N^*_{av} is set equal to 2.4 so as to reflect the broken bond-bending constraint at the oxygen atom bonding sites.

Constraint theory has been shown to provide a remarkably accurate description of network stress in non-ideal CRNs in which $C_{av} > 3$, and its consequences with respect to defect formation (20). The application of constraint theory to bonding defects is based on the simple idea that the bonding forces in a network amorphous solid can be arranged in a hierarchy from stronger-stretching to weaker-bending valence forces. The constraining effects of these forces are a linear function of the average coordination number, N_{av}. For over-constrained networks such as Si_3N_4 for which $N_{av} = 3.43$, Si-atom and N-atom bond-stretching constraints are stronger than the respective Si- and N-atom bond-bending constraints, so that strain energy accumulates along these bending constraints. The most significant accumulation of strain is at the atomic site with the lower coordination number, in this instance nitrogen. Since the bonding geometry of the three-fold coordinated nitrogen atoms is planar, the value of C_{av} as determined from Eq. (5) is 5.0, predicting bonding distortions in the form of bond angle strain at the N-atom bonding sites. This means that the average Si–N–Si bond angle, θ is distorted from the ideal local value of 120° by an amount $\delta\theta$, which is proportional is the difference between N_{av} in the non-ideal strained network and $N^*_{av} = 2.4$ of an ideal, strain free network,

$$\delta\theta \propto [N_{av} - N^*_{av}]. \tag{11}$$

It is further assumed that defect density is associated with broken bonds that relieve local strain build-up. As such this density is expected to be proportional to the strain energy (20), which is proportional to $[\delta\theta]^2$. Therefore, the density of defects, D, in a constrained network is expected to obey the following scaling relationship,

$$D \propto [N_{av} - N^*_{av}]^2. \tag{12}$$

A similar scaling relationship is assumed at the internal interface between two dielectrics. The density of interface defects, designated as N_{def}, is then proportional to the square of the difference in the average number of bonds/atom of the two dielectrics, A and B, that define the interface,

$$N_{def} \propto (N_{av}(A) - N_{av}(B)]^2 = (\Delta(N_{av})]^2, \tag{13}$$

where $[\Delta(N_{av})]$ is given by $N_{av}(A) - N_{av}(B)$. Equation (15) provides the definition for an empirically-defined scaling parameter, N^*_{def}, that will be used to compare the predictions of scaling theory to experiment:

$$N_{def} = N^*_{def}[\Delta(N_{av})]^2, \tag{14}$$

The empirical parameter, N^*_{def}, has been estimated by using the experimental data for SiO_2–Si_3N_4 interfaces (19), this gives a density of fixed charge at this interface of

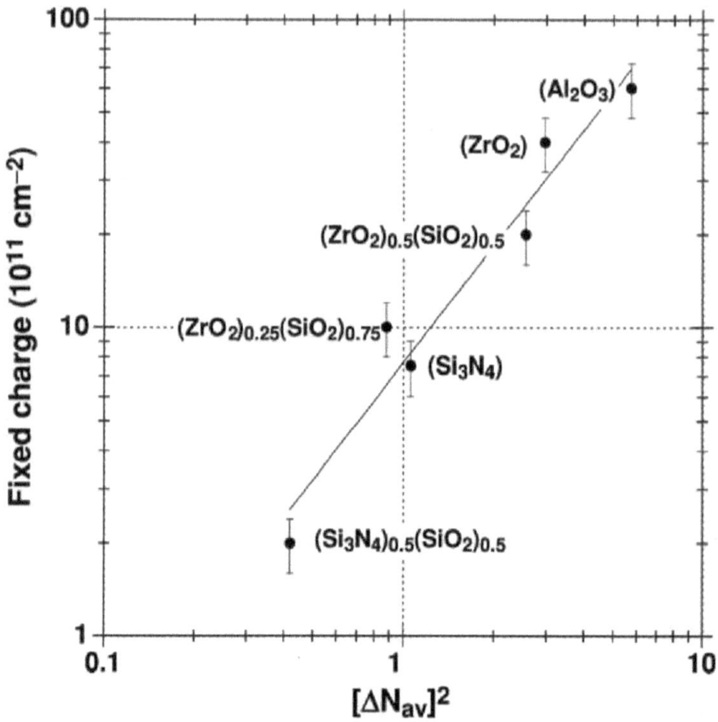

Fig. 40. Log–Log plot of the interfacial defect density, N_{int}, as a function of D^* $[\Delta(N_{av})]^2$. The slope of this plot is 1.25 ± 0.1, differing by a about 25% from the linear modeling value of 1.0.

$7.5 \pm 0.75 \times 10^{11}$ cm^{-2}. Using values of 3.43 for N_{av}(B) for Si_3N_4, and an effective value of 2.4 for N_{av}(A) for SiO_2, and this gives a value for N^*_{def} of $7.1 \pm 0.7 \times 10^{11}$ cm^{-2}.

The validity of the empirically-parameterized scaling relationship in Eq. (14) has been tested by plotting experimentally determined values of fixed charge, N_{def}, as a function of the $[\Delta(N_{av})]^2$, and determining the power law dependence parameter, and the scaling coefficient, N^*_{def}. This yields a value of $7.7 \pm 0.7 \times 10^{11}$ cm^{-2} for N^*_{def}, and a power law factor of 2.5. The plot in Fig. 40 assumes that the bonding discontinuity at the SiO_2–Al_2O_3 interface dominated by Al^{3+} ions, rather than the average bonding that includes a network as well as ionic bonding component. In a similar way the data point for SiO_2–ZrO_2 interfaces with SiO_2, assumes a ZrO_2 rich silicate layer at the internal dielectric interface. Similar results are predicted for HfO_2. Additionally, the insertion of an interfacial silicate is consistent with soft X-ray XPS studies of HfO_2 interfaces with SiO_2 that indicate an interfacial silicate layer forms (94).

Defects at the SiO_2–Si_3N_4, SiO_2–Si silicate alloy interfaces can not be reduced at annealing temperatures up to 1000°C (19), the temperatures required for dopant activation in ion implanted crystalline Si source and drain contacts, and in polycrystalline Si gate electrodes, whereas defects at HfO_2–SiO_2 interfaces can be reduced by at least an order of magnitude after annealing at 900°C (96).

Table 3 also includes calculated values determined from the empirical scaling relationship in fitting the data in Fig. 40. These values have been obtained by comparing measured flat band voltages with those obtained from estimates based on the doping of the Si substrate, and the work function of the gate metal atom (6, 7, 42, 93, 94).

7.3. Strain-Induced Self-organization at Internal Dielectric Interfaces

Several conditions are necessary for self-organization at strained interfaces during a *moderate* temperature anneal at ~600–800°C. These have been addressed above for Si–SiO_2 interfaces. These conditions are (i) that precursor bonding environments are consistent with a self-organization that reduces the total energy with a significant contribution coming from strain-energy reduction, and (ii) that bond-breaking and atomic rearrangements take place at annealing temperatures that are consistent with limitations imposed by other processing constraints, e.g., the melting temperature of Si, the decomposition temperature of SiO_2 into Si and SiO at the Si–SiO_2 interface and/or the chemical and/or micro-structural phase separation of the *bulk* dielectric film. In addition it is important that there be no equilibrium phases with congruent melting points between the end-members of the chemical composition (97), as for example Si and SiO_2 for the chemical phase separation of Si sub-oxides, SiO_x, $x < 2$, at the Si–SiO_2 interface (84). The existence of such a phase would change the end-products in the interfacial self-organization, and either not result in a significant reduction in strain energy, or impede the kinetics for the self-organization. The conditions for chemical phase separation are met for 900°C interfacial anneals for Si–SiO_2 interfaces (84), and the atomic rearrangements associated with these anneals have been confirmed by soft X-ray XPS at the interface bonding level of $1 - 10 \times 10^{14}$ cm^{-2} (77), and the by cathode-luminescence at the defect bonding level of 10^{12} cm^{-2} (78). The temperature of this interfacial relaxation is approximately 100°C lower than the temperature for the onset of visco-elastic relaxation of bulk film elastic strain (78).

These criteria for an interfacial chemical the absence of self-organization is also consistent with the results for SiO_2–Si_3N_4 and SiO_2–Si oxynitride interfaces. These interfaces display no defect reduction for annealing and/or processing temperatures up to 1000°C (19), consistent with the observation that there is a phase, $SiON_2$ with a congruent melting point in excess of 2000°C between SiO_2 and Si_3N_4 (97), that will impede the kinetics for a self-organization, strain-relief mechanism to prevail.

The existence of a relatively low temperature chemical phase separation of SiO_x into Si and SiO_2, and the absence of the low temperature chemical phase separation at internal dielectric interfaces that include either Si_3N_4 or a Si oxynitride in contact with SiO_2, suggest that similar relationships should prevail between (i) strain-drive self-organization of transition metal and lanthanide rare earth silicate and aluminate alloy thin films in contact with SiO_2 and (ii) features in their equilibrium phase diagrams.

This is the case for ZrO_2 and HfO_2 and their respective silicate alloys in contact with SiO_2. In these instances the interface bonding is essentially the same as that in a silicate alloy with up to about 50% ZrO_2 or HfO_2 content. The equilibrium phase diagrams for SiO_2 and ZrO_2, and SiO_2 and HfO_2 indicate a stable silicate phases without congruent melting points, and liquidus curves that are consistent with the spinoidal decomposition of silicate alloys into the end-member oxides (97). Zr and Hf silicate also alloys display a bulk film chemical phase separation at temperatures of at most 900–1000°C, suggesting that an interfacial relaxation could occur at temperatures of at most 800–900°C, and possibly less, especially if the effective ZrO_2 or HfO_2 concentrations are higher than about 50%. This expectation has been realized in devices including HfO_2 and Hf silicate alloys, where fixed charge has been reduced by more than an order of magnitude for annealing temperatures about 700–800°C (85). In contrast devices with Zr and Hf silicate alloys annealed at 500°C do not show defect reduction (2, 93).

In contrast, and also consistent with the equilibrium phase diagram differences between Al_2O_3–SiO_2, and ZrO_2–SiO_2 and HfO_2–SiO_2, there is no reduction of defects at SiO_2–Al_2O_3 interfaces up to temperatures of at least 900°C where the Al_2O_3 films undergo a crystallization. This is consistent with the existence of a compound composition with a congruent melting point in excess of 2000°C between SiO_2 and Al_2O_3 (88). It is also significant that the magnitude of the fixed charge is well above what is anticipated on the base on the $N_{av}(B) - N_{av}(A)$ difference (see Table 3). However, agreement between the scaling is improved if all of the interfacial Al is assumed to be six-fold coordinated, and top exist as Al^{3+} ions. This internal interface is being subjected to additional studies in an attempt to determine the chemical bonding arrangements of the Al atoms.

Compound phases with congruent melting points, such as $(Y_2O_3)_1(SiO_2)_2$, and $(Y_2O_3)_2(SiO_2)_1$ exist in the phase diagrams for the group IIIB silicates of Y and La, and for the lanthanide rare earth elements in trivalent bonding states (96). This means that internal interfaces between interfacial SiO_2, and either the oxides or silicate alloys of the group IIIB and lanthanide rare earth atoms are more than likely not to undergo a self-organization that is driven by bond strain-reduction, and therefore will display levels of fixed charge, typically at the 10^{12} cm^{-2} level. They will therefore not meet targeted performance and reliability as required in aggressively-scaled CMOS devices. This estimate of fixed charge is based on the values of $[\Delta(N_{av})]$ included in Table 3.

Finally, the phase diagrams for ZrO_2–Al_2O_3 and HfO_2–Al_2O_3 do not indicate compound compositions between Al_2O_3 and the respective transition metal oxides (100, 101). Nor due they reveal liquidus curves indicative of stable or incipient liquid immiscibility, and therefore a driving force for spiniodal decomposition. In addition the eutectic compositions in these systems are in the mid-alloy range, and at a temperature at least 100°C higher than in the phase diagrams for the respective phase diagrams with SiO_2. These systems will be investigated by us to determine the magnitude of the interfacial charge, as well as the effectiveness of annealing up to at least 900–1000°C in reducing the magnitude of the interfacial fixed charge. In this regard, studies of Hf aluminate alloys with alloy compositions in the range of 35–50% HfO_2

indicated chemical phase separation at temperatures in excess of 900–1000°C (42). Electrical studies of were made on capacitors prepared from these alloys, subjected to post deposition annealing at 800–900°C prior to Al metallization. These devices showed significant levels of fixed negative charge, similar to those reported for Al_2O_3 devices. These levels were found in both NMOS and PMOS capacitors, confirming that the charge was fixed, and not dominated by electron injection and trapping. This suggests that the interfaces of these devices were Al_2O_3-like, and that the kinetics for a strain-driven self-organization that would reduce fixed charge was not possible for the range of annealing temperatures explored.

8. DISCUSSION

This chapter has applied basic studies of electronic structure and chemical bonding to identify two of the most important considerations for identifying high-k gate dielectrics that have the potential to meet the aggressive scaling targets of semiconductor industry roadmaps, e.g., the ITRS (1). These are (i) anticipated reductions in direct tunneling that are expected from increases in k and hence thickness relative to SiO_2 but are mitigated in large part by reductions in E_B and m^*, and (ii) defects at internal dielectric interfaces, as the composite stacks of Fig. 1.

8.1. Direct Tunneling Reductions: Decreases in Conduction Band Offset Energies and Effective Masses due to d*-State Conduction Bands

This limitation derives from the fundamental electronic structure of transition metal and rare earth lanthanide series oxide dielectrics in which the lowest conduction band states are localized d*-states. The energies of these states relative to the top of the valence band in oxides, and silicate alloys are significantly less than those of the extended s* states in SiO_2 and other non-transition metal and rare earth oxides (e.g., Al_2O_3), and as such, the transition metal and rare earth oxide-based dielectrics have significantly reduced conduction band offset energies, E_B. As such the only candidate oxides, and silicate and aluminate alloys remaining as viable candidates as replacement dielectrics are those of Hf, and Zr, Y, La and the lanthanide trivalent rare earths. This list may be extended to included complex mixed oxides comprised of mixtures of Tm and Re (Tm′) oxides in which d-state mixing promotes conduction band offset energies greater 1 eV, and preferably greater than 1.5 eV. However, this now seems unlikely because of reductions in the local site symmetries of the Tm and Re (Tm′) atoms that may be required to promote local bonding relaxations that reduce the strain energy below that of an ordered bonding arrangement with higher site symmetries.

In addition low conduction band offset energies are also generally accompanied by low tunneling electron masses, since they both derive from the same intrinsic aspects of the electronic structure, i.e., the fact that the lowest conduction band states have anti-bond d*-state character. As reflected in the tunneling figure of merit, Fm, the combination of these mitigates increases in thickness associated with increased k, and results in many order of magnitude smaller decreases in the tunneling than what would be expected from increases in the dielectric constant, and reflected in an increased physical thickness as well.

8.2. Interfacial Defects due to Inherent Differences in Chemical Bonding and Bond Ionicity

This chapter has demonstrated that fixed charge in stacked gate dielectrics comprised of interfacial SiO_2, and alternative gate dielectrics including (i) Si_3N_4 and Si oxynitride alloys, (ii) A_2O_3, (iii) group IVB and IIIB transition metal, and lanthanide rare earth atom oxides, and silicate and aluminate alloys, derives from bond-strain at their internal dielectric interfaces. Constraint theory as applied to dielectrics with CRN and modified CRN amorphous morphologies indicates that the average number of bonding constraints/atom scales linearly with the average number of bonds/atom, and that electronically-active defects result when the average number of bonding constraints/atom is greater than the network dimensionality. Since strain energy scales with the square of the bond-angle strain, $[\delta\theta]^2$, relative defect concentrations in the thin film dielectrics, and at their interfaces scale with bond angle strain energy, and hence are proportional to the square of the difference of the average number of bonds/atom relative to a low defect density standard CRN dielectric, e.g., SiO_2. Following these arguments, it has been shown that defects responsible for fixed charge levels ranging from the mid-10^{11} cm^{-2} to mid-10^{12} cm^{-2} range scale with the square of the step in the average number of bonds/atom between SiO_2, and the alternative dielectric.

This chapter also identifies a mechanism for bond-strain relief by a strain-driven interface self-organization, which has material and interface specific applications. Paralleling results presented in Section 7, self-organization is restricted to systems in which chemical phase separation into SiO_2 and an end-member elemental nitride, or oxide is possible. One condition for this in mixed oxide systems in that there are no silicate or aluminate phases that have congruent melting points. The paper distinguishes between four different internal interfaces. First, Si–Si_3N_4 and Si–Si oxynitride alloy interfaces at which strain relief does not occur up to processing temperatures of at least 1200°C, and at which defect densities are sufficiently low not to degrade performance and reliability in high power applications with EOT extending to about 1.1–1.2 nm. Second, SiO_2–Al_2O_3 at which Al_2O_3 crystallizes at temperature of ~900°C, and at a temperature below which self-organization occur; additionally in which there is a compound aluminio-silicate phase with a congruent melting points that would impede an interfacial self-organization into SiO_2 and Al_2O_3. Defect densities at SiO_2–Al_2O_3 interfaces are in excess of 5×10^{12} cm^{-2}, and require relatively thick interfacial layers of SiO_2 to mitigate the effects of fixed charge on channel transport. Stated differently, these values of fixed charge are too high for device applications in which EOT must be scaled to less than 2 nm. The third group is the group IVB silicates, in particular Zr and Hf silicates. These display a strain-driven self-organization at temperatures less than about 800°C, and densities of interfacial fixed charge have been reduced by more than one order of magnitude permitting EOT scaling to at least 0.8–1.0 nm. This self-organization is consistent with the respective binary oxide equilibrium phase diagrams in which the compound silicate phases, $ZrSiO_4$ and $HfSiO_4$, do not have congruent melting points. In addition, the liquidus curve for the ZrO_2–SiO_2 system displays a stable liquid immiscibility characteristic, and the curve for HfO_2–SiO_2 displays an incipient liquid immiscibility characteristic, each of which are necessary prerequisites for an interface self-organization. Additionally, based on scaling arguments, and a limited set of measurements, group IIIB, Y and La, and lanthanide rare earth atom oxides, and their silicate and aluminate alloys form a fourth groups that are not expected to display strain-driven self-organization at their internal interfaces with SiO_2. The determinant factors are a multiplicity of compound silicate and aluminate phases with congruent melting points that exist for all of the specific systems studied to date.

Finally, results to date on group IV transition metal aluminate systems have not displayed any indication of defect reduction via interfacial strain-relief mechanisms that promote self-organization. In contrast, studies of Ta aluminate devices have displayed significant injection into low lying Ta d*-states, disqualifying them for device applications, and Hf aluminate devices display high-densities of interfacial traps, also raising questions relative to device applications.

The results of this chapter, therefore have served to identify two important limitations for aggressively scaled devices in general. The first is the necessity for an ultra-thin interfacial SiO_2 layer into be contact with the Si substrate in which channel transport occurs. The 900°C interfacial relaxation provides an interfacial bonding structure that is responsible for continuance of scaling metrics, including low densities of D_{it}, excellent channel transport mobilities of electrons and holes, and interface-limited/determined device reliability. This places a limit on attainable scaling of EOT. It has been argued above that this limit may be in range of 0.8–1.0 nm, and not as low as ITRS targets, that extend to at least 0.5 nm (13).

The second limitation is on the actual alternative gate dielectrics that have a chance of working. There are two possibilities. The first is in devices based on the optimized Si oxynitride alloys of Ref. 5, in which the nitrogen profile controlled at the atomic level (30, 31). These devices have the potential to exhibit tunneling leakage currents of (i) <5 A-cm^{-2} at approximately a 1 V of oxide bias for EOT between 1 and 1.2 nm, and (ii) $\sim 10^{-2}$ A-cm^{-2} at approximately a 1V for oxide bias for EOT of $\sim$1.3–1.5 nm. These currents/EOTs are predicated on an interfacial monolayer nitride silicon oxide region that contributes approximately 0.35 nm to EOT, and which is sufficient to preserve channel transport properties, defect densities, and reliability similar to those of SiO_2 devices, i.e., meeting ITRS scaling metrics (1). Although these devices will operate with doped polycrystalline Si gate electrodes, performance in bulk CMOS devices would be considerably improved with dual metal gate electrodes.

The second class of dielectrics includes HfO_2 and ZrO_2 and their silicate alloys. There are two important considerations that must be integrated into process integration approaches. First, like the example presented above for optimized Si oxynitride alloy devices, it is necessary to include an interfacial monolayer nitride silicon oxide region that contributes approximately 0.35 nm to EOT, so that ITRS performance and reliability scaling metrics can be met. Second, it will necessary to subject both dielectric interfaces, the Si–SiO_2 interface, and the internal SiO_2-high-k interface, to thermal annealing at temperature of approximately 900°C to promote strain-induced self-organization at each of these interface.

Finally, it will be necessary to quantify the effects of remote phonon scattering on channel transport (102), and in particular to determine if dual metal gate electrodes can effectively screen channel carrier-phonon coupling and yield current drive meeting ITRS metrics for bulk CMOS devices. Preliminary results by the Intel group are encouraging, suggesting that metal gate electrodes are effective in screening remote phonon scattering, but more complete data, and analyses are required (96).

ACKNOWLEDGEMENTS

The author acknowledges collaborations with Professors Jerry Whitten, Jon-Paul Maria, Robert Nemanich, Harald Ade and Dave Aspnes at North Carolina State University, Jack Rowe of the University of North Carolina at Chapel Hill, and Hisham Massoud of Duke University, and the research contributions of his research assistants and postdoctoral fellows who appear in the citations. The author acknowledges Jim Phillips, retired from AT&T Bell Laboratories for his collaboration and encouragement in the application of constraint theory to Si and dielectric interfaces. Finally, the author acknowledges support from the Office of Naval Research (ONR) the Air Force Office of Scientific Research (AFOSR), the Semiconductor Research Corporation (SRC) the National Science Foundation (NSF) and the SRC/International SEMATECH (ISMT) Front Processes Center.

REFERENCES

1. "*International Technology Roadmap for Semiconductors*" (2001 ed.) at http://public.itrs.net.
2. G. Lucovsky, *Microeletronics Reliability* **43**, 1417 (2003).
3. G. Lucovsky, *IBM J. Res. Develop.* **43**, 301 (1999).
4. H. Yang, G. Lucovsky, *1999 IEDM Digest of Technical Papers*, p. 245.
5. T.P. Ma, *IEEE Trans. Electron Devices* **45**, 680 (1998).
6. G.D. Wilk, R.W. Wallace, J.M Anthony, *J. Appl. Phys.* **87**, 484 (2000).
7. G.D. Wilk, R.W. Wallace, J.M Anthony, *J. Appl. Phys.* **89**, 5243 (2001).
8. H. Niimi, G. Lucovsky, *J. Vac. Sci. Techol. A* **17**, 3185 (1999).
9. H. Niimi, G. Lucovsky, *J. Vac. Sci. Techol. B* **17**, 2610 (1999).
10. R. MacFadden, J. Kavalieros, R. Arghavani, D. Barlage, R. Chau, US Patent 6,610,615, *Plasma nitridation for reduced leakage gate dielectric layers*, issued August 26, 2003.
11. S.V. Hattangady, H. Niimi, G. Lucovsky, *Appl. Phys. Lett.* **66**, 3495 (1995).
12. R. Zallen, *The Physics of Amorphous Solids* (John Wiley and Sons, New York, 1983), Chapter 2.
13. L. Robertson, S. Moss, *J. Non-Cryst. Solids*, **106**, 330 (1988).
14. J. Neufeind, K.-D. Liss, *Bur. Bunsen Phys. Chem.* **100**, 1341 (1996).
15. J.C. Phillips, *J. Non-Cryst. Solids* **34**, 153 (1979).

16. J.C. Phillips, *J. Non-Cryst. Solids* **43**, 37 (1981).
17. P. Boolchand, *Insulating and Semiconducting Glasses* (World Scientific, Singapore, 2000), p. 191.
18. P. Boolchand, D.G. Georgiev, M. Micoulaut, *J. Optoelectronics Adv. Mater.* **4**, 823 (2002).
19. G. Lucovsky, H. Yang, H. Niimi, J.W. Keister, J.E. Rowe, M.F. Thorpe, J.C. Phillips, *J. Vac. Sci. Techol. B* **18**, 1742 (2000).
20. G. Lucovsky, Y. Wu, H. Niimi, V. Misra, J.C. Phillips, *Appl. Phys. Lett.* **74**, 2005 (1999).
21. D.L. Griscom, *The Physics of SiO_2 and its Interfaces*, ed. S.T. Pantelides (Pergammon Press, New York, 1978), p. 232.
22. J.L. Whitten, Y. Zhang, M. Menon, G. Lucovsky, *J. Vac. Sci. Techol. B* **20** 1710 (2002).
23. V. Misra, H. Lazar, Z. Wang, Y. Wu, H. Niimi, G. Lucovsky, J.J. Wortman, J.R. Hauser, *J. Vac. Sci. Techol. B* **17** (4), 1836 (1999).
24. G.B. Rayner, Jr., D. Kang, Y. Zhang, G. Lucovsky, *J. Vac. Sci. Techol. B* **20**, 1748 (2002).
25. R. Kerner, J.C. Phillips, *Solid State Commun.* **117**, 47 (2001).
26. L. Pauling, *The Nature of the Chemical Bond*, 3rd Edition (Cornell University Press, Ithaca, NY, 1936).
27. R.T. Sanderson, *Chem. Bonds and Bond Energy* (Academic Press, New York, 1971).
28. G. Lucovsky, *J. Vac. Sci. Techol. A* **19**, 1553 (2001).
29. B. Rayner, H. Niimi, R. Johnson, R. Therrien, G. Lucovsky, F.L. Galeener, *AIP Conf. Proc.* **550**, 149 (2001).
30. G. Lucovsky, H. Yang, H. Niimi, J.W. Keister, J.E. Rowe, M.F. Thorpe, J.C. Phillips, *J. Vac. Sci. Techol. B* **18**, 1742 (2000).
31. G. Lucovsky, J.C. Phillips, *J. Phys. A* (2004), in press.
32. R.L. Mozzi, B.E. Warren, *J. Appl. Cryst.* **2**, 164 (1969).
33. M. O'Keeffe, B.G. Hyde, *Acta Crystallogr. B* **34**, 27 (1978).
34. G. Lucovsky, T. Mowrer, L.S. Sremaniak, J.L Whitten, *J. Non-Cryst. Solids* **155**, 338–40 (2004).
35. T. Mowrer, G. Lucovsky, L.S. Sremaniak, J.L Whitten, *J. Non-Cryst. Solids* **543**, 338–40 (2004).
36. J.G. Hong, Ph.D. Dissertation, North Carolina State University (2003).
37. J. Robertson, C.W. Chen, *Appl. Phys. Lett.* **74**, 1164 (1999).
38. J. Robertson, *J. Vac. Sci. Techol. B* **18**, 1785 (2000).
39. V.V Afanas'ev, A. Stesmans, *High-k Gate Dielectrics*, ed. M. Houssa (Institute of Physics, Bristol, 2003), p. 217.
40. Appendix A, *High-k Gate Dielectrics*, ed. M. Houssa (Institute of Physics, Bristol, 2003), p. 597.
41. R.S. Johnson, J.G. Hong, G. Lucovsky, *J. Vac. Sci. Techol. B* **19**, 1606 (2001).
42. R.S. Johnson, J.G. Hong, C.L. Hinkle, G. Lucovsky, *J. Vac. Sci. Techol. B* **20**, 1126 (2002).
43. Fulton, unpublished.
44. Yu Zhang, unpublished.
45. P.A. Cox, *Transition Metal Oxides* (Oxford Science Publications, Oxford, 1992), Chapter 2.
46. E.U. Condon, G.H. Shortly, *The Theory of Atomic Spectra* (Cambridge University Press, Cambridge, 1957), Chapter V.
47. G. Lucovsky, G.B. Rayner Jr., D. Kang, G. Appel, R.S. Johnson, Y. Zhang, D.E. Sayers, H. Ade, J.L. Whitten, *Appl. Phys. Lett.* **79**, 1775 (2001).
48. *X-ray Data Handbook*, ed. A. Thompson et al. (Lawrence Berkeley National Laboratory, University of California, Berkeley, CA, 2001).
49. P. Lim, I.-W. Chen, J.E. Penner-Han, *Phys. Rev. B* **48**, 10063 (1993-II).
50. L.A. Grunes, R.D. Leapman, C.N. Nilker, R. Hoffman, A.B. Kunz, *Phys. Rev. B* **25**, 7157 (1983).
51. L.A. Grunes, *Phys. Rev. B* **27**, 2111 (1983).

52. N.A. Stoute. D.E. Aspnes, G. Lucovsky, unpublished.
53. V.V. Afanas'ev, A. Stesmans, *High-k Gate Dielectrics*, ed. M. Houssa (Institute of Physics, Bristol, 2003), p. 217.
54. *X-ray Absorption: Principles, Applications, Techniques of EXAFS, SEXAFS and XANES*, eds. D.C. Koningsberger, R. Prins (Wiley, New York, 1988).
55. S.-G. Lim et al., *J. Appl. Phys.* **91,** 4500 (2002).
56. G.B. Rayner, D. Kang, G. Lucovsky, *J. Vac. Sci. Techol. B* **21**, 1783 (2003).
57. H.H. Tippins, *J. Phys. Chem. Solids* **27**, 1069 (1966).
58. L. Orgel, *Introduction to Transition Metal Chemistry* (Wiley, New York, 1960).
59. D.S. McClure, *Electronic Spectra of Molecules and Ions in Crystals* (Academic Press, New York, 1959).
60. D.E. Raymaker, J.S. Murday, N.H. Turner, C. Moore, M.G. Legally, *The Physics of SiO_2 and its Interfaces*, ed. S.T. Pantelides (Pergammon Press, New York, 1978), p. 99.
61. S. Miyazaki, M. Narasak, M. Ogasawaga, M. Hirose, *Microelectronic Eng.* **59**, 373 (2001).
62. S. Miyazaki, M. Hirose, *AIP Conf. Proc.* **550**, 89 (2000).
63. J.W. Keister, J.E. Rowe, J.J. Kolodziej, H. Niimi,.T.E. Madey, G. Lucovsky, *J. Vac. Sci. Techol. B* **17**, 1831 (1999).
64. H. Sato, T. Nango, T. Miyagawa, T. Katagiri, K.S. Seol, Y. Ohki, *J. Appl. Phys.* **92**, 1106 (2002).
65. N.A. Stoute, D.E. Aspnes, J.G. Hong, G. Lucovsky (unpublished).
66. C.L. Hinkle et al., *Microelectronic Eng.* **72**, 257 (2004).
67. M. Zhu, T.P. Ma, T. Tamagawa, J. Kim, R. Carruthers, M. Gibson, T. Furukawa, 2000 *IEDM Digest of Technical Papers*, p. 463.
68. W. Franz, *Handbuch der Physik*, Vol. XVIII, ed. S. Flugge (Springer, Berlin, 1965), p. 155.
69. J. Maserjian, *J. Vac. Sci. Technol.* **11**, 996 (1974).
70. K.F. Schuegraf, C.C. King, C.-M. Hu, *1992 VLSI Symposium*.
71. H.-Y. Yang, H. Niimi, G. Lucovsky, *J. Appl. Phys.* **83**, 2327 (1998).
72. L.C. Feldman, L. Stensgard, P.J. Silverman, T.E. Jackman, *Proceedings of International Conference on the Physics of SiO_2 and its Interfaces*, ed. S.T. Pantelides (Pergamon Press, New York, 1978), p. 344.
73. D.E. Aspnes, J.B. Theeten, *J. Electrochem. Soc.* **127**, 1359 (1980).
74. F.T. Himpsel, F.R. McFeely, J.A. Yarmoff, G. Hollinger, *Phys. Rev. B* **38**, 6084 (1988).
75. C.H. Bjorkman, T. Yasuda, C.E. Shearon Jr., U. Emmerichs, C. Meyer, K. Leo, H. Kurz, *J. Vac. Sci. Techol. B* **11**, 1521 (1993).
76. J. Schafer, A.P. Young, L.J. Brillson, H. Niimi, G. Lucovsky, *Appl. Phys. Lett.* **73**, 791 (1998).
77. J.W. Keister, J.E. Rowe, J.J. Kolodziej, H. Niimi, H.S. Tao, T.E. Madey, G. Lucovsky, *J. Vac. Sci. Techol. A* **17**, 1250 (1999).
78. J.T.Fitch, C.H. Bjorkman, G. Lucovsky, F.H. Pollak, X. Yim, *J. Vac. Sci. Technol. B* **7**, 775 (1988).
79. G. Luepke, *Surface Sci. Reports* **35**, 75 (1999).
80. C.H. Bjorkman, C.E. Shearon, Jr., Y. Ma, T. Yasuda, G. Lucovsky, U. Emmerichs, C. Meyer, K. Leo, H. Kurz, *J. Vac. Sci. Techol. A* **11**, 964 (1993).
81. U. Emmerichs, C. Meyer, H.J. Bakker, F. Wolter, H. Kurz, G. Lucovsky, C.H. Bjorkman, T. Yasuda, Yi Ma, Z. Jing, J.L. Whitten, *J. Vac. Sci. Techol. B* **12**, 2484 (1994).
82. H. Yang, H. Niimi, J.W. Keister, G. Lucovsky, *IEEE Electron Dev. Lett.* **21**, 76 (2000).
83. G. Lucovsky, Y. Wu, H. Niimi, V. Misra, J.C. Phillips, *APL* **74**, 2005 (1999).
84. B.J. Hinds, F. Wang, D.M. Wolfe, C.L. Hinkle, G. Lucovsky, *J. Vac. Sci. Techol. B* **16**, 2171 (1998).
85. S. Takagi, A. Toriumi, M. Iwase, H. Tanjo, *IEEE Trans. Electron Devices* **41**, 2357 (1994).
86. M.V. Fishchetti, *J. Appl. Phys.* **89**, 1232 (2000).

87. M.V. Fischetti, D.A. Neumayer, E.A. Cartier, *J. Appl. Phys.* **90**, 4587 (2001).
88. R. Helms, E.H. Poindexter, *Rep. Prog. Phys.* **83**, 2449 (1998), and references therein.
89. Y.-M. Lee, Ph.D. Dissertation, NC State University (2003).
90. J.A. Theil, D.V. Tsu, S.S.Kim, G. Lucovsky, *J. Vac. Sci. Techol. A* **8**, 1374 (1990).
91. H. Yang, G. Lucovsky, *J. Vac. Sci. Techol. A* **16**, 1525 (1998).
92. G. Lucovsky, *Extended Abstracts of the 6th Workshop on Formation, Characterization, and Reliability of Ultrathin Silicon Oxides*, January 26–27, 2001, Atagawa Heights, Japan, p. 5.
93. R.S. Johnson, G. Lucovsky, I. Baumvol, *J. Vac. Sci. Techol. A* **19**, 1353 (2001).
94. J.J. Chambers, G.N. Parsons, *J. Appl. Phys.* **90**, 918 (2001).
95. M.D. Ulrich, J.G. Hong, J.E. Rowe, G. Lucovsky, A.S.-Y. Chan, T.E. Madey, *J. Vac. Sci. Techol. B* **21**, 1777 (2003).
96. R. Chau, S. Datta, M. Doczy, J. Kavalieros, M. Metz, *International Workshop on Gate Insulator(s) 2003*, November 6–7, 2003, Tokyo, Japan.
97. J.P Maria, D. Wichakana, J. Parrete, A.I. Kingon, *J. Mater. Res.* **17**, 1571 (2002).
98. H.J. Richter, M. Herrrmann, W. Hermel, *J. Eur. Ceram. Soc.* **7**, 3 (1991).
99. G.J. Ball, M.A. Mignanelli, J.I. Barry, J.A. Gisby, *J. Nucl. Mater.* **20**, 238 (1993).
100. G. Cevales, *Ber. Dtsch. Keram. Ges.* **45**, 216 (1968).
101. V.A. Lysenko, *Neorg. Mater.* (Enlish Trans.) **30**, 930 (1994).
102. F. Gamiz, M.V. Fischetti, *Appl. Phys. Lett.* **83**, 4848 (2003).
103. G. Lucovsky et al., IEEE Transactions on Device and Materials Reliability, (2004), in press.

NOTE ADDED IN PROOF

In Section 4.4, spectra were presented for binary oxides comprised of transition metal and rare earth oxides. A d-state coupling mechanism was proposed to account for the multiplicity of features the XAS spectra for Gd and Dy scandates. Subsequent studies, summarized in Ref. 103, have identified the correct mechanism that accounts for the multiplicity of scandate spectral features, as well as the multiplicity of XAS features in the spectra of TiO_2, ZrO_2 and HfO_2.

The following points summarize the research results of Ref. 103. Thin films of i) TiO_2, ZrO_2 and HfO_2, ii) complex scandate oxides, including $LaScO_3$, $DyScO_3$ and $GdScO_3$, and iii) Zr and Hf titanate alloyss, are nano-crystalline, on deposition when chemically pure. This means that the RCP designation of the last two paragraphs on page 124 is not applicable to oxide dielectrics with large Pauling bond ionicities. The lowest conduction band states in these oxides are Jahn-Teller *term split* d states associated with *intrinsic bonding distortions*. These term-spit states have also been detected in band edge spectroscopy measurements, including spectroscopic ellipsometry and photoconductivity, confirming large differences in matrix elements for optical absorption between octahedrally coordinated Ti and Sc atoms, and eight-fold coordinated Zr and Hf atoms. Localized *band edge traps* resulting from *additional bonding distortions* at grain boundaries are observable in band edge absorption and photoconductivity and contribute to trap assisted tunneling and Frenkel Poole electrical transport for *substrate injection* from n-Si.

Chapter 5

ATOMIC STRUCTURE, INTERFACES AND DEFECTS OF HIGH DIELECTRIC CONSTANT GATE OXIDES

J. ROBERTSON AND P.W. PEACOCK

Engineering Department, Cambridge University, Cambridge CB2 1PZ, UK

ABSTRACT

The properties of gate oxides with high dielectric constant are described. The bulk bonding and electronic structure are described. The band offsets are calculated by various means and compared to their experimental determinations. The bonding at abrupt Si–oxide interfaces are considered in order to obtain an insulating interface. The energy levels of point defects and of interstitial hydrogen are considered as candidates for the substantial fixed charge present in these oxides.

1. INTRODUCTION

The scaling of dimensions of complementary metal-oxide semiconductor (CMOS) transistors has led to the thickness of the silicon dioxide used as the gate insulator to decrease below 1.6 nm. Below this thickness, the leakage current due to direct tunnelling increases above the desired values of about 1 A/cm^2. It becomes necessary to replace the SiO_2 with an alternative, high dielectric constant (κ) oxide as the gate dielectric (1–4). The higher K of the oxide allows us to use a physically thicker layer of oxide but having the same areal capacitance as that required of SiO_2, the so called equivalent oxide thickness or EOT.

The alternative oxides must satisfy various conditions to act as a satisfactory gate oxide:

(1) They should be thermodynamically stable in contact with the Si channel (5).
(2) They must be able to withstand the process conditions of CMOS of 5 seconds at 1000° C. Oxygen diffusion should not be so large that it generates a sizeable SiO_2 interface layer (6).
(3) They must have sufficient band offsets to act as barriers for electrons and holes (7).

A.A. Demkov and A. Navrotsky (eds.), Materials Fundamentals of Gate Dielectrics, 179–214.

(4) They must form high quality interfaces in contact with silicon, in order to minimise the number of interface states and defect states which could trap charge or reduce the channel mobility.
(5) The oxide should have a high electrical reliability.
(6) A suitable gate metal can be found.

Criteria (1–3) restrict the choice mainly to the oxides of Hf, Zr, Al, Y and La. Other lanthanides are also suitable, but Iwai (8) has shown that they have few advantages over La. The leading contenders presently are HfO_2, Hf silicates and their nitrided alloys (1, 9). A figure of merit has been developed to classify oxides in terms of their effectiveness as a tunnel barrier (10).

The interface criterion (4) arises because the performance of a field effect transistor depends fundamentally on the quality of the oxide–Si interface as the current flows in the Si channel next to this interface. However, despite the intensive work on high K oxides, the performance of devices with high K gate oxides is still rather poor compared to those with SiO_2 gate oxides, in terms of their trapped charge and carrier mobility (11–14), so that a deeper understanding of this interface is needed.

There is also an interest in epitaxial oxides on silicon for future functional oxide devices, in which the gate oxide could be ferroelectric, ferromagnetic, superconducting or have giant magneto-resistance. $SrTiO_3$ is the example of such oxides (14–17).

This article describes the behaviour of the interface, in terms of the bulk electronic structure, band offsets, and then describes the behaviour of the defects expected in the oxides.

2. CHEMICAL BONDING

Silicon is a covalently bonded crystal, in which each atom forms four sp^3 bonds to its neighbours (18). The usual oxide SiO_2 also consists of covalent bonds, with each Si being bonded to four oxygens and each oxygen being bonded to two silicons. The bonds are now polar, with the oxygen being negatively charged, but the atom coordinations are still given by so-called 8-N rule, where N is the number of valence electrons on that atom. These coordinations also hold for the amorphous phases.

The bonding in the alternative high K oxides is ionic. This means that their coordinations are larger and they are not determined by the 8-N rule. These coordinations carry over into the amorphous or nano-crystalline phases. The electronic structure and density of states of any material depends on the coordinations and thus the electronic structures carry over from the crystalline phases to the amorphous phase provided that they have the same coordinations.

The electronic structure of each oxide is given in terms of their band structure, partial density of states and ionic charges. The band structures of the various oxides were calculated (19) by the pseudopotential method (20, 21), using Vanderbilt ultrasoft pseudopotentials (22), and using the generalised gradient approximation of the local density approximation (LDA) to give the exchange-correlation energy for the electron gas.

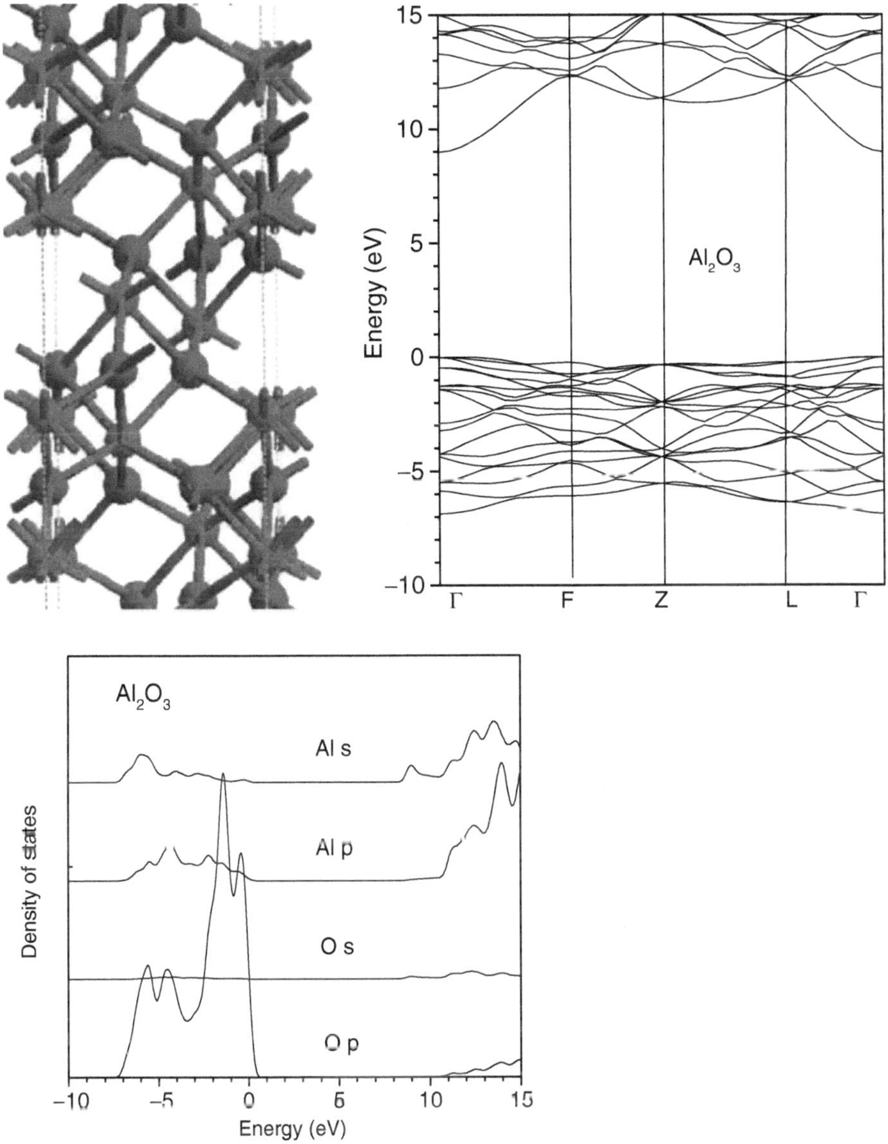

Fig. 1. Crystal structure, band structure and density of states of α-Al_2O_3. The band gap is corrected to experimental value, 8.8 eV.

We first consider Al_2O_3 which is an s,p bonded material. In hexagonal α-Al_2O_3 (sapphire), the Al is primarily six-fold coordinated while the O is four-fold coordinated (Fig. 1). There is also the possibility of the Al being four-fold coordinated. The band structure of α-Al_2O_3 is shown in Fig. 1(b) and the density of states (DOS) is shown

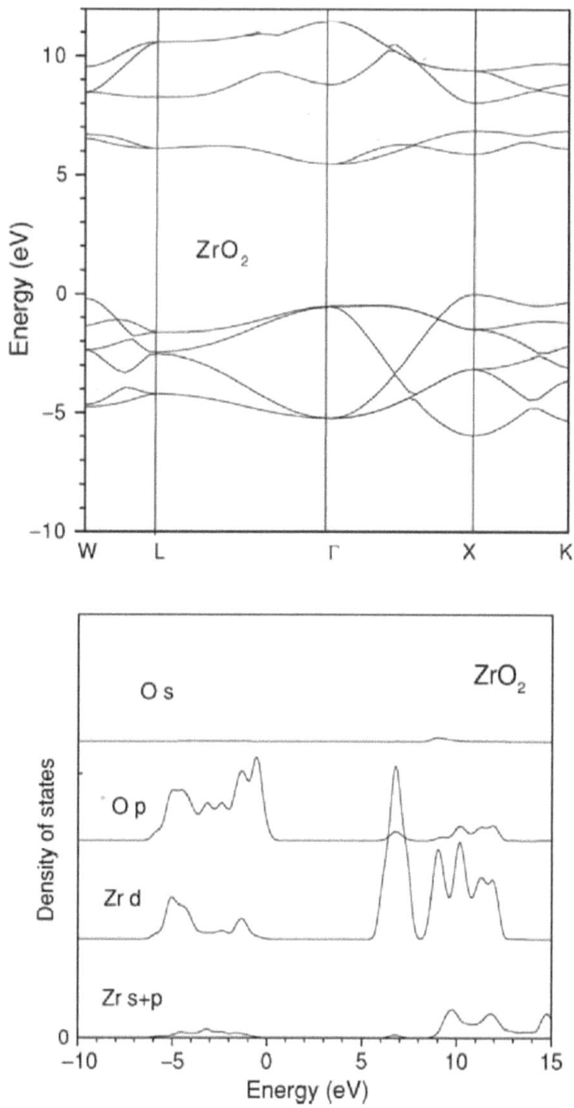

Fig. 2. Band structure and density of states of cubic ZrO_2. The band gap is corrected to experimental value, 5.8 eV.

in Fig. 1(c). Its band gap is 8.8 eV wide (23, 24). (In this article, the band gaps have been adjusted to their experimental values.) The valence band consists of oxygen p states, and the conduction band minimum consists of Al s states.

All the other oxides of interest are transition metal oxides. The simplest high K oxide is ZrO_2. ZrO_2 films are amorphous at room temperatures, but they crystallise relatively easily. ZrO_2 is stable in the monoclinic structure at room temperature, it transforms to the tetragonal structure above 1170° C and to the cubic fluorite structure at 2400° C or by the addition of Y. In cubic ZrO_2, Zr has eight oxygen neighbours and each oxygen has four Zr neighbours with a bond length of 2.20Å, while in monoclinic ZrO_2, each Zr atom has seven oxygen neighbours.

The band structure of cubic ZrO_2 is given in Fig. 2(a) and has an indirect gap of 5.8 eV, the experimental value (25). The valence band is 6 eV wide, and is largely formed from O p states. The conduction band minimum is a Γ_{12} state of Zr 4d orbitals. The Zr d states are split by the crystal field into a lower band of e symmetry (d_{x^2}) states and an upper band of t_2 (d_{xy}) states 5 eV higher (at Γ), with the Zr 4s states lying at 9 eV in between. This crystal field splitting is washed out in the lower symmetry and amorphous phases. The partial DOS shows considerable charge transfer, with the valence band being strongly O p states, and conduction band on Zr d states, with 30% admixture.

HfO_2 is chemically very similar to ZrO_2. The Hf–O bond length is 2.21 Å. Electronically, the band structure of HfO_2 in Fig. 3 is similar to ZrO_2 (26), except that the crystal splitting of the Hf 5d states in the conduction band is larger. Its band gap is 5.8 eV (27).

Crystalline La_2O_3 has the La_2O_3 structure in which La is seven-fold coordinated, with four short La–O bonds (2.30Å) and three longer bonds (2.70 Å). The band structure of La_2O_3 is shown in Fig. 4. The band gap is indirect and is about 6 eV. The valence band maximum is at Γ and the valence band is now only 3.5 eV wide, narrower than in ZrO_2. This is due to its higher ionicity than ZrO_2. The conduction band minimum is due to La d states. The DOS of La_2O_3 in Fig. 4 shows that the valence band is strongly localised on O p states and the conduction band in on La d with some La s,p states starting at 8 eV. La_2O_3 can be taken as a model for other lanthanide oxides.

The group IIIA metal oxide Y_2O_3 has the cubic bixbyite (defect spinel) structure. This has a large unit cell in which there are two types of Y sites, both seven-fold coordinated. We have calculated the bands of Y_2O_3 in the La_2O_3 structure, because it has a smaller unit cell. The gap is direct at Γ and is about 6 eV (28, 29). The valence band is again only 3 eV wide. The partial DOS in Fig. 5 shows the valence band is largely O p states. The major part of the conduction band is due to Y 4d states.

Another large class of oxides are those with the cubic perovskite structure, ABO_3. Many of the perovskites are ferroelectric and so distort into lower symmetry tetragonal or rhombohedral phases, but the cubic phase displays their essential electronic structure. The transition metal ion B occupies the octahedral site coordinated by six oxygens. The oxygens are bound to two B ions, while the A ion is surrounded by 12 oxygen ions. Figure 6 shows the band structure of $SrTiO_3$ and Fig. 6(b) shows the DOS. The direct band gap at Γ is 3.3 eV wide (30, 31). The lowest conduction bands are Ti d_{xy} t_2 states followed by the Ti d_{x^2} states. The states of the A ion (Sr) are well above the band gap, and this ion can be considered to be passive and

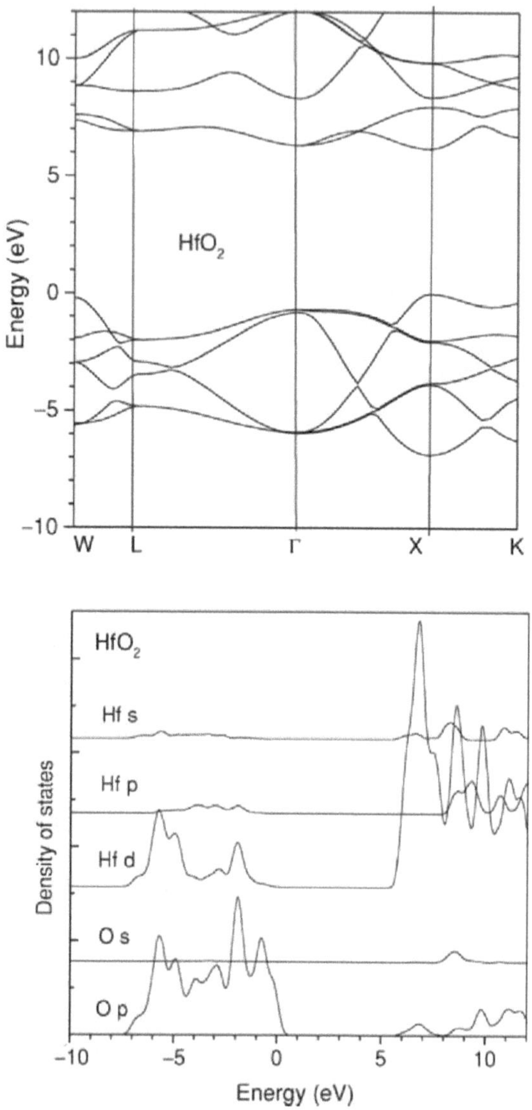

Fig. 3. Band structure and density of states of cubic HfO_2. The band gap is corrected to experimental value 5.8 eV.

essentially fully ionised. On the other hand, the Ti–O bond is polar and about 60% ionic.

In all these transition metal oxides, the minimum band gap is purely ionic, and it is given by the energy difference between the metal d state and the oxygen 2p state. That is, the states at the band extrema are each purely anionic or cationic, although the

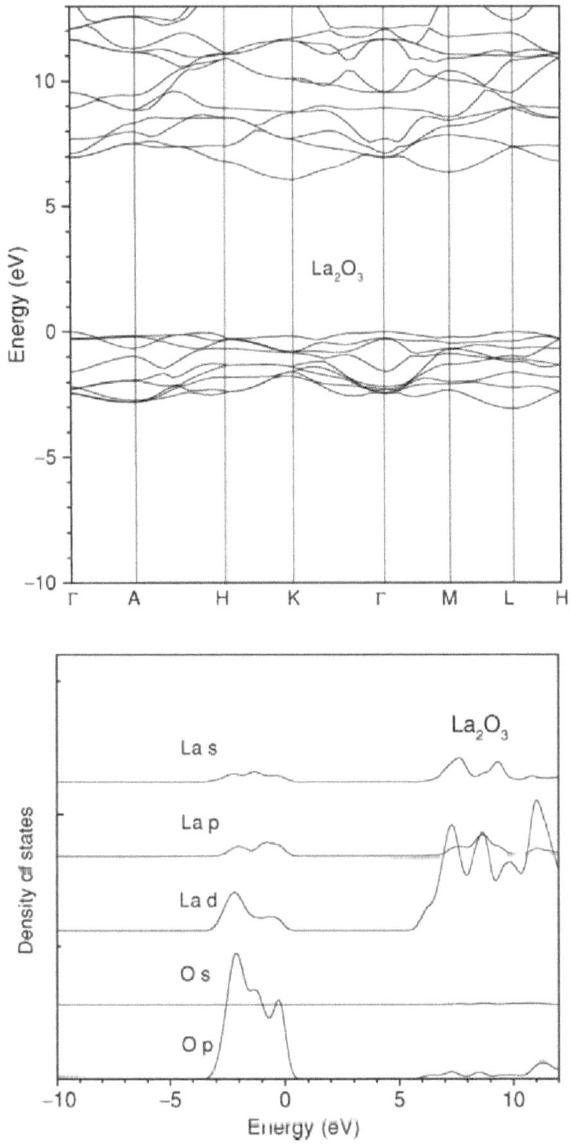

Fig. 4. Band structure and density of states of La_2O_3. The band gap is corrected to experimental value 6 eV.

average band states have more mixed character. The gap increases for early transition metals, and also for groups IV and V compared to group III metals.

$LaAlO_3$ is an unusual perovskite oxide, in that the transition metal La occupies the A site and Al occupies the octahedral B site. $LaAlO_3$ is typical of aluminates, which

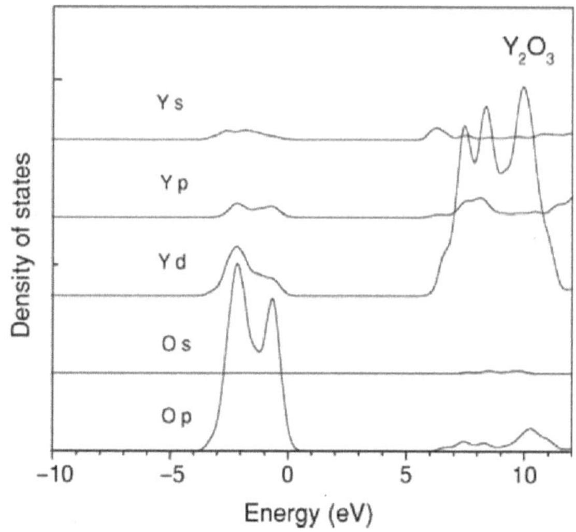

Fig. 5. Density of states of Y_2O_3. The band gap is corrected to experimental value.

are of interest as they have larger dielectric constants than the silicates. It is lattice matched on Si(100). The band structure and partial DOS of $LaAlO_3$ is shown in Fig. 7. The band gap is 5.6 eV (27). The band gap of the amorphous phase is 6.2 eV.

Finally, $ZrSiO_4$ is a typical transition metal silicate. These are of interest as gate oxides because of their greater glass-forming tendency than the simple oxides, despite their lower dielectric constant. $ZrSiO_4$ has the body-centred tetragonal structure. Each Zr atom has eight O neighbours. Each Si has four O neighbours in a tetrahedral arrangement. Each O is bonded to two Zr and one Si atom. Its band structure and partial DOS is shown in Fig. 8. The band gap is taken to be about 6.5 eV. The valence band is 7 eV wide (32). The conduction bands form two blocks. The lower conduction bands due to Zr d states lie between 6.5 and 8 eV, followed by a set of bands mainly due to Si–O antibonding states mixed with further Zr d states. Thus, the conduction band DOS can be considered to be the sum of ZrO_2-like and SiO_2-like components. $HfSiO_4$ is essentially similar to $ZrSiO_4$.

The band gaps of the amorphous $(HfO_2)_x(SiO_2)_{1-x}$ and $(ZrO_2)_x(SiO_2)_{1-x}$ alloys were determined experimentally (33, 34). They follow the '2 band' model in which the lowest band gap is determined by the Hf d states of the HfO_2 component while they are present.

The bonding of heteropolar materials can be analysed in terms of their ionicity or polarity of their bonds. This is done by defining the average energy gap between valence and conduction bands, the Penn gap (35). The Penn gap E consists of both a homopolar or covalent gap (E_h) and an ionic contribution C, which add in quadrature, $E^2 = E_h^2 + C^2$. The ionicity is then defined as the ratio of the ionic gap to the total gap, C^2/E^2. The gap energies are calculated by the empirical method of Levine (36), which is generalisation of the method of Philips and vanVechten (35). We can

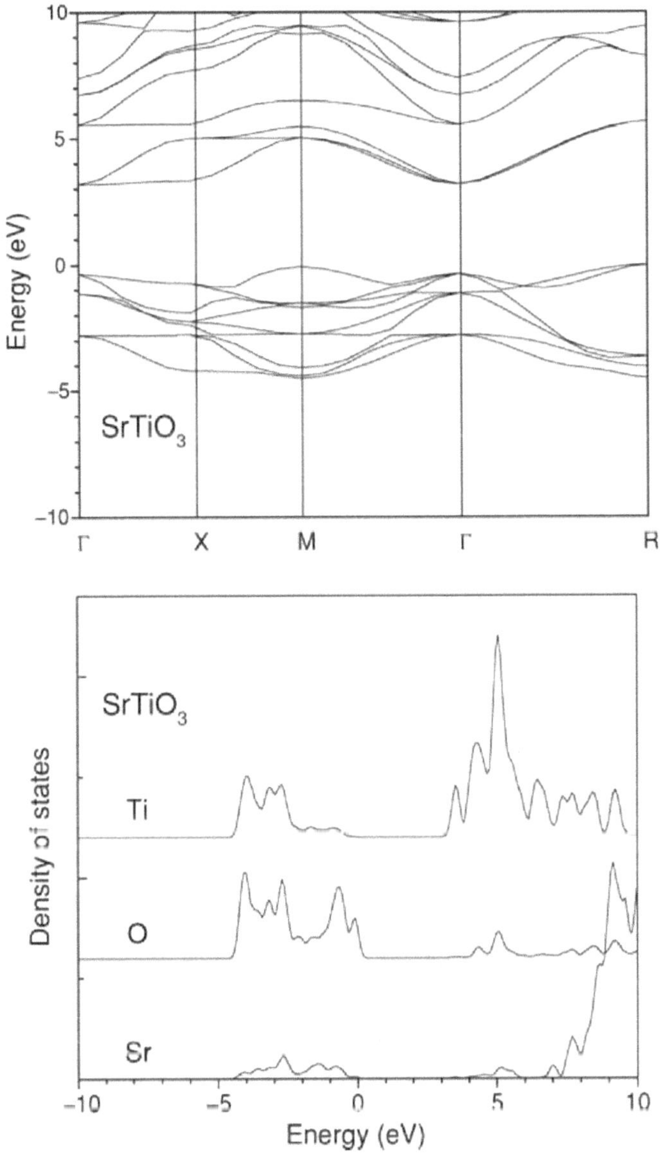

Fig. 6. Band structure and partial density of states of $SrTiO_3$. The band gap is corrected to experimental value 3.2 eV.

plot the ionicity of the bonding in the various oxides by plotting the homopolar and ionic contributions to the Penn gap (Fig. 9) to emphasise the difference between the predominantly covalent SiO_2 and silicon nitride and the predominantly ionic high K oxides (37).

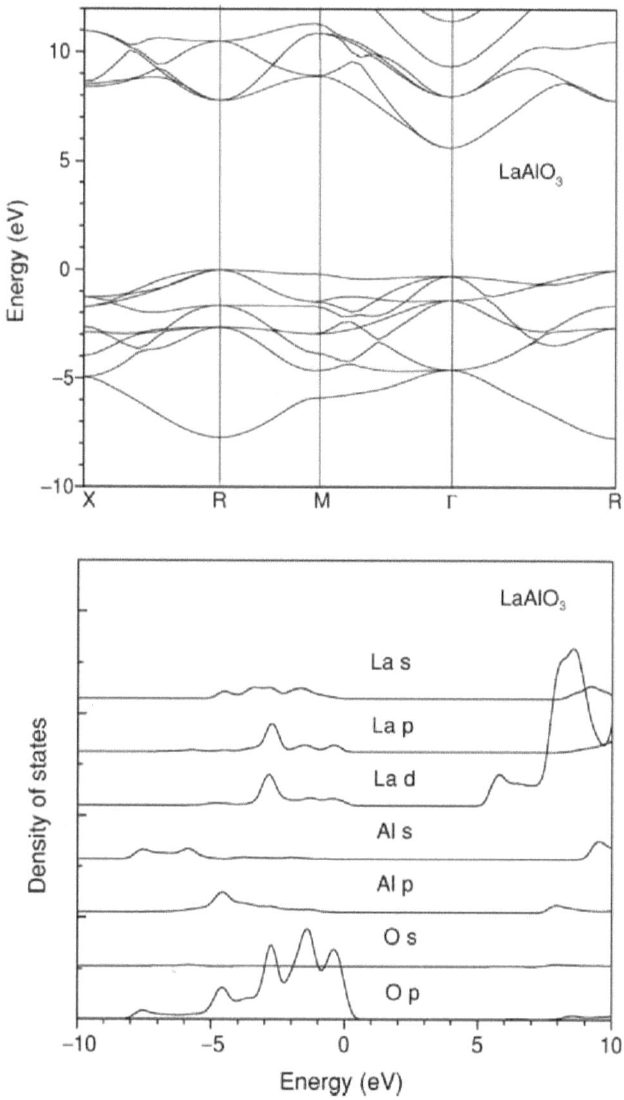

Fig. 7. Band structure and density of states of $LaAlO_3$. The band gap is corrected to 5.8 eV.

3. DIELECTRIC CONSTANT

It is desirable to maximise κ for economic reasons, but κ should not be too high or otherwise it creates high electric field distributions around the drain. Figure 10 plots the static dielectric constant κ against the band gap for the oxides of interest. It is seen that a high κ generally occurs for a smaller band gap.

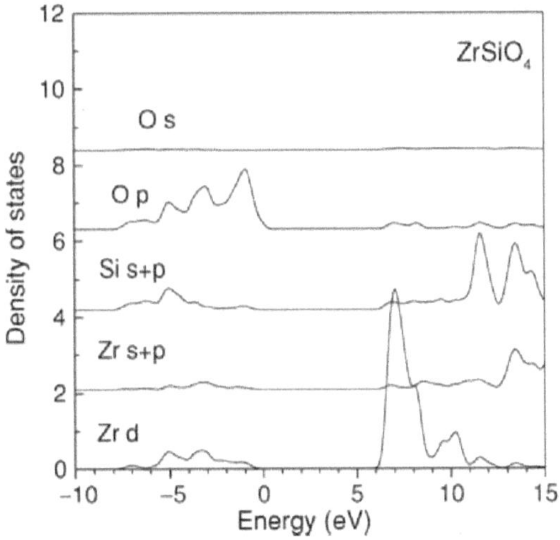

Fig. 8. Band structure and density of states of $ZrSiO_4$. The band gap is corrected to experimental value of about 6.5 eV.

As a number of factors (band offsets, chemical stability) cause us to choose a moderately high band gap of over 5 eV, the K value is often not so high. Consequently, we now consider what controls K at a microscopic level, in order to know how to maximise it.

The low frequency dielectric constant κ is the sum of electronic and lattice contributions,

$$\kappa = \kappa_e + \kappa_l$$

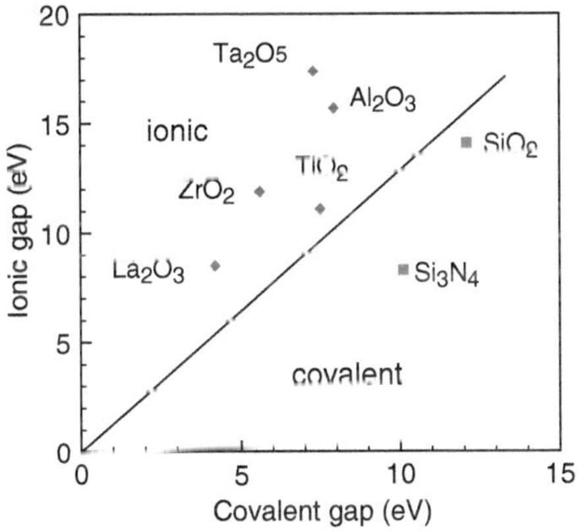

Fig. 9. Covalent and ionic band gaps of oxides, showing the more ionic bonding character of high K oxides.

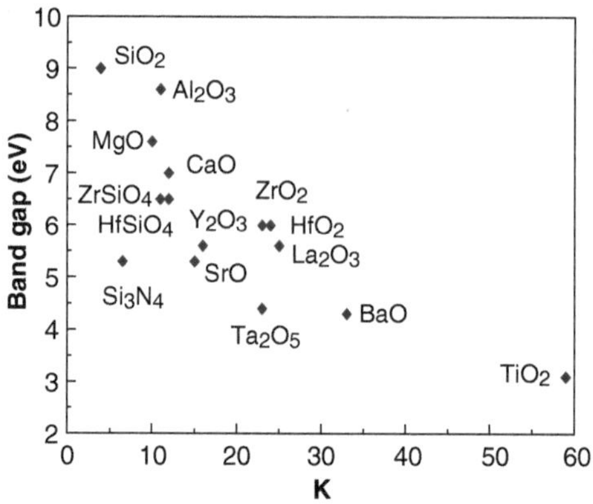

Fig. 10. Correlation of dielectric constant with band gap of candidate oxides.

Here κ_e includes the dielectric constant of free space (1). This equation can also be written as $\kappa = 1 + \chi_e + \chi_l$ where the χ's are the equivalent susceptibilities (38). The electronic component κ_e is also the optical dielectric constant ε_∞ and it is given by refractive index squared, $\kappa_e = \varepsilon_\infty = n^2$. Refractive indices of the oxides of interest were given previously (7). ε_∞ values are typically 4–5 and do not exceed 8. This is small, so the majority of κ must originate from the lattice contribution, $\kappa_l \cdot \kappa_l$ is related to microscopic parameters by (18),

$$\kappa_l = \frac{Ne^2 Z_T^{*2}}{m\omega_{TO}^2} \tag{1}$$

Here, N is the number of ions per unit volume, e is the electronic charge, Z_T^* is the transverse (infra-red) effective charge, m is the reduced ion mass and ω_{TO} is the frequency of the transverse optical phonon. Large values of κ_l occur when Z^* is large and/or ω_{TO} is small. A small ω_{TO} corresponds to a low frequency or 'soft' vibrational mode. A negative value of ω_{TO}^2, that is an imaginary value of ω_{TO}, gives a ferroelectric. This is why the oxides of interest tend to be incipient ferroelectrics.

There are various types of effective charges, which can cause confusion. First there is integer formal charge, which is ± 1 in Na^+, Cl^-. Secondly, there is the static charge Z^*, which describes the degree of charge transfer between the Na and Cl. It is about 0.9 in NaCl as the ionicity is less than 100%. Thirdly, there is the dynamic or infra-red effective charge Z_T^*, which is that used in Eq. (1). Z_T^* is due to the rearrangement of electronic charge as the ion moves. It adds to the static charge, and it can be non-zero even in covalent systems such as Se, if the symmetry is low enough.

The static charge was calculated by Harrison (18) for tetrahedrally bonded solids in terms of the polarity α of the individual bonds as

$$Z^* = 4\alpha - \Delta Z$$

The polarity α is given by

$$\alpha = \frac{V_3}{(V_2^2 + V_3^2)^{1/2}}$$

Here, V_2 is the covalent energy of the bond and V_3 is the ionic energy, and the total band gap is $V_g = (V_2^2 + V_3^2)^{1/2}$. V_3 is given by half the difference of atomic energies across the bond, and is a constant for a given compound. V_2 varies with bond length x as

$$V_2 = V \cdot x^{-n} \tag{2}$$

Here, V is a constant, $n = 2$ for s,p interactions and $n = 7/2$ for p,d interactions of interest in high K oxides. The transverse or 'dynamic' effective charge Z_T^* is given by the change in dipole moment divided by the displacement,

$$Z_T^* = \frac{d(xZ^*)}{dx}$$

where x is the instantaneous bond length. As a bond gets longer, it becomes more ionic, so the dipole gets larger. Thus,

$$Z_T^* = 4\frac{d\alpha}{dV_2} \cdot \frac{dV_2}{dx} \cdot x$$

so

$$Z_T^* = 4\alpha(1 - \alpha^2)n \tag{3}$$

with n from Eq. (2). The variation of Z^* and ω_{TO} with the band gap explains the variation of κ with band gap seen in Fig. 10 for the transition metal oxides.

4. BAND OFFSETS

4.1. Background

The purpose of the gate oxide layer is to prevent current conduction across it. This can occur by tunnelling thought it (either Fowler–Nordheim or direct tunnelling), excitation to the oxide band states (Schottky emission) or by hopping through via defect states (Poole–Frenkel). The barrier at each band or 'band offset' should be over about 1 eV for both the conduction and valence bands in order to inhibit Schottky emission of electrons or holes into their bands (7, 39). The band offset is the alignment of bands between the Si and the oxide, as shown in Fig. 11. There is a band offset at the conduction band and one at the valence band.

It is unclear that all of the proposed oxides in Fig. 10 will act as potential barriers, as some have quite small band gaps. SiO_2 has a wide gap of 9 eV, so it has large band offsets for both electrons and holes. On the other hand, $SrTiO_3$ has a band gap of only 3.2 eV, so its bands must be aligned almost symmetrically with respect to Si for both barriers to be 1 eV. It turns out that for most high K oxides, the conduction

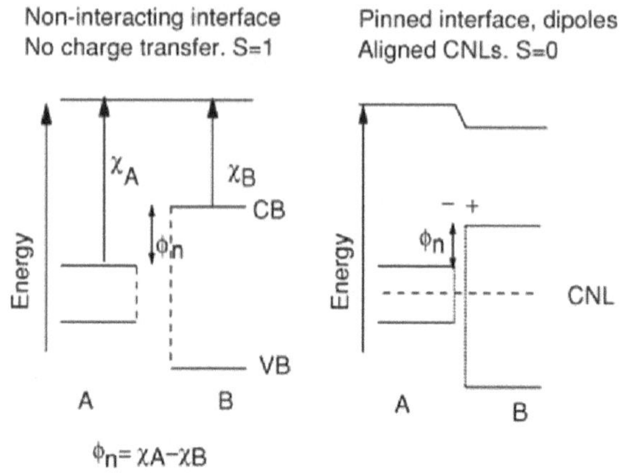

Fig. 11. Schematic of band alignments at a metal–semiconductor interface for (a) no charge transfer, (b) charge transfer and dipole formation.

band offset is smaller than the valence offset, so this offset limits the leakage current. Generally, a band offset is a function of the oxide band gap and the asymmetry of the band alignment. We will see that the asymmetry of alignment depends on details of the electronic structure. The band gap is a fundamental property of the oxide. As the band gap tends to vary inversely with the dielectric constant, as shown in Fig. 10, there is a trade-off between their κ value and their band gap.

The band offsets of an oxide on silicon can be found by treating the oxide as a wide band gap semiconductor. It is then the band offset between two semiconductors. The band offset is closely related to the barrier height between the semiconductor and a metal, which is known as the Schottky barrier height. Both these subjects have been intensively studied in the past.

The band offset between two semiconductors depends on the energy levels of the two semiconductors and the presence of any charge transfer across the interface which would create an interface dipole (Fig. 11).

In the absence of charge transfer, the band offset is given by placing the energies of each semiconductor on a common energy scale as in Fig. 11(a)—usually with respect to the vacuum level. This is called the Electron Affinity rule (40), which says that the conduction band offset is given by the difference in their electron affinities (EAs).

In practice, there is charge transfer across the interface. Consider first two metals in contact. There is charge transfer between the two metals which creates a dipole and equates their Fermi levels. Now consider a metal and semiconductor—the Schottky barrier (41). The semiconductor surface now has states within its band gap which decay into the semiconductor. There can be charge transfer between the metal Fermi level and the interface gap states of the semiconductor, which tends to align the metal Fermi level and its equivalent for the interface states, which is called a charge neutrality level (CNL). The charge transfer at a Schottky barrier tends to align the Fermi level of the metal to the CNL of the semiconductor, as shown in Fig. 11(b).

This leads to an equation for the Schottky barrier height of electrons ϕ_n between the semiconductor S and the metal M of

$$\phi_n = S(\Phi_M - \Phi_S) + (\Phi_S - \chi_s) \tag{4}$$

or

$$\phi_n = S(\Phi_M - \chi_s) + (1 - S)(\Phi_S - \chi_s)$$

Here, Φ_M is the metal work function, Φ_S is the charge neutrality level of the semiconductor and χ_S is the electron affinity (EA) of the semiconductor. All the energies in (1) are measured from the vacuum level, except ϕ_n which is measured from the conduction band edge. S is a dimensionless constant, called the pinning factor, which is given by (42)

$$S = \frac{A}{1 + (e^2 N\delta/\varepsilon\varepsilon_0)} \tag{5}$$

where e is the electronic charge, ε_0 is the permittivity of free space, N is the density of the interface states per unit area and δ is their extent into the semiconductor.

The parameter S in Eq. (1) is a dimensionless pinning factor, which describes the degree of alignment or 'pinning' caused by the interface states. $S = 1$ corresponds to the unpinned Schottky limit, equivalent to the Electron Affinity rule, and $S = 0$ corresponds to the strongly pinned or 'Bardeen' limit.

There have been a number of models of the states causing the pinning. Intrinsic states are a more general source of pinning than defects because there are more of them. Thus we discount defect models of pinning.

We have used the model of virtual gap states (VGS) or metal-induced gap states (MIGS) (43–46). The VGS can be visualised as the dangling bond states of the broken surface bonds of the semiconductor dispersed across its band gap, or alternatively as the evanescent states of the metal wavefunctions continued into the forbidden energy gap of the semiconductor. Φ_S is the charge neutrality level (CNL) of the interface states. The CNL is like a Fermi level for interface states; it is the energy near mid-gap to which the interface states are filled on a neutral surface.

It was originally believed that S depends on the semiconductor ionicity (47). It is now known that S depends on the electronic part of the dielectric constant ε_∞ (48). By plotting the pinning factor S against ε_∞, as in Fig. 12, Monch found that S empirically obeys (49, 50),

$$S = \frac{1}{1 + 0.1(\varepsilon_\infty - 1)^2} \tag{6}$$

This is shown again in Fig. 12(b), while Tersoff (46) suggests $S = 0.5/\varepsilon_\infty$.

The band offset at a semiconductor heterojunction is defined in the same way, as the energy difference between the conduction bands or the valence bands. We can use the VGS model, and for two semiconductors a and b, the electron barrier ϕ_n is the

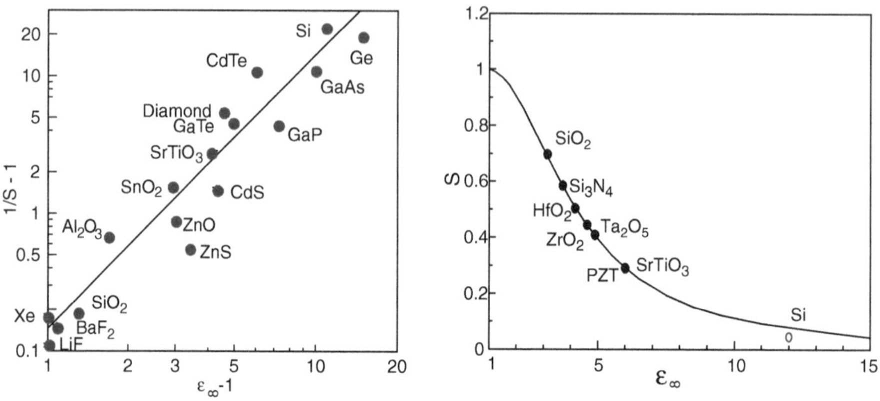

Fig. 12. The variation of the experimental value of pinning factor S with electronic dielectric constant ε_∞ (45), and theoretical variation of S with electronic dielectric constant ε_∞, from Eq. (6).

conduction band offset, given by

$$\phi_n = (\chi_a - \Phi_{S,a}) - (\chi_b - \Phi_{S,b}) + S(\Phi_{S,a} - \Phi_{S,b}) \tag{7}$$

S is the pinning parameter of the wider-gap semiconductor, that is the oxide. In the case of strong pinning $S = 0$, the CNLs of each semiconductor line up, while if there is no pinning $S = 1$, the band offset is given by the Electron Affinity rule. A wide comparison of the band offsets of epitaxial heterojunctions of zinc-blende semiconductors by Yu et al. (51) found that the charge neutrality model with $S = 0$ gives a good description.

Thus, it appears that the CNL model gives a reasonable description of the Schottky barriers and band offsets at heterojunctions. However, this is not strictly true. Firstly, Harrison and Tersoff (52) and Tung (53, 54) note that the density of VGS is also not high enough to pin the metal at a Schottky barrier and certainly not at a semiconductor heterojunction. The band alignment is actually caused by the polarisation of the metal–semiconductor interface bonds, that is states in the valence band, which happens to be reasonably well described by the CNL equations ((4)–(7)).

The second point is that there are in general two components to the charge transfer across an interface, one due the potential difference between the bulk solids, and the other due to the local atomic configuration at the interface. The main part of the band alignment is due to the first term, but the second term can be significant.

The VGS and CNL model gives a reasonable description of the bulk contribution to the charge transfer. The advantage of this method is that the CNLs are properties which are defined by the band structure of the bulk materials, it is applicable to both covalent and ionic bonding, it has been tested over a wide range of band gaps and screening constants, and it does not require an explicit model of the interfacial bonding. This is an advantage because the oxides are often amorphous and whose interface bonding is not known at this stage.

Band offsets of zinc-blende semiconductor interfaces have also been calculated by methods using explicit models of the interface, and then to calculate the offset in terms of the potential step at the interface, as in the work of Baldereschi et al. (55). Alternatively the potential step can be calculated from the bonding using the model solid method of van der Walle (56). These methods need a calculation for each interface structure. These direct calculations allow one to study the variation of interface dipoles with interface structure and oxide termination.

Nevertheless, it is known that Schottky barrier heights do depend on the interface bonding. For example the barrier height of the A and B configurations of the (111)Si:$NiSi_2$ interface differ by 0.17 eV (53). Band offsets at semiconductor heterojunctions is also found to depend on the termination of each face (57, 58). This must ultimately be taken into account, as we do for oxides in the next section.

4.2. Calculations

The method used here is to calculate the band offset between these two semiconductors, Si and oxide, taking the oxide as a wide band gap semiconductor, using the VGS and charge neutrality levels. The CNL is evaluated as the energy at which the Greens function of the band structure, integrated over all bands and over k points in the Brillouin zone, is zero

$$G(E) = \int_{BZ} \int_{-\infty}^{\infty} \frac{N(E')dE'}{E - E'} = 0 \tag{8}$$

This integral can be replaced by a sum over special k points of the zone (59). For tight-binding bands, there is a finite number of bands corresponding to the atomic basis set, whereas for pseudopotential bands we must fix a finite upper limit in integral (8). We take this as the same number of bands as in the tight binding case.

Table 1 gives the calculated energy of the charge neutrality level for each compound with respect to the valence band maximum. The calculation uses Eq. (8) and the LDA band energies, with the conduction band energies shifted upwards by the scissors correction to give the experimental band gaps. An earlier calculation used the tight-binding band structures. These were found by fitting to experimental data, and previous calculations and by scaling of parameters (7, 39).

The CNL values are calculated and given in Table 1. The S parameter is calculated here from Eq. (6) and ε_∞ using the experimental value of refractive index, $\varepsilon_\infty = n^2$. The experimental values of the refractive index (60) are also given in Table 1. The experimental band gaps are used. The offsets also need experimental values of electron affinity. These are taken from photoemission or electrochemical data (61–63).

4.3. Comparison with Experiment

$SrTiO_3$ provides a good test of Schottky barrier models. Figure 13 compares the predicted Schottky barrier heights of various metals on $SrTiO_3$ with the experimental values (64–69). $SrTiO_3$ is the most studied system and the best test of our calculations. The experimental data are quite scattered but they are quite consistent with $S << 1$

Table 1. Tabulation of band gaps (23–31), experiment electron affinities (EA) (61–63, 73), experimental ε_∞, the S factor, and a comparison of the calculated values of charge neutrality levels and conduction band offsets, found by the tight-binding (TB) and the present LDA methods

	Gap (eV) exp	EA (eV) exp	ε_∞	S	CNL, TB (eV)	CNL, LDA (eV)	CB, offset, TB (eV)	CB offset, LDA (eV)
SiO_2	9	0.9	1.5	0.83				
Ta_2O_5	4.4	3.3	4.84	0.4	3.3		0.3	
Al_2O_3	8.8	1*	3.4	0.63	5.5	6.6	2.8	2.4
Y_2O_3	6	2*	4.4	0.46	2.4	2.7	2.3	2.2
La_2O_3	6	2*	4	0.53	2.4	2.5	2.3	2.3
ZrO_2	5.8	2.5*	4.8	0.41	3.6	3.3	1.4	1.6
HfO_2	5.8	2.7	4	0.53	3.7	4	1.5	1.3
$SrTiO_3$	3.3	3.9	6.1	0.28	2.6	1.7	−0.1	0.4
$SrZrO_3$	5.3	2.6	4	0.53	3.7	2.7	0.8	1.6
$LaAlO_3$	5.6	2.5*	4*	0.53		3.8		1.0
$ZrSiO_4$	6.5*	2.4*	3.8	0.56	3.6	4.0	1.5	1.3

*: Estimates.

and our calculated value of 0.28. Note that it is the Ti–O bond which controls ε_∞ and thereby S and not the Sr–O bond, as this Ti–O bond gives states nearest to the gap. In contrast, the Sr states are well above the gap and have little influence on S. Note also that S depends on ε_∞ not ε_0 or otherwise S would be almost zero.

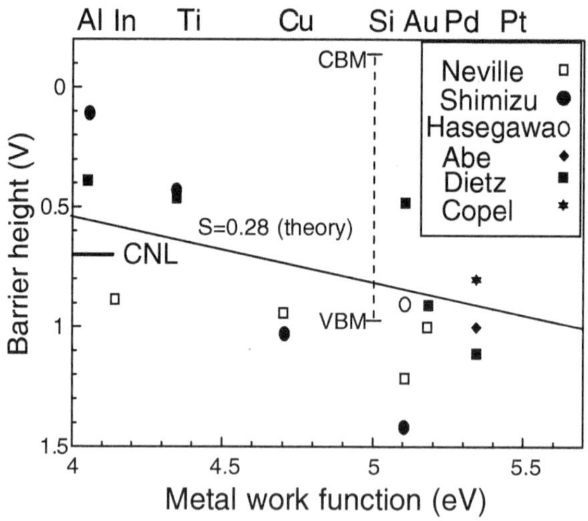

Fig. 13. Comparison of the predicted and experimental (64–69) Schottky barrier heights for various metals on $SrTiO_3$.

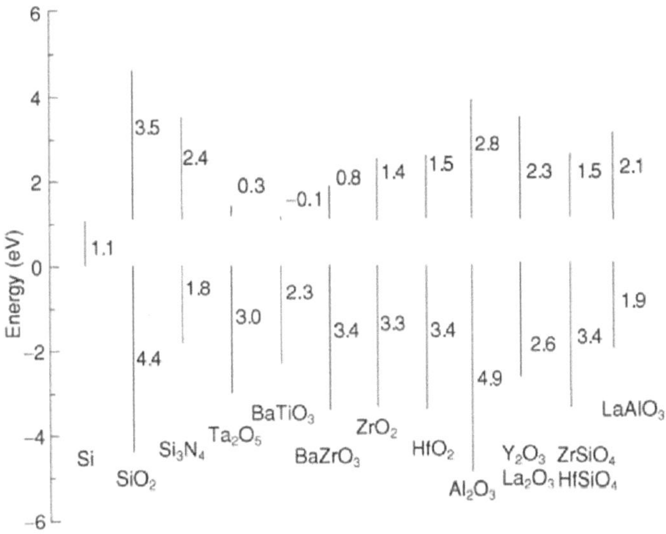

Fig. 14. Summary of predicted band offsets of high K oxides on Si.

The resulting band offsets are given in Table 1 and summarised in Fig. 14. It is seen that the conduction band (CB) offset is always the smaller offset and this limits the oxides which can act as good gate dielectrics. Only Al, Y, La, Zr and Hf based oxides have CB offsets over 1 V, which is the minimum needed to limit electron injection. Earlier preferred oxides such as Ta_2O_5 and $SrTiO_3$ are seen to have too low CB offsets (7).

A number of the band offsets have since been measured experimentally, and were found to be surprisingly close agreement with the predicted values. The calculated and experimental values of some CB offsets are compared in Table 2. The data for SiO_2 is taken from experiment (70, 71). A CB offset of $SrTiO_3$ of 0 eV was found by photoemission by Chambers et al. (72), in close agreement to our calculated value.

For Ta_2O_5, Miyazaki (69) derived a conduction band offset of 0.3 V from a valence band offset measured by photoemission, similar to the tight-binding estimate (35).

The offset of Al_2O_3 on Si was found to be 2.8 eV by DiMaria (74), 2.8 eV by Ludeke (75), and 2.2 eV by Afansev (76). Our calculated value of 2.4 eV from LDA is in good agreement with these values. The lower offset value found by Afanasev (76) was mainly due to the much lower band gap found by them, 7.0 eV than 8.8 eV found by the others. A lower band gap occurs because the Al_2O_3 was prepared by atomic layer deposition (ALD) from organo-metallic precursors. The gap increases when the sample is annealed. This is a common observation with ALD films.

Miyazaki (73) found the conduction band offset of ZrO_2 to be 1.4 eV, using photoemission to find the valence band offset. This is similar to the 1.6 eV calculated here and the 1.4 eV found by tight binding (7). Afanasev (76) found the conduction band offset of ZrO_2 directly, using internal photoemission from the Si valence band

Table 2. Comparison of the calculated conduction band offset (by LDA method) and experimental values for various gate oxides, by various authors

	Calculated, LDA (eV)	Experiment (eV)	Reference
SiO_2		3.1	Keiser (69), Alay (70)
Ta_2O_5	0.35	0.3	Miyazaki (73)
$SrTiO_3$	0.4	0	Chambers (72)
ZrO_2	1.6	1.4	Miyazaki (73)
		2.0	Afanasev (76)
		1.4	Rayner (77)
HfO_2	1.3	1.3	Sayan (63)
		2.0	Afansev (76)
		1.1	Zhu (78)
Al_2O_3	2.4	2.8	Ludeke (75)
		2.2	Afansev (76)
a-$LaAlO_3$	1.0	1.8	Edge (80)
Y_2O_3	2.3	1.6	Miyazaki (29)

to the ZrO_2 conduction band. Their value of 2.0 eV is larger than found by the photoemission method. Rayner et al. (77) found a VB offset of 3.1 eV by photoemission, equivalent to a CB offset of 1.6 eV for a band gap of 5.8 eV.

For HfO_2, Sayan et al. (63) found a CB offset of 1.3 eV assuming a band gap of 5.8 eV. This compares with the 1.3 eV calculated here or the 1.5 eV by tight-binding. Afanasev et al. (76) found 2.0 eV for HfO_2 using internal photoemission. Zhu et al. (78) obtain a CB offset of 1.1 eV from the electrical barrier height, which is close to that found here. However, this measurement is derived from the tunnelling current and uses the Schottky limit to extract a barrier. We know that the Schottky limit is not valid for HfO_2 as $S \sim 0.5$, unlike in SiO_2 where $S \sim 0.9$.

Recently, Xeo et al. (79) experimentally determined that $S = 0.52$ for HfO_2, and ZrO_2 and $S = 0.69$ for Al_2O_3 from CV measurements of various metals on the oxides. The S values are close to those expected from MIGS theory. Afanasev (76) found S values from their barrier heights which were in less good agreement with MIGS values.

There is presently no direct experimental determination of the CB offset for La oxide. Nevertheless, the fact that the leakage current density of La oxide is lower than that of HfO_2 (8) suggests that La oxide has the larger CB offset, as predicted in Table 1. Note that the CB offset was wrongly given as 1.3 eV not 2.3 eV in the diagram in ref (7).

Rayner et al. (77) found a VB offset of about 4 eV for a $ZrSiO_4$ alloy. This is equivalent to a CB offset of 1.4 eV if we take the band gap as 6.5 eV. This CB offset is close to our calculated 1.3 eV by LDA and 1.5 eV by tight-binding.

For amorphous $LaAlO_3$, Edge et al. have measured a valence band offset of 3.2 eV to Si, which gives a CB offset of 1.8 eV for a band gap of 6.2 eV (80). This compares with calculated CB offsets of 1.0–2.1 eV. This large range arises from the stronger CB dispersion in the LDA result.

Overall, there is good agreement between experiment and calculated values for the various alternative gate oxides. Surprisingly, the tight-binding predictions are often closer to experiment than the new LDA values.

Lucovsky (81–83) has recently shown how the CB offsets tends to follow the differences in s–d atomic energies in transition metals, because the band gap between metal s and oxygen 2p states remains rather constant. This gives a method to scan oxides more quickly.

Why are some conduction band offsets so small? The offsets arise mainly by the alignment of the CNL of the oxide and Si, so it depends on the CNL energy in the oxide band gap. The CNL energy is set by competing forces (7). From Eq. (6), we see that a large density of valence states pushes the CNL up, and a large density of conduction states pushes it down. Now, valence states are O 2p states and conduction states are metal d or s,p states. The ratio of these states is just the ratio of oxygen to metal atoms—the oxide stoichiometry. Thus, the CNL energy depends on metal valence. A high metal valence pushes the CNL up and a low valence pushes it down. The CB offset is small in Ta and Ti compounds because the metal valence is too high. On the other hand, La_2O_3 has a lower CNL than ZrO_2 because La's valence is lower.

5. ATOMIC MODELS OF INTERFACES

5.1. Bonding Rules

The ideal situation is an abrupt interface between Si and the high K oxide. So far, this rarely happens because processing leads to oxygen diffusion through the oxide to form a SiO_2-rich interface layer under the high K oxide. Nevertheless, it can happen, and this desirable situation occurs when epitaxial oxides are grown on Si. There are two main classes of epitaxial oxides on Si, the fluorite based systems ZrO_2, CeO_2 and the related cubic bixbyite oxides Y-rich $(Y,La)_2O_3$ and the perovskite structure oxides $SrTiO_3$ and $LaAlO_3$.

The lattice constants of Si and ZrO_2 are 5.43 and 5.07 Å, respectively. This allows Y-stabilised ZrO_2 to grow epitaxially on the Si(100) cube face (84, 85), with the ZrO_2 cube face lying directly on top of the Si cube face. This is expressed as ZrO_2(100)//Si(100), and ZrO_2[001]//Si[001]. There have been many attempts to grow fluorite-structure oxides on Si(100) with this 'cube-on-cube' epitaxy. Generally, this has been difficult, and there are two main successful reports by Wang and Ong (85) for ZrO_2 and by Ami et al. (86) for CeO_2. Often, the oxide grows with (100)Si//(110)oxide orientation. The reason is unclear, but Nagata et al. (87) speculated that it is due to the orientation of the initial oxygen bonding. It has proved easier to grow oxides on (111)Si with the orientation (111)Si//(111)oxide. This has been achieved for CeO_2, and also for the bixbyite $(Y,La)_2O_3$ (88–90). In the latter case, there is a stacking fault at the interface, so that the orientation is (111)Si//(111)oxide and [111]Si//[$\bar{1}\bar{1}\bar{1}$]oxide.

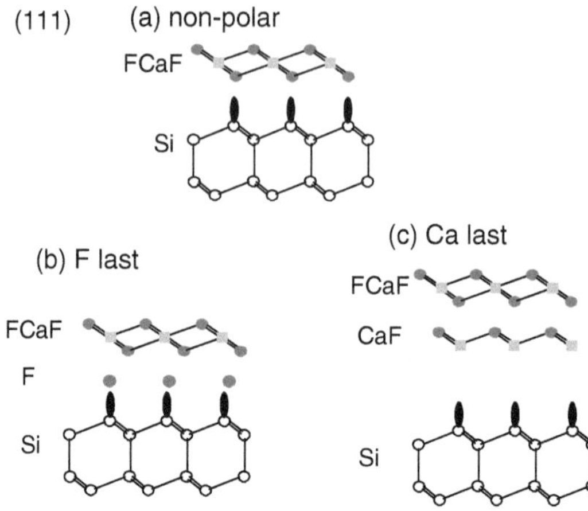

Fig. 15. Schematic of bonding at a (111)Si:CaF_2 interface: (a) with non-polar CaF_2 face, (b) F-terminated face, (c) Ca-terminated face.

$SrTiO_3$ has lattice constant of 3.91 Å. Thus the (001)$SrTiO_3$ lattice can be matched to the (001)Si lattice with the $SrTiO_3$ lattice rotated by 45° with [110]Si// [100]$SrTiO_3$. The lattice constant of $LaAlO_3$ is 3.78 Å and this allows even closer lattice matching to Si in this orientation.

We focus first on the Si:ZrO_2(100) interface. ZrO_2 has the cubic fluorite structure, in which each Zr atom is eight-fold coordinated by oxygens and each O is four-fold coordinated by Zr's. Other fluorite compounds are the metal silicides $NiSi_2$ and $CoSi_2$, and CaF_2. These each form epitaxial interfaces with Si which have been intensively studied (91–94). The most stable interface configuration can be understood in terms of the occupation of bonding states (92, 93).

The CaF_2 interfaces are more complex because CaF_2 has no common element with Si. The ideal (100) and (111) surfaces of CaF_2 are polar, that is they contain only Ca^{2+} or F^- ions. This fixed charge makes them unstable. On the other hand one can think of CaF_2 as FCaF tri-layer units stacked along the [100] or [111] directions, in which alternate F ions are assigned to Ca above or below. Their (100) or (111) faces now contain half the number of F ions, and are effectively non-polar.

On the Si(111) surface, each surface Si atom has one broken or 'dangling' bond (DB). This is half occupied, and it would give a metallic interface if it is left like this. If we make a Si:CaF_2(111) interface by joining Si to a non-polar FCaF unit, as in Fig. 15(a), it would leave the surface Si DB states half occupied in the middle of the Si band gap and a metallic interface (94). Counter-intuitively, this non-polar interface would be bad!

What is needed is to join a *polar* FFCaF unit or a CaF unit to the Si(111), as in Fig. 15(b). The extra F of FFCaF will now form a strong Si–F bond with the Si DB, and sweep the DB state out of the gap, converting into a Si–F bonding state deep in the valence band, to leave an insulating interface with no gap states. This can be

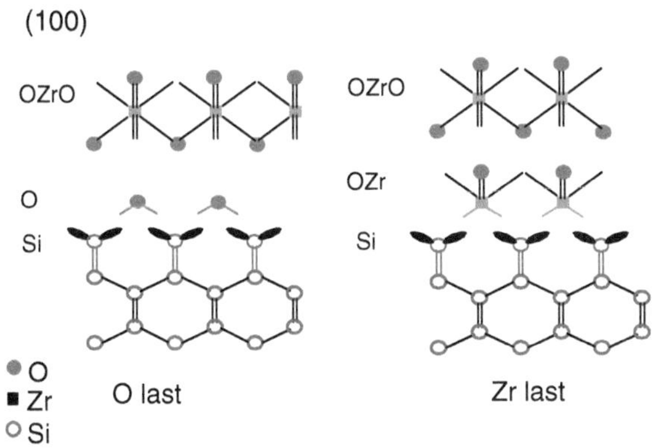

Fig. 16. Schematic of bonding at a (100)Si:ZrO_2 interface: (a) O-terminated face, (b) Zr-terminated face.

considered as a F-terminated interface $\equiv Si^{+}F^{-}F^{-}Ca^{2+}F^{-}$ unit (each dash denotes a Si–Si back-bond). Alternatively, we can have a Ca-terminated interface. A CaF unit donates an electron to the Si DB, to make a Si^{-} dangling bond. The DB state is now filled, giving an insulating interface as in Fig. 15(c). In practice, the Ca-terminated interface is found (94).

Now extend this idea to the Si:ZrO_2(111) interface (95). The ZrO_2 is expressed as $O^{2-}Zr^{4+}O^{2-}$ units. Again, it is no good putting a non-polar OZrO face against the Si(111) surface, as this leaves the Si DB half-filled. We could try with the OOZrO unit. However, this time it does not work. Oxygen needs to get two electrons, but the one Si DB can only give one electron. Thus, the interface is metallic. This is also true for the ZrO unit. The 4-2 valences of Zr and oxygen are not compatible with the single DB per site of Si(111).

Now consider the ideal Si:ZrO_2(100) face, shown in Fig. 16. Here, each surface Si atom has two DBs. This is compatible with ZrO_2. Again we first must form the non-polar OZrO trilayers for the (100) faces. If we put a $O^{2-}O^{2-}Zr^{4+}O^{2-}$ unit on the Si(100), the first O forms two strong Si–O bonds with each silicon, as in Fig. 16(a). Another way of looking at this is that the first O, being divalent, saturates the two DBs of the surface Si's. Then, the non-polar OZrO unit can lie on this. This also works with the ZrO terminating unit in Fig. 16(b). In this case, the Zr donates two electrons to fill the two Si DBs, to give $=Si^{2-}Zr^{4+}O^{2-}$. Thus, the 4-2 valence of ZrO_2 is compatible with (100)//(100) epitaxy.

5.2. Calculations

We have carried out total energy calculations of various atomic models of (100) interfaces to test these ideas (95). Figure 17(a) shows the ideal Si:OOZrO interface discussed above, with a double oxygen layer. Here, the interfacial oxygens are initially four-fold coordinated to two Si's and two Zr's, so we denote this interface as O_4. We find that the interfacial oxygens relax to the structure in Fig. 17(b). One

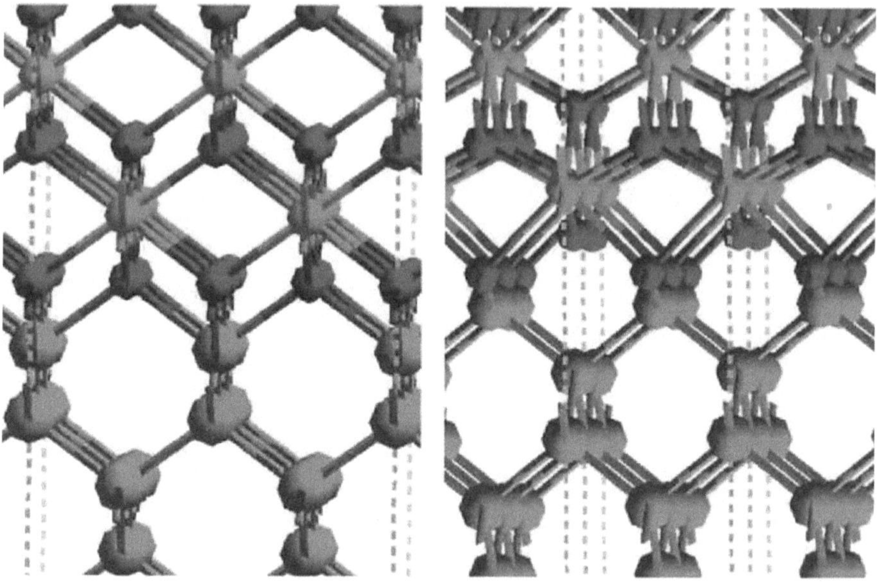
(a) (b)

Fig. 17. Bonding at the O_4-terminated (100)Si:ZrO_2 interface: (a) ideal structure, (b) relaxed structure.

oxygen relaxes downwards towards the silicon layer, to give Si–O–Si bridges. The other oxygen relaxes upwards towards the ZrO_2 layer. This creates a structure like in Fig. 16 (b) which replicates our previous discussion.

Figure 18(a) shows the ideal Si:OZrO interface, with half a layer of four-fold coordinated oxygen sites at the interface. We denote this as O_{4V}, V denoting an oxygen vacancy compared to a full ZrO_2 cell. We find the ideal interface to be metallic with the Fermi level in the conduction band, as expected from the above discussion. This interface relaxes by breaking a Si–O bond as in Fig. 18(b) and this creates a semiconductor.

Figure 19 shows other O-terminated interfaces. The first interface in Fig. 19(a) has oxygens initially three-fold coordinated, to one Si atom and two Zr atoms. This is denoted the O_3 interface. The oxygen bonding is then similar to that in $ZrSiO_4$. This interface structure relaxes to the configuration shown in Fig. 19(a) in which half of the oxygens are bonded to two Si's and one Zr, and the other half are bonded to two Zr's and one Si. The top layer Si's are each five-fold coordinated.

Another O-terminated structure is shown in Fig. 19(b). Here, one of each Si DB is used in a lateral Si–O–Si bridge. This leaves one DB to bond to the ZrO_2 layer. However, this is like the (111)Si, it needs an extra half monolayer to saturate it, to give a $Si^+(O^{2-})_{0.5}OZrO$ configuration overall. This is denoted the O_{3B} interface (B for bridge).

(a)

Fig. 18. Bonding at O-terminated (100)Si:ZrO_2 interface: (a) ideal O_{V4} structure, (b) relaxed structure.

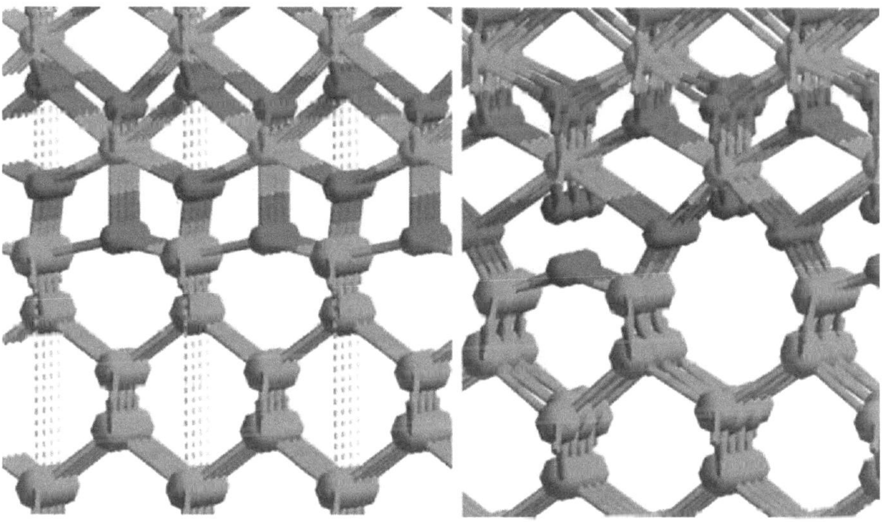

(a)

Fig. 19. Bonding at other O-terminated (100)Si:ZrO_2 interfaces: (a) O_3 interface, (b) O_{3B} structure.

Table 3. Calculated relative total energies (eV per surface Si atom), VB offsets (eV) and insulating character

	Energy (eV/Si)	VB offset (eV)	Metal?
O_4	0.27	2.9	No
O_3	0	2.9	No
O_{3B}	0.39	2.7	No
Zr_6	1.2	3.3	No
Zr_{10}	0	2.6	Yes
Bulk CNL (LDA)		3.1	

Overall, the O_4, O_3 and O_{3B} interfaces have the same number of oxygen atoms. Their total energies are compared in Table 3. The O_3 interface is found to be the most stable structure. The O_4 interface is marginally less stable (95).

Experimentally, Wang and Ong (85) studied the (100) interface configuration by high-resolution transmission electron microscopy (TEM), and modelled the lattice image. They found an atomic configuration similar to O_4, with two oxygen atoms per Si in the last O layer. This suggests that experimentally the most symmetric O-terminated interface forms, despite it not being the most stable. TEM gives evidence about the alignment of atomic planes across the interface. However, the TEM result is not conclusive as to the stoichiometry of the final O layer, because TEM does not see light elements like oxygen. Other interfaces have been studied (96) but these are not found experimentally.

Zr-terminated interfaces are also possible. The simplest has a six-fold coordinated Zr, as shown in Fig. 20(a). This structure relaxes so that the terminal Zr–Si bond lengthens. Another interface has a 10-fold coordinated Zr, with the Zr bonded to four oxygens, four Si's in the top layer and to two more Si in the layer under that. This bonding is similar to in $ZrSi_2$. Our calculation finds that the Zr_{10} is the more stable of the two (Table 3).

The calculations find that the three interfaces, O_4, O_3 and O_{3B} are insulating. They have no states in the Si band gap. However, the Zr_6 interface has states in the gap, while the Zr_{10} interface is metallic. Thus, O-terminated interfaces are more useful in devices.

The band offsets can be calculated for each structure, by comparing the energies of the VB maximum at bulk Si and oxide sites. The CB offset can be derived using the respective band gaps. Note that it is better to calculate the CB offset this way, that directly, because of the underestimation of the band gap within LDA. We find that the O-terminated interfaces have rather similar CB offsets, which are also similar to that found from the bulk CNLs. This indicates that the O-terminated configurations do not introduce significant extra dipole layers at the interface. On the other hand, the two Zr-terminated interfaces are found to have different band offsets.

The Si:$SrTiO_3$(100) interfaces have also been studied in detail (97–100). The $SrTiO_3$ lattice consists of alternating non-polar SrO and TiO_2 atomic layers along the [100] direction. Hence SrO and TiO_2 terminated faces are both non-polar. Recall that

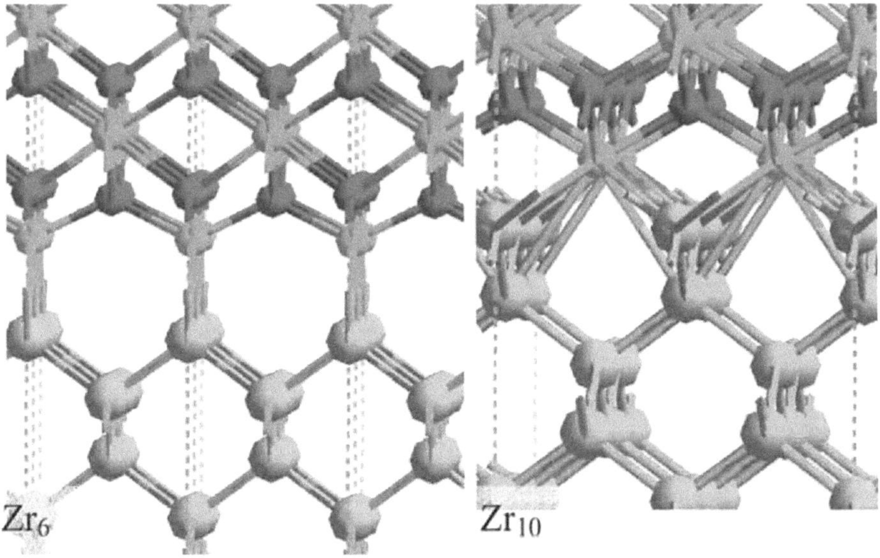

Fig. 20. Bonding at the Zr-terminated (100)Si:ZrO_2 interface: (a) Zr_6 structure, (b) Zr_{10} structure.

the ideal (100)Si face has two dangling bonds. This can reconstruct in a 2 × 1 pattern into surface dimers and leave one DB per Si. Adding a 0.5 ML of Sr will transfer an electron into each Si DB and make this surface insulating (98, 99). We can then build a $SrTiO_3$ lattice on top of this from non-polar SrO and TiO_2 layers, starting with either SrO or TiO_2 first, as in Fig. 21(a) and (b). These interfaces are analogous to Zintl salts with negatively charged Si^- DBs. Both these Sr-terminated interfaces are insulating. The interface of Fig. 21(a) may in fact be the interface formed experimentally by McKee et al. (15) despite their model showing a more complex one.

Alternatively, we can add an oxygen to each Si DB of the dimerised surface. Each oxygen wants to make two bonds, so this still leaves a half-filled oxygen DB. If we add 0.5 ML of Sr, charge transfer fills these DB states to give $-O^-$. This O-terminated interface is insulating and is shown in Fig. 21(c). This interface configuration is analogous to a metasilicate structure with Sr^{2+} being partnered by two negatively charged non-bridging oxygen sites.

Finally, an oxygen can be put into the Si–Si dimer (in either structure) to give a Si–O–Si bridge, as in Fig. 21(d). Zhang et al. (97) and Ashman (98) both find that the CB offset increases as the oxygen-content increases. Fissel et al. (101) found a similar result for Pr_2O_3 interfaces.

6. CARRIER MOBILITY AND DEFECTS

The objective of device scaling is that the smaller device is faster than the larger device. Clearly, it is not useful if a device with a high K gate oxide actually has

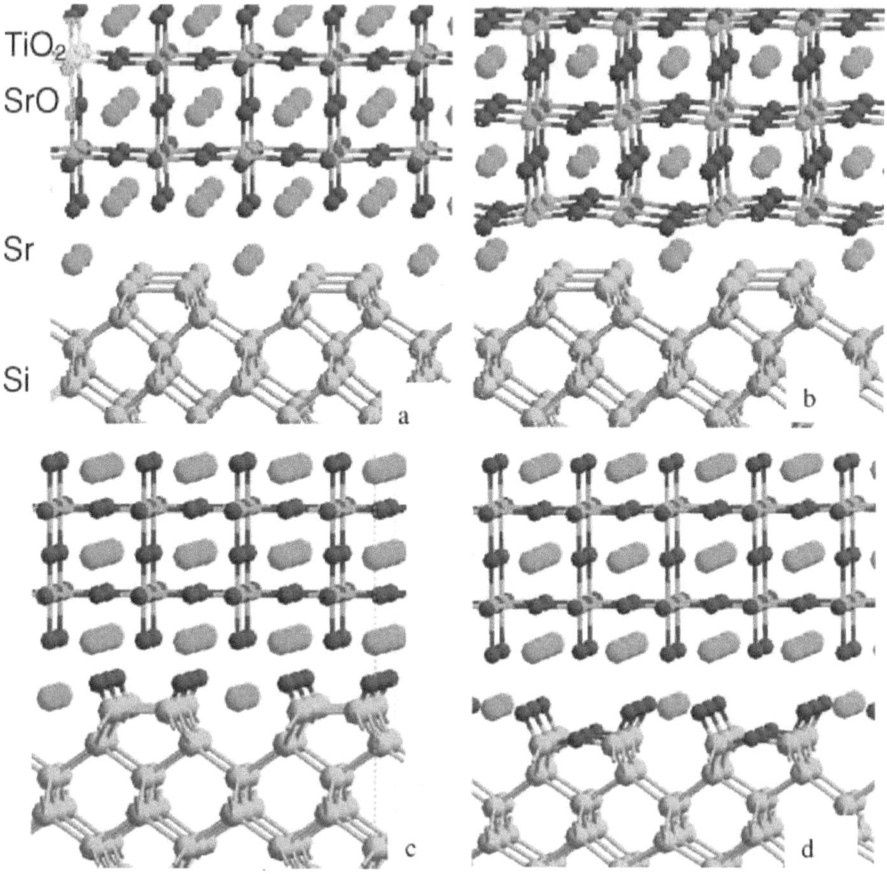

Fig. 21. Four relaxed interface structures of $SrTiO_3$ on (001)Si: (I) with 0.5ML of Sr, (II) with 0.5ML of Sr and 1 ML of oxygen, (III) with Si–O–Si bridge, and (IV) with TiO_2 layer next to interface. Sr atoms have no bonds.

a lower carrier mobility than the equivalent SiO_2 device (102), as this defeats the objective of device scaling.

The electrical performance of the high K oxides has so far been somewhat disappointing. Metal oxide semiconductor (MOS) capacitors made of many high K oxides show displaced capacitance–voltage curves, indicating the presence of a significant trapped charge (11–13, 103, 104). Field effect transistors (FETs) made with high K gate oxides show large shifts in their gate threshold voltages before annealing, which also indicates trapped charge (11–13).

The FETs with high K gate oxides show carrier mobilities less than FETs with a SiO_2 gate oxide, and well below the universal mobility curve (11–13), see Fig. 22. In the universal mobility model of Takagi et al. (105), the mobility varies universally as a function of the gate field. The mobility μ is limited at lower gate fields by Coulombic

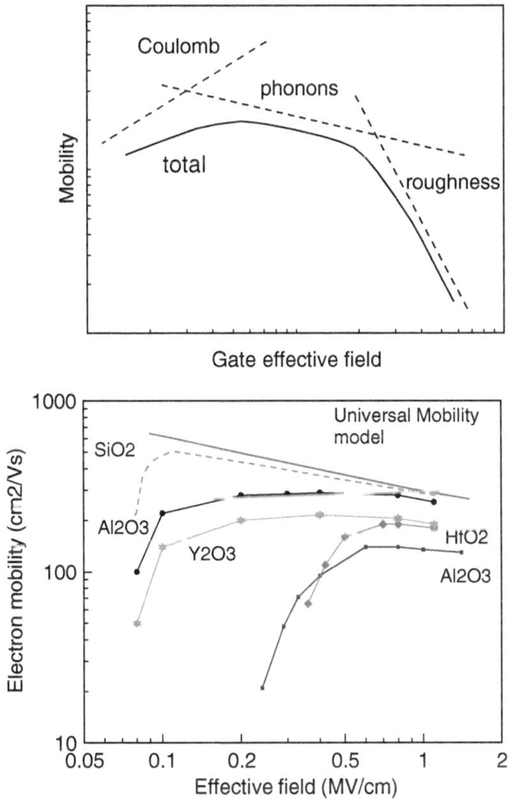

Fig. 22. (a) Scattering mechanisms in the universal mobility model, (b) Electron mobility in devices with high K gate oxides, replotted from Gusev (11) and others.

scattering, at moderate fields by phonon scattering and at high fields by interfacial roughness. The individual components of mobility add according to Matthiessen's rule,

$$\frac{1}{\mu} = \frac{1}{\mu_c} + \frac{1}{\mu_{ph}} + \frac{1}{\mu_r}$$

The carrier mobility has also been measured by Hall effect, which confirmed the lower mobility when high K gate oxides are used (106).

The cause of the lower mobilities in such devices is not fully understood. One contribution could be the remote scattering of carriers by low frequency phonon modes in the oxide, as noted by Fischetti et al. (107). The phonons of SiO_2 lie at moderately high frequencies and are not a problem. However, high K oxides get their high K because they are incipient ferroelectrics, with low-lying soft polar modes. These modes can limit mobility. It has been possible to identify this size of this contribution because phonon scattering is the only scattering mechanism that is temperature dependent

(108, 109). The silicates such as $HfSiO_4$ and $ZrSiO_4$ do not have such modes, which is an advantage for them. It has been noted that all devices in which the mobility is not degraded tend to have an interfacial layer of SiO_2 or a silicate. Degraded mobilities tend to occur when the gate oxide is just the pure high K oxide.

Once the importance of remote phonon scattering is understood, the remaining limit on mobility is the Coulomb scattering by charged defects at the interface and in the oxide. These are the same charges that cause the shifts in CV plots and the gate threshold voltage.

The origin of this trapped charge is uncertain. The first obvious source is intrinsic defects in the oxide or it could be interface traps. CV plots taken on a thickness series of HfO_2 layers have shown that the largest contribution to trapping and mobility lowering is bulk defects (110). Zafar et al. (111) also showed that trapping in HfO_2 and Al_2O_3 occurs by the filling of existing defect levels rather than the creation of new defects. This indicates that bulk defects in high K oxides are the serious problem.

The high K oxides contain more defects than SiO_2. The SiO_2 possessed such a low concentration of defects for three reasons. First, its high heat of formation means that off-stoichiometry defects such as O vacancies are costly and so they are rare. The second is that SiO_2 has covalent bonding with a low coordination. This makes SiO_2 an excellent glass former, so that the SiO_2 is amorphous (a-). The third reason is that the bonding in the a-SiO_2 can relax locally to minimise the defect concentration. The defects are dangling bonds and these can be removed by a rebonding the network. This occurs in particular for defects at the Si:SiO_2 interface.

The high K oxides differ in that their bonding is ionic, and they have higher coordination number (32). The bonding quality was displayed by plotting the average covalent gap and average ionic band gap for each oxide in Fig. 9. The average or Penn gap is the average energy between the centres of the valence and conduction bands, not the minimum gap. The average gap is made up of the some of squares of the covalent and ionic contribution.

The greater ionic character of the bonding and the higher atomic coordination numbers mean that the high K oxides are poor glass formers (32). This is well known, as it is difficult to maintain these oxides as amorphous during their high temperature processing. It is an important reason for preferring HfO_2 or La_2O_3 to ZrO_2, for using silicates instead of pure oxides, and for adding nitrogen, all to inhibit crystallisation.

The effect of poor glass forming ability and high coordination is that the oxides have higher defect concentrations. The oxides have very high heats of formation (112), so the equilibrium concentration of non-stoichiometric defects should be low (except in cases where mixed valence is possible, such as TiO_2). However, the non-equilibrium concentration of defects is high, because the oxide network is not so able to relax and rebond to remove defects.

The experimental study of electronic defects in these oxides is only just beginning. Zafar et al. (111) observed a continuous spectrum of defects across the band gap of Al_2O_3 and HfO_2. However, there is little experimental information on their chemical origin.

The nature of intrinsic defects in ionic oxides differs from those in SiO_2. They are oxygen vacancies, oxygen interstitials, or oxygen deficiency defects due to possible multiple valence of the metal. Detecting defects by electron spin resonance (ESR) is critical as it is the only technique which gives chemical information. So far, most defects found by ESR have been those related to the Si dangling bond on the Si side, called the P_b centre (113). Recently, Lenahan et al. (114) have identified three paramagnetic defects in HfO_2 subjected to corona discharging by ESR; the O vacancy, the Hf^{3+} ion (an electron trapped at Hf^{4+}) and the superoxy radical (or oxygen interstitial). These are the same centres, which were previously identified in ZrO_2 powder used in catalysis (115).

The energetics and energy levels of oxygen vacancies and oxygen interstitials in ZrO_2 and HfO_2 were calculated by Forster et al. (116, 117). They found the oxygen vacancy to be an important defect. They also found that its energy levels lie below mid gap. This is consistent with a more advanced calculation by Louie et al. (118) using the GW approximation, which corrects the LDA problem, who placed the energy level of the O vacancy near midgap in ZrO_2. They only calculated for the unrelaxed vacancy, but the effect of relaxation is small. This is consistent with the fact that the vacancy states lie in metal orbitals, which are part of the conduction band. Thus, the O vacancy is likely to be ionised as V_O^{2+}. Further work is needed. We have recently found similar results using the screened exchange (sX) method, which is much faster computationally than GW.

Another type of defect is hydrogen. Hydrogen is a ubiquitous impurity in SiO_2, where it is believed to be responsible for fixed charge in the oxide, and for many reliability processes at the interface (119, 120). The low concentration of defects at the Si:SiO_2 interface is attributed to the passivation of Si DBs by atomic hydrogen in Si–H bonds. The release of H from these bonds by hot carriers, etc, is held to cause the mobility degradation (120).

Hydrogen is also ubiquitous in high K oxide systems. The oxides are often annealed in forming gas, as were devices with SiO_2. But a major source of hydrogen is the deposition process. The most popular method of producing high K gate oxides is atomic layer deposition (ALD). This usually involves water as oxidant or organometallic precursors that can leave residual hydrogen species in the film.

The question arises as what effect hydrogen has in these oxides. The hydrogen could be tied up as OH groups, especially if introduced during growth from ALD precursors. These OH groups do not form midgap states, so they are deleterious only if they dissociate and liberate H. However, H introduced by post-deposition annealing may be present as an interstitial species. Interstitial atomic hydrogen in most semiconductors acts as a deep amphoteric impurity (121). There, H is stable in all three charge states H^+, H^0 and H^- but it is deep and unreactive. In contrast, in ZnO, H acts as a shallow donor, as the H^0 level lies above the conduction band edge (122). How does hydrogen behave in high K oxides? We have found that hydrogen acts as a shallow donor in many of the oxides, whereas it is deep in the silicates, SiO_2 and Al_2O_3. It is therefore a possible source of positive fixed charge (123).

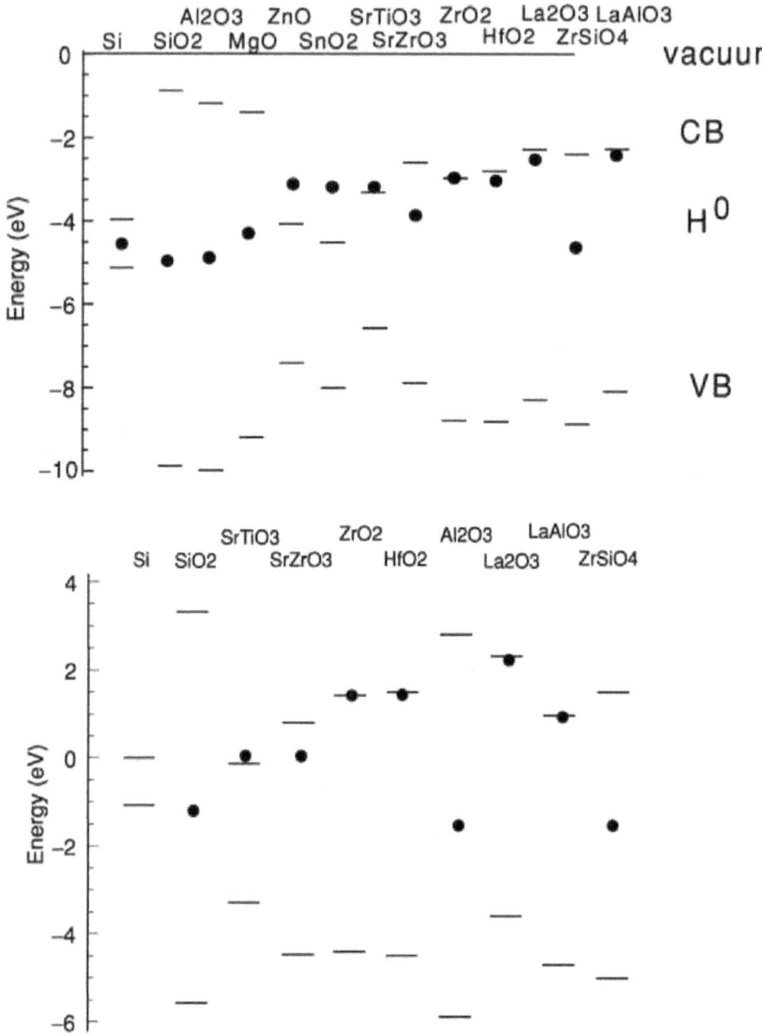

Fig. 23. Variation of H^0 level in oxides: (a) referenced to the vacuum level using electron affinities; (b) using the band alignments of the oxides and Si.

Figure 23 shows the energy level of the interstitial H^0 in the various oxides given with respect to the oxide band edges. The energies are referenced to the vacuum level by using experimental values of the electron affinity, the energy of the conduction band (CB) minimum below the vacuum level. We see that the H^0 level lies above the CB and H acts as a shallow donor in many of the candidate high K oxides, such as ZrO_2, HfO_2, La_2O_3, Y_2O_3, TiO_2, $SrTiO_3$ and $LaAlO_3$. On the other hand, we find H^0 to be deep in SiO_2, Al_2O_3, $SrZrO_3$, and in the silicates $ZrSiO_4$ and $HfSiO_4$. This is consistent with previous calculations for SiO_2 (124). Kilic and Zunger (125) proposed that the H^0 level lay at a roughly constant energy below the vacuum level. Our calculations do not support this. Van der Walle (126) proposed that it should lie at a constant energy with respect to the CNL. We find that in oxides, the level lies at a relatively constant energy with respect to the VB maximum (123).

The energy level of H^0 in the various oxides can also be aligned to the band energy of the underlying Si channel, using the relevant band offset energies of the oxides and Si (7) (Fig. 23(b)). The alignment of oxide and Si bands occurs by aligning the charge neutrality levels in the oxide and the Si. This gives a slightly different alignment than the electron affinity plot in Fig. 23(b). We see that the H^0 level lies above the Si gap in ZrO_2, HfO_2, La_2O_3 and Y_2O_3. Thus, hydrogen in thin layers of these oxides will ionise as H^+ and donate their electrons to the Si channel (even if the H^0 were deep in the bulk oxide). This gives a positive oxide fixed charge. On the other hand, we find that the H^0 level in Al_2O_3 lies in the lower half of its band gap. It would lie below midgap of Si, and therefore H in Al_2O_3 is a possible source of negative fixed charge, accepting an electron from the Si channel. $ZrSiO_4$ gives states in the Si gap region and is amphoteric.

Experimentally, the fixed charge in ZrO_2 and HfO_2 layers is usually found to be positive, while in Al_2O_3 it is negative (1). Gusev et al. (11) observed that Al_2O_3 possessed negative fixed charge whereas HfO_2 possesses positive fixed charge. Houssa et al. (103, 104) have already proposed that hydrogen gives rise to a positive fixed charge in ZrO_2 after annealing, based on the observation that H annealing makes the fixed charge more positive.

In summary, this chapter has discussed the electronic structure of the various high K oxides. It then described how to determine their band offsets first in terms of a general model of bonding, and then in terms of calculations on specific interface structures. Finally, we considered the performance limitations of the high K oxides, in terms of trapped charge at defects and other mechanisms.

ACKNOWLEDGEMENT

The authors thank many colleagues for discussions of the experimental situation.

REFERENCES

1. G. Wilk, R.M. Wallace, J.M, Anthony, *J. Appl. Phys.* **89**, 5243 (2001).
2. A.I. Kingon, J.P. Maria, S.K. Streiffer, *Nature* **406**, 1032 (2000).
3. S.H. Lo, D.A. Buchanan, Y. Taur, W. Wang, *IEEE ED Lett.* **18**, 209 (1997).
4. R.M. Wallace, G. Wilk (eds.), *MRS Bull.* 27 (April 2002).
5. H.J. Hubbard, D.G. Schlom, *J. Mater. Res.* **11**, 2757 (1996).
6. M. Copel, M. Gribelyuk, E. Gusev, *Appl. Phys. Lett.* **76**, 436 (2000).
7. J. Robertson, *J. Vac. Sci. Technol. B* **18**, 1785 (2000).
8. H. Iwai et al., *Tech. Digest IEDM* (IEEE, 2002).

9. M.R. Visokay, J.J. Chambers, A.L.P. Rotondaro, A. Shanware, L. Colombo, *Appl. Phys. Lett.* **80**, 3183 (2002).
10. Y.C. Yeo, T.J. King, C. Hu, *Appl. Phys. Lett.* **81**, 2091 (2002).
11. E.P. Gusev et al., *Tech Digest—International Electron Devices Meeting* (2001).
12. L.A. Ragnarsson, S. Guha, M. Copel, E. Cartier, N.A. Bojarczuk, J. Karasinski, *Appl. Phys. Lett.* **78**, 4169 (2001).
13. M. Copel, E. Cartier, V. Narayanan, M.C. Reuter, S. Guha, N. Bojarczuk, *Appl. Phys. Lett.* **81**, 4227 (2002).
14. K. Eisenbeiser, J.M. Finder, Z. Yu, J. Ramdani, J.A. Curless, J.A. Hallmark, R. Droopad, W.J. Ooms, L. Salem, S. Bradshaw, C.D. Overgaard, *Appl. Phys. Lett.* **76**, 1324 (2000).
15. R.A. McKee, F.J. Walker, M.F. Chisholm, *Phys. Rev. Lett.* **81**, 3014 (1998).
16. R.A. McKee, F.J. Walker, M.F. Chisholm, *Science* **293**, 468 (2001).
17. H. Li, X. Hu, Y. Wei, Z. Yu, X. Zhang, R. Droopad, A.A. Demkov, J. Edwards, K. Moore, W. Ooms, J. Kulik, P. Fejes, *J. Appl. Phys.* **93**, 4521 (2003).
18. W.A. Harrison, *Electronic Structure* (W.A. Freeman, San Francisco, 1979).
19. P.W. Peacock, J. Robertson, *J. Appl. Phys.* **92**, 4712 (2002).
20. M.C. Payne, M.P. Teter, D.C. Allan, T.A. Arias, J.D. Joannopoulos, *Rev. Mod. Phys.* **64**, 1045 (1992).
21. V. Milman, B. Winkler, J.A. White, C.J. Pickard, M.C. Payne, *Int. J. Quant. Chem.* **77**, 895 (2000).
22. D. Vanderbilt, *Phys. Rev. B* **41**, 7892 (1990).
23. R.H. French, *J. Am. Ceram. Soc.* **73**, 477 (1990).
24. Y.N. Xu, W.Y. Ching, *Phys. Rev. B* **43**, 4461 (1991).
25. R.H. French, S.J. Glass, F.S. Ohuchi, Y.N. Xu, W.Y. Ching, *Phys. Rev. B* **49**, 5133 (1994).
26. A.A. Demkov, *Phys. Stat. Solidi. (b)* **226**, 57 (2001).
27. S.G. Lim, S. Kriventsov, T.N. Jackson, J.H. Haeni, D.G. Schlom, A.M. Balbashov, R. Uecker, P. Reiche, J.L. Freeouf, G. Lucovsky, *J. Appl. Phys.* **91**, 4500 (2002).
28. T. Tomiki et al., *J. Phys. Soc. Jpn.* **55**, 4543 (1986).
29. A. Ohta, M. Yamaoka, S. Miyazaki, *Micro Electron* **72**, 154 (2004).
30. L.F. Mattheis, *Phys. Rev. B* **6**, 4718 (1972).
31. D. Redfield, W.J. Burke, *Phys. Rev. Lett.* **28**, 435 (1972).
32. G.M. Rignanese, X. Gonze, A. Pasquarello, *Phys. Rev. B* **63**, 104305 (2001).
33. H. Kato, T. Nango, T. Miyagawa, T. Katagiri, K.S. Seol, Y. Ohki, *J. Appl. Phys.* **92**, 1106 (2002).
34. A. Callegari, E. Cartier, M. Gribelyuk, H.F. Okorn-Schmidt, T. Zabel, *J. Appl. Phys.* **90**, 6466 (2001).
35. J.C. Philips, *Bonds and Bands in Semiconductors* (Academic Press, New York, 1976).
36. B.F. Levine, *J. Chem. Phys.* **59**, 1463 (1973).
37. G. Lucovsky, *J. Vac. Sci. Technol. A* **19**, 1553 (2001).
38. see chapter by R. Waghmare, K. Rabe.
39. J. Robertson, C.W. Chen, *Appl. Phys. Lett.* **74**, 1168 (1999).
40. R.L. Andersen, *Solid. State. Electronics* **5**, 341 (1962).
41. E.H. Rhoderick, R.H. Williams, *Metal Semiconductor Contacts* (Oxford University Press, Oxford, 1988).
42. A.W. Cowley, S.M. Sze, *J. Appl. Phys.* **36**, 3212 (1965).
43. W. Mönch, *Phys. Rev. Lett.* **58**, 1260 (1987).
44. C. Tejedor, F. Flores, E. Louis, *J. Phys. C* **10**, 2163 (1977).
45. J. Tersoff, *Phys. Rev. Lett.* **52**, 465 (1984).
46. J. Tersoff, *Phys. Rev. B* **30**, 4874 (1984); **32**, 6989 (1985).
47. S. Kurtin, T.C. McGill, C.A. Mead, *Phys. Rev. Lett.* **30**, 1433 (1969).
48. M. Schluter, *Thin. Solid. Films* **93**, 3 (1982).
49. W. Mönch, *Surface Sci.* **300**, 928 (1994).

50. W. Monch, *Appl. Surface Sci.* **92**, 367 (1996).
51. E.T. Yu, J.O. McCaldin, T.C. McGill, *Solid. State. Phys.* **46**, 1 (1992).
52. W.A. Harrison, J. Tersoff, *J. Vac. Sci. Technol. B* **4**, 1068 (1986).
53. R.T. Tung, *Phys. Rev. Lett.* **84**, 6078 (2000); R.T. Tung, *Phys. Rev. B* **64**, 205310 (2001).
54. R.T. Tung, *Mater. Sci. Eng.* R **35**, 1 (2001).
55. A. Baldereschi, A. Baroni, R. Resta, *Phys. Rev. Lett.* **61**, 734 (1988).
56. C.G. van de Walle, *Phys. Rev. B* **39**, 1871 (1989).
57. R.G. Dandrea, C.B. Duke, *J. Vac. Sci. Technol. B* **11**, 1553 (1993).
58. C. Berthod, N. Binggeli, A. Baldereschi, *Phys. Rev. B* **68**, 085323 (2003).
59. M. Cardona, N.E. Christensen, *Phys. Rev. B* **35**, 6182 (1987).
60. E.D. Palik, *Handbook of Optical Properties of Solids*, vols. 1–3 (Academic Press, 1985).
61. J.F. Scott, *Ferroelectrics Rev.* **1**, 1 (1998).
62. W. Schmickler, J.W. Schultze, *Modern Aspects of Electrochemistry*, vol. 17, ed. J.M. O'Bockris (Plenum Press, London, 1986).
63. S. Sayan, E. Garfunkel, S. Suzer, *Appl. Phys. Lett.* **80**, 2135 (2002) and private commun.
64. R.C. Neville, C.A. Mead, *J. Appl. Phys.* **43**, 4657 (1972).
65. G.W. Dietz, W. Antpohler, M. Klee, R. Waser, *J. Appl. Phys.* **78**, 6113 (1995).
66. M. Copel, P.R. Duncombe, D.A. Neumayer, T.M. Shaw, R.M. Tromp, *Appl. Phys. Lett.* **70**, 3227 (1997).
67. K. Abe, S. Komatsu, *Jpn. J. Appl. Phys.* **31**, 2985 (1992).
68. H. Hasegawa, T. Nishino, *J. Appl. Phys.* **69**, 1501 (1991).
69. T. Shimizu, N. Gotoh, N. Shinozaki, H. Okushi, *Appl. Surf. Sci.* **117**, 400 (1997).
70. J.W. Keiser, J.E. Rowe, J.J. Kolodziej, H. Niimi, T.E. Madey, G. Lucovsky, *J. Vac. Sci. Technol. B* **17**, 1831 (1999).
71. J.L. Alay, M. Hirose, *J. Appl. Phys.* **81**, 1606 (1997).
72. S.A. Chambers, Y. Liang, Z. Yu, R. Dropad, J. Ramdani, K. Eisenbeiser, *Appl. Phys. Lett.* **77**, 1662 (2000).
73. S. Miyazaki, *J. Vac. Sci. Technol. B* **19**, 2212 (2001).
74. D.J. Maria, *J. Appl. Phys.* **45**, 5454 (1974).
75. R. Ludeke, M.T. Cuberes, E. Cartier, *Appl. Phys. Lett.* **76**, 2886 (2000).
76. V.V. Afanasev, M. Houssa, A. Stesmans, M.M. Heyns, *Appl. Phys. Lett.* **78**, 3073 (2001); *J. Appl. Phys.* **91**, 3079 (2002).
77. G.B. Rayner, D. Kang, Y. Zhang, G. Lucovsky, *J. Vac. Sci. Technol. B* **20**, 1748 (2002).
78. Y.C. Yeo, T.J. King, C. Hu, *J. Appl. Phys.* **92**, 7266 (2002).
79. W.J. Zhu, T.P. Ma, T. Tamagawa, J. Kim, Y. Di, *IEEE. ED. Lett.* **23**, 97 (2002).
80. L.F. Edge, D.G. Schlom, S.A. Chambers, E. Cicerrella, J.L. Freeouf, B. Hollander, J. Schubert, *Appl. Phys. Lett.* (February 2004).
81. G. Lucovsky, Y. Zhang, G.B. Rayner, G. Appel, H. Ade, J.L. Whitten, *J. Vac. Sci. Technol. B* **20**, 1739 (2002)
82. G. Lucovsky, *Microelectronic Reliability* **43**, 1417 (2003).
83. see chapter by G. Lucovsky.
84. S.J. Wang, C.K. Ong, S.Y. Xu, P. Chen, W.C. Tjiu, J.W. Chai, A.C.H. Huan, W.J. Yoo, J.S. Lim, W. Feng, W.K. Choi, *Appl. Phys. Lett.* **78**, 1604 (2001).
85. S.J. Wang, C.K. Ong, *Appl. Phys. Lett.* **80**, 2541 (2002).
86. T. Ami, Y. Ishida, N. Nagasawa, A. Machida, M. Suzuki, *Appl. Phys. Lett.* **78**, 1361 (2001).
87. H. Nagata, T. Tsukahara, S. Gonda, M. Yoshimoto, H. Koinuma, *Jpn. J. Appl. Phys.* **30**, L1136 (1991).
88. G. Apostolopoulos, G. Vellianitis, A. Dimouloas, M. Alexe, R. Scholz, M. Fancuilli, D.T. Dekadjevi, C. Wimer, *Appl. Phys. Lett.* **81**, 3549 (2002).
89. V. Narayanan, S. Guha, N.A. Bojarczuk, F.M. Ross, *J. Appl. Phys.* **93**, 251 (2003).
90. S. Guha, N.A. Bojarczuk, V. Narayanan, *Appl. Phys. Lett.* **80**, 766 (2002).
91. D. Cherns, G.R. Anstis, J.L. Hutchison, J.C.H. Spence, *Philos. Mag. A* **46**, 849 (1982).

92. D.R. Hamann, *Phys. Rev. Lett.* **60**, 313 (1988).
93. P.J. van den Hoek, W. Ravenek, E.J. Baerends, *Phys. Rev. Lett.* **60**, 1743 (1988).
94. S. Satpathy, R.M. Martin, *Phys. Rev. B* **39**, 8494 (1989).
95. P.W. Peacock, J. Robertson, *Phys. Rev. Lett.* **92**, 057601 (2004).
96. V. Fiorentini, G. Gulleri, *Phys. Rev. Lett.* **89**, 266101 (2002).
97. X. Zhang, A.A. Demkov, H. Li, X. Hu, H. We, J. Kulik, *Phys. Rev. B* **68**, 125323 (2003).
98. C.J. Forst, C. Ashman, K. Schwarz, P.E. Blochl, *Nature* **427**, 56 (2004).
99. P.W. Peacock, J. Robertson, *Appl. Phys. Lett.* **83**, 5497 (2003).
100. see chapter by R.A. McKee and Yang.
101. A. Fissel, J. Dabrowski, H.J. Osten, *J. Appl. Phys.* **91**, 8986 (2002).
102. see chapter by Eisenberger et al.
103. M. Houssa, V.V. Afanasev, A. Stesmans, M.M. Heyns, *Appl. Phys. Lett.* **77**, 1885 (2000).
104. M. Houssa, M. Nailli, M.M. Heyns, A. Stesmans, *Appl. Phys. Lett.* **81**, 709 (2002).
105. S.I. Takagi, A. Toriumi, M. Iwase, H. Tango, *IEEE. Trans. Ed.* **41**, 2357 (1994).
106. L.A. Ragnarsson, N.A. Bojarczuk, J. Karasinski, S. Guha, *IEEE. ED. Lett.* 24 689 (2003).
107. M.V. Fischetti, D.A. Neumayer, E.A. Cartier, *J. Appl. Phys.* **90**, 4587 (2001).
108. Z. Ren, M.V. Fischetti, E.P. Gusev, E.A. Cartier, M. Chudzik, *Tech Digest IEDM* (2003) paper 33.2.
109. P. Chau, *Proceedings of International Workshop on Gate Integration*, Tokyo (November 2003).
110. G. Bersuker et al., presently at Motorola workshop, Austin (November 2003).
111. S. Zafar, A. Callegari, E. Gusev, M.V. Frischetti, *J. Appl. Phys.* **93**, 9298 (2003).
112. see chapter by A. Navrotsky.
113. A.Y. Kang, P.M. Lenahan, J.F. Conley, *Appl. Phys. Lett.* **83**, 3407 (2003).
114. P.M. Lenahan, preprint (November 2003).
115. J. Matta et al., *Phys. Chem. Chem. Phys.* **1**, 4975 (1999).
116. A.S. Foster, V.B. Sulimov, F. Lopez Gejo, A.L. Shluger, R.N. Nieminen, *Phys. Rev. B* **64**, 224108 (2001).
117. A.S. Foster, F. Lopez Gejo, A.L. Shluger, R.N. Nieminen, *Phys. Rev. B* **65**, 174117 (2002).
118. B. Kralik, E.K. Chang, S.G. Louie, *Phys. Rev. B* **57**, 7027 (1998).
119. W.L. Warren et al., *Appl. Phys. Lett.* **68**, 2993 (1996).
120. J.H. Stathis, E. Cartier, *Phys. Rev. Lett.* **72**, 2745 (1994).
121. C.G. van de Walle, P.J.H. Denteener, Y. Bar Yam S.T. Pantelides, *Phys. Rev. B* **39**, 10791 (1989).
122. C.G. van de Walle, *Phys. Rev. Lett.* **85**, 1012 (2000).
123. P.W. Peacock, J. Robertson, *Appl. Phys. Lett.* **83**, 2025 (2003).
124. A. Yokozawa, Y. Miyamoto, *Phys. Rev. B* **55**, 13783 (1997).
125. C. Kilic, A. Zunger, *Appl. Phys. Let.* **81**, 73 (2002).
126. C.G. van de Walle, J. Neugebauer, *Nature* **423**, 626 (2003).

Chapter 6

DIELECTRIC PROPERTIES OF SIMPLE AND COMPLEX OXIDES FROM FIRST PRINCIPLES

U.V. WAGHMARE[1] AND K.M. RABE[2]

[1] *Theoretical Sciences Unit, J Nehru Centre for Advanced Scientific Research, Jakkur PO, Bangalore, 560 064, India*

[2] *Department of Physics and Astronomy, Rutgers University, 136 Frelinghuysen Rd, Piscataway, NJ 08854-8019, USA*

ABSTRACT

We review the formalism available in first-principles density functional theory methods for the computation of the dielectric response of periodic insulators. Drawing on previous theoretical work, we discuss the sources and magnitudes of errors in these calculations. For perovskites and related oxide materials, we compare theoretical results with available experimental data on dielectric response and on related properties such as optical absorption by IR-active phonons. Giant dielectric response is related to the soft vibrational modes and the near-divergence of Born effective charges in the vicinity of ferroelectric and metal–nonmetal transitions, respectively. We include discussions of electric field tunability, dielectric loss, and finite-size effects in thin films. This microscopic analysis is used to develop guidelines in the search for new high-dielectric-constant materials. We discuss methods to model and simulate dielectric response of inhomogeneous materials (e.g. composites). The microscopic analysis is used to develop guidelines in the search for new high-dielectric constant materials.

1. INTRODUCTION

Dielectric properties of materials are of great importance to modern technology, with its characteristic trend towards faster and smaller devices. The limits on miniaturization of static and dynamic random access memories, which are based on capacitive components, can be extended by the use of capacitor materials with static dielectric constant ϵ_0 larger than that of SiO_2. Materials with low dielectric constants are also important, for example, for high-speed interconnects in integrated circuit chips.

A.A. Demkov and A. Navrotsky (eds.), Materials Fundamentals of Gate Dielectrics, 215–247.

Sensitivity of the dielectric constant to applied electric field, called tunability, is essential to applications of dielectrics in microwave communications. In most applications, a low value for the dielectric loss is essential.

The dielectric behavior is determined by a number of factors. First and foremost in our discussion will be the "intrinsic" dielectric response of the material in the form of an infinite ideal single crystal. The response can be decomposed into two parts: the dielectric response of the electrons with the nuclei held fixed, and the so-called lattice contribution, the additional screening associated with the motion of the nuclei in response to the applied field. The latter can expressed in terms of zone-center polar phonons and Born effective charges. The intrinsic dielectric loss in the lattice contribution is associated with anharmonic coupling of phonons. In most cases, the intrinsic response is expected to dominate in the range of optical phonon frequencies, typically at frequencies greater than a THz. We refer to other contributions to the total dielectric response of a system as "extrinsic." Specifically, dielectric materials generally have structural features such as point and line defects, grains, and compositional disorder that break the underlying periodicity of their single-crystal form. The total dielectric response then depends on these nonperiodic structures, both directly through their response to applied electric fields, and indirectly through their influence on the phonons and electronic states that determine the dielectric response of the ideal periodic system. In addition, if the material is ferroelectric, the sample may be in a multiple-domain state, and the presence and motion of domain walls in applied fields will contribute to the dielectric response. For a finite sample, such as a thin film or island, the net dielectric response will be modified by finite-size effects, the presence of surfaces and of interfaces, and the change in mechanical and electrical boundary conditions.

For the rational design of new dielectrics, it is necessary to understand the relationships between chemistry, structure and dielectric properties in classes of promising materials. *Ab initio* calculations provide an unbiased, reliable and cost-effective tool for exploring these complex relationships. With controlled access to atomic-scale information and ability to describe accurately the macroscopic behavior of materials, *ab initio* methods play an important role in determining the microscopic mechanisms responsible for dielectric behavior and rationalizing trends in dielectric properties of various oxide materials. In Section 2, we review the first-principles methodology for studying intrinsic dielectric response and its use in developing models that extend its applications to multi-scale and extrinsic response of materials. We include a discussion of some of the fundamental issues that arise in using "ground-state" density functional theory in the prediction of dielectric response. In Section 3, we illustrate the use of *ab initio* methods with applications to classes of oxide materials ranging from binaries to complex multinary compounds. Results from the literature are complemented by calculations, presented here for the first time, for additional compounds in each class. With relevance to reducing the size of device systems in technological applications, we review the use of *ab initio* calculations in exploring finite-size effects in dielectrics. We discuss the origins and trends in dielectric properties of various oxides in Section 4, leading to guidelines for the design of new dielectrics.

2. METHODOLOGY: FIRST-PRINCIPLES CALCULATIONS AND MODELS

For the quantitative investigation of many properties of individual materials, first-principles calculations provide a valuable tool for obtaining information at the atomic level, including the forces and stresses for predicting ground state structures of ideal crystals and various defects, and higher derivatives of the total energy for predicting polarization, Born effective charges, and phonon frequencies. In this section, we present, in some detail, the formalism underlying the first-principles computation of structural parameters and quantities related to the dielectric response, allowing the computation of the intrinsic response of an infinite single crystal. Then, we briefly discuss the construction of first-principles models for the temperature dependence of the intrinsic response and for extrinsic contributions.

2.1. The Born–Oppenheimer Approximation

The starting point for our discussion is the Born–Oppenheimer, or adiabatic, approximation, which allows the decoupling of the solution of the electronic problem from that of the ions (nuclei). In this approximation, electrons, which are much lighter than ions, are assumed to be in their quantum mechanical (QM) ground state, which evolves adiabatically to follow the ionic motion. Given that dielectrics are insulators with a gap in the electronic spectrum, this is quite often a good approximation for time scales longer than a picosecond and temperatures such that $k_B T$ is less than the electronic gap. The motion of ions is generally treated classically, QM effects being potentially important only for very light nuclei or in certain cases at low temperature. Within the Born–Oppenheimer approximation, ionic dynamics are determined by an interatomic potential derived from the total energy of the electron–ion system with stationary ions:

$$E_{\text{tot}}(Z_i, \vec{R}_i) = E_{i-i}(Z_i, \vec{R}_i) + E_{\text{G}}^{\text{el}}(Z_i, \vec{R}_i), \tag{1}$$

where E_{i-i} is the electrostatic potential energy of ions and E_{G}^{el} is the QM ground state energy of electrons, Z_i and $\vec{R}_i$ being the atomic number and position, respectively, of the ith atom. $E_{\text{tot}}(Z_i, \vec{R}_i)$ is thus the central material-specific quantity in determining the crystal structure and lattice contribution to the dielectric response, and forms the objective of first principles calculations. Calculation of the electrostatic energy E_{i-i} is quite straightforward and done using the Ewald summation technique for periodic systems. Calculation of the electron ground state energy E_{G}^{el} is the intellectually and computationally intensive ingredient and is based on density functional theory (1) (DFT), to be described in the following subsection.

2.2. DFT: Ground State and Linear Response

2.2.1. Ground state total energy and polarization

Calculation of the electron ground state energy E_{G}^{el} is based on the variational principle of density functional theory (1), which states that the ground state energy of a system of electrons in a fixed potential $V_{\text{ext}}(r)$ is a unique functional of electron density $\rho(r)$ that is variational with respect to $\rho(r)$ for a fixed total number of electrons. The functional has the following form:

$$E_G[\rho(\vec{r})] = T[\rho(\vec{r})] + \int \rho(\vec{r}) V_{ext}(\vec{r}) dr + E_{int}[\rho(\vec{r})], \tag{2}$$

where T is the kinetic energy, E_{int} is the interaction energy and V_{ext} is the external potential, for example that of fixed ions. In practice, the exact form of this functional is unknown and one uses various approximations. These can be described by expressing the density in terms of single particle (Kohn–Sham (KS)) wave functions:

$$\rho(r) = \sum_i^N |\psi_i(r)|^2,$$

where N is the number of electrons and ψ_i form an orthonormal set of functions. The energy functional then is expressed as (we use atomic units where $e = 1$, $\hbar = 1$)

$$E_G = T_s + \int \rho(\vec{r}) V_{ext}(\vec{r}) dr + \frac{1}{2} \int \frac{\rho(\vec{r})\rho(\vec{r}')}{|\vec{r} - \vec{r}'|} dr\, dr' + E_{xc}[\rho(\vec{r})], \tag{3}$$

where T_s is the noninteracting kinetic energy $-\frac{1}{2}\sum_i \int \psi^*(\vec{r})\nabla^2\psi_i(\vec{r})dr$ and

$$E_{xc}[\rho(\vec{r})] = T - T_s + E_{int} - \frac{1}{2} \int \frac{\rho(\vec{r})\rho(\vec{r}')}{|\vec{r} - \vec{r}'|} dr\, dr',$$

is the exchange-correlation energy functional. Approximations to kinetic and many-body interaction energy are buried in the functional E_{xc}, and its exact form is unknown.

There are three categories of functional forms used in the approximation of E_{xc}: (a) the local density approximation (LDA) treats it as a local function of density $E_{xc} = \int dr \rho(\vec{r}) \epsilon_{xc}(\rho(\vec{r}))$, obtained from quantum Monte Carlo calculations for the homogeneous electron gas (2); (b) the generalized gradient approximation (3) (GGA) treats it as a local function of density and its gradient $E_{xc} = \int dr \rho(\vec{r}) \epsilon_{xc}(\rho(\vec{r}), |\nabla \rho(\vec{r})|)$ and (c) the weighted density approximation (4) (WDA) treats it as a nonlocal functional of density with a short range. The computational effort increases from LDA to GGA to WDA.

In first-principles calculations for atoms, molecules and solids, V_{ext} is the electrostatic potential due to nuclei or ions. In "all-electron methods," it is the full potential due to the nuclei; an alternative is to simplify the problem by observing that only valence electrons are involved in bonding and replace the full coulomb potential of nuclei and the core electrons by a smooth pseudopotential (5). There has been extensive work on development of pseudopotentials that are transferable to different chemical environments (5).

The variational principle for E_G results in Kohn–Sham equations satisfied by wave functions ψ_i:

$$-\frac{1}{2}\nabla^2 + V_{ext} + \int \frac{\rho(\vec{r}')}{|\vec{r} - \vec{r}'|} dr' + \frac{\delta E_{XC}}{\delta \rho(\vec{r})})\psi_i(\vec{r}) = \epsilon_i \psi_i(\vec{r}), \tag{4}$$

ϵ_i being the Kohn–Sham eigenvalues. Solution of the DFT problem can either be based on direct minimization of the energy functional E_G with orthonormality constraints on the wavefunctions (6) or on the iterative self-consistent solution of the Kohn–Sham equations. These methods scale as $O(N^3)$ with system size N and become prohibitively demanding for large system sizes.

The energy E_{tot} can be minimized with respect to strain and atomic positions (often referred to as internal strain) to determine the lowest energy structure. This will be referred to as the theoretical equilibrium structure at $T = 0$ K. The energy of this structure can be used to obtain the cohesive energy (relative to free atom energies), and its second derivative with volume yields the bulk modulus. We point out that the approximation used in the exchange correlation energy leads to some error in the total energy and hence in the derived physical properties. For example, lattice constants are typically slightly underestimated in calculations based on LDA, generally by less than 2%. GGA overcorrects the LDA results and usually gives a lattice constant within 1% of experimental value. Both GGA and WDA give improved cohesive energies, but the bond-lengths and curvatures of energy they yield are not consistently better than LDA (7).

The computation of electric polarization is central to the study of dielectric behavior, yet presents a surprising degree of subtlety. Martin (8) showed that the bulk polarization of an infinite insulating crystal could not be obtained from the electron charge density alone, as one might naively suppose. It was shown by King–Smith and Vanderbilt (9) to be the geometric phase of electrons:

$$P_\alpha^{\text{el}} = \frac{(i)}{\Omega} \sum_{ki} < u_{ki} | \frac{\partial}{\partial k_\alpha} | u_{ki} >, \tag{5}$$

where Ω is unit cell volume, k is the Bloch vector and u_{ki} is the cell periodic part of the Bloch wave function of electrons. In practice, P_α^{el} is obtained using a discretized form of Eq. (5), given in ref. (9).

The difficulties in formulating a correct expression for the polarization were intimately related to those of considering an infinite system in a homogeneous electric field. Below, the polarization will appear as the first derivative of the total energy with respect to homogeneous electric field, yet the Hamiltonian in such a case formally is unbounded from below. This can be resolved by enforcing periodicity on the electron charge density. The corrresponding practical methods for doing first-principles DFT calculations in the presence of *finite* fields have only recently begun to be developed (10, 11). Fortunately, for infinitesimal fields, one can treat the homogeneous field in perturbation theory, and with the methods of density functional theory linear response (described in the next subsection), one can readily study the vibrational excitations (phonons) of the system and response of electrons and ions to homogeneous electric fields necessary to obtain the dielectric tensor.

2.2.2. Linear response: 2nd and 3rd derivatives of total energy

Once the ground state of electrons has been determined for a given V_{ext}, changes in the external potential can be treated perturbatively. The nth order changes in the Kohn–Sham wave functions due to a change in external potential can be written $\{\psi_i^{(n)}\}$. The $2n + 1$ theorem due to Gonze and Vigneron (12) states that knowledge of the

Kohn–Sham wavefunctions up to order n is sufficient to obtain changes in the ground state energy up to order $2n + 1$. Thus, $\{\psi_i^{(0)}\}$ (ground state wavefunctions) yield the first derivatives of total energy, including forces on the atoms in the unit cell (the Hellman–Feynman forces), stresses, and polarization. These quantities can be used greatly to increase the efficiency of structural prediction through minimization of the total energy, and are routinely included in first-principles implementations.

Working at first order to compute the wavefunctions $\{\psi_i^{(1)}\}$, one can obtain second and third derivatives of total energy with respect to the perturbation parameter. For a perturbation parameter λ and perturbing potential $V_{\text{ext}}^{(1)}$:

$$V_{\text{ext}} \rightarrow V_{\text{ext}}^{(0)} + \lambda V_{\text{ext}}^{(1)},$$

KS wavefunctions change as

$$\psi_i \rightarrow \psi_i^{(0)} + \lambda \psi_i^{(1)}.$$

The wavefunctions $\psi_i^{(1)}$ are solutions of the KS equations at first order:

$$\left(H_{\text{KS}}^{(0)} - \epsilon_i^{(0)}\right)\psi_i^{(1)} = -\left(H_{\text{KS}}^{(1)} - \epsilon_i^{(1)}\right)\psi_i^{(0)}, \tag{6}$$

where $H_{\text{KS}}^{(0)}$ is the zeroth order KS hamiltonian and $H_{\text{KS}}^{(1)} = V_{\text{ext}}^{(1)} + \int \frac{\partial^2 E_{\text{HXC}}}{\partial\rho(r)\partial\rho(r')}\rho(r')\mathrm{d}r'$ is the screened first order KS-hamiltonian. These quantities then are used to obtain the second derivative: $\partial^2 E_{\text{tot}}/\partial\lambda^2$. Furthermore, it can be shown that a mixed second derivative of total energy with respect to two different perturbations requires first-order wavefunctions with respect to only one of the perturbations, the other appearing through the derivatives of the potential. Details of such calculations are given in refs. (13, 14).

There are two efficient algorithms for practical linear response calculations: (a) the Green's function method (13) and (b) the variational direct minimization method (14). Two computational codes based on these methods are currently freely distributed: PWSCF code (15) and ABINIT (16), respectively. Both codes have full capability for performing ground state as well as linear response calculations and have been tested for various materials and on a number of different computing platforms.

2.3. Intrinsic Dielectric Response of Crystals at 0 K

The total energy of a material in the presence of a homogeneous electric field can be expanded in powers of the field, and the dielectric constant expressed in terms of the second derivative of E_{tot} with respect to electric field. In the Born-Oppenheimer approximation, we expand E_{tot} as a Taylor expansion in field E_α and atomic displacements $\vec{d}_i$ with respect to the equilibrium structure:

$$\begin{aligned} E_{\text{tot}}(\vec{d}_i, E_\alpha) = E_{\text{eq}} - \sum_\alpha P_\alpha^{\text{s}} E_\alpha - \frac{\Omega}{8\pi}\sum_{\alpha,\beta}\epsilon_{\alpha\beta}^{\infty} E_\alpha E_\beta \\ + \frac{1}{2}\sum_{\alpha,\beta,i,j} K_{ij,\alpha\beta} d_{i\alpha} d_{j,\beta} - \sum_{i\alpha,\beta} Z_{i\alpha,\beta} d_{i,\alpha} E_\beta, \end{aligned} \tag{7}$$

where P^s_α is the spontaneous polarization, $\epsilon^\infty_{\alpha\beta} = 1 + 4\pi \frac{dP_\alpha}{dE_\beta} = \frac{4\pi}{\Omega}\frac{d^2 E_{tot}}{dE_\alpha dE_\beta}$ is the electronic dielectric tensor, $K_{ij,\alpha\beta}$ is the force constant matrix, and $Z_{i\alpha,\beta}$ is the coupling between atomic displacement and electric field $\Omega\frac{\partial P_\beta}{\partial d_{i,\alpha}}$, known as the Born effective charge. Throughout this chapter, we use cgs units for electric field.

If the couplings are expressed as matrices (K is of size $3N \times 3N$, ϵ is a 3×3 matrix and Z is $3N \times 3$), field and atomic displacements as vectors, the energy expansion can be written in a compact form:

$$E_{tot}(\vec{d}_i, \vec{E}) = E_{eq} - \frac{\Omega}{8\pi} E \cdot \epsilon^\infty \cdot E + \frac{1}{2} d \cdot K \cdot d - d \cdot Z \cdot E - \Omega P^s \cdot E. \quad (8)$$

In the presence of an electric field, ions (atoms) feel a force proportional to the field and displace to minimize the total energy:

$$K \cdot d - Z \cdot E = 0.$$

Thus, $d = K^{-1} \cdot Z \cdot E$. We point out that the inverse of the force constant matrix K does not exist due to its three vanishing eigenvalues corresponding to the acoustic modes (or the global translational symmetry). For the sake of formal simplicity, we just lift the eigenvalues of the acoustic modes to a nonzero value; we show below that this does not affect the value of the dielectric response. Using this, one can rewrite the total energy in terms of electric field only:

$$E_{tot}(E_\alpha) = E_{eq} - \frac{\Omega}{8\pi} E(\epsilon^\infty + \frac{4\pi}{\Omega} Z \cdot K^{-1} \cdot Z) \cdot E - \Omega P^s \cdot E. \quad (9)$$

From this, the total dielectric constant (at constant strain) is readily identified:

$$\epsilon_{\alpha\beta} = \epsilon^\infty_{\alpha\beta} + \frac{4\pi}{\Omega} \sum_{i\gamma, j\delta} Z_{j\gamma,\alpha} K^{-1}_{j\gamma,i\delta} Z_{i\delta,\beta}. \quad (10)$$

In terms of eigenvalues K_i and eigenvectors u^i of the force constant matrix, the dielectric constant is expressed as

$$\epsilon_{\alpha\beta} = \epsilon^\infty_{\alpha\beta} + \frac{4\pi}{\Omega} \sum_i \frac{Z^M_{i\alpha} Z^M_{i\beta}}{K_i}, \quad (11)$$

where $Z^M_{i\alpha} = \sum_{j\beta} u^i_{j\beta} Z_{j\beta,\alpha}$ is the mode effective charge. Here we see that shifting the eigenvalue of the acoustic modes to a nonzero value does not change the result as their mode effective charges are exactly zero due to charge neutrality of the unit cell.

While the above derivation is based on the minimization of the interatomic potential energy of the system, it can be shown that the resulting ϵ is the same as the one that arises in the treatment based on the dynamical matrix D; the latter derivation has the advantage that it also provides an expression for the frequency dependence of the dielectric tensor. We introduce a diagonal matrix $S_{ij} = \delta_{ij} \frac{1}{\sqrt{M_i}}$ with dimension $3N \times 3N$, where M_i is the mass of the ith atom. Then the energy in terms of the dynamical matrix is

$$E_{\rm tot}(E_\alpha) = E_{\rm eq} - \frac{\Omega}{8\pi} E \cdot \epsilon^\infty \cdot E + \frac{1}{2}(d \cdot S^{-1}) \cdot D \cdot (S^{-1} \cdot d) - (d \cdot S^{-1}) \cdot (S \cdot Z) \cdot E,$$

With transformation $d \to d \cdot S^{-1}$, and through minimization as done previously, we get

$$\epsilon = \epsilon^\infty + \frac{4\pi}{\Omega}(S \cdot Z) \cdot D^{-1} \cdot (S \cdot Z).$$

Since $D^{-1} = S^{-1} \cdot K^{-1} \cdot S^{-1}$, this can be readily seen to be identical to the expression obtained in Eq. (10). In terms of the normal modes v^i (eigenvectors of the dynamical matrix), we write the static dielectric constant as

$$\epsilon_{\alpha\beta} = \epsilon^\infty_{\alpha\beta} + \frac{4\pi}{\Omega} \sum_i \frac{Z^{*M}_{i\alpha} Z^{*M}_{i\beta}}{\omega_i^2}, \tag{12}$$

where $Z^{*M}_{i\alpha} = \sum_{j\beta} v^i_{j\beta} Z_{j\beta,\alpha}/\sqrt{M_j}$ is the mode effective charge for phonon (note the mass factor) with frequency ω_i. Note that the definitions of the mode effective charges vary in different contexts; in particular the definition used here differs from that in Eq. (11). The frequency dependence of the dielectric constant can be derived by solving the equation of motion at frequency ω:

$$\omega^2 \sum_{i,\alpha} M_i d_{i\alpha} = \frac{\partial E_{\rm tot}}{\partial d_{i\alpha}},$$

which gives a dielectric constant

$$\epsilon_{\alpha\beta}(\omega) = \epsilon^\infty_{\alpha\beta} + \frac{4\pi}{\Omega} \sum_i \frac{Z^{*M}_{i\alpha} Z^{*M}_{i\beta}}{\omega_i^2 - \omega^2}. \tag{13}$$

The expressions up to this point have been derived for a system with fixed strain. The dielectric tensor thus obtained is that expected at frequencies high enough so that the strain degrees of freedom cannot follow the field, although below that of the lowest polar zone-center phonon. Response of a dielectric under fixed stress can be discussed by augmenting the energy functional (17) with piezoelectric and elastic energy terms and the coupling of stress with its conjugate variable, strain η:

$$F(R, \sigma, E) = E_{\rm tot}(R, \eta, E) + E_{\rm piezo}(\eta, E, R) + E_{\rm elastic}(\eta) - \Omega \sum_{\alpha\beta} \sigma_{\alpha\beta}\eta_{\alpha\beta}. \tag{14}$$

For a paraelectric, the expressions for the dielectric tensor are unchanged. For a system with a nonzero polarization at zero field, additional terms appear, and the full energy functional can be used also to compute the piezoelectric response.

In the following subsections, we show how the electron and lattice contributions to dielectric constants can be determined using DFT-based methods to evaluate the quantities that appear in the above expressions. Two alternative approaches are available: (a) linear response methods that directly evaluate the second derivatives and (b) finite-difference formulae, which for phonons are also known as the frozen-phonon method.

2.3.1. Electronic contribution: ϵ_∞

The simplest and most efficient way to compute $\epsilon_{\infty,\alpha\beta}$ is to use a linear-response code. The expression for the perturbation due to the infinitesimal homogeneous electric field is

$$H^{(1)}_{E_\beta}|u_{n\vec{k}}>= i P_c|\frac{\partial}{\partial k_\beta}u_{n\vec{k}}>, \tag{15}$$

P_c being the operator of projection onto unoccupied states (conduction bands). A detailed derivation can be found in ref. (14). In practice, it is calculated through another non-self-consistent linear response calculation similar to the $k \cdot p$ perturbation theory, usually called $\frac{d}{dk}$ linear response.

Without a linear-response code, it is still possible to compute $\epsilon_{\infty,\alpha\beta}$, though it is generally more involved. Here, we briefly mention two classes of methods, the first based on the long-wavelength limit and the second on the modern theory of polarization. In the first,

$$V^{(1)} = 2E_{\rm mac} \lim_{q\to 0} \frac{\sin(q \cdot r)}{q}.$$

A small but finite value of q and a finite value of the field can be treated in a DFT total energy calculation with a supercell which is commensurate with wave vector q. In this method (18), one takes a long wavelength ($q \to 0$) limit of $P(q) = i\rho(q)/q$ arising from the finite field. This is done by calculating $P(q)$ for several values of q and extrapolating it to zero. Yu and Krakauer (19) used the long-wavelength limit in linear response calculations by taking small but finite values of q and obtained ϵ_∞ for $KNbO_3$. In the second class of methods, Bernardini et al. (20) developed a method based on geometric phases in DFT calculations with supercells of polar solids with two slabs and used the interface charge to determine ϵ_∞. Nunes and Vanderbilt (21) developed a method based on localized orbitals and density matrix formalism to obtain change in polarization with electric field. Very recently, the capability of first-principles computations with finite electric fields has been added to ABINIT, so that it is possible to compute ϵ_∞ directly from a finite-difference formula.

In the present work, we throughout present results of LDA linear-response calculations unless otherwise specified. Calculations for a wide variety of semiconductors and insulators reported in the literature show that ϵ_∞, compared to experiment, is generally significantly overestimated, errors being of the order of +10–20% in LDA, GGA slightly improves it (22) and WDA improves it substantially but not consistently (23). It has been suggested that this systematic deviation is linked to a similiar systematic underestimate of the fundamental bandgap. This led to the development of a "scissor" correction technique (24) in which the conduction bands are lifted up to the experimental band gap and ϵ_∞ recalculated using linear response.

There are also efforts to improve the functionals (25) to be used in calculations of polar insulators, in the course of which it has become clear that there are still unresolved fundamental issues. For example, Gonze, Ghosez and Godby (GGG) (26) have recently pointed out the limitations of density functional theory in treating response of a periodic solid to an infinitesimal electric field. In particular, they find that the change in periodic charge density is not sufficient to uniquely determine the potential, which also depends on the polarization. These issues are discussed in more detail in ref. (27).

2.3.2. Lattice contribution

As shown earlier in this section, two physical quantities determine the phonon contribution to dielectric response: Born effective charges and the phonon frequencies at $\vec{q} = 0$. These are routinely calculated in one of two ways: (1) a method based on DFT linear response or (2) frozen phonon calculation with a geometric phase expression for polarization.

In the first method, $\psi^{(1)}_{kl,d_{i,\alpha}}$ is calculated in response to the perturbation produced by the atomic displacement $d_{i,\alpha}$. Each displacement yields one row and column of the force constant matrix $K_{i\alpha,j\beta}$.

$$K_{i\alpha,j\beta} = \sum_{kl} \left\langle \psi^{(0)}_{kl} \left| \frac{\partial V_{\text{ext}}}{\partial d_{j\beta}} \right| \psi^{(1)}_{kl,d_{i,\alpha}} \right\rangle.$$

The same set of first-order wavefunctions $\psi^{(1)}_{kl,d_{i,\alpha}}$ can be used in the computation of the mixed second derivatives with the electric field perturbation to obtain the Born effective charges:

$$Z_{i\alpha,\beta} = \sum_{kl} i \left\langle u^{(1)}_{kl,d_{i,\alpha}} \left| \frac{\partial}{\partial k_\beta} \right| u_{kl} \right\rangle + Z_{\text{ion}}.$$

Thus, with the minimal number of phonon linear response calculations required by the symmetry of the system, one simultaneously obtains the Born effective charges and force constant matrix (hence the phonon frequencies at Γ point).

In the frozen phonon method, an atom i is displaced in direction α by a small amount $d_{i\alpha}$ (about 1% of the lattice constant) and the DFT total energy and forces are calculated. The derivative of the force with respect to atomic displacement is calculated using a finite difference formula:

$$K_{i\alpha,j\beta} = \frac{F_{j\beta}(d_{i\alpha}) - F_{j\beta}(0)}{d_{i\alpha}},$$

where $F_{j\beta}$ is the force on atom j in β direction. Similarly, the polarization with and without the atomic displacement yield, through a finite difference formula, the Born effective charge:

$$Z_{i\alpha,\beta} = \frac{P_\beta(d_{i\alpha}) - P_\beta(0)}{d_{i\alpha}}.$$

These expressions are accurate and reliable if the energy surface is harmonic and the polarization linear with respect to the atomic displacements made. In each case, this assumption needs to be checked with additional calculations.

The inadequacy of DFT for treating a periodic solid with a macroscopic electric field described in the context of ϵ_∞ also affects the calculation of the phonon contribution to ϵ. To argue that the error is of the same order of magnitude as that in ϵ_∞, we make the assumption that the LO–TO splittings in the long wavelength limit (which are proportional to Z^2/ϵ_∞) are in principle accessible within DFT, being the result of a ground-state supercell calculation. However, we point out that there is no clear proof or rigorous justification for this and it is a subtle issue involving the effect of the inhomogeneous electric field. This assumption implies that Z^2 has percentage errors of the same magnitude as that in ϵ_∞. Since the TO phonon frequencies are certainly accessible within DFT, the phonon contribution to the dielectric constant ($\propto Z^2/\omega_i^2$) is expected to have errors which are very similar to those in ϵ_∞.

2.4. Response at Finite Electric Fields: Tunability

The importance of electric field tunability in applications of dielectric materials has prompted recent interest in the development of methods to compute this property from first-principles. In the limit $\omega \to 0$, this requires knowledge of the functional dependence $\vec{P}(\vec{E})$ for finite $\vec{E}$. The problems of performing first-principles calculations for finite electric fields can be circumvented by Taylor expansion of the relevant quantities around $\vec{E} = 0$. A first-principles formalism based on $\vec{P}$ as the fundamental variable was presented in ref. (28). Results presented in this paper, and in subsequent related studies (29) and (30), used approximations regarding the electronic susceptibility and its higher derivatives, the latter being zero. In ref. (31), a formalism was presented that allows the computation of the zero-field tunability correct to zeroth order in the field, without such approximations. It should be noted, however, that this quantity is nonzero only for a system with a zero-field polarization, such as a ferroelectric or pyroelectric.

2.5. Computational Considerations

Most first-principles studies of extended solids use periodic boundary conditions. The computational effort scales rapidly with the volume of the periodic unit cell, with current practical limits in the neighborhood of 50 atoms per unit cell for oxide compounds. This suffices for the first-principles computation of the intrinsic zero-temperature response of a wide variety of oxide compounds. However, there are materials that require very large unit cells, such as solid solutions or compositionally modulated structures. In addition, this limit is far from that needed to realize a unit cell that can exemplify a system with a realistic arrangement of point, line and planar defects for the direct computation of extrinsic response. For finite-temperature calculations, not only the supercell must be large enough to accommodate long-wavelength thermal fluctuations, but also calculations must be performed for a very large number of configurations for thermal averaging in molecular dynamics and Monte Carlo simulations. Finite systems such as thin films and islands can be studied in periodic boundary conditions by periodic replication, however, the volume of the separating vacuum greatly increases the volume of the periodic unit cell and thus the computational effort.

We have found that the most productive approach is to use a model with parameters determined from selected first-principles calculations. A successful phenomenological model is the best starting point for the construction of a first-principles model, which then is easily connected back to experimental observations. A model-based approach also has the advantage of being relatively conceptually transparent.

2.6. Response at Finite Temperature

In the insulating oxides that are the subject of our discussion, the fundamental gap is large enough so that the temperature dependence arises mainly in the lattice contribution to the dielectric response. Examinination of the phonon formalism of Eq. (12) suggests that the main source of temperature dependence should be the anharmonicity of the lattice vibrations.

The simplest treatment of lattice anharmonicity is the quasiharmonic method. In this method, we consider the system to be harmonic, but with phonon frequencies that depend on the lattice parameters, the latter information being available from first-principles calculations. At finite temperature, the structure is determined by minimizing the free energy, taken to be the energy plus the vibrational entropy of the phonon system. This leads to temperature dependence of the structural parameters (e.g., thermal expansion), and thus of the phonon frequencies and, through Eq. (12), the dielectric response (17).

Quantitatively, the quasiharmonic method is limited to systems where the anharmonicity of all relevant phonons thoughout the Brillouin zone is dominated by coupling to the structural parameters, rather than to other phonons. The latter coupling gives rise to changes in effective phonon frequencies with temperature even with fixed structural parameters. Neglect of this effect for phonons at $q \neq 0$ could cause difficulties in correctly predicting the temperature dependence of the structural parameters. If, on the other hand, the dominant coupling of the zone-center polar phonons is to the structural parameters, a semi-empirical analysis can be performed. In this case, the temperature dependence of the structural parameters are taken from experiment, and the zone-center polar phonon frequencies calculated for the varying structure.

Another consideration regarding the quasiharmonic method is that it is limited to systems that are stable at harmonic order, that is, that have no unstable phonons or double well potentials. Thus, the quasiharmonic approximation cannot be applied to the high-temperature phase(s) of a dielectric material that undergoes a symmetry-breaking structural phase transition as a function of temperature. For example, the paraelectric phases of ferroelectrics such as BST are of interest for their high dielectric constants. In these systems, phonon-phonon coupling is essential to stabilizing the high-symmetry phase. In first-principles studies, the needed anharmonicity has been incorporated by construction of an effective Hamiltonian for the phonon branch(es) containing the instability. Rabe and Joannopoulos used inputs from first-principles calculations to model and simulate the phase transition in ferroelectric GeTe (32). Rabe and Waghmare later developed a systematic method based on lattice Wannier functions to construct models for complex materials (33). In this scheme, a subspace Λ_0 of unstable phonons is identified based on the first-principles phonon dispersion for the high symmetry (usually a high-T) phase and a localized symmetrized basis (lattice analog of an electronic Wannier function) is constructed that spans this subspace. With the assumption that the strongest anharmonicity (verifiable from first-principles) is contained in the subspace Λ_0, a model is derived by integrating out the rest of the degrees of freedom. The finite temperature behavior of the model is obtained through finite temperature simulations, usually Monte Carlo. Dielectric constants are determined as the correlation functions from these simulation using the standard techniques of statistical mechanics (34, 35). Applications to real materials (36, 37) appearing in the literature will be discussed below. The effective Hamiltonian approach has been applied also to predicting the behavior of systems with complex structure, such as solid-solution and compositionally-modulated PZT (38) and PSN (39).

2.7. Extrinsic Response

Modeling of the effects of point defects, dislocations, twins and anti-phase boundaries, grain boundaries, free carriers, compositional disorder on the dielectric properties has up to the present been mainly carried out on a phenomenological level. For inhomogeneous dielectrics (such as composites) with grains of size of a micron, there are continuum methods based on simple electrostatics (40) developed to model dielectric properties. These methods can be generalized to study effects of free carriers through the use of Thomas–Fermi models for free electrons. Considerable progress has been made modeling the contributions of domain walls to the dielectric response of ferroelectric perovskites, particularly PZT (41, 42).

There has been relatively little work on incorporating first-principles input into such models. Part of the difficulty is the fact that similar effects do not seem to be of importance in semiconductors, and thus there is no precedent established by work in these systems, generally less computationally demanding, as there is for first-principles calculations of other structural and electronic properties. However, this capability is beginning to be realized; two examples of phenomenological discussion drawing on first-principles results can be found in refs. (43, 44). Further developments will incorporate first-principles calculations into phenomenological models including those mentioned in the previous paragraph.

In dielectrics with significant deviation from periodicity on different length-scales, for example the relaxor ferroelectrics, the dynamical behavior can be remarkably rich, with a very wide range of relaxation time-scales. The frequency dependence of the dielectric response of relaxors obeys the Vogel–Fulcher law, with particularly interesting behavior below 1 GHz. At present, there exist only phenomenological models for these materials. First-principles investigations that link parameters in the models with microscopics are much to be encouraged. A full understanding will require the development of multiscale models and simulation technques, as the exploration of the range of time scales requires much longer simulation times than can be achieved with even simple atomistic models.

2.8. Dielectric Loss

Dielectric losses often are the constraint on applications of dielectrics to technology. Intrinsic losses arise from the anharmonic interactions between phonons (45) and extrinsic features such as impurities, grain boundaries, free carriers in materials. There are phenomenological models for these properties (46, 47). First-principles calculations have been only recently begun to be applied even to the case of intrinsic loss (30).

3. APPLICATIONS TO REAL MATERIALS

In this section, we present results of first-principles calculations for the static dielectric tensors of a wide variety of oxide compounds. We review the previous literature in which results for the dielectric tensor have been reported for specific compounds. We will also refer to reports of computed phonon dispersion relations that do not include explicit calculations of the dielectric tensor.

This paper is, however, more than just a review. Previous investigations have typically focused on one or two compounds at a time. Technical differences between different studies can obscure the comparison between compounds in the same class. Furthermore, in a given class there may be omissions, with certain compounds not being studied at all. Here, we present calculations, many reported for the first time, for all related materials in a given class using a uniform first-principles implementation. To facilitate interpretation of these results, we include Born effective charges of these compounds and frequencies of IR-active phonon modes. We will see as a common theme that a large dielectric constant is in most cases produced by a single very-low-frequency phonon, that will also be associated with large anharmonicity and temperature dependence, and proximity to structural transitions. We will also devote some attention to the effect of d-electrons in constituent elements, both in filled d-shells (as in Cd and Zn) and in low-lying conduction band states (transition-metal oxides).

3.1. Technical Details

We used the plane-wave pseudopotential (PWPP) implementation of density functional theory (DFT) within the local density approximation (LDA) (2). For all the transition metals and other cations (except for Be), we included semi-core electrons as valence electrons. We used extended norm-conserving pseudopotentials available with ABINIT code for Ti, Zr, Sr, Ba, Pb and O and optimized pseudopotentials (5) of Rappe et al. for the rest and for BaO. A scalar relativistic treatment was used in the pseudopotential construction for elements with atomic number greater than 25.

Table 1. Calculated and measured dielectric constants of simple binary oxides that occur in the rocksalt structure

	MgO	CaO	SrO	BaO	CdO
ϵ_∞	3.20	3.89	3.91	3.86	7.32
ϵ	9.98	14.2	18.2	21.5	18.5
Expt (LB)	9.65	11.8	13.3	34	21.9
LDA lattice constant (Å)	4.17	4.7	5.05	5.44	4.64
Expt lattice constants (Å)	4.21	4.8	5.16	5.52	4.72

The experimental values for CdO are quoted from (53).

An energy cutoff ranging from 60 to 120 Ry was used depending on the material to ensure convergence with respect to plane wave basis. We approximated the Brillouin zone integrations with a minimum of $6 \times 6 \times 6$ mesh of k-points. Most of the total energy and linear response calculations were performed with ABINIT code (16) on COMPAC ES 40. All the calculations are performed at the LDA lattice constants (found in most cases to be within −2% of the experimental values), except for the ferroelectric perovskites, where we relaxed the structure internally at the experimental unit cell size and shape.

3.2. $T = 0$ K: Binary Oxides

We first discuss dielectric properties of AO oxides. Oxides of alkaline earth metals (except for Be, which forms its oxide in the wurtzite structure) and Cd occur in the rocksalt structure. In Table 1, we present results of first-principles calculations of dielectric properties of these oxides, along with experimental values taken from ref. (48). The zone-center phonon frequencies compare well with those previously reported (49–51). The electronic contribution to the dielectric response ϵ_∞ gradually increases from Mg to Ba; this is expected as the cations become more polarizable as the number of core electrons increases (51). The value for CdO is about double that of the alkaline earth oxides as the filled high-lying d-electron shell of Cd further contributes to the electronic polarizability. In contrast to ϵ_∞, we note that the lattice contributions in CaO and CdO are very similar. As can be seen from Table 3, both the effective charges and phonon frequencies of the two compounds are quite similar. We note that the lattice constants of CuO and CdO are also very similar, consistent with the Shannon–Prewitt radii of 1.0 and 0.95 Å for Ca and Cd, respectively.

The lattice contribution to the dielectric response increases significantly from MgO to BaO. This is mainly due to the softening of the optical phonon, with some contribution from the increased Born effective charge (see Table 3). For CaO and SrO, our calculated dielectric constants are slightly overestimated, while the phonon frequencies (see Table 3) are within 5% of the experimental values (51). This is partly due to the underestimate of unit cell volume in our calculations. Also, these compounds show a noticeable thermal expansion. The case of ϵ for BaO needs a

Table 2. Calculated dielectric constants of BeO and ZnO in wurtzite structures and of ZnO in the zincblende structure

	BeO	ZnO wurtzite	ZnO zincblende
ϵ_∞	3.10, 3.15	5.90, 4.44	6.79
ϵ	6.62, 7.5	10.1, 8.94	11.3
Expt (LB)	7.35	8.33, 8.84	–
LDA lattice constant (Å)	2.67, 4.33	3.25, 5.20	4.56
Expt lattice constants (Å)	2.70, 4.38	3.26, 5.21	–

separate discussion; our calculated value of 21.5 is much smaller than the experimental value of 34. This is partly due to an overestimate of the phonon frequency ($\omega = 179$ cm^{-1}, compared to the experimental value of $\omega = 155$ cm^{-1}). Calculation of the phonon frequency at the experimental lattice constant gives a significantly lower value of $\omega = 157$ cm^{-1}. Our value for ϵ at the experimental volume of BaO unit cell is 25.4, arising from change in frequency to $\omega = 157$ cm^{-1}. The latter is close to the experimentally measured frequency (50). The fact that dielectric response changed by 18% when the lattice constant changed by 2% suggests that BaO may have a large electrostrictive response. This should not be too surprising, as BaO is known to undergo structural phase transitions as a function of pressure (52).

Oxides of the first-row element in groups IIA (BeO) and IIB (ZnO) occur in the wurtzite structure. In this structure, there are two independent components of the dielectric tensor (ϵ_{xx} and ϵ_{zz}). In Tables 2 and 3, we report the computed values for BeO and ZnO. These results differ slightly from the results in ref. (17) due to better convergence with k-point sampling in the present calculations. Both the effective charge Z^*_{zz} and ϵ_∞ of BeO and of ZnO are the smallest for the oxides of their respective groups. The TO phonon frequencies of BeO and ZnO are the highest in their respective groups, that of ZnO being close to that of the TO phonon of MgO. As in the case of the group IIA and IIB oxides in the rocksalt structure, the electronic dielectric constant of ZnO is much larger than that of BeO, and the lattice contribution to the dielectric response is rather similar. This may initially seem surprising, as the calculated TO phonon frequencies of BeO and ZnO are quite different: 756 and 423 cm^{-1}, respectively (this difference can partly be attributed to the difference in unit cell volumes of BeO and ZnO, the smaller volume tending to increase interatomic force constants and stiffen phonons). However, the phonon-induced polarization is a dipole moment density, and therefore is larger in BeO because of the smaller unit cell volume, resulting overall in a similar contribution to ϵ. Hartree–Fock calculations (55) of dielectric properties of ZnO yielded effective charges and TO phonon frequencies with large error and subsequent overestimation of dielectric constant.

Recently, there has been some progress in epitaxial stabilization of ZnO in the zincblende structure, though no dielectric properties have yet been reported (54). With first-principles calculations, we can investigate how the change in structure would affect the dielectric response. Results are summarized in Table 2. The dielectric constant in the zincblende ZnO (zincblende) is isotropic, as required by symmetry,

Table 3. Calculated effective charges and IR-active phonon frequencies of various oxides. W, Z, T, O, C, m and R in parentheses indicate wurtzite, zincblende, tetragonal, orthorhombic, cubic, monoclinic and rhombohedral crystal structures respectively. "S" for $BaTiO_3$ means $BaTiO_3$ calculated at the unit cell parameters of $SrTiO_3$. xx, yy and zz components of the effective charge tensors are given in order for MO_2; for cubic perovskite ABO_3, effective charges for A, B and two independent components of O are given. For the rutile structure principle values of effective charges along (110), (1-10) and (001) directions are given. For $LaAlO_3$, we give values of Z for different atoms grouped in parentheses. For complex oxides with low symmetry, the ranges of effective charges and phonon frequencies are given.

Oxide	Effective charges	ω of IR-active phonons (cm^{-1})
MgO	1.98	404
CaO	2.38	300
SrO	2.49	222
BaO	2.58	179
CdO	2.47	291
BeO	−1.85	−756
ZnO (W)	2.17, 2.11	423, 397
ZnO (Z)	2.18	414
PbO (T)	3.15, 2.46	275, 399
PbO (O)	2.65, 3.5, 2.83	281–360, 78, 343, 418
ZrO_2	5.74, 5.74, 5.15	154, 334, 437
HfO_2 (C)	5.85	286
HfO_2 (T)	5.84, 5.84, 5.0	117, 384, 536
HfO_2 (m)	5.56, 5.55, 4.74	140–779
CeO_2	5.6	279
SiO_2	3.8, 3.8, 4.1	450–1080
TiO_2	5.34, 7.34, 7.54	165–808
La_2O_3	4.13, 3.77, −2.78, −2.53	204, 408, 221, 444
$LaAlO_3$	(4.26, 3.6), (3.18, 3.5), −2.9, −2.1, −2.4	95–394
Y_2O_2S	3.66, 3.71, −2.52, −2.6, −2.28, −2.2	201–433, 264–461
$ZrSiO_4$	5.41,4.63 (Zr), 3.25 (Si), −4.42−1.16	285–1096, 285–1029
$SrTiO_3$	2.56, 7.23, −5.69, −2.05	90, 184, 573
$BaTiO_3$ (S)	2.82, 7.24, −5.66, −2.19	187, 213, 551
$BaSnO_3$	2.75, 4.49, −3.46, −1.89	126, 209, 663
$BaZrO_3$	2.75, 6.12, −4.8, −2.03	107, 196, 543
$KTaO_3$	1.08, 8.46, −6.29, −1.62	108, 208, 547
$PbTiO_3$ (T)	(3.81,3.51), (6.35,5.58), (−4.6−2.2)	210–533
$BaTiO_3$ (R)	2.81, 6.37, −5.1, −2.04	192–580
$KNbO_3$ (R)	1.1, 8.06, −6.17, −1.49	207, 558
$CaCu_3Ti_4O_{12}$	-	125–563
$CaTiO_3$	2.27–2.59, 6.9–7.06, −1.91−5.38	85–546
Ca $(AlNb)_{1/2}O_3$	2.36–2.62, 4.1–6.26, −1.79–3.87	116–635

and is about 10% larger than that of wurtzite ZnO. In contrast, the anisotropy in wurtzite ZnO is substantial, ϵ_∞ in the *ab*-plane being about 30% larger than that along the *c*-axis. The electronic structure is quite sensitive to the difference in Zn–O bond-lengths along *c*-axis and in the *ab*-plane. This is thought to be related to stronger dispersion in electronic structure in the xy-plane of the Brillouin zone, both near the fermi energy and in the Zn 3d-band. As the local structure is tetrahedral in both cases and the unit cell volumes are within half a percent, vibrational frequencies of the two structures differ only by a few percent, as do the effective charges, and the lattice contribution is very similar in the two structures.

Pb in the +2 oxidation state is known for its interesting stereochemical activity arising from the lone pair of s-electrons. Correspondingly, the crystal chemistry of Pb-based oxides is more complex than the systems considered above. Lead monoxide occurs in two crystal structures: room temperature α-PbO litharge (red, tetragonal) and high-temperature β-PbO massicot (yellow, orthorhombic). Both forms are semiconductors with band gap energies of 1.92 and 2.7 eV, for α-PbO and β-PbO, respectively. Due to the lone pair, the effective charges deviate from the nominal value of 2: we find $Z^\star$ of Pb in α-PbO as high as 3.15 and diagonal components of $Z^\star$ of β-PbO up to 3.5 (see Table 3). Anomalous effective charges often are a clue that the system is in the vicinity of a structural phase transition (56, 57). In β-PbO, $Z^\star = 3.5$ for the yy-component. As we discuss below, in β-PbO there is a very low frequency mode (78 cm^{-1}) at Γ polarized along the y-axis, resulting in a large value of ϵ_{yy}, and in fact our LDA calculations at the experimental lattice constants find this mode to be unstable (i 85 cm^{-1}), with a resulting transition for β-PbO to a lower-symmetry structure. This implies a strong structure and hence temperature dependence in the ϵ_{yy} dielectric response of β-PbO.

In Table 4, we present results for the dielectric response of both forms of PbO. Calculations of α-PbO have been carried out at the LDA structure (compare ref. (58)), while those for β-PbO were done for an LDA structure constrained to have experimental c/a and b/a ratios. The low symmetry of the structures $P4/nmm$ and $Pbcm$ results in an anisotropic dielectric response, with two and three independent diagonal components of the dielectric tensor, respectively. The electronic contributions range between 6.5 and 8.9 for the two structures; this is higher than the oxides of divalent alkaline earth metals and arises from the lone pair of lead s-electrons. The lattice contribution is far more anisotropic than the electron contribution, which is not surprising as phonons are structural excitations and should sensitively reflect the differences in the local structure. In the case of β-PbO, we find $\epsilon_{yy}^{\text{phonon}}$ anomalously

Table 4. Calculated dielectric constants of PbO in tetragonal and orthorhombic structures

	PbO (T)	PbO (O)
ϵ_∞	6.87, 6.50	7, 8.01, 7.56
ϵ	18.3, 9.83	14.2, 224, 12.4
Expt	32	25.9
LDA lattice constant (Å)	3.91, 4.93	5.77, 4.68, 5.38
Expt lattice constants (Å)	3.98, 5.023	5.89, 4.78, 5.49

The experimental ϵ for PbO (T) is taken from ref. (59).

large due to the fact that the Γ-phonon with y-polarization is marginally stable at the LDA lattice constants. In fact, since β-PbO is only stable at high temperature, $\epsilon_{yy}^{\text{phonon}}$ should be quite temperature sensitive, as for other soft-mode systems we discuss below. Thus, if we assume that our values for the ϵ_{xx} and ϵ_{yy} are reliable, the indication from experimental estimates is that ϵ_{yy} is much larger than the other components xx and yy. This is consistent with our results, the precise values being difficult to obtain.

We now turn to oxides with metal ions in the +4 oxidation state. ZrO_2, HfO_2 and CeO_2 occur in the CaF_2 structure or its distorted forms. Dioxides of Zr and Hf (Ce) have the metal ions in the d^0 (f^0) state and the d (d, f) electronic states of the cation form the bottom of conduction bands. These can mix strongly with the p-states of oxygen at the top of the valence bands in response to perturbations and have a significant effect on the properties.

To results on ZrO_2 and HfO_2 quoted from refs. (60–62), respectively, we have added calculations for CeO_2 (Table 5). ϵ_∞ is not available for HfO_2, but is expected to be comparable to that of ZrO_2 and CeO_2 (around 6). While these oxides are often considered ionic, their effective charges deviate substantially from the nominal ionic charges, arising from mixing of d-orbitals of transition metal with the p-orbitals of oxygen. For example, the effective charges of Zr in tetragonal ZrO_2 are 5.15 and 5.74, those of Hf in tetragonal HfO_2 are 5.55 and 4.74, and that of Ce in cubic CeO_2 is 5.6. The static dielectric constants of ZrO_2 and HfO_2 are quite anisotropic, reflecting the fact that they are stable in closely related structures and possibly are close to structural phase transitions. For example, 80% of $\epsilon_\perp$, the large dielectric response perpendicular to the c-axis, of t-ZrO_2 is contributed by a soft E_u mode (153 cm^{-1}). For tetragonal HfO_2, it is interesting to see that the phonon contribution to $\epsilon_\perp$ is close to 100, with only one soft phonon mode of 117 cm^{-1}. For the monoclinic phase of HfO_2, estimates of the dielectric constant within the GGA (62) were found to be about 18% larger than the LDA values. CeO_2 is quite stable in its cubic structure (IR-active TO phonon frequency of 303 cm^{-1}) and its calculated dielectric constant is in good agreement with the experimental value.

For investigation of the role of d-electrons, we now compare dielectric properties of SiO_2 (A with no d-electrons) (64) and TiO_2 (A with d-electrons) (65) in the rutile structure (Table 6). The presence of empty d-conduction bands is clearly reflected in the anomalous Ti effective charges (as large as 7.54), in contrast to nearly nominal values for Si. The p–d hybridization also results in ϵ_∞ for TiO_2 almost twice as large as that of SiO_2. The differences in the lattice contribution to the dielectric response are even more marked. The lattice response of SiO_2 is quite similar in magnitude to that of other sp-bonded oxides discussed earlier. On the other hand, the lattice response of TiO_2 is very large and quite anisotropic. The large magnitude can be attributed to the soft phonons (A_{2u} at 176 cm^{-1} and E_u at 165 cm^{-1}) and the marked anisotropy arises from that of the effective charges. There is strong temperature dependence in the experimental measurements for TiO_2. Parker (66) reported values of $\epsilon = 86$ and 170 at room temperature whereas Samara and Peercy reported values of 115 and 251 at 4 K for a and c axes, respectively. Calculated estimates should have a good

Table 5. Calculated xx and zz components of dielectric constants of binary oxides that occur in CaF_2 structure

	t-ZrO_2	HfO_2	t-HfO_2	m-HfO_2	CeO_2
ϵ_∞	5.74, 5.28	–	–	–	6.93
ϵ	48.1, 20.3	23.9	92.3, 10.7	13.64 , 10.8, 6.98	26.0
Expt	12.5 (CRC, RT)				26
LDA lattice constant (Å)	5.02, 5.09	5.037	5.06, 5.13	5.11, 5.17, 5.28	5.36
Expt lattice constants (Å)	5.05, 5.18	5.08	–	5.12, 5.18, 5.29	5.41

For HfO_2, we also list results for HfO_2 in the high symmetry cubic structure. Results for ZrO_2 are taken from ref. (60) and those for HfO_2 are from ref. (62). Experimental results for CeO_2 are from ref. (63).

agreement with the latter low-temperature measurement (67), but the theoretical value of 165 for the zz response is thus rather low, though this mainly reflects the difficulty in accurately calculating the frequency of a very-low-frequency mode.

In ref. (64), a first-principles investigation of the pressure dependence of dielectric properties of SiO_2 is presented. It is found that the anisotropy of dielectric response decreases with pressure. The electronic contribution changed from 3.3 to 3.13 as pressure was changed from 0 to 114 GPa, whereas the static dielectric constant dropped from 11 to <8.3, indicating that the phonon frequencies, and thus the lattice dielectric response, are much more sensitive to pressure.

To probe effects of change in the coordination number, we calculated dielectric properties of TiO_2 in the CaF_2 structure. These results in comparison with those for rutile TiO_2 are shown in Table 7. As the coordination number is higher in the CaF_2 structure (= 8) than in the rutile (= 6), the p–d hybridization is expected to be

Table 6. Calculated xx and zz components of dielectric constants of SiO_2 and TiO_2 in the rutile structure

	SiO_2	TiO_2
ϵ_∞	3.3, 3.5	7.54, 8.67
Expt		6.84, 8.43
ϵ	11, 9.1	117, 165
Expt	3.9 (Wilk)	86, 170 (RT); 111, 257 (1.6 K)
LDA lattice constant (Å)	4.14, 2.66	4.54, 2.92
Expt lattice constants (Å)	4.18, 2.67	4.59, 2.96

Computed values for SiO_2 and TiO_2 are taken from refs. (64, 65) respectively. Experimental values for SiO_2 are from ref. (68) and for TiO_2 are from refs. (66, 67).

Table 7. Calculated dielectric constants of TiO_2 in rutile (65) and CaF_2 structures

	TiO_2 (rutile)	TiO_2 (CaF_2 structure)
ϵ_∞	7.54, 8.67	9.58
Z^*	5.34, 7.34, 7.54	6.57
ω	165–808	196
ϵ	117, 165	123

stronger, hence a larger ϵ_∞. The effective charge, on the other hand, is slightly lower than the average in rutile. This, as will be discussed next, is also dependent on the connectivity of the structure. The IR phonon frequencies are lower in rutile than in the CaF_2 structure, giving a larger phonon dielectric response.

As materials become more complex (e.g., with increasing number of atoms per formula unit or increasing number of formula units per cell), their structural parameters generally increase in number and tend to become temperature dependent. We consider a sesquioxide, La_2O_3, which occurs in the hexagonal $P\bar{3}ml$ structure, with two internal structural parameters. We calculated dielectric properties of La_2O_3 for both experimental and LDA structures; results are presented in Table 8. The most anisotropic quantity is the effective charge tensor of La (4.11 and 3.77) (see Table 3). The electronic response ϵ_∞ is relatively less anisotropic. The effects of anisotropy in $Z^\star$ are washed out in the overall lattice dielectric response due to a compensating anisotropy in phonon frequencies. Comparison of results calculated with the two different sets of structural parameters indicates that the primary structural sensitivity comes from the lattice contribution, mainly changes in phonon frequencies rather than effective charges. For example, one of the IR-active phonon frequency changes from 180 cm^{-1} (Th) to 203 cm^{-1} (E). Experimental measurement of dielectric response of La_2O_3 yields ϵ of 30, higher than our calculated value, but we expect temperature dependence to account for some if not most of the discrepancy.

Table 8. Calculated dielectric constants of La_2O_3 for two sets of structural parameters: (Th) denotes the theoretical equilibrium structure in LDA and (E) denotes experimental structure

	La_2O_3 (Th)	La_2O_3 (E)
ϵ_∞	4.86, 4.75	4.84, 4.75
ϵ	21.8, 21.1	25.7, 23.8
Expt ϵ		30
Lattice constant (Å)	3.90, 6.07	3.94, 6.13

Experimental values of dielectric constant are quoted from ref. (68).

Table 9. Calculated and measured dielectric constants of selected simple ternary oxides

	$LaAlO_3$	Y_2O_2S	$ZrSiO_4$
	xx, zz	xx, zz	xx, zz
ϵ_∞	4.83, 4.36	5.23, 4.87	4.06, 4.26
Expt	–		3.5, 3.8
ϵ	38.2, 18.3	12.5, 12.2	11.96, 11.53
Expt	–		11.25, 10.69
LDA lattice constant (Å)	3.53, 5.92	3.75, 6.525	6.54, 5.92
Expt lattice constants (Å)	–	3.79, 6.596	6.61, 6.00

The calculated results of Y_2O_2S and $ZrSiO_4$ are quoted from refs. (69, 70), respectively. The experimental results for $ZrSiO_4$ are from refs. (71, 72) for xx and zz components, respectively.

3.3. $T = 0$ K: Simple Ternary Oxides

We have chosen three cases of pseudo-binary oxides for consideration here. $LaAlO_3$ is a hypothetical material (not to be confused with perovskite $LaAlO_3$) obtained by substituting one of the La atoms in La_2O_3 with Al and relaxing the hexagonal structure; Y_2O_2S is a ternary obtained by substituting one of the oxygen atoms with sulphur, and $ZrSiO_4$ is a special case of the solid solution of $(ZrO_2)_x(SiO_2)_{1-x}$), with $x = \frac{1}{2}$. Results are presented in Table 9.

First, we compare the properties of La_2O_3 (Table 8) and $LaAlO_3$. Due to the smaller Al atom, the computed $LaAlO_3$ structure has smaller lattice constants, a by about 10% and c by about 2%. The effect on ϵ_∞ is in the opposite sense, the zz-component being reduced much more than the xx one. The xx component of the lattice dielectric response is almost doubled by Al substitution, whereas the zz-component is reduced by about 15%. There is an overall reduction in the effective charges with Al-substitution (see Table 3). The increase in the lattice contribution to the xx component is related to the soft (95 cm^{-1}) IR-active phonon in $LaAlO_3$.

We discuss properties of Y_2O_2S in comparison with Y_2O_3 (experimental ϵ of 15) and La_2O_3. Calculated values of ϵ_∞ are expected to be overestimated relative to experiemnt (ϵ_∞ of Y_2O_3 is 3.64). These values are higher than those of La_2O_3 and this is consistent with the general trend of increasing ϵ_∞ as one goes down in a metal-group of the periodic table, provided the structure of the oxides remains the same. The experimental value for ϵ of Y_2O_2S is about 18, much higher than the calculated (69) values in Table 9. We note that the dielectric response is much lower than that of La_2O_3; this arises both from smaller effective charges and harder phonons (see Table 3).

Zircon, $ZrSiO_4$, is a member of one of the pseudo-binary systems which shows promise for optimizing properties of high-k dielectric materials. Its electronic part of the dielectric response is quite small, typical of ionic solids, and the first-principles value (70) is well within the LDA errors with respect to the experimental values. Its calculated properties generally agree quite well with experiment. In spite of anomalous effective charges of Zr, the dielectric response more closely resembles that of the parent binary compound SiO_2 than that of ZrO_2. The anisotropy in the effective charges of Zr (Table 3) is washed out in the total lattice contribution to the dielectric response. For both ϵ_{xx} and ϵ_{zz}, one phonon (the mode that involves motion of both Zr and Si, with frequency near 300 cm^{-1}) contributes about 60% of the dielectric response, with little contribution from the other hard IR-active phonons. We speculate that to produce soft phonons and a large dielectric response, it is not enough simply to have anomalous effective charges, but there must also be a continuous geometrical linkage between atoms with anomalous effective charges. This is not the case in $ZrSiO_4$, though it will be in the perovskite oxides to be considered next.

3.4. $T = 0$ K: Perovskite Oxides

Many ternary ABO_3 oxides form in the perovskite structure. The cubic perovskite structure is obtained by corner-linking B-centered oxygen octahedral BO_6 units into a simple-cubic network, with 12-fold coordinated A atoms in the remaining holes. Many interesting materials including ferro- and antiferroelectrics, ferro- and antiferromagnetic oxides are found in this family. Due to the structural frustration inherent in the cubic perovskite structure, the ground state structures in most cases are lower-symmetry structures obtained by various types of distortion of the high-symmetry cubic perovskite structure. As different distortions can be energetically competitive for a given system, many of these oxides undergo structural phase transition(s) as a function of temperature or pressure or doping, often accompanied by electronic phase transitions. Their properties are very sensitive to structure and hence to temperature, especially in the vicinity of phase boundaries.

3.4.1. Cubic perovskites

Only a few ABO_3 oxides are stable against zone-center phonons in the cubic perovskite structure at zero temperature. Here we consider $SrTiO_3$, $BaSnO_3$, $BaZrO_3$ and $KTaO_3$. All of these have a relatively small electronic dielectric response (ϵ_∞ ranging from 4.9 to 6.6). The static dielectric response is especially large if the B site is occupied by an ionized transition metal with empty d-orbital (i.e., $SrTiO_3$, $BaZrO_3$ and $KTaO_3$, with $BaSnO_3$ being significantly lower).

$SrTiO_3$ at the LDA lattice constant is just barely stable, with a very low frequency IR-active phonon and correspondingly large static dielectric constant, in general agreement with experimental measurements. First-principles zone-center phonon frequencies have been variously reported and the low-frequency mode seems to be rather sensitive to details of the calculation. In general, decrease in the lattice constant of the cubic perovskite structure tends to harden the polar zone-center mode. For example, we can compare $SrTiO_3$ with the properties of $BaTiO_3$ computed in the cubic perovskite structure at the $SrTiO_3$ LDA lattice constant at the lattice constant (note that $BaTiO_3$ at zero pressure has a larger unit cell volume and is a rhombohedral ferroelectric at $T = 0$ K). At this lattice constant, $BaTiO_3$ is under compressive stress and we find its phonons to be hardened, as expected. This results in a much smaller

Table 10. Calculated lattice and dielectric constants of cubic perovskite oxides

	a(Å)	ϵ_∞	ϵ	Expt
$SrTiO_3$	3.86 (3.905)	6.22	245.	170, 277
$BaTiO_3$ at the STO volume	3.94 (4.00)	6.56	82.9	–
$BaSnO_3$	4.065 (4.117)	5.05	22.7	18 (LB, CRC)
$BaZrO_3$	4.135 (4.193)	4.91	53.6	43 (CRC)
$KTaO_3$	3.916 (3.989)	5.09	146.	242 (CRC)

Experimental values of dielectric constants for $SrTiO_3$ are from ref. (75); those of $BaSnO_3$, $BaZrO_3$ and $KTaO_3$ are from ref. (74).

(about a factor of 3) response in comparison with $SrTiO_3$ (see Table 10), reminiscent of the pressure dependence of the dielectric properties of SiO_2.

We calculated properties of $BaSnO_3$ and $BaZrO_3$ (see Table 10) and use comparison of the two to investigate the role of B atoms in these oxides. Shannon Prewitt radii of Sn^{4+} and Zr^{4+} are 0.69 and 0.72 Å, respectively, whereas the lattice constants of $BaSnO_3$ and $BaZrO_3$ are 4.12 and 4.19 Å. From this, one would expect to be stronger bonding between Zr and O than between Sn and O. The presence of d-states of Zr and its hybridization with O is responsible for the larger effective charge of Zr (6.03 (56)) in comparison with 5.5 of Sn. The calculated dielectric constants of both $BaSnO_3$ and $BaZrO_3$ are about 20% overestimated relative to the experimental measurements (73, 74). The lattice dielectric response is significantly larger in $BaZrO_3$ than in $BaSnO_3$, owing to the presence of a transition metal at the B-site.

Finally, our calculations for $KTaO_3$ yield a dielectric constant much smaller than the experimental value (74), which we believe is partly due to temperature dependence. Sizeable finite temperature effects are expected, as we find a soft phonon (at 108 cm^{-1}) that will be sensitive to anharmonic effects and alter the dielectric response. The effective charge of Ta is 8.46, which both reflects the strong p–d hybridization and correlates with the presence of a soft mode (56).

3.4.2. Ferroelectric perovskites

In this subsection, we consider three ferroelectric perovskites at $T = 0$ K: $PbTiO_3$, $BaTiO_3$ and $KNbO_3$. While $PbTiO_3$ has a tetragonal structure, $BaTiO_3$ and $KNbO_3$ occur in the rhombohedral structure. Each has two different principal values of the dielectric constant (eigenvalues of the 3×3 dielectric tensor), one parallel to the axis of polarization and another for the two directions perpendicular to this axis. In the rhombohedral structure, the axis is along the (111) direction, while in the tetragonal structure it is along (001). Because of the strong volume dependence of ferroelectric properties, the calculations reported here are not for the fully relaxed LDA structures, but for relaxed structures with lattice parameters taken from experiment (Table 11).

All three systems have electronic dielectric response comparable to simple binary oxides (between 5 and 8). ϵ_∞ of $PbTiO_3$ is larger than that of the others, which is expected due to the lone pair of electrons in 6s-orbital of Pb which hybridizes with the

Table 11. Calculated dielectric constants of ferroelectric perovskite oxides; components for principal axis parallel and perpendicular to the axis of polarization are reported

	$\epsilon_{\infty,\perp}$	$\epsilon_{\perp}$	$\epsilon_{\infty,\parallel}$	$\epsilon_{\parallel}$	Expt
$PbTiO_3$	7.58	134	7.03	30.5	200 (CRC, RT)
$BaTiO_3$	5.91	37.5	6.31	27	80, 2300 (CRC, RT)
$KNbO_3$	5.39	50.6	5.82	30.8	700 (CRC, RT)

p-orbitals of oxygen. We also note that the anisotropy of the electronic dielectric tensor is much greater in $PbTiO_3$ compared to the other two, with the collinear component parallel to the axis being smaller than the component perpendicular to the polarization axis in $PbTiO_3$. This is correlated with a large c/a ratio in the unit cell of t-$PbTiO_3$.

In general, we find the experimental values of the static dielectric constants to be much larger than our calculated estimates (at the experimental structures). As was motivated earlier and will be discussed further in the next section, most if not all this difference is due to the effects of temperature, which lead to softening of modes and divergence of the dielectric response as the ferroelectric transition is approached.

3.4.3. More complex perovskites

There have been only a few first-principles calculations of zero temperature dielectric response of relatively complex perovskite oxides. The complexity can arise either from the complicated distortions of a system with a small formula unit a relatively simple chemical oxide (e.g., antiferroelectric oxides such as $PbZrO_3$ or $CaTiO_3$) or from substitutions on the A site $A_xA'_{1-x}BO_3$, the B site $AB_xB'_{1-x}O_3$, or both. In this section, we consider $CaTiO_3$ from the first category (76) and $CaAl_{1/2}Nb_{1/2}O_3$ (CANO) (77) and $CaCu_3Ti_4O_{12}$ (CCTO) (43) from the second. Results are summarized in Table 12.

For $CaTiO_3$, the crystal structure is orthorhombic, with 20 atoms per unit cell. The computed electronic dielectric response is found to be that typical of perovskites described earlier. Results for static dielectric constants are given for calculations

Table 12. Calculated and measured dielectric constants of perovskite based complex oxides

	$CaTiO_3$	$CaAl_{1/2}Nb_{1/2}O_3$	$CaCu_3Ti_4O_{12}$
Lattice consts	5.28 (5.38)	(5.38)	7.29 (7.384)
	5.41 (5.44)	(5.415)	
	7.54 (7.64)	(7.626)	
ϵ_∞	6.08 (average)	4.87, 5.02, 4.89	–
ϵ	255, 281, 214	24.1, 24.3, 27.6	$40 + \epsilon_\infty$
ϵ (LDA min)	106, 100, 91	–	
Expt	300	23	120–70

performed at both LDA and experimental structures. A substantial difference in found in the two sets, with the response of the LDA structure being much weaker than that at the experimental structure, reflecting the structure dependence discussed earlier. The experimental value is larger than either of the two.

Calculations for CANO resulted in a much better overall agreement with experiment than for $CaTiO_3$. The lattice contribution to the dielectric response is dominant in both $CaTiO_3$ and CANO, though it is much smaller in CANO. A possible explanation can be found in comparison of calculations for La_2O_3 and $LaAlO_3$ presented in the earlier section. In CANO, Al sites break the continuity of linkages of NbO_6 octahedra results in harder phonons as well as a change in the character of the phonon eigenvectors (77). Agreement of theory with experiment is much better for CANO, due to lack of relatively soft phonons.

$CaCu_3Ti_4O_{12}$ proves to be a considerably more complicated case. In this case, an estimate of phonon dielectric response is considerably smaller than the experimental estimate of static dielectric response. An enlightening comparison was made with first-principles results for $CaCu_3Ti_4O_{12}$ in ref. (78). The source of difference in dielectric estimates by theory and by experiment has been conjectured to be due to extrinsic effects such as those of defects, domain or microstructures in the material (79).

Calculations for solid solutions PZT (80) and PSN (81, 82) have been carried out using an effective Hamiltonian method. These show the expected behavior in the vicinity of the morphotropic phase boundaries, with near-diverging responses dominated by a soft mode.

3.5. Finite Temperature Properties

In this section, we review the limited number of first-principles studies of the temperature dependence of the dielectric properties of simple, and then complex oxides. The application of the quasiharmonic method to ZnO, with results for temperature dependent dielectric and piezoelectric properties, can be found in ref. (17). It was found that the static dielectric constant of ZnO changes by about 4% with a change in temperature of about 300 K. Most of this change comes from the change in TO phonon frequency, resulting from the coupling of this phonon with strain. The effective charges or electronic dielectric constants are less temperature dependent. This weak T-dependence of dielectric properties appears to be captured well by a quasi-harmonic analysis, and this approach could be expected to have wider success in systems without soft modes.

The temperature dependence of dielectric properties of ferroelectric perovskite oxides on the other hand is rather strong. This is generally due to the presence of one or more branches of soft modes that are appropriately treated with an effective Hamiltonian method as discussed in Section 2.6. As is readily understood through a phenomenological Landau theory based on free energy (46), the soft mode contribution is expected to diverge $\epsilon \sim \frac{1}{T_c - T}$ near a second-order transition, and thus dominate the response. In most ferroelectric materials, the structural transition is a first-order transition due to coupling of soft modes with strain (83) and the divergence in dielectric response is cut off at a finite value at the transition. Indeed, the temperature dependence of the dielectric constant based on first-principles simulations for $BaTiO_3$ (35), $PbTiO_3$ (34) and $PbSc_{1/2}Nb_{1/2}O_3$ (39) showed a near-divergence near the weakly first-order structural phase transition. Typical simulation cell size in the $BaTiO_3$ (35) and $PbTiO_3$ (35) studies was about $15 \times 15 \times 15$. Effective Hamiltonian simulations on PSN (39) considered a much larger cell ($40 \times 40 \times 40$) to study the effects of disorder and short-range order in the B-cation substitution in $PbSc_{1/2}Nb_{1/2}O_3$ on dielectric response. It was found that the disorder and nano-range order made the transition less sharp and reduced T_c. Fluctuations in the order parameter get stronger with disorder and the peak in the dielectric response gets smeared out. Convergence in the estimate of dielectric response near the transition becomes more demanding as a result and hints at some nontrivial features in the dynamical dielectric response.

3.6. ω-Dependence of Dielectric Response

Experimental measurements of dielectric properties generally involve AC fields and it is important to take this into account in comparing experiemntal and theoretical values. Dielectric response measured at a frequency ω includes the contributions of excitations whose characteristic oscillation frequency is higher than ω. It is usually straightforward to separate electronic response, measured at optical frequencies, from the phonon contribution, which contributes below about 0.1 THz (see the expressions in Section 2). This behavior is illustrated for the example of $BaSnO_3$ in Fig. 1.

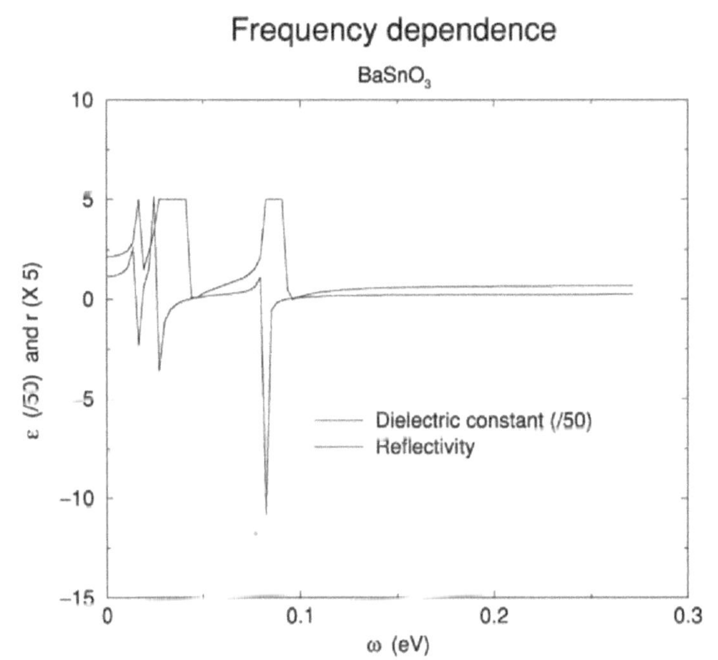

Fig. 1. Frequency dependence of the dielectric constant and reflectivity of $BaSnO_3$ calculated from first-principles.

The ω-dependence of dielectric constant in the frequency range shown have features characteristic of the TO-phonons that contribute to the response. For example, the dielectric constant sharply changes sign at the phonon frequency and the reflectivity (also shown in Fig. 1) remains close to one for the range in which ϵ remains negative.

Dielectric measurements made at frequencies much lower than phonon frequencies (<1 GHz) can be expected to include extrinsic contributions, such as the response of polar point defects or ferroelectric domain walls. This phenomenon was mentioned above in the discussion of CCTO. An important example is of the relaxor ferroelectrics (41, 84), such as $PbMg_{1/3}Nb_{2/3}O_3$. These materials show a very broad and diffused peak in dielectric response with very large magnitude and its frequency dependence exhibits Vogel–Fulcher behavior (85). The microscopic origin of this behavior is at present not clear. An attempt to relate this to nano-scale domain structure motivated by atomistic models is reported in ref. (86). The only first-principles simulations reported are for PSN (39) based on molecular dynamics. There are two challenges in first-principles understanding of these materials, even if one assumes the availability of a good model. First, the disorder and nano-range ordering are known to be crucially responsible for the properties of these materials necessitating simulations of systems with large sizes and secondly, the relevant time scales are in greater than 1 ns necessitating very long simulations. Both are prohibitively expensive with current computational resources, and thus this remains an open question for now. A strategy with multi-scale dynamical model and simulations would be quite effective in addressing these problems.

3.7. Finite Size Effects

The central question of current interest concerning finite size effects is that of the dielectric behavior of thin films, especially the difference between the dielectric constants of films and bulk, and thickness dependence in films. Experimental studies show that generally, the dielectric constants of films are substantially less than bulk values, and that the dielectric constant of a film reduces with thickness. Landau theory analysis suggests that part of the effect comes from clamping, which is especially large because of the coupling of the polar modes to strain (88). Also, details of the sample preparation techniques used to fabricate these films are likely to influence the dielectric behavior, through the introduction of defects.

More generally, soft modes of the bulk are observed to be hardened in films (87). It has been suggested (88) that the observed decrease in the dielectric constant and coercive field for nanoscale ferroelectric systems is due to a fundamental shift in lattice dynamics from a second order displacive transition to a first-order transition and disappearance or weakening of the phonon mode softening. As the temperature is increased toward T_c they do not see a dramatic shifting of modes, but rather a spectral weight transfer from the high frequency mode to the low frequency mode. The absence of mode softening reported here is quite dramatic. This work suggested a clear change in lattice dynamics for nanoscale ferroelectric films that may be highly dependent on sample preparation techniques.

Superlattice systems offer the possibility of studying finite size effects in a reasonably controlled way. $BaTiO_3/SrTiO_3$ superlattices have been rather widely investigated. A recent experimental work that focused on their dielectric properties is reported in ref. (89). A symmetric BTO/STO superlattice of period 4 was fabricated using pulsed laser deposition. The BTO layer showed significant enhancement of the tetragonality (and presumably of the polarization) of ($c/a = 1.076$ for samples thinner than 10 nm). The dielectric constant decreased from 1300 to 150 as the thickness reduced from 1000 to 100 Å, though it was larger than any of the single layer oxide films (BTO: 37, STO: 94). It was suggested that the homogeneous strain is the primary cause for finite size effects on the dielectric response. Surface effects become pronounced at the thickness of 10 nm and measurement of the dielectric constant below 5 nm thickness was reported to be quite difficult.

First-principles calculations on oxide films and superlattices have focused mainly on atomic and electronic structure at the substrate-film interface, reconstruction of the free surface, and field in the interior. In ferroelectric films, substantial attention has also been paid to the switchable polarization and its dependence on mechanical and electrical boundary conditions. First-principles calculations of effective charges and phonons of these systems are becoming feasible, and relevant results can be expected in the near future.

4. DISCUSSION

Throughout the presentation of the first-principles results, we have noted various relationships among chemistry, structure and dielectric properties in oxides. In this section, we summarize these observations and discuss the current status of rational design of dielectric materials from first-principles.

4.1. Microscopic Origins and Trends

For simple s–p bonded oxides (e.g., alkaline earth oxides), the dielectric constant increases as one goes down the column of the periodic table, provided the structure remains the same. This correlates with the increase in covalency reflected in the increasing deviation of the Born effective charges from their nominal values. Based on the experimental information on T-dependence, the phonon-strain-coupling also increases down the column.

The presence of d electrons generally increases dielectric response. Closed d-shell ZnO and CdO have substantially larger electronic dielectric response than the alkaline-earth oxides. Transition metal d-shells, particularly in the conduction bands, hybridizing with oxygen p electrons enhance the dielectric response relative to s–p systems (e.g., rutile structure TiO_2 compared to SiO_2). The electronic part of the response is enhanced by roughly a factor of 2. The anomalously large effective charges and softer phonons lead to a large contribution to the lattice dielectric response.

The dielectric properties appear to be much more strongly sensitive to local structure than to long-range order. For example, wurtzite and zincblende ZnO, with a common tetrahedral local structure, show little difference in the dielectric response. However, the changes in local structure in the cases of the polymorphs of PbO and HfO_2, lead to a nontrivial dependence of dielectric properties, as is also confirmed by our comparison of TiO_2 in the rutile and CaF_2 structures.

The geometrical arrangement of the transition metals also appears to be relevant in producing an anomalous value for the effective charge and hence enhancing dielectric response. In perovskites and related materials, infinite chains of corner-shared MO_6 octahedra are a key structural feature. When these chains are disrupted, the response decrease For example, Al-substitution in $CaTiO_3$ breaks up the infinite MO_6 chains and reduces the $Z^{\star}$ of Ti, as does Si-substitution in $ZrSiO_4$ to the $Z^{\star}$ of Zr. The breaking of the chains in Ruddlesden–Popper phases (90) and by antiphase domain boundaries (44) also decreases the effective charges and lattice contributions to the dielectric response.

In other studies (not presented here) (91), it was found that an oxygen vacancy causes breaks in the MO_6 linkages and substantially reduces the effective charge of the transition metal next to it, correlating with fewer lattice instabilities in Ca-doped $BaTiO_3$. Secondly, in the case of $BaTi_2O_5$, only the Ti atoms at the centers of predominantly corner-shared octahedra contributed to ferroelectricity (92). This is also reflected in the principal values of effective charges of Ti in the case of TiO_2 (Table 3).

4.2. Guidelines for the Design of New Dielectrics

The observations gathered in the previous section can be incorporated into a scheme for the rational design of new dielectric materials. Of course, this is a quite complex and ambitious undertaking, which requires a strong collaboration and iteration of ideas with experimental groups that can carry out controlled synthesis of bulk, thin films, and superlattices, and accurately characterize their dielectric properties. While the difficulties in precisely calculating the dielectric response of a particular materials are evident from the previous discussion (mainly arising from sources including structural errors in LDA, anharmonicity and temperature dependence), input from calculations can give valuable information about chemical and structural trends.

There are two ideas for engineering the dielectric response that we would particularly like to emphasize. One is the strong sensitivity of the lattice dielectric response to homogeneous strain. This can be used to "tune" the dielectric response of a coherent epitaxial film over a wide range through an appropriate choice of substrate. Examples of such variation has been described both in Landau theory (93) and first-principles studies (94).

The second idea is "artificial structuring": the modification of the properties of a compound by ordered substitution on one or more of the cationic sites. This could be used in the application of our observation regarding the geometrical arrangement of transition metals. For example, chains of corner-linked oxygen octahedra can be disrupted in selected directions to achieve a particular dielectric anisotropy. While most such orderings are much easier to imagine and investigate from first-principles than to realize in the laboratory, the configuration dependence of various properties, including dielectric properties, can be substantial (82).

4.3. Conclusions

Fundamental understanding of the physics underlying the dielectric response of insulating oxides, and its manifestations in the properties of individual materials, forms a rich area for discussion. In this article, we have attempted to give a sense of what first-principles calculations can do at present. In addition, there is great promise for continued progress that leads us to expect productive interactions between experimentalists and theorists in the near future for the successful design of new materials.

ACKNOWLEDGEMENTS

U.V.W. acknowldeges kind hospitality of the Department of Physics and Astronomy, Rutgers University and use of the central computing facility at J. Nehru Centre for Advanced Scientific Research, Bangalore. This work was supported in part by ONR N00014-00-0261 and NSF-NIRT DMR-0103354.

REFERENCES

1. P. Hohenberg, W. Kohn, *Phys. Rev.* **136**, 864 (1964); W. Kohn, L.J. Sham, *Phys. Rev.* **140**, 1133 (1965).
2. D.M. Ceperley, B.J. Alder, *Phys. Rev. Lett.* **45**, 566 (1980). J.P. Perdew, A. Zunger, *Phys. Rev. B* **23**, 5048 (1981).
3. J.P. Perdew, K. Burke, M. Ernzerhot, *Phys. Rev. Lett.* **77**, 3865 (1996); 78, 1396 (1997) (E).
4. O. Gunnarsson, M. Jonson, B.I. Lundquist, *Phys. Rev. B* **20**, 3136 (1979).
5. G.B. Bachelet, D.R. Hamann, M. Schluter, *Phys. Rev. B* **26**, 4199 (1982). A.M. Rappe, K.M. Rabe, E. Kaxiras, J.D. Joannopoulos, *Phys. Rev. B* **41**, 1227 (1990). D. Vanderbilt, *Phys. Rev. B* **41**, 7892 (1990).
6. M.C. Payne, M.P. Teter, D.C. Allan, T.A. Arias, J.D. Joannopoulos, *Rev. Mod. Phys.* **64**, 1045 (1992).
7. M. Saad, M.P. Teter, *Phys. Rev. B* **54**, 13643 (1996).
8. R.M. Martin, *Phys. Rev. B* **9**, 1998 (1974).
9. R.D. King-Smith, D. Vanderbilt, *Phys. Rev. B (Rapid Comm.)* **47**, 1651 (1993).
10. I. Souza, J. Iniguez, D. Vanderbilt, *Phys. Rev. Lett.* **89**, 117602 (2002).
11. P. Umari, A. Pasquarello, *Phys. Rev. Lett.* **89**, 157602 (2002).
12. X. Gonze, J.-P. Vigneron, *Phys. Rev. B* **49**, 13120 (1989).
13. S. Baroni, P. Giannozzi, A. Testa, *Phys. Rev. Lett.* **58**, 1861 (1987). S. Baroni, S. de Gironcoli, A. Dal Corso, P. Giannozzi, *Rev. Mod. Phys.* **73**, 515 (2001).
14. X. Gonze, D.C. Allan, M.P. Teter, *Phys. Rev. Lett.* **68**, 3603 (1992). X. Gonze, *Phys. Rev B* **55**, 10337 (1997).
15. S. Baroni, A. Dal Corso, S. de Gironcoli, P. Giannozzi, http://www.pwscf.org.
16. The ABINIT code is a common project of the Université Catholique de Louvain, Corning Incorporated, the Université de Liège, the Commissariat à l'Energie Atomique, Mitsubishi Chemical Corp and other contributors (URL: http://www.abinit.org).

17. N.A. Hill, U.V. Waghmare, *Phys. Rev. B* **62**, 8802 (2000).
18. J.B. McKitterick, *Phys. Rev. B* **28**, 7384 (1983).
19. R. Yu, H. Krakauer, *Phys. Rev. B* **49**, 4467 (1994).
20. F. Bernardini, V. Fiorentini, D. Vanderbilt, *Phys. Rev. Lett.* **79**, 3958 (1997).
21. R.W. Nunes, D. Vanderbilt, *Phys. Rev. Lett.* **73**, 712 (1994).
22. A. Dal Corso, S. Baroni, R. Resta, *Phys. Rev. B* **49**, 5323 (1994).
23. N. Marzari, D.J. Singh, *Phys. Rev. B* **62**, 12724 (2000).
24. Z.H. Levine, D.C. Allan, *Phys. Rev. Lett.* **63**, 1719 (1989).
25. R.M. Martin, G. Ortiz, *Phys. Rev. B* **56**, 1124 (1997).
26. X. Gonze, Ph. Ghosez, R. Godby, *Phys. Rev. Lett.* **74**, 4035 (1995).
27. W.G. Aulbur, L. Jonsson, J.W. Wilkins, *Phys. Rev. B* **54** 8540 (1996). Ph. Ghosez, X. Gonze, R. Godby, *Phys. Rev. B* **56**, 12811 (1997).
28. N. Sai, K.M. Rabe, D. Vanderbilt, *Phys. Rev. B* **66**, 104108 (2002).
29. H. Fu, L. Bellaiche, *Phys. Rev. Lett.* **91**, 057601 (2003).
30. A. Antons, D. Vanderbilt, unpublished.
31. K.M. Rabe, *Mater. Res. Soc. Proc.* **718**, 311 (2002).
32. K.M. Rabe, J.D. Joannopoulos, *Phys. Rev. Lett.* **59**, 570 (1987); *Phys. Rev. B* **36**, 6631 (1987).
33. K.M. Rabe, U.V. Waghmare, *Phys. Rev. B* **52**, 13236 (1995).
34. K.M. Rabe, E. Cockayne, *Amer. Inst. Phys. Conf. Proc.* **436**, 61 (1998).
35. Alberto García, David Vanderbilt, *Appl. Phys. Lett.* **72**, 2981 (1998).
36. W. Zhong, David Vanderbilt, K.M. Rabe, *Phys. Rev. Lett.* **73**, 1861 (1994).
37. U.V. Waghmare, K.M. Rabe, *Phys. Rev. B* **55**, 6161 (1997).
38. L. Bellaiche, A. Garcia, D. Vanderbilt, *Phys. Rev. Lett.* **84**, 5427 (2000).
39. Umesh Waghmare, E.J. Cockayne, B.P. Burton, *Ferroelectrics* **291**, 187 (2003).
40. D. Cule, S. Torquato, *Phys. Rev. B* **58**, R11829 (1998).
41. L.E. Cross, *Ferroelectrics* **76**, 241 (1987).
42. D.J. Kim, J.P. Maria, A.I. Kingon, S.K. Streiffer, *J. Appl. Phys.* **93**, 5568 (2003).
43. L. He, J.B. Neaton, M.H. Cohen, D. Vanderbilt, C.C. Homes, *Phys. Rev. B* **65**, 214112 (2002).
44. H. Li, H. Zheng, L. Salamanca-Riba, R. Ramesh, I. Naumov, K. Rabe, *Appl. Phys. Lett.* **81**, 4398 (2002)
45. V.L. Gurevich, A.K. Tagantsev, *Adv. Phys.* **40**, 719 (1991).
46. M.E. Lines, A.M. Glass, *Principles and Applications of Ferroelectrics and Related Materials* (Clarendon Press, Oxford, 1979).
47. N. McN. Alford et al., *J. Eur. Ceram. Soc.* **21**, 2605 (2001).
48. K.F. Young, H.P.R. Frederikse, *J. Phys. Chem. Ref. Data* **2**, 313 (1973).
49. O. Schutt, P. Pavone, W. Windl, K. Karch, D. Strauch, *Phys. Rev. B* **50**, 3746 (1994).
50. E.J. Wu, G. Ceder, *J. Appl. Phys.* **89**, 5630 (2001).
51. A. Aguado, L. Bernasconi, P.A. Madden, *J. Chem. Phys.* **118**, 5704 (2003).
52. S.T. Weir, Y.K. Vohra, A.L. Ruoff, *Phys. Rev. B* **33**, 4221 (1986).
53. E. Palik (ed.) *Handbook of Optical Constants* (Academic, New York, 1997).
54. S.K. Kim, S.Y. Jeong, C.R. Cho, *Appl. Phys. Lett.* **82**, 562 (2003).
55. T. Kamiya, Japan. *J. Appl. Phys.* **35**, 4421 (1996).
56. W. Zhong, R.D. King-Smith, D. Vanderbilt, *Phys. Rev. Lett.* **72**, 3618 (1994).
57. U.V. Waghmare, N.A. Hill, H. Kandpal, R. Seshadri, *Phys. Rev. B* **67** 125111 (2003).
58. G.W. Watson, S.C. Parker, *J. Phys. Chem. B* **103**, 1258 (1999).
59. H. Mayer, J. Dubois, *Comptes Rendus Hebdomadaires des Seances de L'Academie des Sciences, Serie B (Sciences Physiques)* **277**, 229 (1973).
60. G.-M. Rignanese, F. Detraux, X. Gonze, A. Pasquarello, *Phys. Rev. B* **64**, 134301 (2001).
61. X. Zhao, D. Vanderbilt, *Phys. Rev. B* **65**, 075105 (2002).
62. X. Zhao, D. Vanderbilt, *Phys. Rev. B* **65**, 233106 (2002).

63. T. Inoue, Y. Yamamoto, S. Koyama, S. Suzuki, Y. Ueda, *Appl. Phys. Lett.* **60**, 1381 (1992).
64. C. Lee, X. Gonze, *Phys. Rev. B* **56**, 7321 (1997).
65. C. Lee, Ph. Ghosez, X. Gonze, *Phys. Rev. B* **50**, 13379 (1994).
66. R.A. Parker, *Phys. Rev.* **124**, 1719 (1961).
67. G.A. Samara, P.S. Peercy, *Phys. Rev. B* **7**, 1131 (1973).
68. G.D. Wilk, R.M. Wallace, J.M. Anthony, *J. Appl. Phys.* **89**, 5243 (2001).
69. M. Mikami, S. Nakamura, M. Itoh, *Phys. Rev. B* **65**, 094302 (2002).
70. G.-M. Rignanese, X. Gonze, A. Pasquarello, *Phys. Rev. B* **63**, 104305 (2001).
71. F. Gervais, B. Piriou, F. Cabannes, *J. Phys. Chem. Solids* **34**, 1785 (1973).
72. C. Pecharoman, M. Ocana, P. Tartaj, C.J. Serna, *Mater. Res. Bull.* **29**, 417 (1994).
73. K.-H. Hellwege and A. M. Hellwege, eds., Landolt-Bornstein Numerical Data and Functional Relationships in Science and Technology, Group III: Crystal and Solid State Physics, vol. 11: Elastic, Piezoelectric, Pyroelectric, Electrooptic Constants and Nonlinear Dielectric Susceptibilities of Crystals (Springer-Verlag, Berlin, 1979).
74. D.L. Lide, CRC handbook of Chemistry and Physics, 78th edition 1997–98 (CRC Press, Boca Raton).
75. E.A. Geiss, R.L. Sandstrom, W.J. Gallagher, A. Gupta, S.L. Shinde, R.F. Cook, E.I. Cooper, E.J.M. OA Sullivan, J.M. Roldan, A.P. Segmuller, J. Angilello, *IBM J. Res. Develop.* **34**, 916 (1990)
76. E. Cockayne, B.P. Burton, *Phys. Rev. B* **62** 3735 (2000).
77. E. Cockayne, *J. Appl. Phys.* **90**, 1459 (2001).
78. L. He, J.B. Neaton, D. Vanderbilt, M.H. Cohen, *Phys. Rev. B* **67** 012103 (2003).
79. L. He, J.B. Neaton, M.H. Cohen, D. Vanderbilt, C.C. Homes, *Phys. Rev.* B **65**, 214112 (2002).
80. L. Bellaiche, A. Garcia, D. Vanderbilt, *Phys. Rev. Lett.* **84**, 5427 (2000).
81. A.M. George, J. Iniguez, L. Bellaiche, *Nature* **413**, 54 (2001).
82. J. Iniguez, L. Bellaiche, *Phys. Rev. Lett.* **87**, 095503 (2001).
83. K.M. Rabe, U.V. Waghmare, *Phil. Tran. Roy. Soc. Lond. A* **354** 2897 (1996); K.M. Rabe, U.V. Waghmare, *Ferroelectrics* **194**, 119 (1997).
84. G.A. Smolensky, A.I. Agranovskaya, *Sov. Phys. Sol. State* **1**, 1429 (1959).
85. D. Viehland, S.J. Jang, L.E. Cross, M. Wuttig, *J. Appl. Phys.* **68**, 2916 (1990).
86. S.A. Prosandeev, I.P. Raevski, U.V. Waghmare, *AIP Conference Proceedings* **677**, 41 (2003); S.A. Prosandeev, U. V. Waghmare, I.P. Raevski, L. Jastrabik, *Integrated Ferroelectrics* **58**, 1359 (2003).
87. A.A. Sirenko et al., *Nature* **404**, 373 (2000).
88. D. Wolpert, W. Cox, J. Cerne, A. Markelz, T. Zhao, R. Ramesh, *Technical Proceedings of the 2003 Nanotechnology Conference and Trade Show* Chapter 2, Vol. 2 (2003).
89. Juho Kim, Leejun Kim et al., *Jpn. J. Appl. Phys.* **42**, 5901 (2003).
90. C. Fennie, K.M. Rabe, *Phys. Rev. B* **68**, 184111 (2003)
91. U.V. Waghmare, in preparation.
92. U.V. Waghmare, M.H.F. Sluiter, T. Kimura, T. Goto, Y. Kawazoe, *Appl. Phys. Lett.* **84**, 4917 (2004).
93. N.A. Pertsev et al., *Appl. Phys. Lett.* **77**, 2596 (2000).
94. C. Bungaro, K.M. Rabe, *Phys. Rev.* B **69**, 184101 (2004).

Chapter 7

IVB TRANSITION METAL OXIDES AND SILICATES: AN AB INITIO STUDY

GIAN-MARCO RIGNANESE

Unité de Physico-Chimie et de Physique des Matériaux,
Université Catholique de Louvain,
1 Place Croix du Sud, B-1348 Louvain-la-Neuve, Belgium

ABSTRACT

Using density-functional theory, we investigate the structural, vibrational and dielectric properties of group IVb transition metals (M = Hf, Zr, Ti) oxides and silicates which have drawn considerable attention as alternative high-k materials. For the oxides, three crystalline phases of dioxide are considered. The first two are the cubic and tetragonal structures which exist for hafnia HfO_2 or zirconia ZrO_2, while it is hypothetical for titania TiO_2. The third one is the rutile structure, which on the contrary is the naturally occurring phase of titania, while it is hypothetical for hafnia and zirconia. For the silicates, we analyze first the crystalline phases: hafnon $HfSiO_4$, zircon $ZrSiO_4$ and an hypothetical $TiSiO_4$ structure. Finally, we consider the amorphous silicates. We introduce a scheme which relates the dielectric constants to the local bonding of Si and metal atoms, based on the definition of parameters characteristic of the basic structural units centered on Si and metals atoms and including their nearest O neighbors. This scheme which considerably reduces the computational cost of the calculations allows one to treat much larger systems. Applied to amorphous Zr silicates, it provides a good description of the measured dielectric constants, both of the optical and the static ones.

INTRODUCTION

The key challenge of computational condensed matter physics is to predict the properties of all kinds of materials. In this respect, one of the most successful approaches is the density-functional theory (DFT), which describes very accurately not only standard bulk materials but also complex systems such as proteins and carbon nanotubes.

A.A. Demkov and A. Navrotsky (eds.), Materials Fundamentals of Gate Dielectrics, 249–290.

In the framework of the quest for high-k materials to replace conventional SiO_2 as the gate dielectric in MOS devices, first-principles calculations constitute a valuable tool to understand the behavior of novel materials at the atomic scale without requiring empirical data. This is particularly interesting for the early stages of research when relatively little experimental data are available. In terms of its predictive accuracy, density-functional theory has proven to be very appropriate to study the ground-state properties of the electronic system, such as the structural, vibrational, and dielectric properties on which we will focus in this chapter.

However, DFT calculations have one important drawback associated to their high computational cost, which limits both the length and time scales of the phenomena which can be modeled. Nowadays, it is possible to treat systems containing up to hundreds of atoms within the most widespread DFT approach based on plane-wave basis sets and pseudopotentials. For the high-k materials that we consider here, it is important to note that transition-metal and first-row elements (e.g., oxygen) generally present an additional difficulty when treated with plane-wave basis sets. Namely, their valence wave functions are generally strongly localized around the nucleus and may require a large number of basis functions to be described accurately, thus further limiting the size of the system that can be investigated.

This chapter is dedicated to the first-principles study of the group IVB transition metal oxides and silicates which have drawn considerable attention as alternative high-k materials. Indeed, these systems have shown much promise in overall materials properties (1). On the one hand, the TiO_2 system is attractive due to its anomalously high permittivity (2–5). On the other hand, HfO_2 and ZrO_2 as well as the silicates $HfSiO_4$ and $ZrSiO_4$ in the form of amorphous films are stable in direct contact with Si up to high temperature, which is highly desirable to avoid the degradation of the interface properties by formation of a low-k interfacial layer (6, 7). In fact, the Hf–Si–O and Zr–Si–O phase diagram present a large phase field of stable silicates (a detailed discussion can be found in the chapter by Navrotsky and Ushakov); while, on the contrary, for the Ti–Si–O system, there is little mutual solubility between TiO_2 and SiO_2. Recently, however titanium silicates have also been stabilized and considered as a potential alternative for SiO_2 (8–10). The idea is to increase the static permittivity ϵ_0 with the amount of Hf, Zr, or Ti incorporated into the silicate film. In order to be able to control this process, it is highly desirable to develop an understanding of how the permittivity of Hf, Zr, and Ti silicates are affected by the underlying microscopic structure.

This chapter is organized as follows. In Section 1, we briefly describe the main results of the density functional theory and present the principal equations related to the properties that will be analyzed in the subsequent sections. We also provide some technical details about the calculations. Section 2 is devoted to the study of structural, vibrational and dielectric properties of hafnia (HfO_2), zirconia (ZrO_2), and titania (TiO_2). Three crystalline phases are considered: the cubic, the tetragonal, and the rutile structures. The differences and the analogies between the three phases and between hafnia, zirconia, and titania are presented in details. In Section 3, the structural and electronic properties of the crystalline silicates $MSiO_4$ (with M = Hf, Zr, Ti) are investigated. We discuss their Born effective charge tensors and compare the phonon frequencies at the Γ point. A detail analysis of the dielectric permittivity tensors is presented. Section 4 is dedicated to the study of amorphous silicates. For this purpose, a scheme is introduced which relates the dielectric constants to the local bonding of Si and M (= Hf, Zr, Ti) atoms. The central idea is to define of characteristic parameters for the basic structural units (SUs) formed by Si and M (= Hf, Zr, Ti) atoms and their nearest neighbors. With this scheme, heavy large-scale calculations, which are beyond current computational capabilities, are avoided. Applied to amorphous Zr silicates, our scheme provides a good description of the measured dielectric constants, both of the optical and the static ones. Finally, in Section 5, we present our conclusions.

1. THEORETICAL BACKGROUND

1.1. Ground State Properties

The key concept of density functional theory is to describe an interacting system of electrons through the density rather than through the many-body wavefunction. For a solid with N electrons obeying the Pauli principle and interacting via the Coulomb potential, the complexity of the problem is reduced from $3N$ degrees of freedom for the many-body wavefunction to only three (the spatial coordinates x, y, and z).

Firstly, Hohenberg and Kohn (11) showed that the ground state of the electron system is completely defined by the electron density which minimizes the total energy. Furthermore, they demonstrated that all the other ground state properties of the system (e.g., the lattice constant, the cohesive energy, etc.) are functionals of the ground state electron density. Consequently, once the ground state electron density is known, all the other ground state properties follow (in principle, at least).

Later on, Kohn and Sham (12) proved that this variational approach is equivalent to equations of a very simple form:

$$(T + v_{\rm KS}[n])\,|\psi_\alpha\rangle = (T + v_{\rm ext} + v_{\rm H}[n] + v_{\rm xc}[n])\,|\psi_\alpha\rangle = \epsilon_\alpha|\psi_\alpha\rangle, \tag{1}$$

known today as the Kohn–Sham equations. These effectively single-particle eigenvalue equations are formally similar to the time-independent Schroedinger equation, T being the kinetic energy operator and $v_{\rm KS}$ the potential experienced by the electrons. The latter is usually decomposed into a part which is external to the electronic system $v_{\rm ext}$, for instance the electron–ion interaction, and a part describing the electron–electron interactions. For convenience, the last part is further split into the Hartree potential $v_{\rm H}$ and the exchange-correlation potential $v_{\rm xc}$, whose form is, in general, unknown.

The ground state energy of the electronic system is given by:

$$E_{\rm el}\{\psi_\alpha\} = \sum_\alpha^{\rm occ}\langle\psi_\alpha|T + v_{\rm ext}|\psi_\alpha\rangle + E_{\rm Hxc}[n], \tag{2}$$

where E_{Hxc} is the Hartree and exchange-correlation energy functional of the electron density $n(\mathbf{r})$ with $\delta E_{\mathrm{Hxc}}/\delta n = v_{\mathrm{H}} + v_{\mathrm{xc}}$, and the summation runs over the occupied states α. The occupied Kohn–Sham orbitals are subject to the orthonormalization constraints,

$$\int \psi_\alpha^*(\mathbf{r})\psi_\beta(\mathbf{r})d\mathbf{r} = \langle \psi_\alpha | \psi_\beta \rangle = \delta_{\alpha\beta}, \tag{3}$$

where α and β label occupied states. The density is obtained from

$$n(\mathbf{r}) = \sum_\alpha^{\mathrm{occ}} \psi_\alpha^*(\mathbf{r})\psi_\alpha(\mathbf{r}). \tag{4}$$

Presently, DFT is considered as the method of choice for simulating solids and molecules from first-principles. The interested reader may find a collection of some interesting DFT applications in the review article of Pickett (13). For more technical details about DFT, we recommend the review article of Payne et al. (14).

1.2. Response Properties

In this brief overview, we will only present the *responses* of solid systems to two types of perturbations: (a) collective displacements of atoms characterized by a wavevector $\mathbf{q}$ (phonons) and (b) homogeneous static electric fields. These responses can also be obtained in the framework of DFT using various methods, which can be found in the nice review article by Baroni et al. (15).

The method that is adopted in the calculations described here is based on a variational approach to density-functional perturbation theory, which is presented in details in refs. (16, 17). The first paper (16) is devoted to the computation of the first-order derivatives of the wavefunctions, density and self-consistent potential with respect to the perturbations mentioned above; while the second paper (17) presents the second-order derivatives. We adopt the same notations as in those references to introduce the properties that are studied in the following sections. In particular, κ and α run over the atoms in the unit cell and over the three cartesian directions, respectively; $\tau_{\kappa\alpha}$ denote the equilibrium positions.

The squares of the phonon frequencies $\omega^2_{m\mathbf{q}}$ at $\mathbf{q}$ are determined as eigenvalues of the dynamical matrix $\tilde{D}_{\kappa\alpha,\kappa'\beta}(\mathbf{q})$, or as solutions of the following generalized eigenvalue problem:

$$\sum_{\kappa'\beta} \tilde{C}_{\kappa\alpha,\kappa'\beta}(\mathbf{q}) U_{m\mathbf{q}}(\kappa'\beta) = M_\kappa \omega^2_{m\mathbf{q}} U_{m\mathbf{q}}(\kappa\alpha), \tag{5}$$

where M_κ is the mass of the ion κ, and the matrix $\tilde{C}$ is related to the dynamical matrix $\tilde{D}$ through:

$$\tilde{D}_{\kappa\alpha,\kappa'\beta}(\mathbf{q}) = \frac{\tilde{C}_{\kappa\alpha,\kappa'\beta}(\mathbf{q})}{(M_\kappa M_{\kappa'})^{1/2}}. \tag{6}$$

The matrix $\tilde{C}_{\kappa\alpha,\kappa'\beta}(\mathbf{q})$ is the Fourier transform of the matrix of the interatomic force constants. It is connected to the second-order derivative of the total energy with respect to collective atomic displacements (17).

The limit $\mathbf{q} \rightarrow \mathbf{0}$ must be performed cautiously (17) by the separate treatment of the macroscopic electric field associated with phonons in this limit. A bare dynamical matrix at $\mathbf{q} = \mathbf{0}$ is first calculated, then a non-analytical part is added, in order to reproduce correctly the $\mathbf{q} \rightarrow \mathbf{0}$ behavior along different directions:

$$\tilde{C}_{\kappa\alpha,\kappa'\beta}(\mathbf{q} \rightarrow \mathbf{0}) = \tilde{C}_{\kappa\alpha,\kappa'\beta}(\mathbf{q} = \mathbf{0}) + \tilde{C}^{\mathrm{NA}}_{\kappa\alpha,\kappa'\beta}(\mathbf{q} \rightarrow \mathbf{0}). \tag{7}$$

The expression of the non-analytical part will be presented later on in this section.

For insulators, the dielectric permittivity tensor is defined as the coefficient of proportionality between the macroscopic displacement field and the macroscopic electric field, in the linear regime:

$$\mathcal{D}_{\mathrm{mac},\alpha} = \sum_{\beta} \epsilon_{\alpha\beta} \mathcal{E}_{\mathrm{mac},\beta}. \tag{8}$$

It can be obtained as

$$\epsilon_{\alpha\beta} = \frac{\partial \mathcal{D}_{\mathrm{mac},\alpha}}{\partial \mathcal{E}_{\mathrm{mac},\beta}} = \delta_{\alpha\beta} + 4\pi \frac{\partial \mathcal{P}_{\mathrm{mac},\alpha}}{\partial \mathcal{E}_{\mathrm{mac},\beta}}. \tag{9}$$

In general, the displacement $\mathcal{D}_{\mathrm{mac}}$, or the polarization $\mathcal{P}_{\mathrm{mac}}$, will include contributions from ionic displacements. In the presence of an applied field of high frequency, the contribution to the dielectric permittivity tensor resulting from the electronic polarization, usually noted $\epsilon^{\infty}_{\alpha\beta}$, dominates. This ion-clamped dielectric permittivity tensor is related to the second-order derivatives of the energy with respect to the macroscopic electric field (17). Later on in this section, we will consider the supplementary contributions to the polarization coming from the ionic displacements.

For insulators, the Born effective charge tensor $Z^*_{\kappa,\beta\alpha}$ is the proportionality coefficient relating, at linear order, the polarization per unit cell, created along the direction β, and the displacement along the direction α of the atoms belonging to the sublattice κ, under the condition of zero electric field. The same coefficient also describes the linear relation between the force on an atom and the macroscopic electric field:

$$Z^*_{\kappa,\beta\alpha} = \Omega_0 \frac{\partial \mathcal{P}_{\mathrm{mac},\beta}}{\partial \tau_{\kappa\alpha}(\mathbf{q} = \mathbf{0})} = \frac{\partial F_{\kappa,\alpha}}{\partial \mathcal{E}_{\beta}} \tag{10}$$

where Ω_0 is the volume of the primitive unit cell. The Born effective charge tensors are connected to the mixed second-order derivative of the energy with respect to atomic displacements and macroscopic electric field (17).

Finally, we present two phenomena that arise from the same basic mechanism: the coupling between the macroscopic electric field and the polarization associated with the $\mathbf{q} \rightarrow \mathbf{0}$ atomic displacements. In both cases, the Born effective charges are involved.

On the one hand, in the computation of the low-frequency (infrared) dielectric permittivity tensor, the response of the ions must be include. Their motion will be triggered by the force due to the electric field, while their polarization will be created by their displacement.

At the lowest order of approximation in the theory, the macroscopic frequency-dependent dielectric permittivity tensor $\epsilon_{\alpha\beta}(\omega)$ is calculated as follows:

$$\epsilon_{\alpha\beta}(\omega) = \epsilon_{\alpha\beta}^{\infty} + \frac{4\pi}{\Omega_0} \sum_m \frac{S_{m,\alpha\beta}}{\omega_m^2 - \omega^2}, \tag{11}$$

where the mode-oscillator strength $S_{m,\alpha\beta}$ is defined as:

$$S_{m,\alpha\beta} = \Big(\sum_{\kappa\alpha'} Z^*_{\kappa,\alpha\alpha'} U^*_{m\mathbf{q}=\mathbf{0}}(\kappa\alpha')\Big)\Big(\sum_{\kappa'\beta'} Z^*_{\kappa',\beta\beta'} U_{m\mathbf{q}=\mathbf{0}}(\kappa'\beta')\Big). \tag{12}$$

A damping factor might be added to Eq. (11) in order to take into account anharmonic effects, and fit frequency-dependent experimental data. For our purpose, such a damping factor can be ignored.

At zero frequency, the static dielectric permittivity tensor is usually noted $\epsilon^0_{\alpha\beta}$; it is obtained by:

$$\epsilon^0_{\alpha\beta} = \epsilon^{\infty}_{\alpha\beta} + \sum_m \Delta\epsilon_{m,\alpha\beta} = \epsilon^{\infty}_{\alpha\beta} + \frac{4\pi}{\Omega_0} \sum_m \frac{S_{m,\alpha\beta}}{\omega_m^2}. \tag{13}$$

In parallel to this decomposition of the static dielectric tensor, one can define a mode-effective charge vector:

$$Z^*_{m,\alpha} = \frac{\sum_{\kappa\beta} Z^*_{\kappa,\alpha\beta} U_{m\mathbf{q}=\mathbf{0}}(\kappa\beta)}{\Big(\sum_{\kappa\beta} U^*_{m\mathbf{q}=\mathbf{0}}(\kappa\beta) U_{m\mathbf{q}=\mathbf{0}}(\kappa\beta)\Big)^{1/2}}. \tag{14}$$

This vector is related to the global polarization resulting from the atomic displacements for a given phonon mode m. The non-zero components reveal the directions in which the mode is infrared active.

On the other hand, for phonons in the long-wavelength limit, a macroscopic polarization and electric field can be associated with the atomic displacements. At the simplest level of theory, the phonon eigenfrequencies then depend on the direction along which the limit is taken as well as on the polarization of the phonon. This gives birth to the LO–TO splitting, and to the Lyddane–Sachs–Teller relation (17).

For insulators, the non-analytical, direction-dependent part of the dynamical matrix $\tilde{C}^{\mathrm{NA}}_{\kappa\alpha,\kappa'\beta}(\mathbf{q} \to \mathbf{0})$ is given by:

$$\tilde{C}^{\mathrm{NA}}_{\kappa\alpha,\kappa'\beta}(\mathbf{q} \to \mathbf{0}) = \frac{4\pi}{\Omega_0} \frac{\Big(\sum_\gamma q_\gamma Z^*_{\kappa,\gamma\alpha}\Big)\Big(\sum_{\gamma'} q_{\gamma'} Z^*_{\kappa',\gamma'\beta}\Big)}{\sum_{\alpha\beta} q_\alpha \epsilon^{\infty}_{\alpha\beta} q_\beta}. \tag{15}$$

Hence, once the dynamical matrix at $\mathbf{q} = \mathbf{0}$ as well as $\epsilon^{\infty}_{\alpha\beta}$ and the Born effective charges tensors are available, it is possible to compute the LO–TO splitting of phonon frequencies at $\mathbf{q} = \mathbf{0}$.

1.3. Technical Details

We have performed all the calculations using the ABINIT package, developed by the authors and collaborators (18). The exchange-correlation energy is evaluated within the local density approximation (LDA) to density-functional theory, using Perdew–Wang's parameterization (19) of Ceperley–Alder electron-gas data (20).

Only valence electrons are explicitly considered using pseudopotentials to account for core-valence interactions. We use norm-conserving pseudopotentials (21,22) with Hf(5s, 5p, 5d, 6s), Zr(4s, 4p, 4d, 5s), Ti(3s, 3p, 3d, 4s), Si(3s, 3p), and O(2s, 2p) levels treated as valence states. The following atomic valence configurations are used to generate the pseudopotentials: for Hf, $5s^2 5p^6 5d^2 6s^2$; for Zr, $4s^2 4p^6 4d^2 5s^0$; for Ti, $3s^2 4p^6 3d^2 4s^0$; for Si, $3s^2 3p^2$; and for O, $2s^2 2p^4$. In the case of Hf, we take core radii of 1.50, 2.85, 2.45, 3.50 a.u. for describing angular waves from s to f. The corresponding values (up to the d wave) are 1.75, 1.55, and 1.70 a.u. for the Zr pseudopotential, and 1.25, 1.25, and 1.65 a.u. for the Ti pseudopotential. For the Si pseudopotential, the same cutoff radius of 2.00 a.u. is used for the three lowest angular-momentum waves. For the O pseudopotential, we use a cutoff radius of 1.50 a.u. for both s and p waves. We adopted a separable form for the pseudopotentials (23) treating the following angular-momentum waves as local: f for Hf, d for Zr, d for Ti, d for Si, and p for O.

The wavefunctions are expanded in plane waves up to a kinetic energy cutoff of 30 Ha for Hf- and Zr-based systems and of 45 Ha for Ti-based systems. For each crystalline system, the Brillouin zone is sampled by a Monkhorst–Pack (24) mesh of k-points. For the cubic and tetragonal phases of the oxides as well as for the crystalline silicates, our mesh corresponds to a $4 \times 4 \times 4$ grid in the conventional unit cell, leading to 10, 12, and 15 special k-points in the irreducible Brillouin zone, respectively. For the rutile phase of the oxides, we use a $4 \times 4 \times 6$ grid that results in nine special k-points in the irreducible Brillouin zone. For the amorphous system, we only use the Γ-point to sample the Brillouin zone. The chosen kinetic energy cutoff and k-point sampling of the Brillouin zone ensure convergence of all the calculated properties.

2. CRYSTALLINE OXIDES

2.1. Introduction

Titania (TiO_2) is by far the most important compound formed by the elements of group IVB, its importance arising predominantly from its use as a white pigment. Three forms exists at room temperature: rutile, anatase, and brookite. Each of them occurs naturally. The rutile is the most common form, both in nature and as produced commercially. It is also the most stable phase: the other transform into it on heating. Note also that all three forms contain sixfold coordinated titanium atoms.

Hafnia (HfO_2) and zirconia (ZrO_2) undergo polymorphic transformations with changes in external parameters. At high temperature, the compounds are highly defective and their structure is fluorite type ($Fm\bar{3}m$). The decreasing temperature induces a cubic to tetragonal ($P4_2/nmc$) phase transition ($c - t$) at about 2650°C for

HfO_2 (25) and about 2350°C for ZrO_2 (26). This transition is followed by a tetragonal to monoclinic ($P2_1/c$) martensitic phase transition ($t - m$) at about 1650°C for hafnia (27) and about 1150°C for zirconia (28). In the cubic and tetragonal phase, the metal atoms are eightfold coordinated while in the monoclinic phase they are sevenfold coordinated. None of the existing phases have sixfold coordinated atoms as in the rutile structure.

Hafnia and zirconia have many similar physical and chemical properties, but they differ considerably from titania (e.g., difference in the stable phase). This can be related to the very close chemical homology between Hf and Zr, compared to Ti.

Considering the valence electrons only, the electron configurations of Hf, Zr, and Ti differ only by the principal number of the occupied orbitals it is $5d^2 6s^2$ for hafnium, $4d^2 5s^2$ for zirconium, and $3d^2 4s^2$ for titanium. Thus, in principle, they should be characterized by decreasing electronegativities and increasing atomic and ionic radii from Ti to Hf. However, in the periodic table, the inner transition (rare-earth) elements immediately preceding Hf add electrons to the inner 4f shell from element N° 58, cerium, to N° 71, lutetium (it would actually be more correct to write that the electron configuration of hafnium is $4f^{14} 5d^2 6s^2$). Because the nuclear charge increases while no additional outer shells are filled, there is a contraction in the atomic size. Consequently, the element N° 72, hafnium, has a slightly smaller atomic size than element N° 40, zirconium, the group IVB element in the preceding row. This results in the so-called lanthanide contraction.

As a result, while the atomic radius of Ti (1.40 Å) is indeed smaller than for Zr (1.55 Å), the atomic radius of Hf (1.55 Å) is identical to that of Zr (29). The ionic radii (M^{4+}) also presents the same anomaly: it is smaller for Ti (0.61 Å) than for Zr (0.84 Å), but it is essentially the same for the latter and Hf (0.84 Å) (30). Finally, their electronegativities also show an anomalous trend with values of 1.23 for hafnium and 1.22 for zirconium, compared to 1.32 for Ti (31). All this explains the origin of the close similarity between HfO_2 and ZrO_2 with respect to TiO_2.

The structural, electronic and dynamical properties of hafnia, zirconia, and titania have been the object of several first-principles studies (32–46). For sake of brevity, we here only present here our results for the cubic, tetragonal, and rutile phases.

2.2. Structural Properties

To model the crystalline oxides, we consider the cubic, tetragonal, and rutile structures which are illustrated in Fig. 1. For HfO_2 and ZrO_2, the cubic and tetragonal phases exist in nature, while the rutile phase is purely hypothetical. On the contrary, for TiO_2, the rutile occurs naturally, while the other two are hypothetical.

The cubic phase takes the fluorite structure (space group $Fm\bar{3}m$, N° 225), which is fully characterized by a single lattice constant a. The M (= Hf, Zr) atoms are in a face-centered-cubic structure and the O atoms occupy the tetrahedral interstitial sites associated with this fcc lattice. The primitive unit cell contains one formula unit of MO_2 with M (= Hf, Zr, Ti), while the conventional unit cell has four of them.

The tetragonal phase (space group $P4_2/nmc$, N° 137) can be viewed as a distortion of the cubic structure obtained by displacing alternating pairs of O atoms up and down

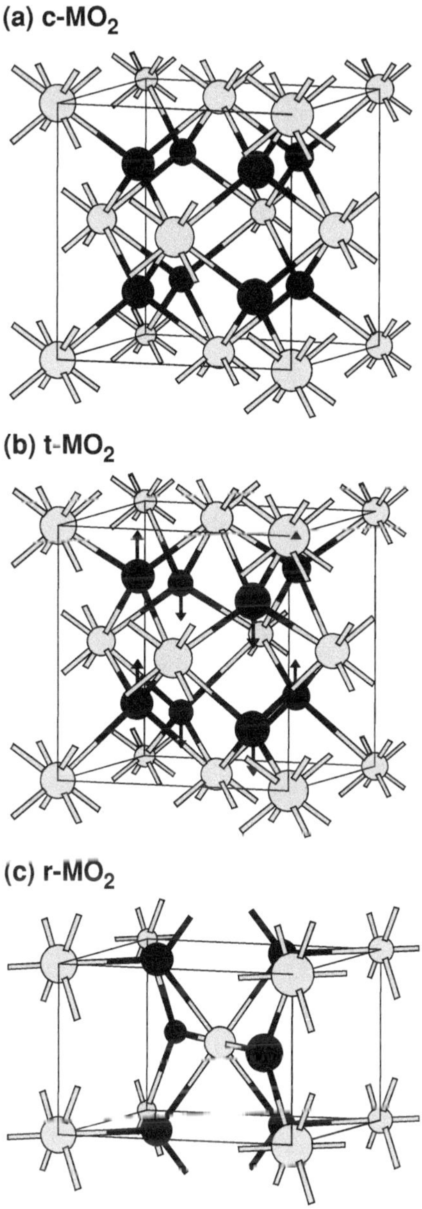

Fig. 1. Structures of the cubic and tetragonal phases of HfO_2, ZrO_2, and TiO_2. A ball and stick representation is adopted where M (= Hf, Zr, Ti) and O atoms are colored in light and medium grey, respectively. For the tetragonal phase, the arrows indicate the displacements of oxygen pairs relative to the cubic structure.

by an amount Δz along the z direction, as marked by the arrows in Fig. 1, and by applying a tetragonal strain. The resulting primitive cell is doubled compared to the cubic phase, including two formula units of MO_2. The conventional unit cell, which is reproduced in in Fig. 1(b), has four formula units of MO_2 with M = (Hf, Zr, Ti). The tetragonal structure is completely specified by two lattice constants (a and c) and the dimensionless ratio $d_z = \Delta z/c$ describing the displacement of the O atoms. The cubic phase can be considered as a special case of the tetragonal structure with $d_z = 0$ and $c/a = 1$ (if the primitive cell is used for the tetragonal phase, $c/a = \sqrt{2}$).

The rutile structure (space group $P4_2/mnm$, N° 136) has a tetragonal unit cell with two formula units of MO_2 with M (= Hf, Zr, Ti). The metal atoms occupy the body-centered-cubic positions and the O atoms are at $(u, u, 0)$, $(1-u, 1-u, 0)$, $(\frac{1}{2}-u, \frac{1}{2}+u, \frac{1}{2})$, and $(\frac{1}{2}+u, \frac{1}{2}-u, \frac{1}{2})$, as reported in Fig. 1(c). The rutile structure is completely specified by two lattice constants (a and c) and the internal parameter u related to the position of O atoms.

For hafnia and zirconia, the tetragonal phase is found to be the most stable with $E_t < E_c < E_r$, at variance with titania for which it is the rutile phase which is energetically favored with $E_r < E_t < E_c$. It is interesting to note that in the rutile phase the M (= Ti, Zr, Hf) atoms are sixfold coordinated while in the cubic and tetragonal phases they are eightfold coordinated. This is a first clear difference between Ti atoms on the one hand, and Zr and Hf atoms on the other. It can be related to the smaller ionic radius of Ti^{4+} (0.61 (47)) compared to Zr^{4+} and Hf^{4+} (0.72 and 0.71 Å, respectively (47)).

Our calculated structural parameters for the cubic, tetragonal, and phases of HfO_2, ZrO_2, and TiO_2 are reported in Table 1. For the naturally occurring, the agreement with the experimental values (28, 48). is very good: the errors on the lattice constants and the volumes are smaller than 2%, as is typical for LDA calculations. The largest discrepancy is for d_z in t-ZrO_2 (the small displacement from the cubic phase localization): our value is about 30% smaller than the experimental data, but it is in excellent agreement with the results of other first-principles calculations. The discrepancy with experiment is probably due to the fact that our calculations are performed at zero-temperature.

While the structural parameters for Zr and Hf based oxides are very similar, the values for Ti based materials can differ by about 5–10% from the former two. The largest difference is observed in the tetragonal phase in which the Ti–O distance is found to be 10 and 13% smaller than the Zr–O and Hf–O distances, respectively. This is another evidence of the different chemistry of 3d metals with respect to 4d and 5d metals.

2.3. Born Effective Charge Tensors

In Table 2, we report non-vanishing components of the calculated Born effective charge tensors for M (= Hf, Zr, Ti) and O atoms in the cubic, tetragonal, and rutile phases of hafnia, zirconia, and titania. These values should be compared with the nominal ionic charges $Z = +4$ for M (= Hf, Zr, Ti) atoms and $Z = -2$ for O atoms.

Table 1. Structural parameters for the cubic (c), tetragonal (t), and rutile (r) phases of HfO_2, ZrO_2, and TiO_2

	HfO_2	ZrO_2	TiO_2
c			
a	5.11	5.01	4.72
Volume	33.36	31.44	26.29
d(M–O)	2.21	2.17	2.04
t			
a	5.11	5.02	4.71
c	5.17	5.09	4.93
d_z	0.0310	0.0400	0.0701
Volume	33.75	32.07	27.34
d(M–O)	2.13	2.07	1.89
	2.32	2.31	2.29
r			
a	4.90	4.80	4.53
c	3.27	3.22	2.92
u	0.3051	0.3054	0.3033
Volume	78.51	74.19	60.12
d(M–O)	2.11	2.07	1.93
	2.12	2.08	1.94

The lengths are expressed in Å.

Due to the symmetry of the cubic phase, the Born effective charge tensors of M (= Hf, Zr) and O atoms are diagonal and isotropic. The value of Z^* is anomalously large for M (= Hf, Zr, Ti) atoms compared to the nominal ionic charge $Z = 4$. This behavior has also been observed in the case of $PbZrO_3$ (49). A detailed analysis of the physics of Born effective charges in the case of perovskite ferroelectrics (like $PbZrO_3$) ascribed this effect to a mixed covalent–ionic bonding (50).

In the tetragonal structure, the symmetry imposes that the Born effective charge tensor of M (= Hf, Zr, Ti) atoms is diagonal and only has two independent components: parallel ($Z^*_{\parallel}$) and perpendicular ($Z^*_{\perp}$) to the *c* axis. For hafnia and zirconia, the tensors are quite isotropic. The value of $Z^*_{\perp}$ is identical to the one calculated for the cubic phase, while $Z^*_{\parallel}$ is 6 and 10% smaller for HfO_2 and ZrO_2, respectively. For titania, the tensor is very anisotropic: the value of $Z^*_{\perp}$ is 3% larger than the one calculated for the cubic phase, while $Z^*_{\parallel}$ is more than 30% smaller. This is another manifestation of the different behavior of Ti atoms compared to Zr and Hf atoms. The Born effective charge tensor of O atoms is also diagonal, but with three independent components. It is quite anisotropic for all three systems. In t-TiO_2, it is even more anisotropic than the tensor of Ti atoms: the ratio between the largest and the smallest components is about 2.5. This ratio is only 1.6 for t-ZrO_2 and 1.4 for t-HfO_2. Such a strong anisotropy of the Born effective charge tensor for O atoms has already been observed in SiO_2-stishovite (51).

Table 2. Non-vanishing components of the calculated Born effective charge tensors for M (= Hf, Zr, Ti) and O atoms in the cubic (c), tetragonal (t), and rutile (r) phases of HfO_2, ZrO_2, and TiO_2

Atom	HfO_2	ZrO_2	TiO_2
c			
M	$(+5.58\ +5.58\ +5.58)$	$(+5.74\ +5.74\ +5.74)$	$(+6.40\ +6.40\ +6.40)$
O	$(-2.79\ -2.79\ -2.79)$	$(-2.87\ -2.87\ -2.87)$	$(-3.20\ -3.20\ -3.20)$
t			
M	$(+5.57\ +5.57\ +5.24)$	$(+5.74\ +5.74\ +5.15)$	$(+6.63\ +6.63\ +4.42)$
O	$(-3.22\ -2.35\ -2.62)$	$(-3.51\ -2.24\ -2.57)$	$(-4.76\ -1.94\ -2.21)$
r			
M	$(+5.38\ +0.66\ +6.31)$	$(+5.58\ +0.69\ +6.51)$	$(+6.36\ +1.00\ +7.52)$
	$[+6.04\ +4.72\ +6.31]$	$[+6.27\ +4.89\ +6.51]$	$[+7.36\ +5.36\ +7.52]$
O[a]	$(-2.69\ -1.37\ -3.15)$	$(-2.79\ -1.46\ -3.25)$	$(-3.18\ -1.81\ -3.76)$
	$[-4.06\ -1.32\ -3.15]$	$[-4.25\ -1.33\ -3.25]$	$[-4.99\ -1.37\ -3.76]$

For the cubic and tetragonal phases, the tensors are diagonal, only the principal elements are given. For the rutile phase, the three independent components ($Z^*_{xx} = Z^*_{yy}$, $Z^*_{xy} = Z^*_{yx}$, and Z^*_{zz}) of the tensors are given, and the principal components are indicated between brackets.
[a] For the rutile phase, the components for O atoms in the rutile phase refer to the atom located at $(u, u, 0)$. The corresponding values for the other oxygen atoms can be obtained using the symmetry operations.

In the rutile structure, the Born effective charge tensors of M (= Hf, Zr, Ti) or O atoms have only three independent components: Z^*_{xx}, Z^*_{xy}, and Z^*_{zz}. Indeed, Z^*_{yy} and Z^*_{yx} are equal to Z^*_{xx} and Z^*_{xy}, respectively, while the other components are zero. When we take a coordinate system whose axes are along the [110], $[1\bar{1}0]$, and [001] directions, the tensors are diagonalized. The principal values obtained in this coordinate system are reported between brackets in Table 2. For the metal atoms, the ratio between the largest and the smallest components is about 1.3 in r-HfO_2, 1.3 for r-ZrO_2, and 1.4 for r-TiO_2. For O atoms, this ratio is 3.1 for hafnia, 3.2 for zirconia, and 3.6 for titania. Thus, the anisotropy is even larger than in the tetragonal phase, especially for O atoms in hafnia and zirconia.

It is interesting to note that the Born effective charges of c-HfO_2 are about 3% smaller (in absolute value) than those of c-ZrO_2 which are in turn more than 11% smaller than those of c-TiO_2. The comparison between the Z^* values in the tetragonal phases of hafnia, zirconia, and titania is also very instructive. In directions perpendicular to the c axis, the Born effective charges of the M (= Hf, Zr, Ti) atoms compare in the same way as for the cubic phase: the values of t-HfO_2 are about 3% smaller than those of t-ZrO_2, which are 15% smaller than of t-TiO_2. The Born effective charges of O atoms show an increasing anisotropy from t-HfO_2 to t-ZrO_2: the values of $Z^*_\perp$ for t-HfO_2 are comprised between those of t-ZrO_2, which in turn are surrounded by those of t-TiO_2. In the direction parallel to the c axis, the Born effective charges in t-HfO_2 are larger than in t-ZrO_2 by about 2% and than in t-TiO_2 by more than 15%, showing an opposite trend with respect to the comparison for the cubic phase.

To summarize, the largest components of the Born effective charge tensors are found in r-TiO_2 and in the rutile phase in general. For the metal atoms, the strongest anisotropy is found for the Ti atoms in t-TiO_2, while for Zr and Hf atoms the strongest anisotropy appears in the rutile phase. For the oxygen atoms, the strongest anisotropy is found in r-TiO_2 and in the rutile phase in general.

2.4. Phonon Frequencies

The theoretical group analysis (see Appendix) predicts the following irreducible representations of optical and acoustical zone-center modes for the cubic phase:

$$\Gamma = \underbrace{F_{2g}}_{\text{Raman}} \oplus \underbrace{F_{1u}}_{\text{IR}} \oplus \underbrace{F_{1u}}_{\text{Acoustic}}, \tag{16}$$

for the tetragonal phase:

$$\begin{aligned}\Gamma = &\underbrace{A_{1g} \oplus 2B_{1g} \oplus 3E_g}_{\text{Raman}} \oplus \underbrace{A_{2u} \oplus 2E_u}_{\text{IR}} \\ &\oplus \underbrace{A_{2u} \oplus E_u}_{\text{Acoustic}} \oplus \underbrace{B_{2u}}_{\text{Silent}},\end{aligned} \tag{17}$$

and for the rutile phase:

$$\begin{aligned}\Gamma = &\underbrace{A_{1g} \oplus B_{1g} \oplus B_{2g} \oplus E_g}_{\text{Raman}} \oplus \underbrace{A_{2u} \oplus 3E_u}_{\text{IR}} \\ &\oplus \underbrace{A_{2u} \oplus E_u}_{\text{Acoustic}} \oplus \underbrace{A_{2g} \oplus 2B_{1u}}_{\text{Silent}}.\end{aligned} \tag{18}$$

Although the space group differ for the tetragonal and rutile phases, their point group ($4/mmm$) is the same. Hence, the same notations appear in the irreducible representations of the zone-center modes.

Because of the non-vanishing components of the Born effective charge tensors, the dipole–dipole interaction must be properly included in the calculation of the interatomic force constants (17, 52, 53). In particular, the dipole–dipole contribution is found to be responsible for the splitting at the Γ point between the longitudinal and transverse optic (LO and TO, respectively) modes F_{1u} in the cubic phase, and E_u (perpendicular to the c axis) and A_{2u} (parallel to c axis) in the tetragonal phase.

Our calculated phonon frequencies are reported in Table 3. Our results are in very good agreement with experimental data (54–58) and previous first-principles calculations (32, 44) when available. Note that in t-TiO_2, one E_u mode has an imaginary frequency. This corresponds to a negative curvature of the total energy dependence on the atomic positions, hence to a saddle point. This vibration mode tends to break the symmetry imposed in the calculation. Hence, the tetragonal structure is found to be unstable.

Table 3. Fundamental frequencies of the cubic (c), tetragonal (t), and rutile (r) phases of HfO_2, ZrO_2 and TiO_2 (in cm^{-1}) with their symmetry assignments

		Mode		HfO_2	ZrO_2	TiO_2
c						
	Raman	F_{2g}	O*	579	596	619
	Infrared	F_{1u} (TO)	O	285	280	177
		F_{1u} (LO)		630	677	686
t						
	Raman	A_{1g}	O*	218	259	382
		$B_{1g}(1)$	M	244	331	351
		$B_{1g}(2)$	O	582	607	669
		$E_g(1)$	M	110	147	130
		$E_g(2)$	O	479	474	435
		$E_g(3)$	O	640	659	731
	Infrared	A_{2u} (TO)	O	315	339	429
		A_{2u} (LO)		621	664	678
		E_u (TO1)	O	185	153	$116i$
		E_u (LO1)		292	271	166
		E_u (TO2)	O	428	449	496
		E_u (LO2)		669	734	850
	Silent	B_{2u}	O*	665	673	660
r						
	Raman	A_{1g}	O*	638	626	636
		B_{1g}	O*	91	92	116
		B_{2g}	O*	792	800	844
		E_g	O*	486	483	481
	Infrared	A_{2u} (TO)	O	308	301	204
		A_{2u} (LO)		670	721	779
		E_u (TO1)	O	193	197	180
		E_u (LO1)		214	274	354
		E_u (TO2)	M	222	302	404
		E_u (LO2)		296	333	448
		E_u (TO3)	O	478	462	502
		E_u (LO3)		733	765	825
	Silent	A_{2g}	O*	353	368	419
		$B_{1u}(1)$	M	92	124	126
		$B_{1u}(2)$	O	428	424	423

In t-TiO_2, the E_u mode with an imaginary frequency tends to break the symmetry imposed in the calculation and implies an instability of the structure. The letter (M or O) in the second column indicates the atoms (metal or oxygen) whose motion dominates in the vibrational mode, a star superscript showing that the other atoms are fixed by symmetry.

The atomic motions associated to the various vibrational modes have been described in detail in the literature. The interested reader will refer to refs. (57, 59, 60) for the cubic and tetragonal phases, and ref. (32) for rutile.

It is very interesting to compare the phonon frequencies calculated for HfO_2, ZrO_2, and TiO_2 (see Table 3). There are several possible origins for the variations that are observed between Hf, Zr, and Ti oxides: structural changes (e.g., the volume), change of the mass ratio Hf/Zr = 1.96 and Zr/Ti = 1.90, and differences in interatomic force constants.

The structural changes reported in Table 1 are relatively small, in particular between hafnia and zirconia. We suspect that their effect should not be the most important origin for the variations observed in the phonon frequencies. In order to check this, we compute the phonon frequencies for hafnia and titania assuming that the interatomic force constants are the same as those for zirconia, while the volume is allowed to vary. In hafnia, we find that the frequencies are decreased by at most 2%; where as in titania, they are increased by at most 6% in the cubic and tetragonal phase, and 11% in the rutile phase. This analysis shows that the structural changes play a relatively minor role in agreement with our intuition. Their effect is slightly more important in titania since the structural changes are larger (in particular in the rutile structure).

As for the role of the mass ratio, it is interesting to focus on the modes in which the M (= Hf, Zr, Ti) atoms are not much involved (either they are fixed, or they move significantly less than O atoms) and on those in which on the contrary the M (= Hf, Zr, Ti) atoms move significantly more than O atoms. In the former case (modes indicated by the letter O in the second column of Table 3), the phonon frequencies should not be affected by the change between Hf, Zr, or Ti; whereas, in the latter case (modes indicated by the letter M in the second column of Table 3), the variation should be very important.

In the cubic phase, the F_{2g} is the only mode in which the M (= Hf, Zr, Ti) atoms are fixed (as indicated by the letter O with star superscript in Table 3), and we observe indeed that the phonon frequencies do not vary very much (at most 4% with respect to zirconia). In the tetragonal phase, there are two such modes. A_{1g} and B_{2u}. While for the latter, we again do not observe any significant variation of the phonon frequencies (at most 2% with respect to zirconia); for the former, the changes are quite important: 16% decrease and 47% increase with respect to t-ZrO_2 for hafnia and titania, respectively. This is a case where the effects due to differences in the interatomic force constants are dominant. In the rutile phase, there are five modes in which the metal atoms are fixed by symmetry. Between hafnia and zirconia, the frequencies vary by at most 1% indicating that the effect of the interatomic force constants is negligible. When comparing titania to zirconia, three modes (A_{1g}, B_{2g}, and E_g) do not show significant variations, while for the other two (B_{1g} and A_{2g}) the effect of the interatomic force constants is well pronounced (25 and 14% increase of the frequencies, respectively). For the modes in which the oxygen atoms move significantly more than the metal atoms and the latter are not fixed by symmetry (as indicated by the letter O only in Table 3), the effect of the interatomic force constants is more pronounced. In particular, the mode $E_u(1)$ in the tetragonal phase even becomes unstable in titania. However, in some of these modes, this effect remains negligible (e.g., the $B_{1u}(2)$ mode in the rutile phase).

Finally, for the modes in which the M (= Hf, Zr, Ti) atoms move significantly more than O atoms (as indicated by the letter M in Table 3), it is also possible to evidence the influence of the interatomic force constants. For this purpose, we compute the phonon frequencies for hafnia and titania assuming that the interatomic force constants are the same as those for zirconia, while the mass of the metal atom is changed to that of Hf or Ti. These simple calculations lead to frequencies which are reduced by roughly 28% for hafnia and increased by about 34% for titania with respect to zirconia. When these results compare well with those of Table 3, it can be considered that the effect of the interatomic force constants is negligible. This is actually what we find in all cases for hafnia compared to zirconia as well as for the $E_u(2)$ mode in r-TiO_2. On the contrary, the effect of the interatomic force constants is found to be very important for the B_{1g} and $E_g(1)$ modes in t-TiO_2 and for the $B_{1u}(1)$ mode in r-TiO_2.

In conclusion, while the interatomic force constants in hafnia and zirconia are very similar, they differ considerably in titania. As a result, while the differences in the phonon frequencies in Hf and Zr oxides can mostly be explained by the ratio between the masses of the two metals, the most important origin for the variation in Ti oxides is the interatomic force constants.

2.5. Dielectric Permittivity

In the cubic phase, the electronic (ϵ_∞) and static (ϵ_0) permittivity tensors are diagonal and isotropic. Due to the symmetry of the tetragonal and rutile crystals, these tensors are still diagonal, but have two independent components $\epsilon_\parallel$ and $\epsilon_\perp$, parallel and perpendicular to the c axis, respectively. In Table 4, the calculated values of ϵ_∞ and ϵ_0 are reported for the cubic, tetragonal, and rutile phases of hafnia, zirconia, and titania.

In the tetragonal phase, the ϵ_∞ tensor is only slightly anisotropic with about 5 and 10% difference between the parallel and perpendicular values for hafnia and zirconia. For titania, it is a bit more anisotropic with about 25% difference between these values. On the contrary, the ϵ_0 tensor is highly anisotropic: the value of ϵ_0 in the direction parallel to the c axis is 1.6 and 2.4 times smaller than that in the perpendicular direction for t-HfO_2 and t-ZrO_2, respectively. For t-TiO_2, the static dielectric tensor cannot be calculated due to the instability of the phase (the E_u mode with an imaginary frequency tends to break the symmetry imposed in the calculation).

In the rutile phase, the ϵ_∞ tensors present the same relatively small difference (12%) between $\epsilon_\parallel$ and $\epsilon_\perp$ for the three oxides. For hafnia and zirconia, the anisotropic character is reduced for the static dielectric permittivity tensor with 4 and 1% difference between its parallel and perpendicular components, respectively. For titania, ϵ_0 is more anisotropic than ϵ_∞.

The calculated dielectric tensors can only be compared with experimental values for the cubic and tetragonal phases of hafnia and zirconia, and for the rutile phase of titania. Moreover, a direct comparison is very difficult since there are very few data available in the literature, especially for hafnia. The main problem encountered in the experimental determination of the dielectric properties is that good quality single crystals are not available. For the tetragonal phase, the results obtained for undoped powders stabilized by their small particle size must be analyzed in the framework of effective medium theory (61). As a result, a unique value of ϵ is found without distinction between the directions parallel and perpendicular to the c axis. In order

Table 4. Electronic and static dielectric tensors for the cubic (c), tetragonal (t), and rutile (r) phases of HfO_2, ZrO_2, and TiO_2 (for t-TiO_2, the static dielectric tensor cannot be calculated due to the instability of the phase, see discussion in the text)

	HfO_2		ZrO_2		TiO_2	
c						
ϵ_∞	5.37		5.74		9.11	
$\Delta\epsilon$	20.80		27.87		128.36	
ϵ_0	26.17		33.61		137.47	
t						
	$\parallel$	$\perp$	$\parallel$	$\perp$	$\parallel$	$\perp$
ϵ_∞	5.13	5.39	5.28	5.74	6.66	8.81
$\Delta\epsilon_1$	14.87	22.34	15.03	35.48		
$\Delta\epsilon_2$		5.08		6.91		
ϵ_0	20.00	32.81	20.31	48.13		
r						
	$\parallel$	$\perp$	$\parallel$	$\perp$	$\parallel$	$\perp$
ϵ_∞	5.19	4.59	5.54	4.93	8.57	7.49
$\Delta\epsilon_1$	19.38	11.64	26.27	19.69	116.16	81.65
$\Delta\epsilon_2$		2.74		1.30		2.90
$\Delta\epsilon_3$		4.65		5.72		5.23
ϵ_0	24.58	23.62	31.81	31.64	124.74	96.28

In all the cases, the tensors are diagonal. For the cubic phase, it is also isotropic, while for the other three phases, the tensors have different components parallel ($\parallel$) and perpendicular ($\perp$) to the c axis. The contributions of the different phonon modes to the static dielectric tensor are also indicated. For the cubic phase, the contribution originates from the IR-active F_{1u} mode. For the tetragonal and rutile phases, the phonon mode contributions to $\epsilon_0^\parallel$ come from the IR-active A_{2u} mode, while the contributions to $\epsilon_0^\perp$ come from the two IR-active E_u modes.

to compare our results with experimental data, we average the values parallel and perpendicular to the c axis:

$$\bar{\epsilon} = \frac{2\epsilon_\perp + \epsilon_\parallel}{3}.$$

This average does not really have any physical meaning, and therefore the comparison is rather qualitative.

For hafnia, we are only aware of measurements of ϵ_0. Our calculated values of 26.17 for the cubic phase, and $\bar{\epsilon}_0 = 28.54$ for the tetragonal phase significantly overestimate the values of 16 (62) and 20 (63) obtained in recent measurements. This overestimation is significantly higher than what can be expected from our density functional approach and the origin of this difference remains poorly understood. For the cubic phase, our results agree within 1% with those obtained by by Zhao and Vanderbilt (43) using similar methods. However, for the tetragonal phase, our calculations disagree significantly with those of Zhao and Vanderbilt (43). In fact, we find a ratio of 1.6 between the values of the ϵ_0 tensor in directions parallel and perpendicular to the c axis, to be compared with the value of 8.6 reported by Zhao and Vanderbilt (43). We note that the value of $\bar{\epsilon}_0 = 70$ proposed by the latter authors appears excessively high in view of the dielectric constant of the cubic phase (~26.17) and the trends observed for zirconia (see below).

For zirconia, an experimental value of $\epsilon_\infty = 4.8$ is reported in the literature for c-ZrO_2 (64,65), while measured values for t-ZrO_2 range between 4.2 (57) and 4.9 (66). Our theoretical values ($\epsilon_\infty = 5.74$ and $\bar{\epsilon}_\infty = 5.59$ for the cubic and tetragonal phases, respectively) are larger than the experimental ones by about 10–15%, as often found in the LDA to density-functional theory. For ϵ_0, the experimental values found in the literature vary from 27.2 (67) to 29.3 (68) for c-ZrO_2, and from 34.5 (67) to 39.8 (68) for t-ZrO_2. For the cubic phase, our calculated value $\epsilon_0 = 33.61$ is somewhat larger than experimental estimates, whereas, for the tetragonal phase, our calculated average $\bar{\epsilon}_0 = 38.86$ falls in the range of the experimental data.

For titania, the experimental values of the electronic permittivity tensor for the rutile phase are 6.84 and 8.43 in the directions perpendicular and parallel to the c axis (54). Our corresponding theoretical values of 7.49 and 8.57 also present the usual 10% overestimation of the LDA. For the static dielectric permittivity tensor, our theoretical values are on the same order of magnitude as the experimental results, which show quite large discrepancies: from 86 and 170 (69) to 115 and 251 (70) for the components perpendicular and parallel to the c axis, respectively.

For a deeper analysis of the static dielectric tensor, we can rely not only on the frequencies of the IR-active modes, but also on the corresponding eigendisplacements and Born effective charges. Indeed, the static dielectric tensor can be decomposed in the contributions of different modes as indicated in Eq. (13).

The contribution of the individual modes $\Delta\epsilon_{\mathrm{m}}$ to the static dielectric constants are presented in Table 4 (except for the t-TiO_2 phase). For each IR-active mode, the relevant component of the oscillator strength tensor is reported in Table 5. This tensor is isotropic for the F_{1u} mode in the cubic phase, while in the tetragonal and rutile phases we indicate the parallel–parallel component for the A_{2u} mode, and the perpendicular–perpendicular component for the E_u modes. We also give the magnitude of the mode-effective charge vector defined by Eq. (14) which is parallel and perpendicular to the tetragonal axis for A_{2u} and E_u modes, respectively, while it has an arbitrary orientation for the F_{1u} mode. The atomic motions for these vibrational modes have been described in detail in the literature (32,57,59,60).

In Table 5, the lowest frequency modes provide the largest contributions to ϵ_0, even if their oscillator strength (S_m) is relatively small. For instance, the $E_u(1)$ mode in the tetragonal and rutile phases of hafnia and zirconia contributes much more to the static dielectric permittivity than the $E_u(2)$ mode in the tetragonal phase and the $E_u(3)$ mode in the rutile phase, which present however larger values of S_m. This emphasizes the crucial role of the frequency factor in Eq. (13). That is particularly true for the cubic phase for TiO_2 compared to ZrO_2 and HfO_2. In this case, the frequency of the F_{1u} mode in the titanium oxide is more than 35% smaller than in the other two oxides. This reduced frequency as well as the increased Born effective charge (see discussion in Section 2.3) leads to a static dielectric constant more than four times larger in c-TiO_2.

Table 5. Components of mode-effective charge vectors Z^*_m and oscillator strength tensor S_m for each of the IR-active modes of the cubic (c), tetragonal (t), and rutile (r) phases of HfO_2, ZrO_2, and TiO_2

	HfO_2		ZrO_2		TiO_2	
	Z^*_m	S_m	Z^*_m	S_m	Z^*_m	S_m
c						
F_{1u}	5.82	6.31	6.42	7.65	7.55	11.69
t						
A_{2u}	7.71	11.10	8.14	12.28	7.38	11.18
$E_u(1)$	5.75	5.76	5.95	5.91	6.04	6.20
$E_u(2)$	5.91	7.03	6.99	9.95	9.53	20.91
r						
A_{2u}	9.29	16.10	10.29	19.61	12.55	32.33
$E_u(1)$	4.81	3.80	6.29	6.30	9.57	17.62
$E_u(2)$	3.13	1.18	2.61	0.99	3.56	3.17
$E_u(3)$	6.18	9.31	6.39	10.09	4.88	7.12

The description of the reported vector and tensor components corresponding to the different modes is given in the text. The components of the mode-effective charge vectors are given in units of $|e|$, where e is the electronic charge. The oscillator strengths are given in 10^{-4} atomic units (1 a.u. = 253.2638413 m^3/s^2).

In the tetragonal phase of hafnia and zirconia, the same argument holds to rationalize why the ϵ_0 tensor is highly anisotropic, while the ϵ_∞ tensor is only slightly anisotropic. Indeed, in these materials, the A_{2u} has the largest oscillator strength (about twice the one of $E_u(1)$ mode) and the largest mode-effective charge. However, its frequency is about twice larger than that of the $E_u(1)$ mode, and its contribution to the static dielectric constant is thus roughly twice smaller than that of the $E_u(1)$ mode.

In Table 5, it can be observed that the oscillator strengths and the mode-effective charges essentially increase from HfO_2 to ZrO_2 and from ZrO_2 to TiO_2. This can be related to the behavior of the Born effective charges $Z^*_{\kappa,\alpha\alpha'}$ and the eigendisplacements $U_m(\kappa\alpha)$, the two quantities that appear in the definitions of $S_{m,\alpha\beta}$ and $Z^*_{m,\alpha}$ given in Eqs. (12) and (14). On the one hand, as discussed in Section 2.3, the Born effective charges show globally the following trend: $Z^*(HfO_2) \leq Z^*(ZrO_2) \leq Z^*(TiO_2)$. On the other hand, the displacements of Hf atoms are smaller than those of Zr atoms, which in turn are smaller than those of Ti atoms, simply because the mass increases from Ti to Hf (as discussed in Section 2.4).

If one now considers the contributions to the static dielectric constant reported in Table 4, it appears clearly that $\Delta\epsilon(HfO_2) \leq \Delta\epsilon(ZrO_2) \leq \Delta\epsilon(TiO_2)$ for almost of all the modes. For a few modes, however, despite the fact that the oscillator strengths are smaller, the corresponding contribution presents exactly the opposite trend due to an increase of the corresponding phonon frequencies. For instance, for $E_u(3)$ mode in the rutile phase, the frequency for ZrO_2 is much smaller than for TiO_2. As a result, the increase by 220% of the oscillator strengths is completely compensated by the raise of 34% in the frequency: in the end, the contribution for TiO_2 is 9% larger than the one for ZrO_2.

3. CRYSTALLINE SILICATES

3.1. Introduction

Due to the chemical homology of Hf and Zr discussed in Section 2.1, hafnon ($HfSiO_4$) and zircon ($ZrSiO_4$) resemble each other in many physical and chemical properties. Their similarities are such that there is complete miscibility between hafnon and zircon (71). In addition to their importance as potential alternative gate dielectrics, hafnon and zircon are of geological significance. They both belong to the orthosilicate class of minerals, which can be found in igneous rocks and sediments. Zircon is used as a gemstone, because of its good optical quality, and resistance to chemical attack. In the earth's crust, hafnon and zircon are host minerals for the radioactive elements uranium and thorium. They have therefore widely been studied in the framework of nuclear waste storage.

In a recent paper (72), we have studied the structural, electronic and dynamical properties of zircon using first-principles calculations. In this section, we present a comparison between the naturally occurring $ZrSiO_4$ and $HfSiO_4$ crystals, and we also consider the $TiSiO_4$ crystal defined by similarity with hafnon and zircon. By lexical analogy, we will also refer to this hypothetical structure as titanon.

3.2. Structural Properties

The $MSiO_4$ with M (= Hf, Zr, Ti) crystals, which are represented in Fig. 2, have a conventional unit cell which is body-centered tetragonal (space group $I4_1/amd$, N° 141) and contains four formula units of $MSiO_4$, as illustrated by the dashed lines in Fig. 2(b). A primitive cell containing only two formula units of $MSiO_4$ can also be defined, as indicated by the heavy lines in Fig. 2(b).

The structure of hafnon, zircon, and titanon crystals may be viewed as consisting of $(SiO_4)^{4-}$ anions and M^{4+} cations with M (= Hf, Zr, Ti), as illustrated by the medium grey tetrahedra and the light grey spheres in Fig. 2(b). This is consistent with the larger bond length (about 25%) of the M–O compared to the Si–O bond.

Alternatively, as presented in Fig. 2(c), a different view may be adopted in which $HfSiO_4$, $ZrSiO_4$, and $TiSiO_4$ consist of alternating (discrete) SiO_4 tetrahedra and MO_8 units, sharing edges to form chains parallel to the c-direction. Note that in these MO_8 units four O atoms are closer to the M atom than the four other ones (about 6, 6, and 11% difference in the M–O bond length in hafnon, zircon, and titanon, as reported in Table 6).

The positions of the M (= Hf, Zr, Ti) and Si atoms are imposed by symmetry: they are located at $(0, \frac{3}{4}, \frac{1}{8})$ and $(0, \frac{1}{4}, \frac{3}{8})$ on the 4a and 4b Wyckoff sites, respectively. The O atoms occupy the 16h Wyckoff sites $(0, u, v)$, where u and v are internal parameters.

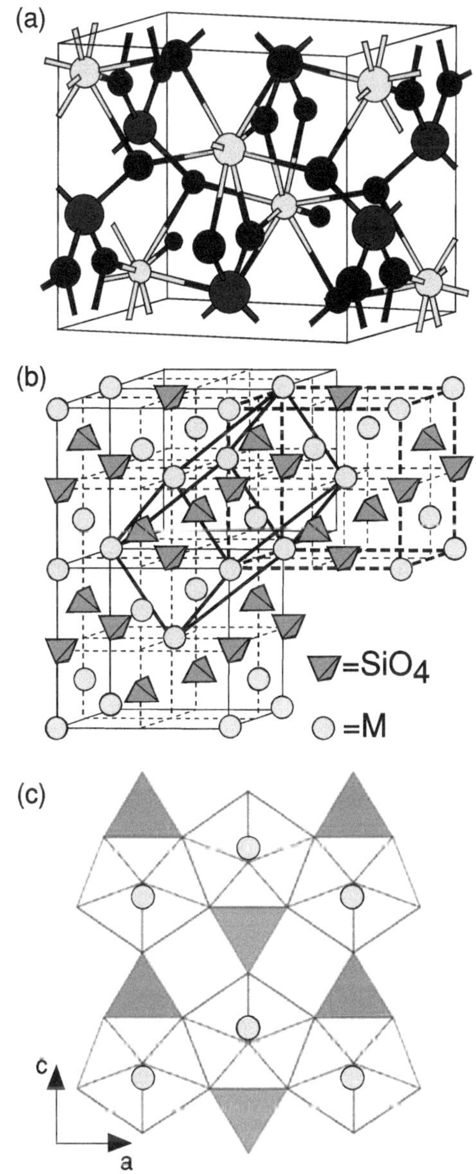

Fig. 2. Structure of $HfSiO_4$, $ZrSiO_4$, and $TiSiO_4$. (a) A ball and stick representation is adopted for the body-centered-tetragonal unit cell, where the M (= Hf, Zr, Ti), O, and Si atoms are colored in light, medium, and dark grey respectively. (b) The individual SiO_4 units are represented schematically by the medium grey tetrahedra, while M atoms are indicated by light grey spheres. The two sets of dashed lines and heavy lines outline the body-centered-tetragonal unit cell and the primitive cell, respectively. (c) Besides the SiO_4 units, the MO_8 triangular dodecahedra with the M atoms in their center are also drawn.

Table 6. Structural parameters of $HfSiO_4$, $ZrSiO_4$, and $TiSiO_4$

	$HfSiO_4$		$ZrSiO_4$		$TiSiO_4$
	Th.	Expt.	Th.	Expt.	Th.
a	6.61	6.57	6.54	6.61	6.21
c	5.97	5.96	5.92	6.00	5.81
u	0.0672	0.0655	0.0645	0.0646	0.0591
v	0.1964	0.1948	0.1945	0.1967	0.1892
Volume	130.42	128.63	126.60	131.08	112.03
d(Si–O)	1.62	1.61	1.61	1.62	1.60
d(M–O)	2.14	2.10	2.10	2.13	1.95
	2.27	2.24	2.24	2.27	2.19
∠(O-Si-O)	97°	97°	97°	97°	96°
	116°	117°	116°	116°	117°

The lengths are expressed in Å. The experimental data are taken from Speer and Cooper (71) for $HfSiO_4$, and from Mursic et al. (73) for $ZrSiO_4$.

Table 6 summarizes our results obtained after structural and atomic relaxation. The calculated lattice constants a and c, as well as the internal parameters u and v are found to be in excellent agreement with their corresponding experimental values for hafnon (71) and zircon (73). Interatomic distances and angles are within 1 or 2% of the experimental values. This accuracy is to address in a meaningful way the dynamical and dielectric properties.

The structural parameters for Hf- and Zr-based silicates are very close, whereas those for Ti-based materials can be 5–10% larger than for the two other silicates. The largest difference is found for the shorter Ti–O bond which is 9 and 10% smaller than the Zr–O and Hf–O distances, respectively. This is further confirmation of the different chemistry of 3d metals with respect to 4d and 5d metals.

3.3. Electronic Structure

In Fig. 3, we present the calculated electronic density of states (DOS) for hafnon, zircon, and titanon. The complete electronic band structure for $ZrSiO_4$ along several directions in the Brillouin zone can be found elsewhere (72). For $HfSiO_4$ and $TiSiO_4$, the electronic band structure is very similar apart from the position of the Hf 5s and 5p bands in hafnon and the Ti 3s and 3p bands in titanon, as explained hereafter.

We clearly distinguish four groups in the DOS of the valence bands, of which the three lowest ones are rather peaked (small dispersion of the bands), indicative of a weak hybridization. The DOS of hafnon, zircon, and titanon exhibit a very sharp peak at −60.2, −47.1, and −56.5 eV, respectively, which are attributed to the Hf 5s, Zr 4s, Ti 3s states, respectively. Each of these peaks corresponds to two flat bands in the band structure (72). The peaks at −29.8 eV for hafnon, −25.5 eV for zircon, and −33.1 eV for titanon are related to the Hf 5p, Zr 4p, and Ti 3s states, respectively. Each of these peak includes six electrons per unit cell. Finally, the O 2s peak (8 electrons per unit cell) is located between −18.0 and −16 eV for hafnon, zircon, and titanon.

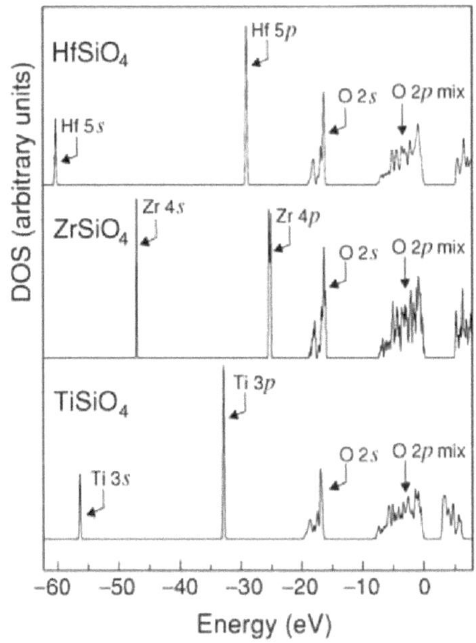

Fig. 3. Electronic density of states (DOS) for $HfSiO_4$, $ZrSiO_4$, and $TiSiO_4$.

By contrast, the fourth group (24 electrons per unit cell), has a much wider spread of 8 eV. These states have mainly an O 2p character with some mixing of Si and M (= Hf, Zr, Ti) orbitals. This mixed covalent–ionic bonding of $HfSiO_4$ and $ZrSiO_4$, appearing in this group of valence bands, should be kept in mind when interpreting the Born effective charge tensors.

3.4. Born Effective Charge Tensors

In the hafnon, zircon, and titanon structures, the local site symmetry of M (= Hf, Zr, Ti) and Si atoms is rather high ($\bar{4}m2$). The Born effective charge tensors of M (= Hf, Zr, Ti) and Si atoms are diagonal and have only two independent components; parallel and perpendicular to the tetragonal axis, $Z^*_\parallel$ and $Z^*_\perp$, respectively. The Born effective charge tensors of M (= Hf, Zr, Ti) and Si atoms are reported in Table 7.

We note that $Z^*_\perp$ for M (= Hf, Zr, Ti) is anomalously large compared to the nominal ionic charge of the hafnium, zirconium, titanium ions $Z = +4$. A similar behavior was also observed in the case of $PbZrO_3$ (49) and of hafnia and zirconia, as discussed in Section 2. A detailed analysis of the physics of Born effective charges in the case of perovskite ferroelectrics (like $PbZrO_3$) ascribed this effect to a mixed covalent–ionic bonding (50). In Section 3.3, we have seen the occurrence of M–O 2p hybridization. Thus the physical interpretation of this phenomenon is likely similar to the case of perovskite ferroelectrics. The other component of the M (= Hf, Zr, Ti) Born effective charge tensor ($Z^*_\parallel$) is also larger than the nominal ionic charge, although the effect is not as pronounced (in the case of Ti, it is basically equal to the nominal charge).

Table 7. Non-vanishing components of the calculated Born effective charge tensors for M (= Hf, Zr, Ti), Si, and O atoms in $HfSiO_4$, $ZrSiO_4$, and $TiSiO_4$.

Atom	$HfSiO_4$	$ZrSiO_4$	$TiSiO_4$
M	$\begin{pmatrix} +5.28 & +5.28 & +4.68 \end{pmatrix}$	$\begin{pmatrix} +5.41 & +5.41 & +4.63 \end{pmatrix}$	$\begin{pmatrix} +5.91 & +5.91 & +4.01 \end{pmatrix}$
Si	$\begin{pmatrix} +3.18 & +3.18 & +4.35 \end{pmatrix}$	$\begin{pmatrix} +3.25 & +3.25 & +4.42 \end{pmatrix}$	$\begin{pmatrix} +3.56 & +3.56 & +4.83 \end{pmatrix}$
O[a]	$\begin{pmatrix} -1.15 & 0 & 0 \\ 0 & -3.08 & -0.19 \\ 0 & -0.35 & -2.26 \end{pmatrix}$	$\begin{pmatrix} -1.15 & 0 & 0 \\ 0 & -3.17 & -0.16 \\ 0 & -0.34 & -2.25 \end{pmatrix}$	$\begin{pmatrix} -1.20 & 0 & 0 \\ 0 & -3.54 & -0.11 \\ 0 & -0.45 & -2.22 \end{pmatrix}$
	$\begin{bmatrix} -1.15 & -3.16 & -2.18 \end{bmatrix}$	$\begin{bmatrix} -1.15 & -3.23 & -2.19 \end{bmatrix}$	$\begin{bmatrix} -1.20 & -3.58 & -2.18 \end{bmatrix}$

For M (= Hf, Zr, Ti) and Si atoms, the tensors are diagonal and only the principal elements are given. For O atoms, the full tensor is reported and the principal elements of its symmetric part are indicated between brackets.

[a] The tensors reported in the table for O atoms refer to the atom located at $(0, u, v)$. The Born effective charge tensors for the other oxygen atoms can be obtained using the symmetry operations.

For the silicon atom, there are also some (weaker) deviations with respect to the nominal value ($Z = +4$), one component being larger, and one being lower. While for hafnon and zircon, the deviations are the largest for the perpendicular components; for titanon, it is the parallel component that differs the most from the nominal charge. These deviations are not very different from those observed in tetrahedrally bonded silica polymorphs, like quartz (74), in which each O atom is strongly bonded to two Si atoms, or in the more compact polymorph of silica, stishovite (51), in which each O atom has three close Si neighbors.

Note that $Z^*_\perp$ is about 3% smaller for hafnium in $HfSiO_4$ than for zirconium in $ZrSiO_4$, which in turn is about 10% smaller than for titanium in $TiSiO_4$. This is very similar to what is observed in hafnia, zirconia, and titania as discussed in Section 2.3. The Born effective charge of Si atoms for directions perpendicular to the tetragonal axis shows a very similar behavior: it is about 2% smaller in hafnon than in zircon, and 10% smaller in the latter than in titanon. For the Born effective charge in a direction parallel to the c axis, we find for Si atoms the same trend as for perpendicular directions, but the opposite one for M (= Hf, Zr, Ti) atoms: the Born effective charges for Hf in hafnon is about 1% higher than for Zr in zircon, which in turn is about 15% higher than for Ti in titanon.

The local site symmetry of the O atoms has only a mirror plane. As a consequence, the Born effective charge tensors of O atoms are not diagonal, and depend on five independent quantities. We examine the tensor for the O atom located at $(0, u, v)$, which is reported in Table 7. The Born effective charge tensors of the other oxygen atoms can be obtained using the symmetry operations. For this particular atom, the mirror plane is perpendicular to x. Note that Z^*_{yz} and Z^*_{zy} are different, but rather small, making the Born effective charge tensor almost diagonal. They appear in the mirror plane, where one O–Si bond and two O–M bonds (one long and one short) are present. One can compute the projection of the Born effective charge on these directions. For the O–Si bond, the projection is −3.00 in $HfSiO_4$ −3.01 in $ZrSiO_4$, and −3.22 in $TiSiO_4$. For the O–M bonds, the projections on the shorter and longer ones are −3.15 and −2.25 in hafnon, −3.23 and −2.29 in zircon, and −3.60 and −2.36 for titanon, respectively. In this plane, the magnitude of the Born effective charge components is larger than the nominal ionic charge of oxygen ($Z = -2$). Following an alternative approach to the characterization of the anisotropy of this tensor, we select its symmetric part and diagonalize it. The principal values are given in Table 7 and the principal direction associated to the largest principal value forms an angle of about 20° with respect to the y axis. Both analyses give the same type of anisotropy.

Such a strong anisotropy of the Born effective charge tensor for O atoms, with one component of magnitude much smaller than 2 and much smaller than the two others, was already observed in SiO_2-stishovite (51). By contrast, in tetrahedrally bonded silica, there are *two* components of magnitudes much smaller than 2. Thus, at the level of the Born effective charges, the ionic-covalent bonding of O atoms to M (= Hf, Zr, Ti) and Si atoms in $HfSiO_4$, $ZrSiO_4$, and $TiSiO_4$ is closer to stishovite than to quartz, in agreement with a naive bond-counting argument. Models of amorphous silicates MSi_xO_y should take into account this difference, and might be classified according to the anisotropy of the O Born effective charges. One expects that, for a small content of M (= Hf, Zr, Ti), the quartz-like behavior dominates, while, for M atomic fractions closer to that of hafnon, zircon, and titanon, the stishovite-like behavior becomes stronger.

Note finally that the Born effective charges for O atoms are very similar in $HfSiO_4$, $ZrSiO_4$, and $TiSiO_4$. The first principal component is the same in $HfSiO_4$ and $ZrSiO_4$, while it is 4% larger (in absolute value) in $TiSiO_4$. The last principal component is basically the same for all three silicates. The only significant difference is for the second principal value, which is 2% smaller in hafnon and 13% in titanon with respect to zircon.

3.5. Phonon Frequencies

We also compute the phonon frequencies at the Γ point of the Brillouin zone for hafnon and zircon. The theoretical group analysis (see Appendix) predicts the following irreducible representations of optical and acoustical zone-center modes:

$$\Gamma = \underbrace{2A_{1g} \oplus 4B_{1g} \oplus B_{2g} \oplus 5E_g}_{\text{Raman}} \oplus \underbrace{3A_{2u} \oplus 4E_u}_{\text{IR}} \oplus \underbrace{A_{2u} \oplus E_u}_{\text{Acoustic}} \oplus \underbrace{B_{1u} \oplus A_{2g} \oplus A_{1u} \oplus 2B_{2u}}_{\text{Silent}}.$$

Because of the non-vanishing components of the Born effective charge tensors, the dipole–dipole interaction must be properly included in the calculation of the interatomic force constants (17, 52, 53). In particular, the dipole–dipole contribution is found to be responsible for the splitting between the longitudinal and transverse optic (LO and TO, respectively) modes E_u (perpendicular to c) and A_{2u} (parallel to c) at the Γ point.

Table 8. Fundamental frequencies of $HfSiO_4$, $ZrSiO_4$, and $TiSiO_4$ (in cm^{-1}) with their symmetry assignments

	Mode		$HfSiO_4$ Th.	$HfSiO_4$ Expt.	$ZrSiO_4$ Th.	$ZrSiO_4$ Expt.	$TiSiO_4$ Th.
Raman	$A_{1g}(1)$	O*	462	450	442	439	383
	$A_{1g}(2)$	O*	970	984	971	974	1011
	$B_{1g}(1)$	M	162	157	225	214	259
	$B_{1g}(2)$	O	395	401	397	393	418
	$B_{1g}(3)$	O	638	620	632	–	627
	$B_{1g}(4)$	O	1016	1020	1017	1008	1047
	B_{2g}	O*	247	267	252	266	263
	$E_g(1)$		161	148	194	201	194
	$E_g(2)$		204	212	225	225	242
	$E_g(3)$		369	351	375	357	430
	$E_g(4)$	O	530	–	536	547	544
	$E_g(5)$	O	923	–	923	–	319
Infrared	A_{2u}(TO1)		312		348	338	319
	A_{2u}(LO1)		423		476	480	482
	A_{2u}(TO2)	O	598		601	608	606
	A_{2u}(LO2)		656		646	647	631
	A_{2u}(TO3)	O	983		980	989	1000
	A_{2u}(LO3)		1095		1096	1108	1106
	E_u(TO1)		252		285	287	303
	E_u(LO1)		313		341	352	374
	E_u(TO2)	O	395		383	389	374
	E_u(LO2)		409		420	419	414
	E_u(TO3)	O	420		422	430	433
	E_u(LO3)		461		466	471	497
	E_u(TO4)	O	873		867	885	877
	E_u(LO4)		1023		1029	1035	1048
Silent	B_{1u}	O*	107		120		125
	A_{2g}	O*	233		242		250
	A_{1u}	O*	383		392		418
	$B_{2u}(1)$	O*	573		566		547
	$B_{2u}(2)$	O*	945		943		969

The letter (M or O) in the second column indicates the atoms (metal or oxygen) whose motion dominates in the vibrational mode, a star superscript showing that the other atoms are fixed by symmetry. The experimental values are taken from Hoskin and Rodgers (76) for $HfSiO_4$ (Raman modes only), and from Dawson et al. (77) for $ZrSiO_4$.

The calculated phonon frequencies are reported in Table 8. Our results are in excellent agreement with the values reported in experiments for hafnon (75, 76), and for zircon (77–79), with a rms absolute deviation of 4.1 cm^{-1} for $HfSiO_4$ (9.4 cm^{-1} for $ZrSiO_4$), and a rms relative deviation of 4.2% (2.5%).

In Section 2.4, we have pointed out three origins for the variations of the frequencies in Hf and Zr based oxides: the structural changes, the mass ratio equal to 1.96

for Hf/Zr and 1.90 for Zr/Ti, and the differences in interatomic force constants. It is quite interesting to compare on the same basis the phonon frequencies calculated for $HfSiO_4$, $ZrSiO_4$, and $TiSiO_4$ (see Table 8).

By performing a similar analysis as for the oxides, we find that the structural changes play a very minor role, in agreement with the intuition resulting from the very small variations observed in Table 6. With respect to zircon, we find that the frequencies in hafnon are decreased by about 1%; where as in titanon, they are increased by at most 6%.

As for the oxides, the mass ratio should play a major role for the modes in which the M (= Hf, Zr, Ti) atoms move significantly more than O atoms (as indicated by the letter M in the second column of Table 8). This is the case for the $B_{1g}(1)$ mode. The effect of the mass ratio on the phonon frequencies can be roughly estimated by assuming that the interatomic force constants for hafnon and titanon are the same as those for zircon, while the mass of the metal atom is changed to that of Hf or Ti. These simple calculations lead to frequencies which are reduced by roughly 28% for hafnon and increased by about 33% for titanon with respect to zircon. In Table 8, we observe that, for hafnon, the estimation above is very good; while, for titanon, the frequency only increases by 15% indicating an important change in the interatomic force constants.

On the contrary, the frequencies should not vary much between the three silicates for modes in which the M (= Hf, Zr, Ti) atoms are fixed by symmetry as indicated by the letter O with a star superscript in Table 8, as well as for those in which the O atoms move significantly more than the M (= Hf, Zr, Ti) atoms (as indicated by the letter O). In most of these cases, this is indeed what is observed; in a few cases, however, the differences in the interatomic force constants dominate (for instance, for the $A_{1g}(1)$ mode in $TiSiO_4$ or the B_{1u} mode in $HfSiO_4$ for which the frequencies decrease by 11 and 14% with respect to $ZrSiO_4$.

In summary, the effect of the interatomic force constants is less pronounced for the silicates than for the oxides. As a result, the differences in the phonon frequencies in Hf, Zr, and Ti silicates can mostly be explained by the ratio between the masses of the three metals.

3.6. Dielectric Permittivity

Due to the tetragonal symmetry of the hafnon, zircon, and titanon crystals, the electronic (ϵ_∞) and static (ϵ_0) permittivity tensors have two independent components $\epsilon_\parallel$ and $\epsilon_\perp$ parallel and perpendicular to the c axis, respectively. The calculated values of ϵ_∞ and ϵ_0 are reported in Table 9.

For zircon, values of 10.69 (3.8) (78) and 11.25 (3.5) (79) are reported for the static (electronic) dielectric permittivity in the directions parallel and perpendicular to the tetragonal axis, respectively. Our theoretical values are larger than the experimental ones by about 10%, as often found in the LDA to density functional theory. For hafnon, we were not able to find accurate measurements in the literature: for hafnium silicates, values ranging from 11 to 25 have been reported.

Table 9. Electronic and static dielectric tensors of $HfSiO_4$, $ZrSiO_4$, and $TiSiO_4$

	$HfSiO_4$		$ZrSiO_4$		$TiSiO_4$	
	$\parallel$	$\perp$	$\parallel$	$\perp$	$\parallel$	$\perp$
ϵ_∞	4.11	3.88	4.26	4.06	5.52	5.56
$\Delta\epsilon_1$	4.93	4.38	5.90	5.16	9.90	11.54
$\Delta\epsilon_2$	0.81	0.75	0.52	1.31	0.31	0.00
$\Delta\epsilon_3$	0.80	0.35	0.85	0.05	1.01	0.46
$\Delta\epsilon_4$		1.27		1.38		1.88
ϵ_0	10.65	10.63	11.53	11.96	16.73	19.44

The contributions of individual phonon modes to the static dielectric tensor are indicated. The tensors are diagonal and have different components parallel ($\parallel$) and perpendicular ($\perp$) to the c axis. The phonon mode contributions to $\epsilon_0^{\parallel}$ come from the three IR-active A_{2u} modes, while the contributions to $\epsilon_0^{\perp}$ come from the four IR-active E_u modes.

The contribution of the individual modes $\Delta\epsilon_{\mathrm{m}}$ to the static dielectric constant, as defined in Eq. (13), are also indicated in Table 9. The largest contribution comes from the lowest frequency mode. The decomposition of the static dielectric tensor can further be analyzed using the mode-effective charge vectors and the oscillator strength tensors, defined by Eqs. (14) and (12), respectively. In Table 10, we present for each IR-active mode, the magnitude of its mode-effective charge vectors (this vector is parallel and perpendicular to the tetragonal axis for A_{2u} and E_u modes, respectively), as well as the relevant component of the oscillator strength tensor (the parallel–parallel component for A_{2u} modes, and the perpendicular–perpendicular component for E_u modes).

Table 10. Components of mode-effective charge vectors Z^*_{m} and oscillator strength tensor S_{m} for each of the IR-active modes for $HfSiO_4$, $ZrSiO_4$, and $TiSiO_4$

	$HfSiO_4$		$ZrSiO_4$		$TiSiO_4$	
	Z^*_m	S_m	Z^*_m	S_m	Z^*_m	S_m
$A_{2u}(1)$	6.85	7.39	7.68	10.06	7.73	12.53
$A_{2u}(2)$	3.78	4.24	2.76	2.64	2.05	1.41
$A_{2u}(3)$	6.60	11.22	6.71	11.50	7.10	12.49
$E_u(1)$	5.93	4.05	6.79	5.91	8.45	13.22
$E_u(2)$	2.94	1.70	3.51	2.71	0.90	0.00
$E_u(3)$	1.69	0.91	0.28	0.12	2.66	1.08
$E_u(4)$	7.21	14.02	7.37	14.69	8.10	18.05

The description of the reported vector and tensor components corresponding to the two types of modes is given in the text. The components of the mode-effective charge vectors are given in units of $|e|$, where e is the electronic charge. The oscillator strengths are given in 10^{-4} atomic unit (1 a.u. = 253.2638413 m^3/s^2).

For each symmetry representation (A_{2u} and E_u), the lowest and highest frequency modes exhibit the largest mode effective charges and the largest oscillator strengths. Despite their similar oscillator strengths, the modes of lowest frequency contribute much more to the static dielectric constant than the modes of highest frequency, the frequency factor in Eq. (13) playing a crucial role. The other modes contribute significantly less to the static dielectric constants.

For the lowest and highest frequency modes, the oscillator strengths and the mode-effective charges increase from hafnon to zircon and from zircon to titanon. The origin of this difference can be traced back to the Born effective charges and the eigendisplacements. Indeed, as discussed in Section 3.4, the Born effective charges of M (= Hf, Zr, Ti) and Si atoms show the following behavior: $Z^*(\mathrm{HfOSi_4}) \leq Z^*(\mathrm{ZrSiO_4}) \leq Z^*(\mathrm{TiSiO_4})$. Moreover, due to their heavier weight, the displacements of Hf atoms are smaller than those of Zr atoms, which in turn are smaller than those of Ti atoms.

Coming back to the contributions to the static dielectric constant reported in Table 9, we observe that the contributions of the lowest and highest frequency modes show the following increasing trend: $\Delta\epsilon(\mathrm{HfSiO_4}) \leq \Delta\epsilon(\mathrm{ZrSiO_4}) \leq \Delta\epsilon(\mathrm{TiSiO_4})$. This behavior is essentially related to the increase in the oscillator strengths, since the phonon frequencies do not change significantly in all three silicates.

4. AMORPHOUS SILICATES

The dielectric properties of transition metal amorphous silicates constitute an issue of great practical importance. Early experimental measurements tend to show a supra-linear dependence of the static dielectric constant ϵ_0 on the metal concentration (6, 7). While several phenomenological theories address this behavior (80,81), a close to linear dependence seem to prevail based on more recent (82, 83). In a recent paper (84), we have used DFT simulations to tackle this particularly relevant problem by analyzing how the permittivity of Zr silicates is affected by the underlying microscopic structure.

In addressing this technological issue, we face the more general problem of predicting the dielectric properties of amorphous alloys using first-principles calculations. Brute force investigation of numerous large supercells is beyond present computational capabilities. To overcome this difficulty, we explore the relationship between the dielectric properties of Zr silicates and their underlying microscopic structure. Using density-functional theory, we determine optical and static dielectric constants for various model structures of Zr silicates, both ordered and disordered. We introduce a scheme which relates the dielectric constants to the local bonding of Si and Zr atoms. This scheme relies on the definition of parameters characteristic of the basic structural units (SUs) formed by Si and Zr atoms and their nearest neighbors.

Applied to amorphous Zr silicates, our scheme provides a good description of measured dielectric constants, both optical (83, 85) and static (82, 83), and shows the important contribution of $\mathrm{ZrO_6}$ SUs to the static dielectric constant. In a very similar way, our scheme can also be used to investigate Hf and Ti silicates. We here only briefly indicate how these systems compare with each other.

We set up a series of model structures of $(ZrO_2)_x(SiO_2)_{1-x}$ with x ranging from 0 to 0.5, nine crystalline and one amorphous, and describe them in terms of cation-centered SUs. We start with three different SiO_2 polymorphs ($x = 0$):

- C_0: α-cristobalite with four SiO_4 SUs per unit cell
- Q_0: α-quartz with three SiO_4 SUs
- S_0: stishovite with two SiO_6 SUs

We derive three new crystal structures by replacing one of the Si atoms by a Zr atom for each of these models:

- C_1: Zr^{Si} in α-cristobalite with three SiO_4 and one ZrO_4 SUs per unit cell ($x = 0.25$)
- Q_1: Zr^{Si} in α-quartz with two SiO_4 and one ZrO_4 SUs ($x = 0.33$)
- S_1: Zr^{Si} in stishovite with one SiO_6 and one ZrO_6 SUs

We also consider the zircon crystal, as well as two other models generated by substituting Zr by Si:

- Z_2: zircon which contains two SiO_4 and two ZrO_8 SUs per unit cell ($x = 0.5$)
- Z_1: Si^{Zr} in zircon with two SiO_4, one SiO_6, and one ZrO_8 SUs ($x = 0.25$)
- Z_0: fully Si-substituted zircon with two SiO_4 and SiO_6 SUs ($x = 0$)

Finally, only a single disordered structure could be afforded because of the noticeable computational cost associated:

- A: amorphous structure, generated using classical molecular dynamics with empirical potentials (84), with 3 ZrO_4 and 17 SiO_4 SUs ($x = 0.15$)

The atomic coordinates and the cell parameters of all our model structures are fully relaxed within the local density approximation (LDA) to DFT. The calculated optical and static dielectric constants for our model structures are given in Table 11. Due to the well-known limitations of the LDA, the theoretical values are larger than the experimental ones (when available) by about 10%.

In order to analyze the dependence of the optical dielectric constant (ϵ_∞) on the underlying atomic nanostructure, we use the Clausius–Mosotti relation (81, 83) that connects it to the electronic polarizability $\bar{\alpha}$:

$$\frac{\epsilon_\infty - 1}{\epsilon_\infty + 2} = \frac{4\pi}{3}\frac{\bar{\alpha}}{\bar{V}}, \tag{19}$$

where $\bar{V}$ is the average SU volume. The polarizability $\bar{\alpha}$ can be considered as a local and additive quantity, in contrast with ϵ_∞. Hence, we define α_i values for each SU i, where $i \equiv SiO_n$ (with $n = 4$ or 6) or ZrO_n (with $n = 4$, 6, or 8), in such a way that:

$$\bar{\alpha} = \sum_i x_i \alpha_i, \tag{20}$$

Table 11. Composition (x), optical (ϵ_∞) and static (ϵ_0) dielectric constants, volume ($\bar{V}$) in bohr^3, polarizability $\bar{\alpha}$ in bohr^3, characteristic dynamical charge ($\bar{Z}$), and characteristic force constant ($\bar{C}$) in hartree/bohr^2 for the various model systems

Model	x	ϵ_∞	ϵ_0	$\bar{V}$	$\bar{\alpha}$	$\bar{Z}$	$\bar{C}$
C_0	0.00	2.38	4.30	264.77	19.92	4.21	0.4391
C_1	0.25	2.76	5.25	273.21	24.12	4.59	0.3895
Q_0	0.00	2.54	4.83	240.34	19.46	4.28	0.4169
Q_1	0.33	2.91	5.84	275.28	25.56	4.85	0.3661
S_0	0.00	3.36	10.33	153.74	16.16	4.81	0.2716
S_1	0.50	4.44	24.20	201.88	25.74	6.14	0.1188
Z_0	0.00	3.37	10.11	167.80	17.68	4.76	0.2512
Z_1	0.25	3.94	18.36	189.74	22.42	5.29	0.1287
Z_2	0.50	4.13	11.81	213.28	26.00	5.58	0.2385
A	0.15	3.24	8.92	213.12	21.75	4.83	0.2424

The reported dielectric constants correspond to orientational averages.

where x_i is the molecular fraction. In Table 12, we report the five α_i values that we determine by solving in a least square sense the overdetermined system based on the calculations of ϵ_∞ for the nine crystalline models. The optical dielectric constants derived from these α_i values using Eqs. (19) and (20) are in very good agreement with those computed from first-principles, showing average and maximal errors smaller than 1 and 2.5%, respectively. For the amorphous model, which was not used to determine the α_i values, the calculated value $\epsilon_\infty = 3.25$ compares very well with the first-principles result $\epsilon_\infty = 3.24$. These results give an *a posteriori* motivation for the use of Eqs. (19) and (20) to model the optical dielectric constant.

For the static dielectric constant (ϵ_0), the description in terms of a single local and additive quantity as the electronic polarizability is precluded by the phonon contributions. To overcome this difficulty, we consider the difference between dielectric constants ($\Delta\epsilon$):

$$\Delta\epsilon = \epsilon_0 - \epsilon_\infty = \frac{4\pi}{\Omega_0}\sum_m \frac{S_m}{\omega_m^2} = \frac{4\pi}{\bar{V}}\frac{\bar{Z}^2}{\bar{C}}, \tag{21}$$

Table 12. Polarizability (α in bohr^3), characteristic dynamical charge (Z), and characteristic force constant (C in hartree/bohr^2) for various structural units, extracted from the calculations for the nine crystalline models

	SiO_4	SiO_6	ZrO_4	ZrO_6	ZrO_8
α	19.68	16.14	37.37	35.35	32.69
Z	4.29	4.92	5.66	7.16	6.73
C	0.3597	0.2176	0.4202	0.0817	0.1153

where ω_m and S_m are the frequency and the oscillator strength of the mth mode. The volume of the primitive unit cell Ω_0 is related to the volume $\bar{V}$ and to the number of SUs $\bar{N}$ by $\Omega_0 = \bar{N}\bar{V}$. The characteristic dynamical charge $\bar{Z}$ and characteristic force constant $\bar{C}$ are defined by:

$$\bar{Z}^2 = \frac{1}{\bar{N}} \sum_{\kappa} Z_{\kappa}^2 \quad \text{and} \quad \bar{C}^{-1} = \frac{1}{\bar{N}} \sum_{m} \frac{S_m}{\omega_m^2 \bar{Z}^2}, \tag{22}$$

where Z_κ are the atomic Born effective charges.

A detailed analysis of the variation of $\Delta\epsilon$ due to a Si $\rightarrow$ Zr substitution is given in ref. (84), where the contribution from sixfold coordinated atoms has been highlighted. In fact, these configurations are very similar to those in ABO_3 perovskites. The enhancement of $\Delta\epsilon$ finds its origin in very low frequency modes in which the cations (A or B) move in opposition with the O atoms while carrying opposite effective charges.

By analogy with the polarizability, we define Z_i and C_i values for each SU such that:

$$\bar{Z}^2 = \sum_i x_i Z_i^2 \quad \text{and} \quad \bar{C}^{-1} = \sum_i x_i C_i^{-1}, \tag{23}$$

though the locality and the additivity of these parameters is not guaranteed *a priori*. We determine the optimal values Z_i and C_i in the same way as for α_i (Table 12).

For the nine crystalline models, the values of $\Delta\epsilon$ obtained by introducing these parameters in Eqs. (21) and (23) match quite well those calculated from first-principles (84), though the agreement is not as impressive as for ϵ_∞. Differences result primarily from the determination of $\bar{C}$. By contrast, the values of $\bar{Z}$ given by Eq. (23) agree very well with those computed from first principles, showing an average and maximal error smaller than 2 and 3%, respectively. *A posteriori*, $\bar{C}$ appears to be less local and additive. In fact, it can be demonstrated that the locality of $\bar{C}$ is closely related to the dynamical charge neutrality of the SUs (84).

For the amorphous model, which was not used to determine the Z_i and C_i values, the agreement between the model and the first-principles $\Delta\epsilon$ is excellent with an error smaller than 1% (84). Indeed, our scheme is more accurate for disordered systems, where the localization of vibrational modes is enhanced and the dynamical charge neutrality appears better respected.

The parameters in Table 12 fully determine the dielectric constants of Zr silicates of known composition in terms of SUs. It is important to note the following two points. On the one hand, the three parameters of Zr-centered SUs all contribute to enhancing the dielectric constants over those of Si-centered ones of corresponding coordination.[1] This is clearly at the origin of the increase of ϵ_∞ and ϵ_0 with increasing Zr concentration. Second, while the polarizability α_i of a given SU (Si- or Zr-centered) steadily decreases with increasing coordination, such a regular behavior is not

[1] In Table 12, the value of C for SiO_4 apparently leads to a higher contribution to $\Delta\epsilon$ than that for ZrO_4. This is an artifact of the approach we used to determine the Z_i and C_i.

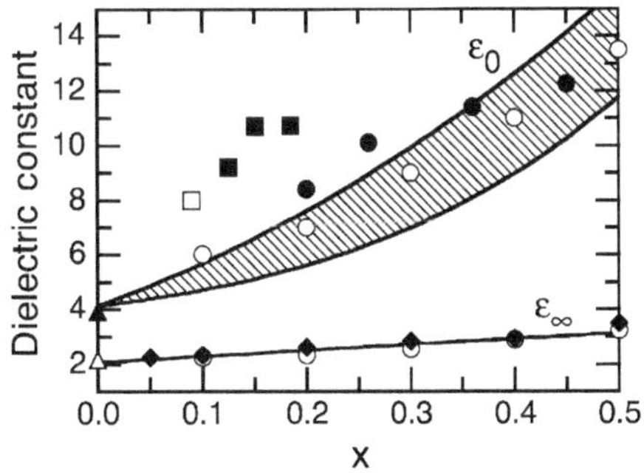

Fig. 4. Dielectric constants (ϵ_∞ and ϵ_0) as a function of composition x for amorphous $(ZrO_2)_x(SiO_2)_{1-x}$. The hatched region corresponds to results derived from our model scheme and reflects the indetermination of the number of ZrO_6 units. The upper curve delimiting the band corresponds to structures entirely composed of ZrO_6 units, while the lower curve represents a smooth transition from a structure composed of ZrO_4 units at $x = 0$ to one composed of ZrO_8 units at $x = 0.5$, without the occurrence of any ZrO_6 unit. The references for the experimental data are: ◆ (85), ● (82), ○ (83), ■ (6, 7), □ (86), ▲ (87), and △ (30).

observed for the parameters Z_i and C_i determining $\Delta\epsilon$. On the other hand , Z_i and C_i concurrently vary to enhance the contribution of ZrO_6 units, which are the SUs giving the largest contribution to $\Delta\epsilon$ in amorphous Zr silicates.

Using the scheme given by Eqs. (19), (20), (21) and (23), we can now obtain an estimate of ϵ_∞ and ϵ_0 for amorphous $(ZrO_2)_x(SiO_2)_{1-x}$ as a function of Zr composition ($0 < x < 0.5$). Using measured densities for Zr silicates (85), we first calculate ϵ_∞ as a function of x. In this case, the effect of Zr coordination is negligible since the various Zr-centered units have close α values compared to SiO_4 (Table 12). As plotted in Fig. 4, our theoretical values agree very well with available experimental data (83, 85).

In order to apply our scheme for $\Delta\epsilon$, additional information on the cationic coordination is required. We assume that the Si atoms are fourfold coordinated. The coordination of Zr atoms is less well determined. Recent EXAFS measurements (80) tend to show that the average Zr coordination increases from about 4 to about 8 for Zr concentrations increasing from $x \sim 0$ to 0.5. In Fig. 4, we also plot the calculated ϵ_0 for amorphous $(ZrO_2)_x(SiO_2)_{1-x}$ as a function of x, together with the available experimental data (6, 7, 82, 83, 86).

The characteristic parameters used to calculate ϵ_0 change noticeably with the local environment of Zr atoms. Therefore, the indetermination with respect to their coordination leads to a range of possible values for our theoretical values as represented by the dashed band in Fig. 4. We consider several suitable suitable distributions of three representative structural units (ZrO_4, ZrO_6, and ZrO_8). The upper curve delimiting the band in Fig. 4 corresponds to structures entirely composed of ZrO_6 units. The lower curve is for amorphous systems which do not contain any ZrO_6 units. The average Zr coordination varies linearly from four to eight between $x = 0$ and 0.5, with concentrations of ZrO_4 and ZrO_8 SUs varying at most quadratically. Note that the

Table 13. Polarizability (α in bohr^3), characteristic dynamical charge (Z), and characteristic force constant (C in hartree/bohr^2) for MO_6 and MO_8 structural units with M (= Hf, Zr, Ti), extracted from the calculations for the r-MO_2 structures on the one hand, and for the c-MO_2 and $MSiO_4$ structures on the other hand[a]

	HfO_6	ZrO_6	TiO_6	HfO_8	ZrO_8	TiO_8
α	35.28	34.54	33.49	32.21	31.66	32.87
Z	7.19	7.45	8.59	6.77	6.95	7.70
C	0.1278	0.1050	0.0437	0.1630	0.1418	0.0778

[a]The parameters extracted from the $MSiO_4$ structures take into account the values already obtained for SiO_4 SUs which are reported in Table 12.

upper part of the band matches well the recent experimental data (82, 83). The earlier data (6, 7, 86) cannot be explained. Figure 4 shows that, for a sufficient amount of ZrO_6 units, values of ϵ_0 at intermediate x can indeed be larger than estimated from a linear interpolation between SiO_2 and $ZrSiO_4$. However, in agreement with recent experiments (82, 83), our theory indicates that the extent of this effect is more limited than previously assumed (6, 7, 80).

Our scheme could also be applied to Hf and Ti silicates which are very similar to Zr silicates, provided that the value of the characteristic parameters are adapted. In this respect, the comparison between the various crystalline oxides and silicates carried out in the preceding sections provides very useful informations. Indeed, it is possible to extract the characteristic parameters of MO_6 and MO_8 SUs from the results obtained for the r-MO_2 structures on the one hand, and for the c-MO_2 and $MSiO_4$ structures on the other hand. These values are reported in Table 13.

Note the results in Tables 12 and 13 for ZrO_6 and ZrO_8 structural units are in good agreement despite the fact they have been obtained using almost completely different sets of crystalline systems. The only common system is the zircon crystal. In Table 12, the parameters are extracted from the results for crystalline systems that all include Si-centered SUs; whereas, in Table 13, the reference crystals do not include such structural units (apart from zircon). This increases our confidence in the validity of the scheme.

Basically, all the parameters in Table 13 show a similar trend. On the one hand, the enhancement of the dielectric permittivities (both electronic and static) will be larger for the Ti-centered SUs than for Hf- and Zr-centered ones. On the other hand, the MO_6 units produces a larger enhancement than MO_8 units. Hence, for the amorphous silicates, the same kind of considerations should apply. In particular, for Ti amorphous silicates, the dielectric constants should be considerably larger than for Hf and Zr amorphous silicates. Firstly, for Ti, the MO_6 SUs tend to be more stable (as in rutile) than the MO_8 ones; whereas, for Hf and Zr, the MO_8 SUs prevail. Secondly, all the characteristic parameters of Ti-centered SUs produce a larger enhancement than Hf and Zr ones.

5. CONCLUSIONS

Using density-functional theory, we have investigated the structural, electronic, dynamical, and dielectric properties for a series of high-k materials belonging to the Hf–Si–O, Zr–Si–O, and Ti–Si–O systems. We have considered three different structures (cubic, tetragonal, and rutile) for hafnia, zirconia, and titania (the crystalline oxides), as well as hafnon, zircon, and titanon (the crystalline silicates), and finally the amorphous silicates.

In all the investigated systems, we have found a very good agreement between the parameters of the relaxed atomic structures and the experimental ones (when available). The structural parameters are found to be very similar in Hf- and Zr-based materials. On the contrary, the Ti-based materials show some significant differences with respect to the corresponding Hf and Zr oxides and silicates.

The phonon frequencies at the center of the Brillouin zone, the Born effective charge tensors, and the dielectric permittivity tensors have been obtained using density-functional perturbation theory. For the crystalline systems, the agreement between the calculated phonon frequencies and their corresponding experimental values (when available) was found to be excellent. In all the cases, the differences between the Hf-, Zr-, and Ti-based systems have been analyzed in detail and interpreted in terms of structural changes, the mass ratio, and variations of interatomic force constants.

The Born effective charge tensors present an important anisotropy. For some directions, we have found that these effective charges are larger than the nominal ionic charge, indicating a mixed covalent–ionic bonding between M (= Hf, Zr, Ti) and O atoms, and between Si and O atoms. We have also discussed the effective charges focusing on the changes between the systems containing hafnium, zirconium, and titanium.

We have computed the electronic and static dielectric permittivity constants, and have proposed a detailed analysis of the contributions of individual vibrational modes. The discussion is based on the calculation of mode-effective charges and oscillator strengths. For the tetragonal systems (t-HfO_2, t-ZrO_2, r-HfO_2, r-ZrO_2, r-TiO_2, $HfSiO_4$, $ZrSiO_4$, $TiSiO_4$), it was observed, for directions both parallel and perpendicular to the tetragonal axis, that a single mode contributes for more than 60% of the full ionic contribution. The corresponding eigenvectors, which could be obtained in our first-principle approach, show clearly that the displacement is characterized by M (= Hf, Zr, Ti) and O atoms moving in opposite directions. In the silicates, the displacement of Si atoms in these modes is more than twice smaller than those of the other species, inducing a substantial distortion of the SiO_4 tetrahedra in contradiction to what was previously thought. For all systems, the modifications related to the presence of Zr rather than Hf have been rationalized in terms of the difference in mass between these atoms, variations of interatomic force constants, and changes in structural parameters (minor effect).

For hafnon, zircon, and titanon, the electronic density of states have been calculated. The contributions from Hf 5s and 5p, Zr 4s and 4p, Ti 3s and 3p, and O 2s and

2p are clearly distinguishable. The spread of the latter indicates hybridization with atomic M (= Hf, Zr, Ti) and Si orbitals.

Finally, the dielectric properties of amorphous silicates have been investigated. We have proposed a simple scheme which connects the optical and static dielectric constants of Zr silicates to their underlying microscopic structure. Our theory supports recent experiments which find a close to linear dependence of ϵ_0 on the Zr fraction x, and shows that higher dielectric constants can be achieved by increasing the concentration of ZrO_6 structural units. We have extended these results to Hf and Ti amorphous silicates. We have proposed that the dielectric constants should be considerably larger for Ti-based systems than for Hf and Zr ones because of the predominance of MO_6 in the former and the larger enhancement produced by Ti-centered structural units.

ACKNOWLEDGMENTS

The author wishes to thank Drs. F. Detraux, A. Bongiorno, G. Jun, and X. Rocquefelte, as well as Profs. K. Cho, X. Gonze, and A. Pasquarello who took an active part in the research leading to the results presented in this Chapter. He is also grateful to R. B. van Dover for providing us his results prior to publication. Support is acknowledged from the FNRS-Belgium, the FRFC project (N° 2.4556.99), and the Belgian PAI-5/1/1.

APPENDIX: SYMMETRIES OF THE PHONON MODES

In this appendix, we present the procedure to determine the symmetries of the phonon modes $U_{m\mathbf{q}}(\kappa\alpha)$ at a wave vector $\mathbf{q}$ using group theory. For pedagogic purposes, we focus on the practical aspects without giving the formal justifications of the formulas, which may be found in the literature (88). As an example, we will consider the phonon modes at the Γ point for the fluorite structure (space group $Fm\bar{3}m$, N° 225). This corresponds to the cubic structure for the oxides studied previously.

In the following, we adopt the Seitz notation for the symmetry operations of the crystal:

$$\{\mathbf{S} \mid \mathbf{v}(S)\}, \tag{A.1}$$

where $\mathbf{S}$ is 3×3 a real orthogonal matrix representation of a rotation and $\mathbf{v}(S)$ is a vector which is smaller than any primitive translation vector of the crystal. Applied to the equilibrium position vector of atom κ relative to the origin of the cell τ_κ, this symmetry operation transforms it according to the rule:

$$\{\mathbf{S} \mid \mathbf{v}(S)\}\, \tau_\kappa = \mathbf{S}\tau_\kappa + \mathbf{v}(S) = \tau_{\kappa'} + \mathbf{R}(a), \tag{A.2}$$

where $\mathbf{R}(a)$ is a translation vector of the crystal. The second equality expresses the fact that, because the symmetry operation $\{\mathbf{S} \mid \mathbf{v}(S)\}$ is one which sends the crystal into itself, the lattice site κ must be sent onto an equivalent site which we label κ'.

Table A1. Character table for space group $Fm\bar{3}m$ (N° 225) at the Γ point

	E	4	2	3	2′	I	−4	m	−3	m'	Functions
Mult.	1	6	3	8	6	1	6	3	8	6	
A_{1g}	1	1	1	1	1	1	1	1	1	1	$x^2+y^2+z^2$
A_{1u}	1	1	1	1	1	−1	−1	−1	−1	−1	
A_{2g}	1	−1	1	1	−1	1	−1	1	1	−1	
A_{2u}	1	−1	1	1	−1	−1	1	−1	−1	1	
E_g	2	0	2	−1	0	2	0	2	−1	0	$(2z^2-x^2-y^2, x^2-y^2)$
E_u	2	0	2	−1	0	−2	0	−2	1	0	
T_{2u}	3	−1	−1	0	1	−3	1	1	0	−1	
T_{2g}	3	−1	−1	0	1	3	−1	−1	0	1	(xy, xz, yz)
T_{1u}	3	1	−1	0	−1	−3	−1	1	0	1	(x, y, z)
T_{1g}	3	1	−1	0	−1	3	1	−1	0	−1	(J_x, J_y, J_z)

The starting point is determine the space group $G_{\mathbf{q}}$ of the wave vector **q**. It consists of the symmetry operations of the crystal $\{\mathbf{S} \mid \mathbf{v}(S)\}$ whose purely rotational part $\{\mathbf{S}\}$ have the property:

$$\mathbf{Sq} = \mathbf{q} + \mathbf{G}, \tag{A.3}$$

where **G** is a translational vector of the reciprocal lattice. It is clear that **G** vanishes if **q** lies inside the first Brillouin zone, and that it can only be non-zero if **q** lies on the boundary of the zone. For our example, since we consider the Γ point, the space group is simply the space group of the crystal.

The next step is to obtain the character table for the space group $G_{\mathbf{q}}$. The various tables for all space groups and special points can be found in books such as ref. (89). Alternatively, the Bilbao Crystallographic Server (90) provides all the tables for the Γ point. For our example, the character table is given in Table A1.

The symmetries of phonon modes are related to the transformation properties of the displacement vectors $U_{m\mathbf{q}}(\kappa\alpha)$. In group theoretical terms, this implies to take direct product of the irreducible representations for the vector (x, y, z) with those of the various atomic sites.

First, we need to find the irreducible representations χ_V of the vector (x, y, z). For our example, we see in Table A1 that:

$$\chi_V = F_{1u}. \tag{A.4}$$

Second, each atomic site is also characterized by one or more symmetry operations that map the atomic site onto itself. The collection of these symmetry operations define the site group (usually labeled using Wyckoff notation). The site group can be one of the 32 crystallographic point groups and must be a subgroup of the space group. For each site, the characters $\chi_{\text{atomic site}}$ represent the number of atoms that are invariant under the symmetry operations of the group. The irreducible representations need to be found for each site. For our example, the Zr and O atoms occupy 4a and 8c

Wyckoff sites, respectively. We find that:

$$\begin{aligned} \chi_{\text{atomic site } 4a} &= A_{1g} \\ \chi_{\text{atomic site } 8c} &= A_{1g} \oplus A_{2u} \end{aligned} \tag{A.5}$$

Finally, the symmetries of the phonon modes are found by taking the direct product of chi_V with $\chi_{\text{atomic sites}}$. For our example, we have that:

$$\begin{aligned} \chi_V \otimes \chi_{\text{atomic site } 4a} &= F_{1u} \otimes A_{1g} = F_{1u} \\ \chi_V \otimes \chi_{\text{atomic site } 8c} &= F_{1u} \otimes (A_{1g} \oplus A_{2u}) = F_{1u} \oplus F_{2g} \end{aligned} \tag{A.6}$$

Hence, the theoretical group analysis predicts the following irreducible representations of optical and acoustical zone-center modes for the cubic phase:

$$\Gamma = \underbrace{F_{2g}}_{\text{Raman}} \oplus \underbrace{F_{1u}}_{\text{IR}} \oplus \underbrace{F_{1u}}_{\text{Acoustic}} . \tag{A.7}$$

The final step is to assign each phonon mode m individually, that is to find its characters $\chi_{m\mathbf{q}}(\{\mathbf{S} \mid \mathbf{v}(S)\})$ with respect to the various symmetry operations of $G_{\mathbf{q}}$. For non-degenerate modes, it can be demonstrated (88) that:

$$\begin{aligned} \chi_{m\mathbf{q}}(\{\mathbf{S} \mid \mathbf{v}(S)\}) &= \sum_{\kappa\kappa'} \sum_{\alpha\beta} U_{m\mathbf{q}}(\kappa\alpha)\ T_{\alpha\beta}\ U_{m\mathbf{q}}(\kappa'\beta)\delta(\kappa, F_0(\kappa'; R)), \\ \text{with } T_{\alpha\beta} &= S_{\alpha\beta} e^{i\mathbf{q}\cdot[\tau_\kappa - \mathbf{S}\tau_\kappa]}, \end{aligned} \tag{A.8}$$

where $F_0(\kappa'; S)$ is the label of the atom to which the atom κ' is brought by the symmetry operation $\{\mathbf{S} \mid \mathbf{v}(S)\}$. The δ expresses that we only need to taken into account those atoms κ' that map on to atom κ by the symmetry operation. In case of degeneracy, the sum in Eq. (A.8) must be extended to all the degenerate modes.

By comparing the characters $\chi_{m\mathbf{q}}(\{\mathbf{S} \mid \mathbf{v}(S)\})$ with Table A1, it is then quite straightforward to assign the various phonon modes. In the end, we can double check the assignments obtained in this way agree with the irreducible representations of $\chi_V \otimes \chi_{\text{atomic sites}}$.

REFERENCES

1. G.D. Wilk, R.M. Wallace, J.M. Anthony, High-k gate dielectrics: Current status and materials properties considerations, *J. Appl. Phys.* **89**(10), 5243–5275 (2001).
2. S.A. Campbell, D.C. Gilmer, X.C. Wang, M.T. Hsieh, H.S. Kim, W.L. Gladfelter, J.H. Yan, MOSFET transistors fabricated with high permitivity TiO_2 dielectrics, *IEEE Trans. Electron Devices* **44**(1), 104–109 (1997).
3. C.J. Taylor, D.C. Gilmer, D.G. Colombo, G.D. Wilk, S.A. Campbell, J.Roberts, W.L. Gladfelter, Does chemistry really matter in the chemical vapor deposition of titanium dioxide? Precursor and kinetic effects on the microstructure of polycrystalline films, *J. Am. Chem. Soc.* **121**(22), 5220–5229 (1999).
4. J. Zhang, J.S. Yuan, Y. Ma, A.S. Oates, Design optimization of stacked layer dielectrics for minimum gate leakage currents, *Solid State Electron.* **44**(12), 2165–2170 (2000).

5. M. Kadoshima, M. Hiratani, Y. Shimamoto, K. Torii, H. Miki, S. Kimura, T. Nabatame, Rutile-type TiO_2 thin film for high-k gate insulator, *Thin Solid Films* **424**(2), 224–228 (2003).
6. G.D. Wilk, R.M. Wallace, Stable zirconium silicate gate dielectrics deposited directly on silicon, *Appl. Phys. Lett.* **76**, 112–114 (2000).
7. G.D. Wilk, R.M. Wallace, J.M. Anthony, Hafnium and zirconium silicates for advanced gate dielectrics, *J. Appl. Phys.* **87**, 484–492 (2000).
8. P.H. Giauque, H.B. Cherry, M.A. Nicolet, Thermal stability of amorphous $Ti_3Si_1O_8$ thin films, *Microelectron Eng.* **55**(1–4), 183–188 (2001).
9. D.K. Sarkar, E. Desbiens, M.A. El Khakani, High-k titanium silicate dielectric thin films grown by pulsed-laser deposition, *Appl. Phys. Lett.* **80**(2), 294–296 (2002).
10. R.C. Smith, N. Hoilien, C. Dykstra, S.A. Campbell, J.T. Roberts, W.L. Gladfelter, Cvd of $Ti_xSi_{1-x}O_2$ films: Precursor chemistry impacts film composition, *Chem. Vapor Deposition* **9**(2), 79–86 (2003).
11. P. Hohenberg, W. Kohn, Inhomogeneous electron gas, *Phys. Rev.* **136**, B864–B871 (1964).
12. W. Kohn, L.J. Sham, Self-consistent equations including exchange and correlation effects, *Phys. Rev.* **140**, A1133–A1138 (1965).
13. W.E. Pickett, Pseudopotential methods in condensed matter applications, *Comput. Phys. Rep.* **9**, 115–197 (1989).
14. M.C. Payne, M.P. Teter, D.C. Allan, T.A. Arias, J.D. Joannopoulos, Iterative minimization techniques for ab initio total-energy calculations: Molecular-dynamics and conjugate gradients, *Rev. Mod. Phys.* **64**, 1045–1097 (1992).
15. S. Baroni, S. de Gironcoli, A. Dal Corso, P. Giannozzi, Phonons and related crystal properties from density-functional perturbation theory, *Rev. Mod. Phys.* **73**, 515–562 (2001).
16. X. Gonze, First-principles responses of solids to atomic displacements and homogeneous electric fields: Implementation of a conjugate-gradient algorithm, *Phys. Rev. B* **55**, 10337–10354 (1997).
17. X. Gonze, C. Lee, Dynamical matrices, born effective charges, dielectric permittivity tensors, and interatomic force constants from density-functional perturbation theory, *Phys. Rev. B* **55**, 10355–10368 (1997).
18. X. Gonze, J.M. Beuken, R. Caracas, F. Detraux, M. Fuchs, G.M. Rignanese, L. Sindic, M. Verstraete, G. Zerah, F. Jollet, M. Torrent, A. Roy, M. Mikami, P. Ghosez, J.Y. Raty, D.C. Allan, First-principles computation of material properties: the ABINIT software project, *Comp. Mater. Sci.* **25**, 478–492 (2002). http://www.abinit.org.
19. J.P. Perdew, Y. Wang, Accurate and simple analytic representation of the electron–gas correlation-energy, *Phys. Rev. B* **45**, 13244–13249 (1992).
20. D.M. Ceperley, B.J. Alder, Ground state of electron gas by stochastic method, *Phys. Rev. Lett.* **45**, 566–569 (1980).
21. N. Troullier, J.L. Martins, Efficient pseudopotentials for plane-wave calculations, *Phys. Rev. B* **43**, 1993–2006 (1991).
22. M. Teter, Additional condition for transferability in pseudopotentials, *Phys. Rev. B* **48**, 5031–5041 (1993).
23. L. Kleinman, D.M. Bylander, Efficacious form for model pseudopotentials, *Phys. Rev. Lett.* **48**, 1425 (1982).
24. H.J. Monkhorst, J.D. Pack, Special points for Brillouin-zone integration, *Phys. Rev. B* **13**, 5188–5192 (1976).
25. P. Duran, C. Pascual, Phase equilibria and ordering the system HfO_2-Yb_2O_3, *J. Mater. Sci.* **19**, 1178–1184 (1984).
26. G. Teufer, The crystal structure of tetragonal ZrO_2, *Acta Cryst.* **15**, 1187–1187 (1962).
27. R. Ruh, H.J. Garrett, R.F. Domagala, N.M. Tallen, System zirconia-hafnia, *J. Am. Ceram. Soc.* **51**, 23–27 (1968).

28. P. Aldebert, J.P. Traverse, Structure and ionic mobility of zirconia at high temperature, *J. Am. Ceram. Soc.* **68**, 34–40 (1985).
29. J.C. Slater, Atomic radii in crystals, *J. Chem. Phys.* **39**, 3199–3204 (1964).
30. D.R. Lide, *CRC Handbook for Chemistry and Physics*, 76th Edition (CRC Press, Boca Raton, FL, 1995).
31. E.J. Little, M.M. Jones, A complete table of electronegativities, *J. Chem. Educ.* **37**, 231–233 (1960).
32. C. Lee, Ph. Ghosez, X. Gonze, Lattice-dynamics and dielectric-properties of incipient ferroelectric TiO_2 rutile, *Phys. Rev. B* **50**, 13379–13387 (1994).
33. C.Y. Lee, X. Gonze, Dielectric-constants and born effective charges of TiO_2 rutile, *Phys. Rev. B* **49**(20), 14730–14731 (1994).
34. K. Parlinski, Z.Q. Li, Y. Kawazoe, First-principles determination of the soft mode in cubic ZrO_2, *Phys. Rev. Lett.* **78**(21), 4063–4066 (1997).
35. B. Králik, E.K. Chang, S.G. Louie, Structural properties and quasiparticle band structure of zirconia, *Phys. Rev. B* **57**, 7027–7036 (1998).
36. M.W. Finnis, A.T. Paxton, M. Methfessel, M. van Schilfgaarde. Crystal structures of zirconia from first principles and self-consistent tight binding, *Phys. Rev. Lett.* **81**(23), 5149–5152 (1998).
37. G. Jomard, T. Petit, A. Pasturel, L. Magaud, G. Kresse, J. Hafner, First-principles calculations to describe zirconia pseudopolymorphs, *Phys. Rev. B* **59**(6), 4044–4052 (1999).
38. J.E. Lowther, J.K. Dewhurst, J.M. Leger, J. Haines, Relative stability of ZrO_2 and HfO_2 structural phases, *Phys. Rev. B* **60**(21), 14485–14488 (1999).
39. R. Asahi, Y. Taga, W. Mannstadt, A.J. Freeman, Electronic and optical properties of anatase TiO_2, *Phys. Rev. B* **61**(11), 7459–7465 (2000).
40. M. Mikami, S. Nakamura, O. Kitao, H. Arakawa, X. Gonze, First-principles study of titanium dioxide: Rutile and anatase, *Jpn. J. Appl. Phys. 2*, **39**(8B), L847–L850 (2000).
41. G.-M. Rignanese, F. Detraux, X. Gonze, A. Pasquarello, First-principles study of dynamical and dielectric properties of tetragonal zirconia, *Phys. Rev. B* **64** 1–7:134301, (2001).
42. A.A. Demkov, Investigating alternative gate dielectrics: A theoretical approach, *Phys. Status Solidi B* **226**, 57–67 (2001).
43. X. Zhao, D. Vanderbilt, First-principles study of structural, vibrational, and lattice dielectric properties of hafnium oxide, *Phys. Rev. B* **65**(1–4) :233106, (2002).
44. X. Zhao, D. Vanderbilt, Phonons and lattice dielectric properties of zirconia, *Phys. Rev. B* **65**(1–10) :075105, (2002).
45. J. Muscat, V. Swamy, N. M. Harrison, First-principles calculations of the phase stability of TiO_2, *Phys. Rev. B* **65**(22)(1–15):224112, (2002).
46. M. Mikami, S. Nakamura, O. Kitao, H. Arakawa, Lattice dynamics and dielectric properties of TiO_2 anatase: A first-principles study, *Phys. Rev. B* **66**(15):155213, 1–6 (2002).
47. N.N. Greenwood, A. Earnshaw, *Chemistry of Elements*, 2nd Edition (Butterworth-Heinemann, Oxford, UK 1997).
48. J. Wang, H.P. Li, R. Stevens, Hafnia and hafnia-toughened ceramics, *J. Mater. Sci.* **27**, 5397–5430 (1992).
49. W. Zhong, D. King-Smith, D. Vanderbilt, Giant LO–TO splitting in perovskite ferroelectrics, *Phys. Rev. Lett.* **72**, 3618–3621 (1994).
50. Ph. Ghosez, J.-P. Michenaud, X. Gonze, Dynamical atomic charges: The case of ABO_3 compounds, *Phys. Rev. B* **58**, 6224–6240 (1998).
51. C. Lee, X. Gonze, Lattice-dynamics and dielectric-properties of SiO_2 stishovite, *Phys. Rev. Lett.* **72**, 1686–1689 (1994).
52. X. Gonze, J.-C. Charlier, D.C. Allan, M.P. Teter, Interatomic force-constants from first principles—the case of alpha-quartz, *Phys. Rev. B* **50**, 13035–13038 (1994).
53. P. Giannozzi, Gironcoli. S. de, P. Pavone, S. Baroni, Ab initio calculation of phonon dispersions in semiconductors, *Phys. Rev. B* **43**, 7231–7242 (1991).

54. J.G. Traylor, H.G. Smith, R.M. Nicklow, M.K. Wilkinson, Lattice dynamics of rutile, *Phys. Rev. B* **3**(10), 3457–3472 (1971).
55. A. Feinberg, C.H. Perry, Structural disorder and phase transitions in ZrO_2-Y_2O_3 system, *J. Phys. Chem. Solids* **42**, 513–518 (1981).
56. T. Hirata, E. Asari, M. Kitajima, Infrared and raman-spectroscopic studies of ZrO_2 polymorphs doped with Y_2O_3 or CeO_2, *J. Solid State Chem.* **110**, 201–207 (1994).
57. C. Pecharromán, M. Ocaña, C.J. Serna, Optical constants of tetragonal and cubic zirconias in the infrared, *J. Appl. Phys.* **80**, 3479–3483 (1996).
58. P. Bouvier, G. Lucazeau, Raman spectra and vibrational analysis of nanometric tetragonal zirconia under high pressure, *J. Phys. Chem. Solids* **61**, 569–578 (2000).
59. K. Negita, Lattice vibrations and cubic to tetragonal phase transition in ZrO_2, *Acta. Metall.* **37**, 313–317 (1989).
60. K. Negita, H. Takao, Condensations of phonons at the tetragonal to monoclinic phase transition in ZrO_2, *J. Phys. Chem. Solids* **50**, 325–331 (1989).
61. C. Pecharromán, J.E. Iglesias, Effective dielectric-properties of packed mixtures of insulator particles, *Phys. Rev. B* **49**, 7137–7147 (1994).
62. K. Kukli, J. Ihanus, M. Ritala, M. Leskela, Tailoring the dielectric properties of HfO_2–Ta_2O_5 nanolaminates, *Appl. Phys. Lett.* **68**, 3737–3739 (1996).
63. E.P. Gusev, E. Cartier, D.A. Buchanan, M. Gribelyuk, M. Copel, H. Okorn-Schmidt, C. D'Emic, Ultrathin high-k metal oxides on silicon: processing, characterization and integration issues, *Micorelectron. Eng.* **59**, 341–349 (2001).
64. D.W. Liu, C.H. Perry, R.P. Ingel, Infrared spectra in nonstoichiometric yttria-stabilized zirconia mixed crystals at elevated temperatures, *J. Appl. Phys.* **64**, 1413–1417 (1988).
65. D.L. Wood, K. Nassau, Refractive index of cubic zirconia stabilized with yttria, *Appl. Opt.* **21**, 2978–2981 (1982).
66. R.H. French, S.J. Glass, F.S. Ohuchi, Y.-N. Xu, W.Y. Ching, Experimental and theoretical determination of the electronic-structure and optical-properties of 3 phases of ZrO_2, *Phys. Rev. B* **49**, 5133–5142 (1994).
67. M.T. Lanagan, J.K. Yamamoto, A. Bhalla, S.G. Sankar, The dielectric properties of yttria-stabilized zirconia, *Mater. Lett.* **7**, 437–440 (1989)
68. A. Dwivedi, A.N. Cormack, A computer-simulation study of the defect structure of calcia-stabilized zirconia, *Phil. Mag.* **61**, 1–22 (1990).
69. R.A. Parker, Static dielectric constant of rutile (TiO_2), 1.6-1060 k, *Phys. Rev.* **124**(6), 1719–1722 (1961).
70. G.A. Samara, P.S. Peercy, Pressure and temperature dependence of the static dielectric constants and raman spectra of TiO_2 (rutile), *Phys. Rev. B* **7**(3), 1131–1148 (1973).
71. J.A. Speer, B.J. Cooper, Crystal structure of synthetic hafnon, $HfSiO_4$, comparison with zircon and the actinide orthosilicates, *American Mineralogist* **67**, 804–808 (1982).
72. G.-M. Rignanese, X. Gonze, A. Pasquarello, First-principles study of structural, electronic, dynamical, and dielectric properties of zircon, *Phys. Rev. B* **63**;104305, 1–7 (2001).
73. Z. Mursic, T. Vogt, H. Boysen, F. Frey, Single-crystal neutron-diffraction study of metamict zircon up to 2000 k, *J. Appl. Crystallogr.* **25**, 519–523 (1992).
74. X. Gonze, D.C. Allan, M.P. Teter, Dielectric tensor, effective charges, and phonons in alpha-quartz by variational density-functional perturbation-theory, *Phys. Rev. Lett.* **68**, 3603–3606 (1992).
75. J.H. Nicola, H.N. Rutt, A comparative study of zircon ($ZrSiO_4$) and hafon ($HfSiO_4$) raman spectra, *J. Phys. C: Solid State Phys.* **7**, 1381–1386 (1974).
76. P.W.O. Hoskin, K.A. Rodgers, Raman spectral shift in the isomorphous series $(Zr_{1-x}Hf_x)SiO_4$, *Eur. J. Solid State Inorg. Chem.* **23**, 1111–1121 (1996).
77. P. Dawson, M.M. Hargreave, G.R. Wilkinson, The vibrational spectrum of zircon ($ZrSiO_4$), *J. Phys. C: Solid State Phys.* **4**, 240–256 (1971).

78. F. Gervais, B. Piriou, F. Cabannes, Anharmonicity in silicate crystals: Temperature dependence of A_u type vibrational modes in $ZrSiO_4$ and $LiAlSi_2O_6$, *J. Phys. Chem. Solids* **34**, 1785–1796 (1973).
79. C. Pecharromán, M. Ocaña, P. Tartaj, C.J. Serna, Infrared optical-properties of zircon, *Mater. Res. Bull.* **29**, 417–426 (1994).
80. G. Lucovsky, G.B.Jr. Rayner, Microscopic model for enhanced dielectric constants in low concentration SiO_2-rich noncrystalline Zr and Hf silicate alloys, *Appl. Phys. Lett.* **77**, 2912–2914 (2000).
81. H.A. Kurtz, R.A.B. Devine, Role of bond coordination and molecular volume on the dielectric constant of mixed-oxide compounds, *Appl. Phys. Lett.* **79**, 2342–2344 (2001).
82. W.-J. Qi, R. Nieh, E. Dharmarajan, B.H. Lee, Y. Jeon, L. Kang, K. Onishi, J.C. Lee, Ultrathin zirconium silicate film with good thermal stability for alternative gate dielectric application, *Appl. Phys. Lett.* **77**, 1704–1706 (2000).
83. R.B. van Dover, L. Manchanda, M.L. Green, G. Wilk, E. Garfunkel, B. Busch, unpublished, (2001).
84. G.-M. Rignanese, F. Detraux, X. Gonze, A. Bongiorno, A. Pasquarello, Dielectric constants of Zr silicates: A first-principles study, *Phys. Rev. Lett.* **89**:117601, 1–4 (2002).
85. M. Nogami, Glass preparation of the ZrO_2–SiO_2 system by the sol-gel process from metal alkoxide, *J. Non-Cryst. Solids* **69**, 415–423 (1985).
86. V. Misra, unpublished, (2001).
87. A.K. Varshneya, *Fundamental of Inorganic Glasses*, (Academic Press Inc., San Diego, CA (1994).
88. A.A. Maradudin, S.H. Vosko, Symmetry properties of the normal vibrations of a crystal, *Rev. Mod. Phys.* **40**, 1–37 (1968).
89. S.C. Miller, W.S. Love, *Tables of Irreducible Representations of Space Groups and Co-Representations of Magnetic Space Groups* (Pruett Press, Boulder, CO, 1967).
90. E. Kroumova, J.M. Perez-Mato, M.I. Aroyo, S. Ivantchev, G. Madariaga, H. Wondratschek, The bilbao crystallographic server: a web site with crystallographic tools using the international tables for crystallography, in: *18th European Crystallographic Meeting* (1998). http://www.cryst.ehu.es.

Chapter 8

THE INTERFACE PHASE AND DIELECTRIC PHYSICS FOR CRYSTALLINE OXIDES ON SEMICONDUCTORS

RODNEY MCKEE

Oak Ridge National Laboratory, PO Box 2008, Oak Ridge, TN 37831, USA

1. INTRODUCTION

At the writing of this book, "scaling" of SiO_2-based transistors is still the overarching performance issue in our semiconductor hungry economy. However, while the methodology of increasing speed by scaling down device dimensions dominates manufacturing, its performance dominance is on the wane. Our SiO_2 scaling era is actually entering a twilight, a twilight that may signal the end of a developmental timeline for solid-state electronics (see Fig. 1) that has been over 70 years in the making. The advent of SiO_2 as a gate dielectric on this timeline certainly enabled our transistor technology, but, unlike many of the other events, it was not physics based. SiO_2 was fortuitously available as the native oxide on silicon (1). As we search for a possible way out of the performance twilight at the end of this scaling era and hopefully begin a new age of functionality in semiconductor device physics, we could do well to re-examine several of the timeline events and the dielectric physics issues that underpin them as the predecessors of the SiO_2 gate choice.

The first to consider is Julius Lilienfeld's seminal conjecture for an electrostatic field effect in a surface thin-film (2); it not only initiated the solid-state electronics timeline, it was truly a profound scientific and technological contribution. His field effect transistor (FET) patent contains a claim that describes a variable resistance, thin-film: "*the thickness of the film, moreover, is minute and of such a degree that the electrical conductivity therethru would be influenced by applying thereto an electrostatic force*". This thin-film was, in fact, field effect inversion charge at a semiconductor surface that was "switchable" in response to an applied electric field. It was indeed the charge gain that is logic-switched in every FET in modern integrated circuits. This field effect gain is crucial to keep, but SiO_2 is not.

Certainly several major developments followed in the solid-state electronics timeline, but Lilienfeld's inversion charge concept reached a significant new level in May of 1957 with the particularly notable mark of 4 patents issued to researchers at Bell

A.A. Demkov and A. Navrotsky (eds.), Materials Fundamentals of Gate Dielectrics, 291–312.

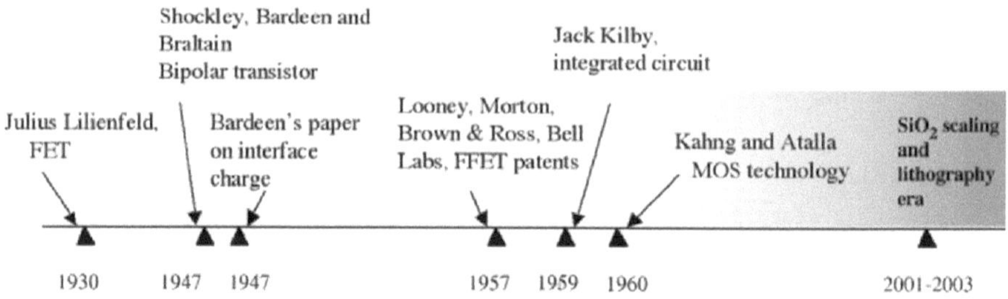

Fig. 1. The last 70 years of an electronics timeline in which field effect and semiconductor surface phenomena have evolved. This time line started with Lilienfeld's (2) "FET" patent; it includes the bipolar transistor (3), the integrated circuit (4), Kahng and Atalla's first metal-oxide-silicon (MOS) device (5), and subsequently all of modern silicon and silicon–germanium transistor technology.

Labs (6). These patents described an invention that was similar to Lilienfeld's FET, but they dramatically added functionality to the field effect concept by "*altering the conductivity of a path through a semiconducting body by polarizing a ferroelectric maintained in proximity to the body to alter the surface charge...* ". The semiconductor in this early device was germanium and the dielectric displacement altering the surface charge would evolve in response to ferroelectric polarization in a thin $BaTiO_3$ crystal, glued to the germanium surface. These notions of coupling ferroelectric polarization from a crystalline oxide with the field effect charge in a semiconductor were clearly fundamental in their insight: these were expectations for entirely new device physics. This ferroelectric field effect transistor (FFET) idea was attempting to add device functionality to a semiconductor technology even before there was an integrated circuit. Dielectric physics associated with cooperative phenomena like ferroelectricity is crucial to keep, but SiO_2 cannot provide it.

As the early work in 1957 showed, crystalline oxides have the potential for use as both field effect dielectrics as well as active functional components in the FET. The short coming of these early results was the highly imperfect interface and trapped charge between the dielectric and the semiconductor (the crystals were simple glued together). The defect structure required almost 40 V to switch the active ferroelectric component, hardly a value with commercial appeal. However, the advent of crystalline oxides grown commensurately on semiconductors (7, 8) along with the identification of a functional interface phase at oxide/semiconductor junctions (9) is giving the semiconductor community a major opportunity to overcome the extrinsic interface charge problem and expand device physics into fundamentally new directions. Importantly, this opportunity is not evolutionary; it is entirely new and is centered in a truly physics-based view of the dielectric system. This physics-based view is amenable to experimental and theoretical characterization, control and analysis that has never been possible within the aging SiO_2-based science and technology of gate oxides (see Demkov et al. (10)). Crystalline oxides along with field effect phenomena associated with interface phase physics in these systems may well be a materials physics way out of the SiO_2 twilight that we now face.

In the remainder of this chapter we will discuss crystalline oxides on semiconductors (COS) as the ideal physical system for solid-state electronics. It is a physical system based on dielectric epitaxy in a monolithic thin-film structure with a semiconductor. The physical interface in such a structure can be made perfectly commensurate, and it thus provides us with the systematics of crystalline periodicity to manipulate interface band structure as well as interface charge. Unlike SiO_2 that is thermally grown on silicon, COS oxides can be adapted to silicon, any silicon–germanium alloy and to pure germanium. Moreover, we will bring out the idea that an interface phase exists at the oxide/semiconductor junction that is a natural consequence of thermodynamic equilibrium in these structures and that thermodynamic/electrodynamic consequences of the interface phase concept provide a strong indicator for how we can achieve the continued development towards high performance and new functionality that we need for the continued progress on the solid state electronics timeline.

2. CRYSTALLINE OXIDES ON SEMICONDUCTORS—THE PHYSICAL STRUCTURE

Figure 2 is a two-panel construction of Z-contrast images of $Ba_{0.725}Sr_{0.275}O$ and $SrTiO_3$ grown on pure silicon using molecular beam epitaxy (MBE) techniques. The BaSrO compound imaged in Fig. 2(a) has a 5 eV band gap and is alloyed to match the lattice parameter of the (001) face of silicon. The overlay in the left side of the image shows a simple model of the epitaxial cube-on-cube NaCl-type oxide structure of the alkaline earth oxide on silicon. While the oxygen atoms are not imaged, the bright contrast of the heavy alkaline earth metal atoms and the [110] symmetry of the epitaxial structure is clear. Figure 2(b) illustrates the case where $SrTiO_3$ has been grown and strained commensurate to silicon, but rotated 45°.

These lattice images are members of a COS structure series that can be generically written as $(AO)_n(A'BO_3)_m$. The subscripts n and m in this structure series are integer repeats of atomic planes and unit cells of constituent crystalline layers. While this structure series can be quite broadly applied, we will discuss it here for cases where A and A′ are elements or combination of elements out of Group IIA of the periodic table (i.e., Ba, Sr, Ca and Mg) and B is a Group IVA transition metal like Ti or Zr.

Figure 3 is a collection of lattice parameter data upon which we base our ideas for generalization of our structure series to germanium and thus to silicon–germanium heterojunction technology. Figure 3 shows lattice parameters and their temperature dependent changes for silicon and germanium and three perovskite oxides (11–15). In analogy to III–V GaAs alloy heteroepitaxy (15) our oxide MBE synthesis technique (7, 17, 18) has shown that lattice matched oxides can be formed in our structure series by source shuttering. The semiconductor alloy series from pure silicon to pure germanium has its room temperature cubic lattice parameter varying from 5.43 to 5.65 Å. The perovskites $CaTiO_3$, $SrTiO_3$ and $BaTiO_3$ are simple cubic "2–4" perovskite structures in which the 2+ alkaline earth metal ions occupy the cube corners, the 4+ transition metal is in the center of the cube, octahedrally coordinated with oxygen ions in the face-centered sites of the unit cell (19). These three perovskites are mutually soluble

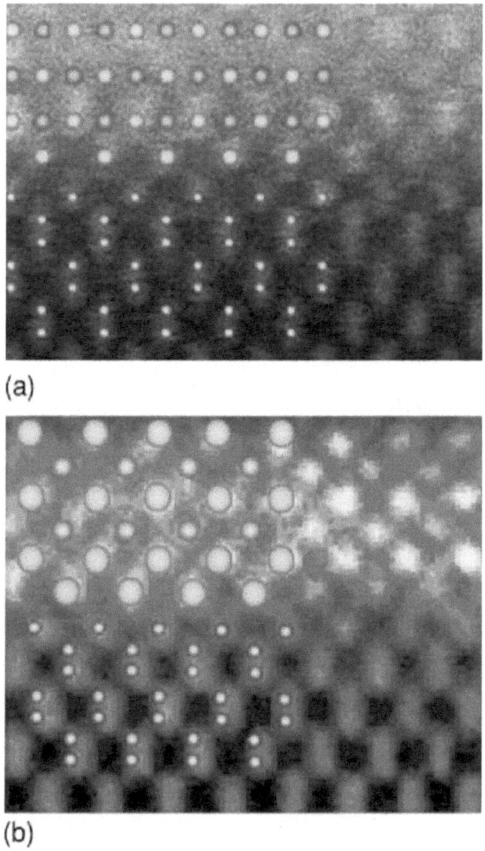

Fig. 2. Alkaline earth and perovskite oxide heteroepitaxy on silicon. Panels (a) and (b) illustrate our ability to manipulate interface structure at the atomic level using our $(AO)_n(A'BO_3)_m$ structure series. The n/m ratio defines the electrical characteristics for this new physical system of COS in a MOS capacitor. In panel (a) $n = 3$, $m = 0$; in panel (b) $n = 1$, $m = 2$.

in each other and by mixing Ca and Sr for instance in a 60/40 ratio, the 5.43 Å lattice parameter of pure silicon can be obtained at room temperature since the perovskites rotate 45° so that the [100] direction in the oxide is parallel with [110]Si and its 3.84 Å spacing (4). The lattice parameters of two pure perovskites are of particular note in Fig. 3: pure $BaTiO_3$ matches the 5.65 Å lattice parameter of pure germanium at room temperature and pure $CaTiO_3$ matches the 5.44 Å lattice parameter of pure silicon at 515°C; no alloying is required for either of these perovskites for growth on the two end members of a Si–Ge substrate series. Results for $BaTiO_3$ on germanium are treated below.

Figure 4 is a Z-contrast image of the epitaxy and structural perfection obtained when pure $BaTiO_3$ is grown on pure germanium using the precepts of layer-by-layer energy minimization developed in Ref 20. This is the ferroelectric $BaTiO_3$/Ge

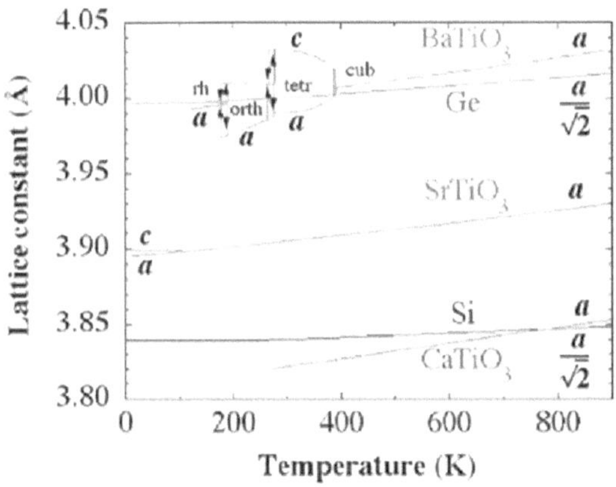

Fig. 3. Lattice parameter vs. temperature for Si, Ge, and perovskite oxides. The perovskite oxides can be grown on any Si–Ge alloy composition and lattice matched or intentionally strained so as to not disrupt the in-plane translational symmetry at the interface.

structure that was envisioned by the Bell group in 1957 (6). In this early work, the researchers looked for dielectric displacement to alter the surface charge set up by ferroelectric polarization in a thin $BaTiO_3$ crystal, glued to the surface of germanium. The "glue" is now a layered, heteroepitaxial structure that is thermodynamically stable and is commensurate at the atomic level. We shall show in the next section that this $BaTiO_3$/Ge structure is electrically perfect as well and completely avoids the trapped

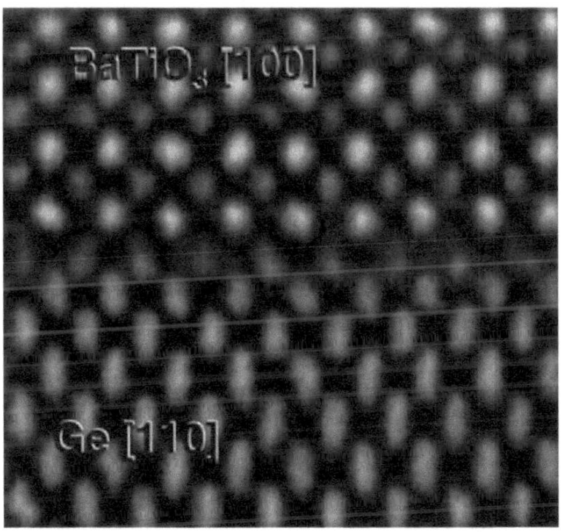

Fig. 4. Z-contrast STEM image of the $BaTiO_3$/Ge interface. The image is a real-space image of the interface that is commensurate and atomically abrupt with the "bright" row of barium at the interface, followed by the "dim" titanium row, one plane up. The perovskite structure then continues: Ba/Ti/Ba .... The oxygen atoms are not imaged.

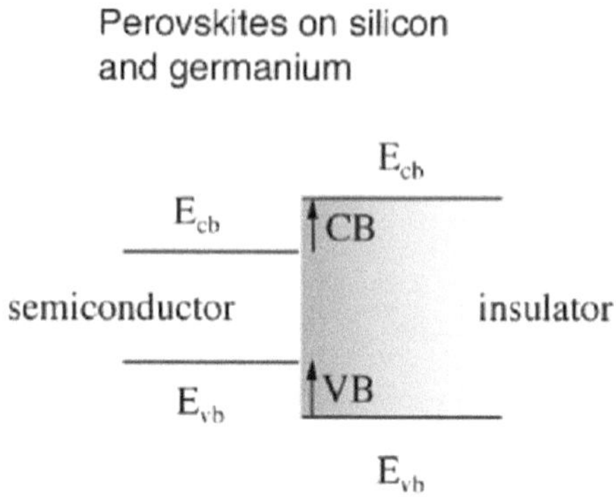

Fig. 5. Schematic band diagram illustrating positive conduction (CB) and valence (VB) offsets. While this diagram is in this case only to demonstrate generalities, the tunability of the interface electrical structure is a principle facet of the physics base of the COS system described here.

charge that induced the discontinuity in dielectric displacement and inversion charge that hampered its early implementation by the Bell group.

3. CRYSTALLINE OXIDES ON SEMICONDUCTORS—THE ELECTRICAL STRUCTURE

COS offers us an interface physics approach to the development of new gate dielectrics for transistor technology through epitaxial growth of the oxides. Moreover, this heteroepitaxial approach allows us to systematically manipulate interface band structure as well as interface charge. We will demonstrate these points directly with experimental data obtained from MOS capacitors (21).

The surface potential, interface charge and inversion charge in the semiconductor of the MOS capacitor are all functions of bias voltage and frequency. Device function for the MOS capacitor requires that the oxide dielectric act as a Schottky barrier (22) with no free charge and support the dielectric displacement that sets up inversion charge in the underlying semiconductor. This coupling of dielectric displacement to inversion charge is critically dependent on the details of the interface band structure of the MOS capacitor.

In a recent treatment of this problem for a number of candidate oxides being considered for alternative dielectrics on silicon, Robertson and Chen (23) and Robertson (24) have predicted that while a number of the oxides in the perovskite class have the desired high dielectric constants, band offset and alignment are highly unfavorable; namely the barrier height to electron transfer across a perovskite dielectric on silicon can be small or even non-existent. Chambers et al. (25) have demonstrated that this is indeed a problem in a recent spectroscopic study of the $SrTiO_3$/Si interface.

Figure 5 is a schematic of the band structure at a semiconductor/dielectric interface. This diagram specifies the valence and conduction band offsets, VB and

Table 1. Valence and condition band offset parameters for COS structures on silicon and germanium

	CNL	Band gap	CB_{Ge}	VB_{Ge}	CB_{Si}	VB_{Si}
Si	0.36 (26)	1.1				
Ge	0.18 (26)	0.6				
$SrTiO_3$	2.6	3.3	+0.18	+2.42	−0.14	+2.4
BaO	2.5	5	+2.1	+2.3	+1.26	+2.1

CB, for hole and electron states at a semiconductor/insulator heterojunction. The charge neutrality level, CNL, referenced to the valence band is the energy level in the semiconductor (or the insulator) density of states above which allowable states are empty. In an n-type semiconductor, for instance, the probability for an electron escaping from the conduction band of the semiconductor to unfilled insulator states is related to CB and conversely, in a p-type material the probability for a hole escaping from the semiconductor to the valence band of the insulator is related to VB. In either case, for the insulator to support field effect inversion of charge in the semiconductor, CB and VB must be positive for the electron or hole-conducting channel.

Table 1 gives representative values reported by Robertson and estimated here for the alkaline earth oxide, BaO. Robertson's conclusions for $SrTiO_3$ (typical of the simple perovskites $SrTiO_3$, $CaTiO_3$ and $BaTiO_3$) show that p-channel field effect transistors on germanium or silicon would have positive values for VB. However, CB is only slightly positive for germanium and is negative for silicon; an n-channel FET would not switch.

This asymmetry in band structure is attributable to the fact that the band gap for transition metal perovskites lies between the valence band of filled oxygen 2p states and the conduction band of empty transition metal d states. The weighting of the transition metal d-states that the integrated CNL drives is thus responsible for the asymmetry. Chambers et al. (25) confirmed this Roberston/Chen prediction with measurements for $SrTiO_3$ on silicon reporting CB values between 0.0 and 0.1 ± 0.1 eV but VB values in excess of 2 eV. It is here that the physics-base for the COS approach to transistor gate dielectrics can be clearly illustrated.

Our COS structure series, $(AO)_n(A'BO_3)_m$, allows the band structure to be systematically manipulated. Unlike the transition metal perovskites, the alkaline earth oxides like BaO are strongly ionic and the charge neutrality level is in the middle of the band gap (27). We have estimated the entries in Table 1 for BaO assuming the mid-gap CNL and clearly show that band offsets, both CB and VB, are substantially positive for germanium and for silicon. Therefore, if two or more BaO planes were inserted between a perovskite and germanium or silicon, then the asymmetry of the band structure would, according to the numbers in Table 1, adjust and support dielectric displacement across the junction. This is a simple and striking prediction for our heteroepitaxial approach.

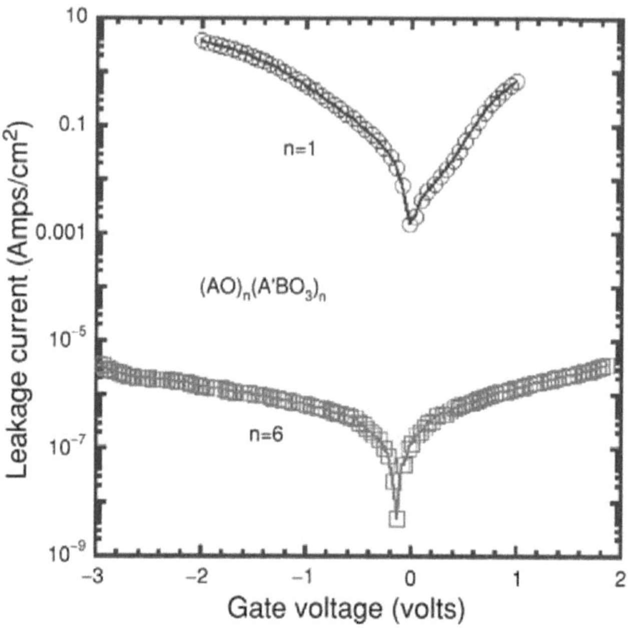

Fig. 6. Leakage Current data for $(AO)_n(A'BO_3)_m$ on germanium. The data are plotted as absolute values of leakage current.

Figure 6 is a collection of leakage current data obtained for $BaTiO_3$ on germanium with two values of the AO repeat, $n = 1$ and $n = 6$. The open circles show the $n = 1$ data for $BaTiO_3$ directly on germanium (see Fig. 4). While this heteroepitaxial structure is perfectly commensurate, and bulk $BaTiO_3$ has a band gap of 3.4 eV (28), a $BaTiO_3$ thin-film grown directly on germanium is not an effective barrier to electron transfer. This is entirely consistent with Robertson's and Chen's predictions. However, if as few as six atomic planes of BaO are grown against germanium first, then the leakage current (open squares) drops *6 orders of magnitude*. This is dramatic evidence for physical structure dominance of the electrical structure in our heteroepitaxial oxide series. Within the context of the generality of our structure series and its n/m configurations, interface band structure can be adjusted for many of the perovskite and transition metal oxides analyzed by Robertson (24).

We turn now to the final part of our discussion of the electrical structure of COS dielectrics. Interface charge, as early as 1947 (29), was recognized as a significant problem for field effect charge inversion in a transistor. Interface charge can completely screen the semiconductor from an applied field and even result in a discontinuity in dielectric displacement, as was the case for the ferroelectric field of $BaTiO_3$ crystals glued to germanium. The thesis of our heteroepitaxial approach is that oxide dielectrics can be grown as monolithic, single crystal structures on semiconductors tying up the dangling bonds typical of the amorphous SiO_2/Si structure. This would in turn, eliminate extrinsic interface charge. We will use the capacitance data in Fig. 7 to prove this point.

Figure 7(a) shows high frequency (1 MHz) and low frequency (10 Hz) capacitance data taken from a 250 Å thick $BaTiO_3$ film on p-type Ge. Figure 7(b) is an expanded view of the data in the bias range where the germanium surface potential varies from zero to its value at the Fermi level. We can extract the density of interface charge (our measure of interface perfection) from the ΔC of C_{LF}–C_{HF} in this bias range. This is done following an analysis developed by Nicollian and Brews (21).

The capacitance of the MOS capacitor can be broken down into its component parts: C_{ox} and C_{Ge} as identified in the equivalent circuit inset in Fig. 7(a). C_{Ge} is dependent on the germanium surface potential and interface and inversion charge. The hashed region in Figs. 7(a) and (b) is the bias range where the field effect is "depleting" the oxide semiconductor interface of majority carrier charge and initiating the process of charge carrier inversion (the upward turn in C_{LF} Fig. 7(a) is signature of field-effect-driven electron–hole pair generation and charge carrier inversion in the underlying p-type germanium). Dielectric displacement via gate charge must overcome the screening effect of any interface-trapped charge before germanium inversion charge can even respond. ΔC in this depletion region provides a measure of any extra capacitance that might be associated with the charging dynamics of interface traps.

The electron–hole recombination rate in this depletion region is frequency dependent as well as is the rate that electrons can move into and out of interface traps. The majority carrier electron–hole contribution to C_{Ge} can keep up at high frequencies, but electron trapping at the interface cannot. Therefore, by sweeping the frequency of a small-signal ac bias, $dV_g(\omega)$, $C_{it}(\omega) \Rightarrow 0$ as $\omega \Rightarrow \infty$; the contribution of interface charge can be separated.

Figure 7(b) provides the data for the determination of the interface trap density, D_{it}; $D_{it} \propto \Delta C$. As these data show, ΔC and hence trapped charge for our commensurate interface is negligible. D_{it} is indistinguishable from zero through the entire depletion region. To our measurement uncertainty, $D_{it} < 10^{10}/cm^2$-eV; this is an electrically perfect interface, free of extrinsic defect interface charge. Germanium inverts from its majority carrier p-type behavior to minority, n-type behavior over a narrow 1-V range.

To our knowledge, this is the first demonstration of field effect charge inversion for a gate oxide on germanium. Its basis is the generalization of our structure series, $(AO)_n(A'BO_3)_m$ in this new physical system of COS oxides. The flexibility that this structure series gives us to manipulate the physical and electrical structure at the atomic level is thus applicable to silicon or germanium and any silicon–germanium alloy.

4. CRYSTALLINE OXIDES ON SEMICONDUCTORS—THE INTERFACE PHASE AND A COULOMB BUFFER

When Schottky (30) and Mott (31) formulated the barrier height theory for a metal/semiconductor junction, and later when Anderson (32) formulated the

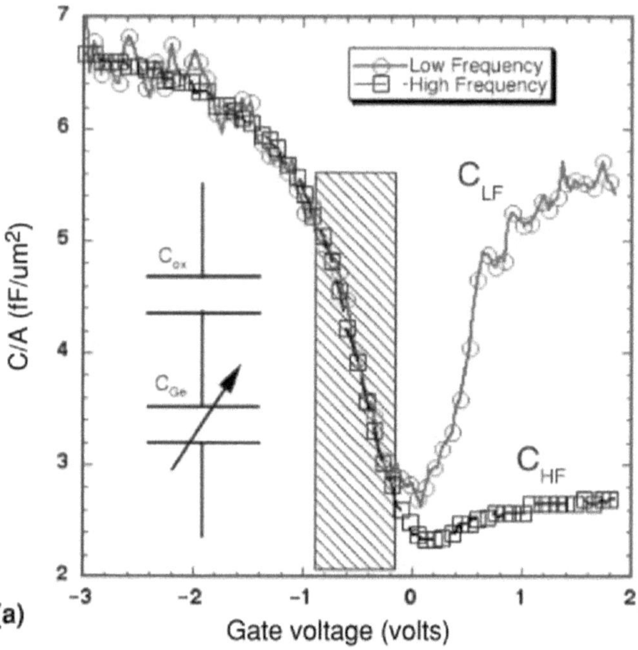

(a)

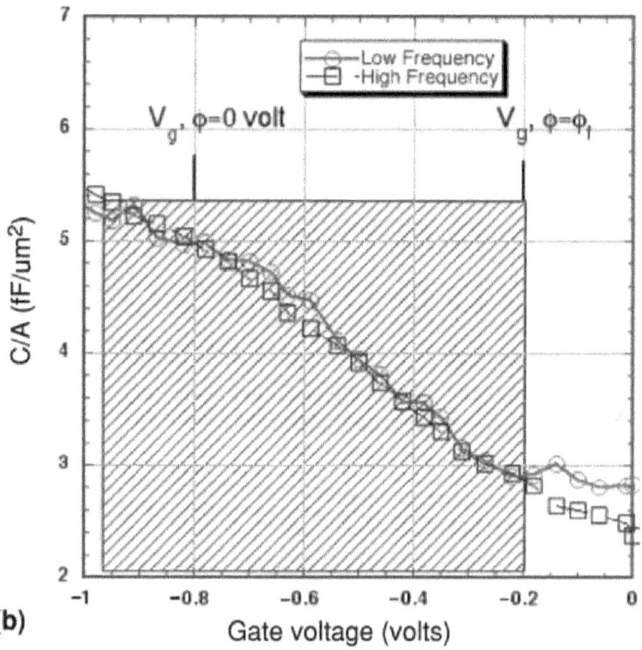

(b)

Fig. 7. High-frequency–low frequency capacitance data for $BaTiO_3$/Ge. Data taken with aluminum top and bottom electrodes. The doping is p-type, $10^{17}/cm^3$. The measured flat band voltage is -0.8 V. With an interface state density of $10^{10}/cm^2$, the flatband shift indicates a fixed positive charge of $10^{12}/cm^2$.

band-edge offset problem for semiconductor/semiconductor junctions, there was no consideration given to interface states as contributions to the electrostatic boundary conditions. The charge distribution at the interface was treated simply as a superposition of the bulk-terminated junction. While these theories have certainly been insightful, they consistently misrepresent the barrier height or band-edge offsets because real interfaces, apparently from interfacial structure variations, modify the intrinsic band alignment (33–35).

The bulk-termination view of the problem has been enhanced over the years with an ever-increasing formalization of theoretical techniques (36–43), but Tung (44–46) has recently argued, from the perspective of molecular systems in chemical physics, that these techniques and their refinements are not adequately accounting for bond polarization and chemical bonding-induced charge transfer. While this issue can be debated, what is certain is that whether first principles theory is used or whether the molecular systems approach is applied, a lack of detailed knowledge about the interface structure often leaves us in a quandary. The quandary is that we do not know how to predict, *a priori*, the evolution of the physical and electrical structure of the interface when two materials are joined. This is clearly apparent for even the simplest of systems (38) and we have no guidance on this issue for the important problem of barrier height adjustment for alternative gate oxides on silicon (47).

If we consider Tung's treatment of the barrier height problem, he identifies an interface specific region in which the wave functions of the junction interact (be they originating in metal/semiconductor, semiconductor/semiconductor or dielectric/semiconductor junctions). His bond polarization arguments suggest that this interaction should in general set up an electronic structure that is distinct from that which is on either side of the interface. In this sense, Tung identifies the crux of the problem, i.e., the interface itself should be considered as a fundamental part of the physical and electrical structure of the junction. Beyond this however, his molecular systems approach, does not give us direction as to how to predict or really even how to expect this interface region to evolve. Here we will present a generalization of Tung's view by expanding it to show that the interface, even at monolayer thicknesses, should be identified as a distinct phase both in a thermodynamic and electrodynamic sense. First principles theory then shows us that the interface phase does two things: it constrains the physical structure of the junction, and it sets the electrical boundary conditions that establish the junction electrostatics.

In what follows we will introduce and analyze energy band diagrams that are part of the barrier height problem for crystalline oxides on semiconductors (COS). These data allow us to draw a clear distinction between a bulk termination view of the problem and our interface phase thesis. An immediate outcome of this distinction is functionalization of the barrier height concept itself. This functionalization (Fig. 8) is via a "Coulomb Buffer" that is identified with charge transfer and ionic bonding between alkaline earth metal atoms in an interfacial silicide and oxygen in the oxide dielectric. This buffer is apparent here from the structure specifics of heteroepitaxy, but it is an electrodynamic concept with far-reaching implications.

Energy band diagrams and barrier heights for two of the Group IIA silicide variants (Sr, Be) in the interface phase for our alkaline earth oxide (AO)/Si junction illustrate

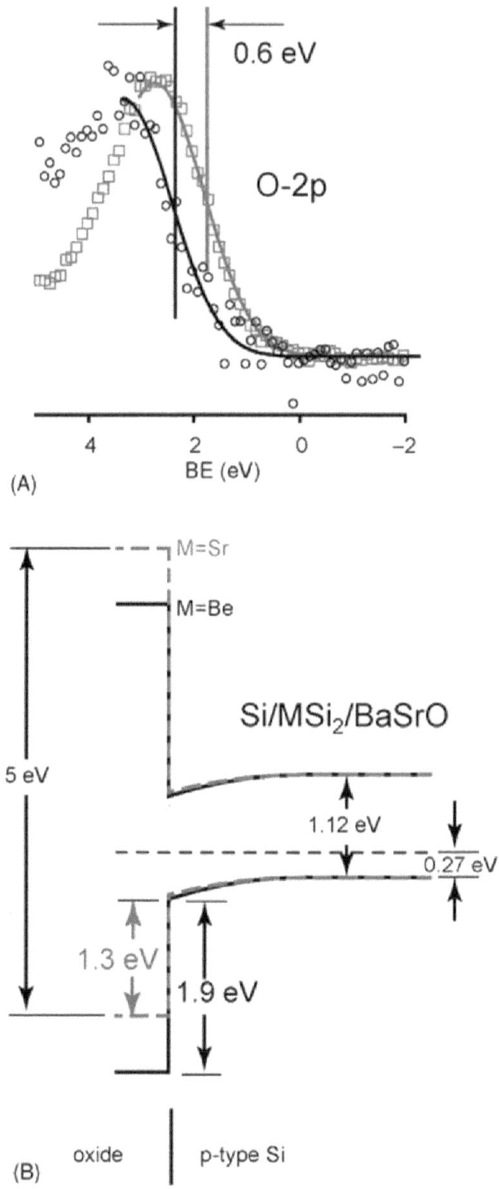

Fig. 8. Band diagrams—Coulomb Buffer with interface variants. (A) O 2p valence band edge data for the silicide variants of the interface phase with $SrSi_2$ in the red data and $BeSi_2$ in the black data with BaSrO as the dielectric. (B) Band diagram for silicide variants on (001)Si with 3 unit cells of lattice matched $Ba_{0.725}O_{0.275}$. The junction is Si/MSi_2/BaSrO where M denotes the alkaline earth metal in a 1 monolayer silicide interface phase. The band gap of the oxide is taken to be 5 eV (22). The band bending is 0.50 eV for $SrSi_2$ and 0.46 eV for $BeSi_2$.

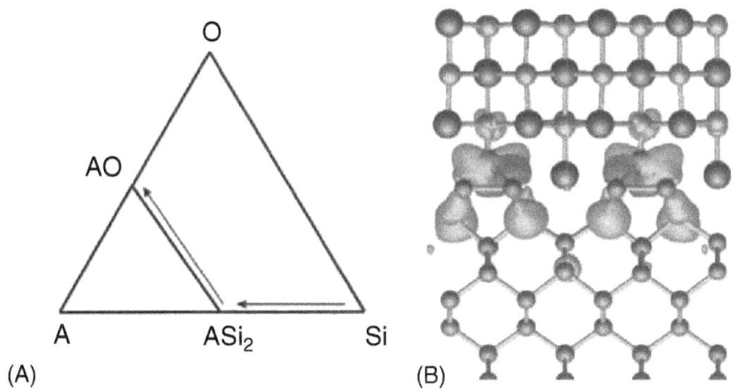

Fig. 9. Thermodynamics and electrodynamics of the Coulomb Buffer. (A) Three-component phase equilibria with stable tie lines between the bulk terminations and the interface phase, ASi_2; A is an alkaline earth metal. (B) The image illustrates three layers of the alkaline earth oxide on the (001) face of silicon observed in cross-section at the [110] zone axis (blue = alkaline earth metal; yellow = oxygen; green = silicon). A distinct interface phase can be identified as a monolayer structure between the oxide and the silicon in which charge density in interface states is strongly localized around the silicon atoms in the interface phase. The dipole in the ionic A–O bond between the alkaline earth metal in the silicide and the oxygen in the oxide buffers the junction against the electrostatic polarization of the interface states localized on silicon. The electron density of this valence surface state at the center of the Brillouin zone (γ) is shown with the purple isosurface ($0.3 \times 10^{-3}e$).

a 0.6 eV valence band shift when an aliovalent substitution of Be is made for Sr (−1.3 to −1.9 eV, Fig. 8(A) and (B)). Neither the magnitude nor asymmetry of this shift can be explained with the classical bulk-termination view of the barrier height problem (48).

We develop our interface phase thesis with surface and interface thermodynamics, in-situ spectroscopic characterization of band-edge alignment and first principles, self-consistent electronic structure calculations. The data we present come from a Si/AO member of the COS dielectric structures (7, 48). This system develops a cube-on-cube commensurate epitaxy via an interfacial silicide in which any of the Group IIA alkaline earth metals can be systematically substituted (7); real-space Z-contrast electron microscopy imaging techniques have been used to deduce the physical interface structure (7).

The primary demonstration of the interface phase thesis is done using the lattice-matched $Ba_{0.725}Sr_{0.275}O$ composition, but we will supplement our findings with data from strained films of pure SrO and BaO on silicon.

An almost intuitive picture of the Coulomb buffer concept can be had by examining the physical structure and the charge localized in interface states on silicon atoms in the ASi_2 interface phase for this system (Fig. 9(B)). The interface states (their distribution and charge density are illustrated in purple) are consistent with the classical view of the problem; the charge occupying these states polarizes the dielectric and shifts the relative electrostatic potential at the interface. However the physical dipole associated

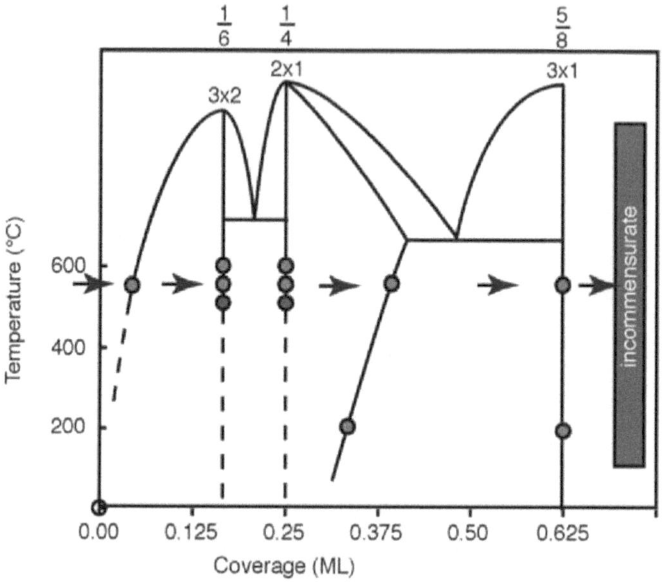

Fig. 10. Surface phase description of heteroepitaxy for the transition from silicon. Upper panel describes the surface structure evolution with data taken using condensation kinetics and RHEED; the filled circles are experimentally measured terminal compositions along with their diffraction symmetry.

with the ionic A–O bond between the alkaline earth metal in the silicide and the oxygen in the oxide buffers the junction against the electrostatic polarization from the interface states. This A–O buffer is not anticipated from the bulk termination of silicon or the alkaline earth oxide. However it is an obvious expectation if heteroepitaxy and phase equilibrium are considered for this junction.

Heteroepitaxy for the Si/AO junction develops with a silicide interface phase under the constraints of a surface phase diagram (Fig. 10); this surface phase diagram specifies equilibrium along the Si–alkaline earth metal tie line (the two-component tie line shown at the bottom of the triangle in Fig. 9). Equilibrium evolves along this tie line through a layered sequence of ordered structures (47) to a 2×1 $SrSi_2$ at 1/4 ML (one monolayer $= 6.78 \times 10^{14}$ sites/cm^2 on (001) silicon). We have determined that stoichiometric compounds (line-compounds) are stable along this tie line at three specific coverages, 1/6, 1/4 and 5/8 ML. These line compounds constrain, or bound, the 1/4 ML interface phase stability of the ASi_2 composition. Experimentally, the terminal line-compound compositions are straightforwardly determined from observations of phase-condensation kinetics using reflection high energy electron diffraction (RHEED) *(the method is described in ref. 7)*. Transitioning to the NaCl-type structure of the alkaline earth oxide in our junction is then accomplished from this 1/4 ML surface silicide along the ASi_2–AO tie line (Fig. 9), satisfying the three-component equilibrium requirements for thermodynamic stability in the junction. While layer sequencing in this Si/AO junction is an essential part of its heteroepitaxial structure, layer-sequenced thermodynamic constraint is a general concept (48) and is fundamental to energy minimization at a junction boundary.

The distinct structural arrangement of Sr and Si in our interface phase is not intuitive. However, what is obvious from examining the physical and electronic characteristics of the interface structure (Fig. 9 and ref. 48), is that it supports the symmetry and coordination of the bulk (001)BaSrO plane, thus minimizing the electrostatic energy of the junction. Moreover, it facilitates the heteroepitaxy to homoepitaxy transition for growth of subsequent oxide overlayers in the flat fcc structure as observed by both RHEED and X-ray photoelectron spectroscopy (XPS) in the growth sequence. The interface phase has both a density (1/2 that of (001)Si) and site occupation of silicon atoms (48) that establishes the electronic stability of the junction. The Sr ion in the $SrSi_2$ interface phase is atop a surface "valley" site of the underlying bulk terminated silicon and acts as an electrostatic (Coulomb) buffer of the crystal potentials on either side of the interface in the junction. This Coulomb buffer is sensitive to electronegativity variations of the aliovalent silicide variants in Group IIA; moving down the group from Be to Ba, dramatically influences the electrodynamics of the junction. With both our first principles calculations and XPS data, we have deduced energy band diagrams for this system to demonstrate this point.

The energy band diagrams and barrier heights (Fig. 8) for two of the Group IIA silicide variants (Sr, Be) in the interface phase are experimentally determined from XPS data (50). The XPS data were developed in a sequenced fashion, starting with clean, reconstructed silicon, followed by growth interruptions and characterization at monolayer additions up to the 6 monolayer, 16.3 Å BaSrO film; this methodology is described in detail (9). The X-ray excitation of the underlying core level of silicon was observed continually as reference. At this total thickness, the valence band spectrum from the oxide overlayer is fully developed as the O 2p state. The valence band edge is shown to be 1.8 eV below the Fermi level of the system with Sr at the interface and 2.4 eV below the Fermi level of the system with Be at the interface. These data present clear examples of Tung's interface specific region and its modification of junction electrostatics.

We can quantify this modification straightforwardly; the bulk-termination approach with its charge neutrality level alignment argument (47) was used recently to set an estimate of the valence band offset for this Si/AO system (8). The bulk-termination estimate for the valence band offset for the Sr variant, assuming that the charge neutrality level is in the center of the gap for the oxide, is -2.1 eV. This estimate is thus significantly in error; $\sim$1 eV (Fig. 8). This discrepancy in the valence band offset originates in the monolayer level electrostatics of the ASi_2 interface phase. We will use first principles quantum mechanical techniques to clarify the dielectric polarization that is responsible for this discrepancy in valence band offsets.

We have calculated the relaxed coordinates and electronic structure of our $Si/Ba_{0.725}Sr_{0.275}O$ system by starting from the $SrSi_2$ interface structure deduced from Z-contrast imaging (8). Since we have the basics of the physical structure as a start, the emphasis of the calculation has been to elucidate the electronic structure and its interface specifics as contributions to the barrier height electrodynamics. The calculation is done with first principles, self-consistent total energy calculations within standard Density Functional Theory (DFT) in the Local Density Approximation (LDA). While the details will be reported elsewhere (50), we have used nonlocal, norm-conserving ultrasoft pseudopotentials within the Vanderbilt scheme (52) and have explicitly incorporated the semicore 4s4p and 5s5p states of Sr and Ba to account for the valence states in the atomic pseudopotential accurately.

In principle, the valence band offset is simply understood; primarily there are two distinct contributions: the first, ΔE_v, is a band structure part, purely the bulk termination, and the second, ΔV, contains all of the interface physics. The valence band offset can be defined as $VB_{offset} = \Delta E_v + \Delta V$. ΔE_v is the difference between the energies of the valence band edges of the bulk terminated oxide and semiconductor structures, and ΔV is the shift in the average electrostatic potentials of the two sides of the interface when the junction is formed.

Unlike the classical, continuum view of the junction electrostatics, where the interface structure of the junction is ignored, a heteroepitaxial, layer-sequenced structure like our oxide/semiconductor model presents an electrostatic potential that contains strong oscillations due to the positions of the atoms. While it is the structure in these oscillations that will give us the details of the interface interaction, we must carefully account for them if an accurate picture is to be obtained of how the anisotropy of chemical bonding and charge transfer influence the potential line-up. We follow a planar averaging method that has been developed earlier (41, 42) for treating such oscillations by defining a planar average of the potential for the atoms in each plane parallel to the interface. Within our pseudopotential approach we get VB_{offset} directly, but we can only obtain a value for ΔV separately that can be interpreted as a physical dipole if we normalize our values of ΔE_v to the "Shottky-Limit" for the problem (the Shottky-Limit of the barrier height being the difference in bulk electron affinities for the oxide and semiconductor (46)). This normalization gives us values for ΔV (the missing part of the barrier height problem) at the junction that are direct measures of a macroscopic dipole moment originating in the interface phase (49).

As stated earlier, we have chosen the Si/AO system for study. The alkaline earth metals in Group IIA develop a significant variation in Pauling's electronegativities, and give us a wide flexibility to investigate the charge transfer characteristics of interface phase formation. Within Table 2 is a compilation of calculated values for offset parameters with variations both in interface and oxide compositions that take advantage of solubility and solution chemistry in the Group IIA alkaline earth silicides and oxides. The last row of the table contains the values of ΔV and VB_{offset} that we interpolate from the SrO and BaO endpoints for direct comparison to the lattice-matched BaSrO data for our model system; the first principles theoretical values are respectively, -1.12 and -1.46 eV. The experimental and theoretical values for VB_{offset} in our commensurate system are now in remarkable agreement for this dielectric/semiconductor junction (-1.3 eV vs. -1.46 eV).

The next and perhaps most striking result shown in Table 2 (graphically in Fig. 11(B)), is in both the magnitude of ΔV and its change ($\sim$0.6 eV) as we move

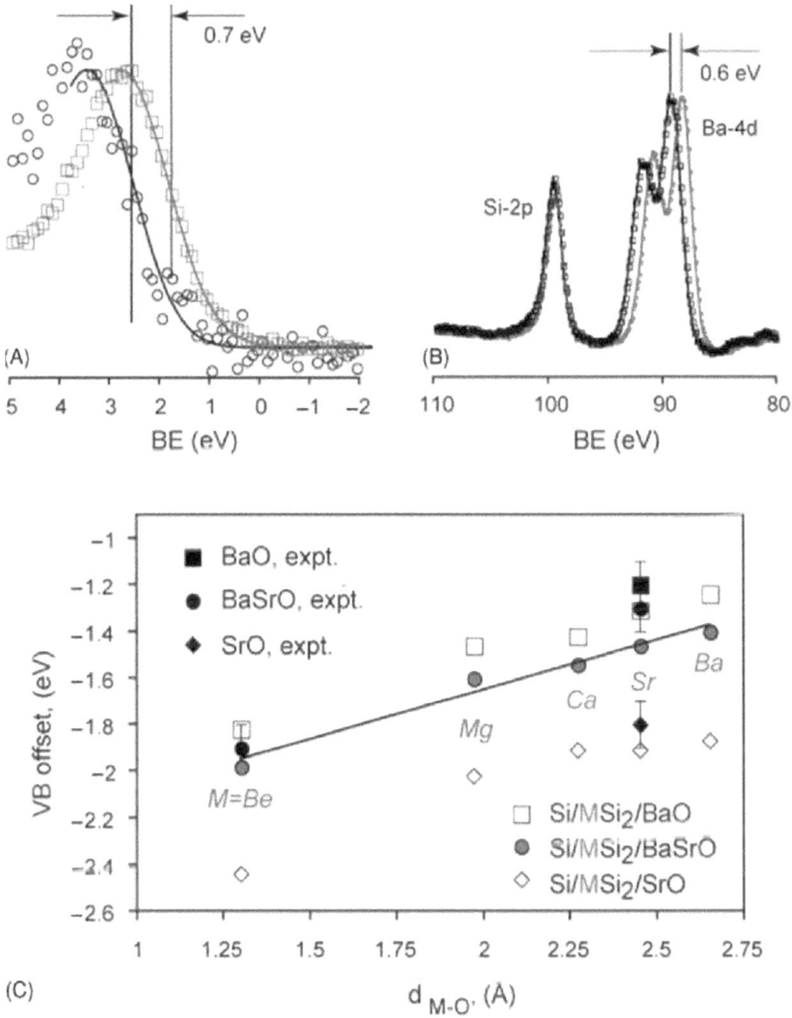

Fig. 11. Valence band offsets for alkaline earth oxides on (001)Si. Panel A is experimental data for the oxide dielectric effect on the O 2p valence band edge with SrO in the red data and BaO in the black data both with the Sr variant in the interfacial silicide; Panel B is the core level shift of the Ba 4d peak for the silicide variants of the interface phase with $SrSi_2$ in the red data and $BeSi_2$ in the black data with BaSrO as the dielectric. Panel C is a theoretical/experimental comparison for both the dielectric effect and silicide variants in the interface phase. Theoretical values for the BaSrO case are linear interpolations of the endpoint SrO and BaO values.

with increasing atomic number down the Group II alkaline earth metal column for the metal in the interface phase while holding the dielectric, SrO or BaO, constant. The measured shift in VB_{offset} for the Sr and Be silicide variants is 0.6 eV (Fig. 8 and 11(B)). With a constant dielectric, $\delta VB_{offset} = \delta\Delta V$; the theoretical estimate from the

Table 2. Interface phase electrodynamics for Si/AO junctions

	ΔV (eV); values normalized to Schottky-Limit	Valence band offset (eV)	Alkaline earth silicide ion displacement from oxygen in the oxide (Å)
$Si/BeSi_2/SrO$	−0.49	−2.44	1.44
$Si:MgSi_2:SrO$	−0.07	−2.02	2.11
$Si:CaSi_2/SrO$	0.04	−1.91	2.33
$Si:SrSi_2:SrO$	0.04	−1.91	2.46
$Si:BaSi_2:SrO$	0.08	−1.87	2.57
$Si/BeSi_2/BaO$	−2.02	−1.82	1.30
$Si/MgSi_2/BaO$	−1.66	−1.46	1.97
$Si/CaSi_2/BaO$	−1.62	−1.42	2.27
$Si/SrSi_2/BaO$	−1.51	−1.31	2.45
$Si/BaSi_2/BaO$	−1.48	−1.24	2.65
$Si/SrSi_2/Sr_{0.25}Ba_{0.75}O$	−1.12	−1.46	

Table 2 is 0.52 for SrO and 0.51 for BaO. As can be seen by examining the displacement column in Table 2, this effect correlates with the displacement of the alkaline earth ion in the silicide relative to the oxygen ion site in the overlayer oxide (this is our Coulomb Buffer). As we have shown, the silicide interface phase (Fig. 9(B)) is a natural consequence of and a requirement for the overall thermodynamic stability for our $Si/B_{0.725}Sr_{0.275}O$ junction. The electrodynamics of the junction gives us the Coulomb Buffer concept via dipole phenomena that correlates with bond distances between alkaline earth metal ions in the silicide interface phase and the oxygen ions in the BaSrO. When the structure of the interface phase is examined in detail we see that the silicon ion in the silicide moves upwards toward the oxygen in the first oxide layer, relaxes to the equilibrium Si–O distance of 1.75 Å and remains at this position to within better than 1% for all of the silicide variants. The alkaline earth ions however, systematically shift away from the oxygen ion as we move from Be down the group to Ba. From the chemist's view, the alkaline earth ions move away from the oxygen as expected as they become more electropositive. The consequence of this displacive process can be understood (see Fig. 9(B)) as a buffering of the oxide electrostatic potential from the charge localized on the silicon atoms in the interface phase. This buffer is thus proportional to the dipole strength, or bond length, of the A–O bond. As the dipole strength increases, the valence band offset is increasingly distinct from the bulk termination Shottky-Limit term, ΔE_v.

With both the silicide variant in the ASi_2 phase and dielectric constant changes in the AO phase (Fig. 11(C)) we find a measure of the Coulomb Buffer in the slope of VB_{offset} plotted against the alkaline earth metal–oxygen ion spacing. Correlating the Coulomb Buffer with the A–O dipole leads us to a determination that this slope is simply the dipole charge over the dielectric constant of the interface phase. The slope (Fig. 11(B) and (C)) of this curve is 0.43 eV/Å. Equating this value to Q/ε and taking Q as the unit cell charge density, n, times the specific charge q, we deduce a value of $+2$ for q given that n is the number of alkaline earth ions/unit cell in the interface phase (1/4 ML). This simple deduction hinges only on an assumption that the dielectric constant of the interface phase is about that of silicon, i.e., 10. We believe that this is a reasonable assumption given the hybridization of the electron charge around the silicon atoms in the interface region (see Fig. 9(B)). This simplified view will be expanded (28) but it serves well to elucidate the basic electrodynamic picture of the Coulomb buffer in the interface phase.

One final characteristic of these data and calculations can be seen in the change in the ΔV and VB_{offset} values as the dielecric itself is changed. BaO has a dielectric constant of 34 with a band gap of 4.8 eV (52). SrO on the other hand has a dielectric constant of 13 and a band gap of 6 eV (52). As the valence band offset is measured, changing from SrO to BaO with the Sr variant in the silicide interface phase, we find a δVB_{offset} of 0.7 eV (see Fig. 11(A) and (C)). From Table 2, the theoretical value of δVB_{offset} is 0.6 eV. In this case however, the A–O bond length does not change (A is Sr in both) so the buffer influence of the Sr–O dipole makes no contribution to the valence band shift. While δVB_{offset} changes by 0.6 eV, $\delta\Delta V$ changes by 1.55 eV. This observation suggests a coupling between the dielectric constant of the oxide and the charge density (Fig. 9(B)) that is localized around the silicon atoms in the interface phase. This coupling may be as simple as image charge displacement, but given that the energy levels of the entire system change when BaO is substituted for SrO, it is prudent to consider this issue in more detail (50).

We have addressed the quandary left by Tung in his notion of the interface specific region of the Shottky barrier problem and shown that this interface specific region should be interpreted within the constraint of thermodynamic equilibrium between the two components of a dielectric junction. This equilibrium constraint can be understood as phase equilibrium at the junction that even at monolayer levels requires an interface phase; this interface phase controls the overall junction electrostatics via a Coulomb Buffer. This Coulomb Buffer is fundamentally distinct from wave function decay of interface states that comes from the classical bulk termination view of the barrier height problem. Moreover, this thermodynamic/electrodynamic view of the problem provides a unifying concept for understanding and designing barrier height function within the barrier offset problem that is general to all of semiconductor physics.

5 SUMMARY

We have addressed two of the primary questions for viability of a gate oxide in MOS transistor technology. We have shown that band offset and alignment can be adjusted by atomic level structural and chemical changes, and we have demonstrated that a highly perfect electrical interface between a polar oxide and a semiconductor can be obtained free of interface charge. In a broader sense, we have taken Kahng and Atalla's MOS device to a new and prominent position in the solid-state electronics timeline. It can now be extensively developed utilizing an entirely new physical system, the commensurate COS interface.

Not only will our COS approach be able to address the near-term needs for alternative gate dielectrics in MOSFET transistor technology, it will enable extremely promising new device physics opportunities that go well beyond anything envisioned for the SiO_2/Si electronics era.

Ahn et al. (53, 54) are adapting atomic force microscopy (AFM) techniques to show that non-contact ferroelectric writing of nanowires in a semiconductor surface is possible. This notion relies on the ferroelectric field effect induced by poling thin-films of $Pb(Zr_{0.52}Ti_{0.48})O_3$ supported on silicon. Levy (55) is proposing to develop a quantum information processor using ferroelectrically coupled Ge/Si quantum dots. Levy's architecture mediates spin interactions between nearest neighbor electrons by nonlinear optical rectification in 10 nm-wide channels in the underlying semiconductor. These channels are potential wells induced by the ferroelectric field effect associated with static polarization of an epitaxial ferroelectric on Si–Ge. Kane (56) is developing ideas for a quantum computer architecture in which spins associated with donors in silicon function as qubits and quantum operations are gated with a voltage induced field effect. Preserving the phase of the electron qubits in Kane's architecture requires that interface defects be separated at least at a μm scale. Only the physical perfection of a crystalline dielectric interface can meet such a requirement. These ideas are the beginnings of an entirely new device physics that is enabled by the crystalline perfection and anisotropic response of this new physical system of a COS-based MOS device.

The revolution envisioned by the men at Bell labs holding crystal chunks in their hands and considering integration of crystalline oxides and semiconductors to develop new device functionality is here.

ACKNOWLEDGEMENTS

Research sponsored the Division of Materials Sciences and Engineering, Office of Basic Energy Sciences, U.S. Department of Energy at Oak Ridge National Laboratory under contract DE-AC05-00OR22725 with UT-Battelle, LLC.

REFERENCES

1. C.J. Froshch, L. Derick, *Proc Electrochem. Soc.* **547** (1957).
2. J. Lilienfeld, A method and apparatus for controlling electric currents, U.S. Patent No. 1,745,175 (January 28, 1930).
3. J. Bardeen, W.H. Brattain, *Phys. Rev.* **71**, 230 (1947).

4. J. St. Clair Kilby, patent filed in February 1959. Issued in 1964, Patent No. 3,138,743 for Miniaturized Electronic Circuits.
5. D. Kahng, M.M. Atalla, Silicon–silicon dioxide field induced surface devices, in: *IRE Solid-State Device Research Conference* (Carnegie Institute of Technology, Pittsburgh, PA., 1960).
6. D.H. Looney, Semiconducting translating device, US Patent # 2,791,758 (1957); J.A. Morton, Electrical swithching and storage, US Patent # 2,791,761 (1957); I.M. Ross, Semiconducting translating device, US Patent # 2,791,760 (1957); W.L. Brown, Semiconductive device, US Patent # 2,791,759 (1957).
7. R.A. McKee, F.J. Walker, M.F. Chisholm, *Phys. Rev. Lett.* **81**, 3014 (1998); R.A. McKee, F.J. Walker, $CaTiO_3$ interfacial template structure on semiconductor-based material and the growth of electroceramic thin-films in the perovskite class, US Patent No. 5,830,270 (1998).
8. R.A. McKee, F.J. Walker, M.F. Chisholm, *Science* **293**, 468 (2001).
9. R.A. McKee, F.J. Walker, M. Buongiorno Nardelli, W.A. Shelton, G.M. Stocks, *Science* **300**, 1726 (2003).
10. A.A. Demkov, Private communication (2004); X. Zhang, A.A. Demkov, H. Li, X. Hu, Y. Wei, J. Kulik, *Phys. Rev. B* **68**, 125323 (2003).
11. D. Taylor, Thermal expansion data VIII. Complex oxides, ABO_3, the perovskites, *Trans. J. Br. Ceram. Soc.* **84**, 181–188 (1985).
12. Landolt-Börnstein, *Numerical Data and Functional Relationships in Science and Technology*, eds. K.-H. Hellwege, A.M. Hellwege (Springer-Verlag, Berlin, 1981), New Series, Group III, Vol. 16, Part a, p. 330.
13. K.G. Lyon, F.L. Salinger, C.A. Swenson, G.K. White, Linear thermal expansion measurements on silicon from 6 to 340 K, *J. Appl. Phys.* **48**, 865–868 (1977).
14. Landolt-Börnstein, *Numerical Data and Functional Relationships in Science and Technology*, ed. O. Madelung (Springer-Verlag, Berlin, 1987), New Series, Group III, Vol. 22, Part a, p. 18.
15. Y.S. Touloukian, R.K. Kirby, R.E. Taylor, P.D. Desai, *Thermal Expansion: Metallic Elements and Alloys, Vol. 12, Part 1 of Thermophysical Properties of Matter* (Plenum, New York, 1975), p. 116.
16. M.R. Melloch, D.D. Nolte, J.M. Woodall et al., *Crit. Rev. Solid State* **21**, 189 (1996).
17. R.A. McKee, F.J. Walker, J.R. Conner, R. Raj, *Appl. Phys. Lett.* **63**, 2818 (1993).
18. R.A. McKee, F.J. Walker, J.R. Conner, E.D. Specht, D.E. Zelmon, *Appl. Phys. Lett.* **59**, 782 (1991).
19. R.W.G. Wyckoff, *Crystal Structures*, Vol. 1, Chapter VII, a5 and Fig VIIA, 6a (Interscience Publishers, Inc., New York, 1951).
20. R.A. McKee, F.J. Walker, E.D. Specht, G.E. Jellisen, L.A. Boatner, *Phys. Rev. Lett.* **72**, 2741 (1994).
21. An excellent treatment of MOS dielectric theory and field effect phenomena in such a device can be found in Nicollian and Brews (see pg. 332 for discussion of D_{it} and ΔC); E.H. Nicollian, J.R. Brews, *MOS (Metal Oxide Semiconductor) Physics and Technology* (John Wiley & Sons, New York, 1982).
22. W. Schottky, *Z. Phys.* **118**, 539 (1942).
23. J. Robertson, C.W. Chen, *Appl. Phys. Lett.* **74**, 1168 (1999).
24. J. Robertson, *J. Vac. Sci. Technol. B* **18**, 1785 (2000).
25. S.A. Chambers, Y. Liang, Z. Yu, R. Droopad, J. Ramdani, K. Eisenbeiser, *Appl. Phys. Lett.* **77**, 1662 (2000).
26. J. Tershoff, *Phys. Rev. Lett.* **52**, 465 (1984).
27. N.F. Mott, R.W. Gurney, *Electronic Processes in Ionic Crystals* (Clarendon Press, Oxford, 1940).

28. G.E. Jellison Jr., L.A. Boatner, D.H. Lowndes, R.A. McKee, M. Godbole, *Appl. Optics* **33**, 6053 (1994).
29. J. Bardeen, *Phys. Rev.* **71**, 717 (1947).
30. W. Schottky, *Phys. Z.* **113**, 367 (1940).
31. N.F. Mott, *Proc. Cambridge Philos. Soc.* **34**, 568 (1938).
32. R.L. Anderson, *Solid-State Electron.* **5**, 341 (1962).
33. For a review of these issues, see the monograph by W. Monch, *Semiconductor Surfaces and Interfaces*, 3rd Edition (Springer, Berlin, 2001).
34. A. Franciosi, C.G. Van de Valle, *Surf. Sci. Repts.* **25**, 1 (1996).
35. A.A. Demkov, O.F. Sankey, *Phys. Rev. Lett.* **83**, 2038 (1999).
36. S.G. Louie, M.L. Cohen, *Phys. Rev. B* **13**, 2461 (1976).
37. W.R. Frensley, H. Kroemer, *Phys. Rev. B* **16**, 2642 (1977).
38. W.A. Harrison, E.A. Kraut, J.R. Waldrop, R.W. Grant, *Phys. Rev. B* **18**, 4402 (1978).
39. W.A. Harrison, J. Tersoff, *J. Vac. Sci. Technol. B* **4**, 1068 (1986).
40. W. Mönch, *Appl. Phys. Lett.* **72**, 1899 (1998).
41. A. Balereshi, S. Baroni, R. Resta, *Phys. Rev. Lett.* **61**, 734 (1988).
42. M. Peressi, S. Baroni, R. Resta, A. Balereschi, *Phys. Rev. B* **43**, 7347 (1991); C.G. Van de Walle, R.M. Martin, *Phys. Rev. B* **35**, 8154 (1987); C.G. Van de Walle, *Phys. Rev. B* **39**, 1871 (1989).
43. F. Leonard, J. Tersoff, *Phys. Rev. Lett.* **84**, 4693 (2001).
44. R.T. Tung, *Phys. Rev. Lett.* **84**, 6078 (2000).
45. R.T. Tung, *Phys. Rev. B* **20**, 205310 (2001).
46. R.T. Tung, *Mater. Sci. Eng.* **35**, 1 (2001).
47. J. Robertson, *J. Vac. Sci. Technol. B* **18**, 1785 (2000).
48. A.P. Sutton, R.W. Balluffi, *Interfaces in Crystalline Materials* (Clarendon Press, Oxford, 1995), pp. 349–394.
49. E.A. Kraut, R.W. Grant, J.R. Waldrop, S.P. Kowalczyk, *Phys. Rev. Lett.* **44**, 1623 (1980).
50. M. Buongiorno Nardelli, W.B. Shelton, G.M. Stocks, F.J. Walker, R.A. McKee, to be published. Calculations in this work have been done using the PWscf package (S. Baroni, A. Dal Corso, S. de Gironcoli, P. Giannozzi, http://www.pwscf.org/).
51. D. Vanderbilt, *Phys. Rev. B* **41**, 7892 (1990).
52. A.M. Stoneham, J. Dhote, A compilation of crystal data for halides and oxides, http://www.cmmp.ucl.ac.uk/~ahh/research/crystal/homepage.htm, University College London, London, and references contained therein (2002).
53. C.H. Ahn, T. Tybell, L. Antognazza, K. Char, R.H. Hammond, M.R. Beaseley, Ø. Fischer, J.-M. Triscone, *Science* **276**, 1100 (1997).
54. A. Lin, X. Hong, V. Wood, A. Verevkin, C.H. Ahn, R.A. McKee, F.J. Walker, E.D. Specht, *Appl. Phys. Lett.* **78**, 2034 (2001).
55. J. Levy, *Phys. Rev. A* **64**, 052306 (2001).
56. B.E. Kane, *Fortschr. Phys.* **48**, 1023 (2001).

Chapter 9

INTERFACIAL PROPERTIES OF EPITAXIAL OXIDE/SEMICONDUCTOR SYSTEMS

Y. LIANG[1] AND A.A. DEMKOV[2]

[1]*Freescale Semiconductor, Inc., 6501 William Cannon Dr. West, Austin, TX 78735, USA*

[2]*Department of Physics, The University of Texas at Austin, Austin, TX USA*

1. INTRODUCTION

To ensure continuous downscaling of CMOS technology, the semiconductor industry must make a transition from the Si/SiO_2/poly-Si triad to a more complex Si/high-k/metal system (1). The integration of this new stack into the current CMOS flow is one of the most urgent tasks of today's electronics. A gate insulator with a high dielectric constant (high-k) enables a physically thick but dielectrically thin insulating layer that ensures continuous downscaling without compromising gate leakage due to quantum mechanical tunneling. Incorporating a metal gate eliminates the depletion at the dielectric/poly Si interface thus further increasing the gate capacitance. The oxide's gate action depends on, among other factors, the barrier height at the oxide–semiconductor and oxide–metal interfaces. In order to have a sufficiently low leakage current, the band offset between the dielectric and Si must be greater than one electron volt (eV). During the past decade, a significant effort was devoted to understanding issues such as dielectric properties of various high-k oxides, defect density at the dielectric-Si interface, and thermodynamic stability of high-k dielectrics in contact with Si. The interfacial electronic structures and band alignment at the Si/dielectric interface, however, have received less attention. This is due, in part, to difficulties in quantitative determination of band offsets at the Si/dielectric interface.

In this chapter we discuss several key factors that dictate the band offset and electronic structure at the oxide/semiconductor interface, and methods of determining the band offset from experimental and theoretical points of view. We use epitaxial $SrTiO_3$ on Si and GaAs as model systems for our discussions.

This chapter is organized as follows: Section 2.1 discusses several key issues important for the band alignment at the oxide/semiconductor interface. These issues include reference level, the dipole at the oxide/semiconductor interface, and the

A.A. Demkov and A. Navrotsky (eds.), Materials Fundamentals of Gate Dielectrics, 313–348.

effect of interfacial atomic structure on band offsets. Sections 2.2 and 2.3 discuss electronic structure of transition-metal oxides and the complex band structure and charge neutrality level of $SrTiO_3$. Section 2.4 outlines the principles of photoemission spectroscopy and its application for band offset determination. Section 3 presents experimental findings on the early stages of the $SrTiO_3$ growth on semiconductors and valance band and conduction band offsets at $SrTiO_3$/Si and $SrTiO_3$/GaAs interfaces. Epitaxial $SrTiO_3$ is chosen in our discussion because it provides a well-defined interface compared to its amorphous or polycrystalline analogue. This enables us to have a closer and more rigorous comparison between experimental results and ab initio theoretical calculations, which is the topic of Section 4.

2. BAND DISCONTINUITY AT OXIDE/SEMICONDUCTOR INTERFACE

2.1. Reference Level for Band Alignment

Understanding and predicting the band lineup between two dissimilar materials has been a long-standing problem in solid state physics (2, 3). Although there are many models and empirical rules to explain band alignments, a key factor for the band alignment is a correct reference level providing a common energy scale for two materials.

Historically, we first meet the problem of band alignment or band discontinuity in the case of the so-called Schottky barrier (4). The barrier forms when a heterojunction is created between a metal and an insulator or a semiconductor. According to Schottky the energy barrier encountered by an electron in the metal at such a heterojunction is simply the difference between the semiconductor *electron affinity* (the energy distance from the bottom of the conduction band to the vacuum level) and the *work function* of the metal (the energy distance from the Fermi level to the vacuum level). Unfortunately, this intuitive picture does not agree with experiment very often. The barrier seems to be only weakly dependent on the metal! In essence the Schottky model is the oldest form of the electron affinity rule where the vacuum level is used as a reference for the band lineup (5). This rule in effect suggests that the charge transfer does not affect the band discontinuity at the interface when two materials are brought together to an intimate contact. While this method has been useful to describe band offsets between some wide band-gap materials, it fails to predict the band offsets between a metal and a semiconductor and between most conventional semiconductors.

In 1947 Bardeen (6) suggested that the so-called surface states newly introduced by Tamm and Schockley (7, 8) altered the charge distribution at the semiconductor surface and fixed or "pinned" the Fermi level, thus fixing the barrier height. Bardeen's argument for an n-type semiconductor goes as follows. Electrons from the impurity level see unoccupied surface states and start filling them, thus charging the surface and leaving an uncompensated positive charge behind. The positive charge attracts the electrons back. Once equilibrium is reached there is positive space charge in the interior of the semiconductor and negative charge at its surface, and thus a double layer causing the band bending at the surface. If the density of surface states is high, the metal will not change the charge distribution in a significant way, and the barrier is determined by the preexisting band bending. The level up to which the surface band is filled is a charge neutrality level. Although this picture includes surface states in band alignment consideration, it suffers several conceptual difficulties. For example, if the contact with metal is intimate, there are no surface states. This difficulty was resolved by Heine who introduced the concept of interface states which play a very similar role (9). Experimentally, the Schottky barrier height often falls between the Schottky and Bardeen limits (4, 6).

Aside from the electron affinity rule and Bardeen model, there are also other empirical approaches such as the common anion rule which states that the valence band offsets is small between materials with the same anion (10). Most of these rules, however, only explain band offsets of a small subset material system.

Theoretically the band alignment is often estimated within the so-called metal-induced gap states (MIGS) model (3, 9). The MIGS model describes both the Bardeen (6) and Schottky (4) limits and interpolates between the two in a linear fashion, provided that electron affinities, charge neutralities, and the pinning factor are known. The theory was successfully used to describe the band discontinuity in heterojunctions between covalent semiconductors. Recently, Robertson (11) used the MIGS model to predict conduction band offsets between Si and a variety of novel dielectric materials. In this model the conduction band offset is given by:

$$\phi = (\chi_a - \Phi_a) - (\chi_b - \Phi_b) + S(\Phi_a - \Phi_b) \tag{1}$$

Here χ_a, χ_b, Φ_a, Φ_a are the electron affinities and charge neutrality levels of Si and dielectric measured from the vacuum level, respectively, and S is an empirical dielectric pinning parameter describing the screening by the interfacial states. This particular form of Eq. (1) is due to (12). If $S = 1$ the offset is given by the difference in electron affinities as was originally proposed by Schottky (4). Alternatively, for $S = 0$ we get strong pinning or the Bardeen limit (6). The pinning parameter can be estimated by an empirical formula (11):

$$S = \frac{1}{1 + 0.1(\varepsilon_\infty - 1)^2} \tag{2}$$

where ε_∞ is the high frequency (electron) component of the dielectric constant.

It is worth noting that, according to Eq. (2), materials with a larger electronic component of the dielectric constant should pin stronger. This observation agrees with experiment (13). However, as we show in the next section, the correlation to the dielectric constant is secondary. The fundamental reason is the smaller forbidden gap in the electronic spectrum that governs both the electronic polarization and decay rate of evanescent states. Transition metal oxides such as $SrTiO_3$, hafnia, zirconia, and other high-k materials considered as gate dielectrics for CMOS all have band gaps smaller (and dielectric constants higher) than that of SiO_2, and thus suffer the same intrinsic disadvantage with respect to the Fermi level pinning.

Modern electronic structure computation techniques such as the method of van de Walle and Martin (14) produce reliable valence band offsets, and if the band gaps are known from experiment, the conduction band offset can be inferred. However, these calculations are rather time consuming, and sensitive to the exact atomic structure of the interface, which in general is not known. An epitaxial oxide on a semiconductor offers a system unique in the sense that the interfacial structure is well defined, and in principle, can be manipulated in a controlled manner (15). Though the atomic arrangement at the $SrTiO_3/Si$ interface is still under debate (16–19), certain features have been well established. Still it is useful to have a simple back-of-the-envelope model to estimate the discontinuity. In what follows, we apply the simple MIGS model to the $SrTiO_3/Si$ interface. We use the complex band structure to determine the charge neutrality level. We compare these simple estimates of the band offset with those obtained experimentally and *via* density functional theory (DFT) calculations discussed in Sections 3 and 4.

2.2. Electronic Structure of $SrTiO_3$ and Complexity of Transition Metal Oxides

Due to the technological importance of SiO_2, its electronic structure has been thoroughly studied both theoretically and experimentally. The crystal structure and the density of states of α-quartz are shown in Fig. 1. The near edge valence band comes from the oxygen–silicon bonding orbitals, while the top of the valence band originates from the non-bonding p-state of oxygen. The bottom of the conduction band can be viewed as coming from the anti-bonding Si hybrid states (20, 21). Thus the large band gap of SiO_2 is a reflection of the large difference in the electronegativity between silicon and oxygen. The Si sp^3 hybrid energy is -8.27 eV, and the p-state of oxygen is at -14.13 eV, the hybridization pushes the anti-bonding hybrid state up in energy, while the non-bonding oxygen orbital stays roughly the same (22).

Figure 2 shows the crystal structure and the density of states of $SrTiO_3$. Note that the top of the valence band is still oxygen-derived, however, the bottom of the conduction band is coming from the d-states of Ti. This is true for most transition metal oxides, where the d-states of the metal come lower in energy than states of non-transition metal (the M–O bond is about 20% longer than Si–O bond and results in a smaller overlap) and form the bottom of the conduction band. Moving from a lighter to a heavier metal in the same column, i.e., from Ti to Zr to Hf, would open up the band gap of the corresponding MO_2 oxide (TiO_2, ZrO_2 and HfO_2) as one goes from 3d to 4d to 5d states (the difference between hafnia and zirconia is not very pronounced due to the lanthanide contraction). An extended discussion of the electronic structure of transition metal oxides can be found in the chapter by J. Robertson of this book.

2.3. Complex Band Structure and Charge Neutrality Level

The charge neutrality level, as introduced by Bardeen in 1947, plays a role of the surface Fermi level, but in general is unknown (6). In his 1984 paper, using ideas previously developed by Appelbaum and Hamann (23), Tersoff suggested that the charge neutrality level can be associated with the branch point of the complex band structure in the fundamental gap (24, 25). The wave function at this energy can be written as equally weighted linear combination of conduction and valence band states. To estimate the position of the branch point, Tersoff extended to three dimensions results proven in one dimension by Allen (26). The branch point in the fundamental

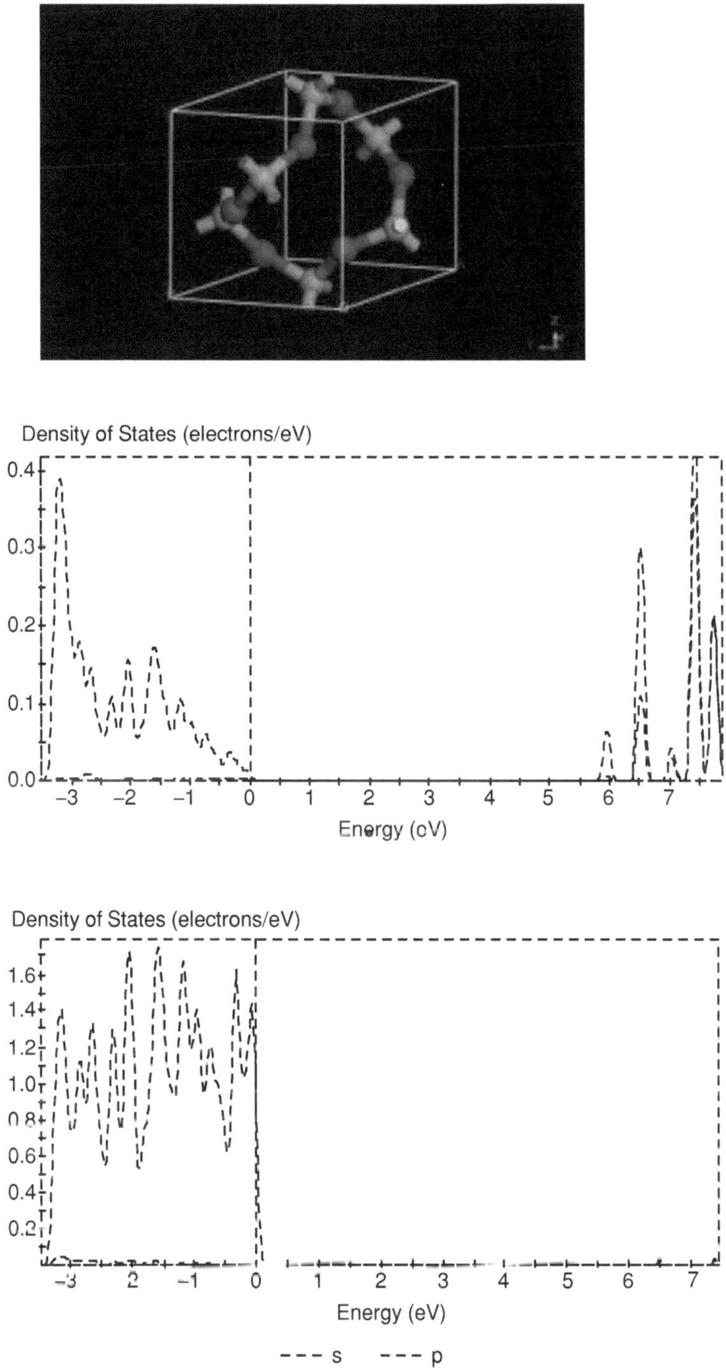

Fig. 1. The crystal structure of SiO_2 α-quartz and its site projected density of states (top panel: Si; bottom panel: oxygen). The zero energy is set at the top of the valence band.

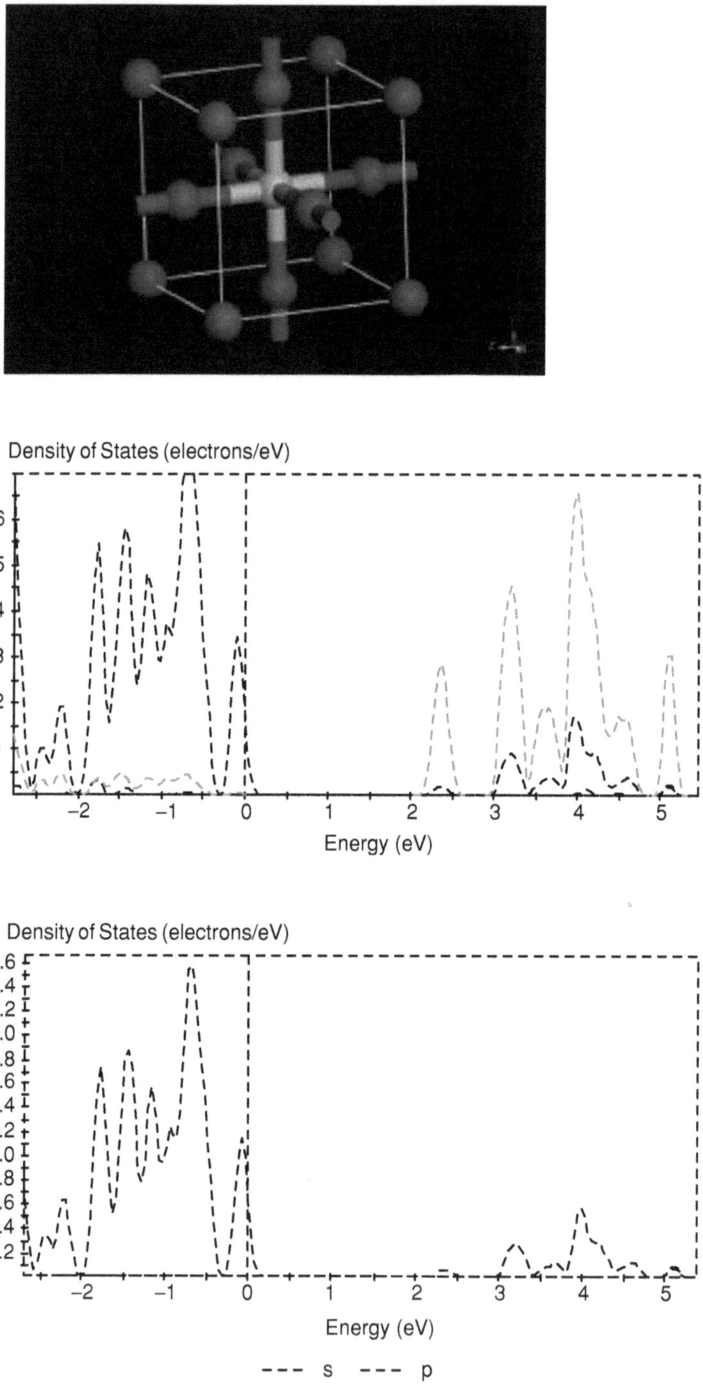

Fig. 2. The crystal structure and the site projected density of states of $SrTiO_3$ (top panel shows s (blue), p (red), and d (green) states of Ti, bottom panel shows s and p states of oxygen).

gap coincides with the zero of the cell-averaged real-space Green's function calculated along a judiciously chosen crystallographic direction:

$$G(\vec{x}, E) = \sum_{n,k} \frac{e^{i\vec{k}\cdot\vec{x}}}{E - E_{n,k} + i\eta} = 0 \tag{3}$$

Here E is the energy in the fundamental gap, and a small imaginary term $i\eta$ in the denominator insures convergence.[1] The direction $\vec{x}$ should be chosen to give the slowest decaying evanescent[2] state.

Here we find the branch point from the actual calculation of the complex band structure. The analytical properties of Bloch functions and energies have been originally studied by Kohn and co-workers (27). They considered the band energy $E_n(\vec{k})$ as a multi-valued function $E(\vec{k})$ of a complex wave vector $\vec{k} = \vec{g} + i\vec{h}$. The usual band structure is then the $Re(E) - g$ cross-section of the Riemann surface. Starting at the lower energy surface (e.g., the valence band) and going into the complex $\vec{k}$-plane around the branch point and back we end up on the next energy surface (i.e., the conduction band). Solutions of the Schrödinger equation with the energy in the band gap thus have complex wave vectors, and are therefore spatially decaying. The character of the solution continuously changes from that of the lower energy band to the higher energy band, with the branch point serving as a point of cross-over from donor-like states to acceptor-like states (28).

The physical connection between the wave vector at a branch point and the interfacial dipole was first made by Heine (9), who used its inverse (the penetration depth of the evanescent gap state) to estimate the separation of the positive charge in the metal and negative charge in the surface states. The dipole is $D = 4\pi\sigma t/\varepsilon$, where σ is the charge density per unit area, ε is the dielectric constant, and $t = \frac{1}{q}$ is the mean separation between the negative charge in the surface states and the positive charge in the metal. Here q is the imaginary wave vector describing the complex band structure in the forbidden energy gap of the semiconductor. When the wave functions are matched at the metal–semiconductor interface, the evanescent wave describes the exponential decay of the metal wave function inside the semiconductor. In other words the metal effectively charges the imaginary wave vector states rather than induces them. Note that the complex band structure is a bulk property of a material, and thus can be calculated without a detailed interface model. More generally, Heine's arguments are made within the effective mass theory. And the wave function decay is a reflection of the macroscopic boundary conditions such as the plane of the interface, and thus the charge transfer is conceptually different from that occurring in the chemical bond formation. The latter charge variation occurs on the length scale comparable with the lattice constant and is beyond the applicability of the effective mass theory.

[1] Recently, Robertson used a slightly different formula, which is appropriate for a tight-binding model since the energy spectrum has an upper bound (11); however in principle Eq. (6) of reference (11) is divergent.
[2] There are no propagating solutions (Bloch waves) of the Schrödinger equation in the band gap energy region. However, solutions with imaginary wave vectors do exist, and those decaying away from the defect, surface or interface are known as *evanescent waves*.

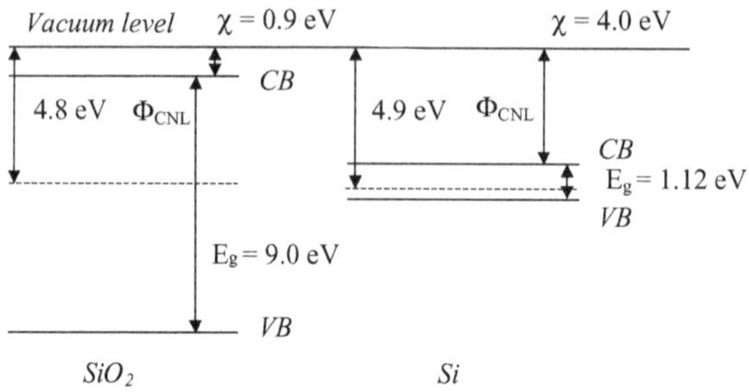

Fig. 3. The band alignment at the Si–SiO_2. A large value of the pinning parameter and a small difference in the charge neutrality level position make the Schottky and Bardeen limits almost identical.

To calculate the conventional band structure, one chooses a set of points along a particular wave vector $\vec{k}$ (e.g., (001)), constructs the corresponding Hamiltonian at each point and diagonalizes it. Since the eigenstates are ordered (this result follows from the Wronskian theorem for local operators) one can connect the nth eigenvalues at each k-point, thus generating $E_n(\vec{k})$ branch of the band structure. To calculate the complex band structure in some sense the opposite is required, one chooses energy and solves the Schrödinger equation for any $\vec{k}$ whether it is real, imaginary or complex. We use the algorithm proposed for this purpose by Boykin (29) as implemented by Tomfohr and Sankey (30). We employ LDA-DFT Hamiltonians to generate the complex band structure. Because the band gap is underestimated (this is typical of LDA calculations), the energy position of the branch point is somewhat uncertain. We find that a simple scaling with respect to the experimental gap value is sufficient to obtain a consistent picture.

We first consider the simple Si/SiO_2 interface. For β-crystobalite, the complex band structure gives the rescaled charge neutrality level 5.1 eV above the valence band maximum (this places it 4.8 eV below the vacuum, assuming an electron affinity χ of 0.9 eV) (31, 32). The imaginary wave vector along the c-axis of the tetragonal cell has a length of 0.67 Å^{-1} at the branch point. That means that an evanescent state at that energy penetrates only about 1.5 Å inside the oxide. The electron affinity and charge neutrality level of Si with respect to vacuum are 4.0 and 4.9 eV, respectively. The pinning parameter S of SiO_2 is 0.9 (this large value makes a contact to SiO_2 almost ideal Schottky case) (11). The conduction band offset calculated using Eq. (1) is 3.1 eV in rather good agreement with experiment and the previous density functional calculations using the reference potential method (33). The corresponding band diagram is shown in Fig. 3.

Figure 4 shows the complex band structure of $SrTiO_3$ calculated using the Hamiltonian computed with the local orbital code SIESTA (34). The branch point is 0.73 eV above the valence band edge. However, since the value of the band gap is significantly underestimated, we rescale the energy associated with the branch point by the ratio of the theoretical and experimental band gaps. The rescaled charge neutrality level is

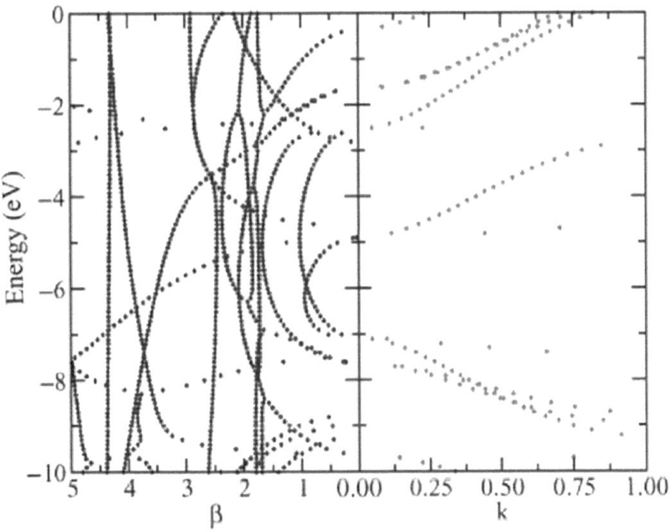

Fig. 4. The complex band structure of $SrTiO_3$ in the near gap region, here $\beta = 2\mathrm{Im}(\vec{k})$. The charge neutrality level is 0.73 eV above the band valence band top as calculated. The band gap is calculated to be 2.1 eV, therefore the rescaled value of the charge neutrality level is estimated to be 6.4 eV with respect to the vacuum level.

6.4 eV for $SrTiO_3$ and 4.9 eV for Si (again both are given with respect to the vacuum level). The electron affinities of Si and $SrTiO_3$ are 4.0 and 3.9 eV, respectively. Thus within this simple theory we expect a 1.6 eV conduction band offset in the Bardeen limit (strong pinning), and a 0.1 eV offset in the Schottky limit (no pinning). The latter value agrees well with the experimental results discussed in Section 3.2. The length of the imaginary wave vector at the branch point along the (001) direction is 0.5 Å^{-1}, and the evanescent state at the branch point penetrates about 2 Å inside the $SrTiO_3$ (roughly the distance between the Sr and Ti planes along the (001) direction of the perovskite structure). This decay is slower than in SiO_2 indicating a larger interface dipole and thus higher pinning strength. Intuitively, this is related to the smaller band gap of $SrTiO_3$ than of SiO_2. It also suggests that the band alignment is sensitive to the quality of the interface, since at least two atomic layers of the oxide are involved.

The smaller band gap also results in a larger electronic susceptibility that scales approximately as the square of the ratio of the plasma frequency ω_p (which measures the electron density) over the so called Penn gap E_{PG} (which measures the average energy gap of the electronic spectrum) (35):

$$\varepsilon = 1 + \left(\frac{\hbar\omega_p}{E_{PG}}\right)^2. \tag{4}$$

This analysis should be considered as only qualitative, however. It is interesting to note that despite a smaller band gap the evanescent states die off much faster in $SrTiO_3$

than in monoclinic HfO_2, but similarly to the decay in the cubic HfO_2 polymorph (32). However, the pinning parameter of hafnia is almost twice that of $SrTiO_3$ and formally does not change from polymorph to polymorph since they have similar band gaps and ε_∞. This indicates that despite being physically intuitive the MIGS theory is missing some important aspects of the problem (36). That is primarily due to the fact that MIGS ignores the atomic or chemical details of the interfacial bonding and dipole formation. Being an "effective mass" theory it breaks down if the spatial extent of the "defect" state is similar to the lattice spacing, which is often the case when chemistry is involved! In Section 4 we show how modern electronic structure theory describes the interface charge density redistribution and its effect on the band discontinuity without using the language of MIGS.

2.4. Photoemission Method for Determination of Band Alignment

Experimentally the band offsets can be determined by a number of techniques including optical (37), transport (38), and photoemission (39). The first two are generally model-dependent and less direct. In comparison photoemission provides a more direct measurement on the band offset because it measures valance band and core level positions of a material and change of the valance band and core level position when the respective material is brought in contact with other. Additionally, photoemission can provide other information such as band bending and reactions at an interface.

Photoemission involves absorption of photons (X-ray or ultraviolet) by atoms near the specimen surface and subsequent emission of photoelectrons. The kinetic energy of an emitted photoelectron is governed by the Einstein relationship: $E_\mathrm{k} = \hbar\nu - E_\mathrm{b} - \mu$, where E_k is the kinetic energy of the photoelectron, $\hbar\nu$ is the photon energy, E_b is the binding energy, and μ is the work function, which equals the work function of the analyzer unless the specimen is biased instead of being grounded to the analyzer. The photoemission intensity can be described as (40):

$$I \propto cN_i\sigma_i(E_\mathrm{k})n_iT(E_\mathrm{k}) \tag{5}$$

where I is the photoemission intensity, n_i is the density of atoms of the element i, σ_i is the photo-ionization cross-section for the atomic orbital of interest, T is the spectrometer transmission function, E_k is the kinetic energy of photoelectron, and c is a geometrical factor depending on the emission angle and the area from which photoelectrons are detected.

The key factor in the photoemission process is the photo-ionization cross-section $\sigma(E_\mathrm{k})$, which involves initial and final states of the photoemission process and can be described in a golden-rule form (41):

$$\sigma(E_\mathrm{k}) \sim \sum_i |< f \mid \Delta \mid i >|^2\, \delta(E_\mathrm{k} - E_i + \hbar\nu) \tag{6}$$

where $|i>$ relates to the initial, occupied electron states of the specimen, $< f|$ is the unbounded final state propagating in the vacuum continuum, and Δ is the interaction operator associated with electromagnetic radiation of photons. For core levels excited by highly energetic photons, the initial localized states dominate the observed

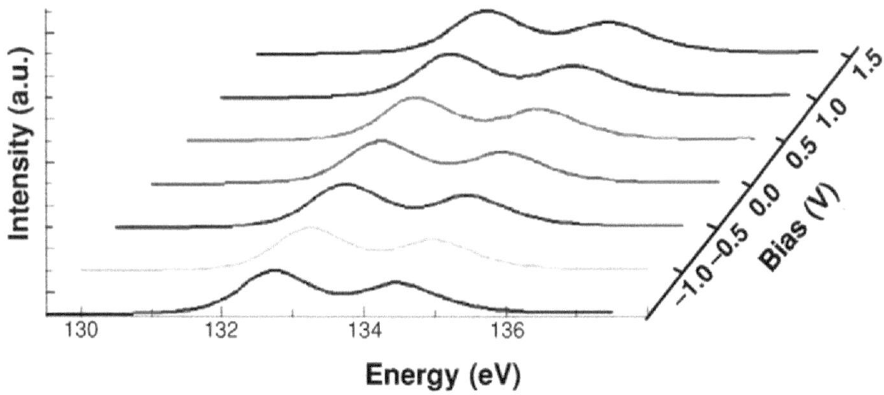

Fig. 5. XPS Sr-3d core level acquired at different sample biases on a 40 Å thick $SrTiO_3$ epitaxially grown on n-Si. The quantitative one-to-one correlation between Sr-3d peak position and bias suggests that surface charging is negligible.

photoemission process. Thus, Eq. (6) allows interpretation of a measured spectrum in terms of intrinsic electronic structure of the specimen. The extension of this method to valance bands requires inclusion of the electron occupancy and an estimation of the integral band intensities after subtraction of the inelastic background (42).

Quantitative application of the photoemission technique to dielectrics can be complicated by surface charging due to their insulating nature. When this occurs, it results in erroneous binding energy and band offset measurements. The surface charging can be identified by measuring core level shifts at different specimen biases. A linear relationship of the two is expected if surface charging is negligible. Alternatively, one can examine the surface charging by measuring core level positions with significantly different photon intensities. Figure 5 shows a shift of the Sr-3d peak position under different biases on a 40 Å thick $SrTiO_3$ film epitaxially grown on Si. The linear, one-to-one relationship suggests that surface charging is negligible.

Photoemission has been used extensively to determine band lineups in semiconductor heterojunctions (43–46). Recently this technique has also been used to determine the band offsets at oxide/oxide and oxide/semiconductor interfaces (47–50). According to Kraut et al. (44), the band offset determination using photoemission involves successive measurements on a clean substrate surface, on a thin film from which core levels of the film and the substrate can both be measured, and finally on a thick film so that only photoelectrons from the film are detectable. The valence band maximum (VBM) is determined by fitting the XPS valence band edge to the Gaussian broadened theoretical density of states. The relationship of core levels and VBMs of the substrate and film are illustrated in Fig. 6. Inspection of the figure allows determination of the valence band offset (44):

$$\Delta E_{\mathrm{v}} = \left(E_{\mathrm{cl}}^{\mathrm{s}} - E_{\mathrm{v}}^{\mathrm{s}}\right)^{\mathrm{s}} - \left(E_{\mathrm{cl}}^{\mathrm{f}} - E_{\mathrm{v}}^{\mathrm{f}}\right)^{\mathrm{f}} + \left(E_{\mathrm{cl}}^{\mathrm{f}} - E_{\mathrm{cl}}^{\mathrm{s}}\right)^{\mathrm{int}} \tag{7}$$

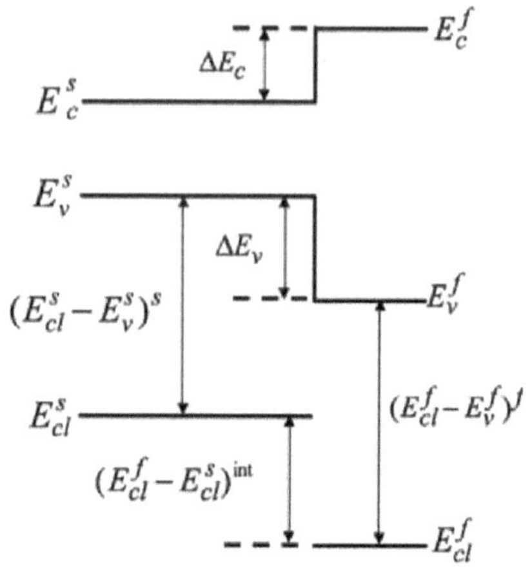

Fig. 6. Energy diagram showing band offsets and their relationship with core levels, VBMs, and band gaps of a substrate and a film.

Here $\Delta E_{\rm v}$ is the valence band offset, $(E^{\rm s}_{\rm cl} - E^{\rm s}_{\rm v})^{\rm s}$ and $(E^{\rm s}_{\rm cl} - E^{\rm f}_{\rm v})^{\rm f}$ are the difference of a core level and the VBM of the substrate and the thick film, respectively, and $(E^{\rm f}_{\rm cl} - E^{\rm s}_{\rm cl})^{\rm int}$ is the difference of core levels between the thin film and the substrate. The conduction band offset can be obtained once the valence band offset is known by $\Delta E_{\rm c} = \Delta E_{\rm v} - \Delta E_{\rm g}$, where $\Delta E_{\rm g}$ is the difference of the band gaps between the substrate and the film. It is worth noting that during the second step of the photoemission measurement, the film has to be thin enough in order to measure the core levels of the film and the substrate, but sufficiently thick so that the band gap and electronic properties of the film are representative of the properties of the thick film. Since accurate determination of the theoretical valence band edge of oxides is time consuming and often difficult, linear extrapolation of the valence band edge is often used to determine the VBM (47). This method was found accurate within hundredths of an eV of the VBM as determined by the Kraut method (51).

In addition to band offset, photoemission has also been used to study band bending and reactions at an interface (39, 52–54). Compared with UPS, XPS can probe deeper into a surface layer or a buried interface so that core level shifts related to the band bending at the surface or the buried interface can be extracted. Considering an interface covered by a film with thickness t, as photoelectrons emitted at distance z from the interface ($z = 0$) are attenuated exponentially, the observed core level binding energy at the interface is given by (54):

$$E_{\rm cl}^{\rm ob} = \frac{\int_t^{\infty} [\phi(z-t) + E_{\rm cl}] \exp(-z/\lambda)\, {\rm d}z}{\int_t^{\infty} \exp(-z/\lambda)\, {\rm d}z} \tag{8}$$

where E_{cl} is core level binding without band bending, and $\phi(z)$ is the real electrostatic potential at distance z. Comparing the observed and the real binding energy allows one to obtain information on band bending at an interface or a surface ($t = 0$). In addition to states at surface or interface, the observed binding energy also depends on the doping level of the semiconductor (54).

3. EXPERIMENTAL INVESTIGATION OF INTERFACIAL PROPERTIES OF OXIDE/SEMICONDUCTOR SYSTEMS

3.1. Early Stages of Epitaxial Growth of $SrTiO_3$ on Si

Several methods were reported for epitaxial growth of $SrTiO_3$ on Si using molecular beam epitaxy (MBE) (16, 55–58). MeKee et al. used a layer-by-layer deposition and subsequent recrystallization method to grow epitaxial $SrTiO_3$ on Si (16, 55). More discussion on this method can be found in McKee's chapter in this book. Yu et al. reported the Sr-induced de-oxidation and subsequent co-deposition of Sr and Ti to form epitaxial $SrTiO_3$ film on Si (56). Li et al. showed a step-growth process for two-dimensional $SrTiO_3$ films on Si (57), and Liang et al. reported Sr-facilitated oxidation for epitaxial growth $SrTiO_3$ on Si (58). While there are different methods for epitaxial growth of $SrTiO_3$ on Si, these methods all involve a Sr mediated template layer on Si. In this section we focus on the role of this template layer.

Prior to the $SrTiO_3$ deposition, the native oxide layer on the Si(001) substrate has to be removed. This can be accomplished using either a Sr-induced de-oxidation method or a thermal de-oxidation method (59). Since the Sr-induced de-oxidation leaves residual Sr atoms on Si and consequently modifies the electronic structure of the Si surface, the thermal de-oxidation should be used for accurate band offset measurements on $SrTiO_3$/Si.

The clean Si(001) surface exhibits a well-known two-domain $(2 \times 1) + (1 \times 2)$ reconstruction. A subsequent deposition of Sr on the clean Si(001) yields a series of reconstructions such as (3×2), (2×1), and (5×1) depending upon the amount of Sr deposited on Si. At approximately half monolayer coverage, Sr forms a well-defined two-domain (2×1) reconstruction. Low-energy-electron-diffraction (LEED) and scanning tunneling microscopy (STM) show that the Sr–Si(001)–(2×1) surface exhibits structural and morphological characteristics similar to those of the clean Si(001) surface. For example, both exhibit the two-domain (2×1) and (1×2) reconstruction and the regular step-terrace morphology. One noticeable difference between the two is that the addition of Sr significantly straightens the step edges, indicating increase of the kink energy upon Sr adsorption. Figure 7 shows LEED and STM images of the clean and Sr-covered Si surfaces.

While both LEED and STM show similar structural characteristics for the clean Si(001) and Sr–Si(001)–(2×1) surfaces, the photoemission reveals a significant difference in their electronic properties. Figure 8 compares the UPS spectrum of the clean Si(001) with that of the Sr–Si(001)–(2×1). Prior to the Sr deposition, a high density of surface-states is clearly visible near the valence band edge. These states

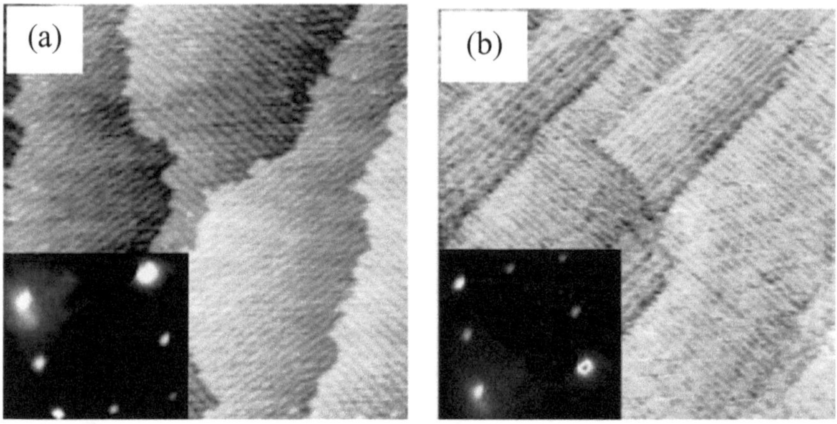

Fig. 7. STM images and LEED patterns showing similar structural characteristics of the clean Si(001) – (2 × 1) (a) and Sr covered Sr–Si(001) – (2 × 1) (b) surfaces. The sizes of the STM images are both 300 Å × 300 Å.

disappear after the Sr adsorption, indicating that Sr stabilizes the Si(001) surface by eliminating the states derived from the Si surface dimers. XPS of the Si-2p core level on the clean and Sr–Si(001)–(2 × 1) surfaces show an increase of the binding energy by a few tenths of an eV on n-type Si after the Sr deposition, indicating less band bending at the Sr/n-Si(001) surface. In contrast, a decrease of the Si-2p binding

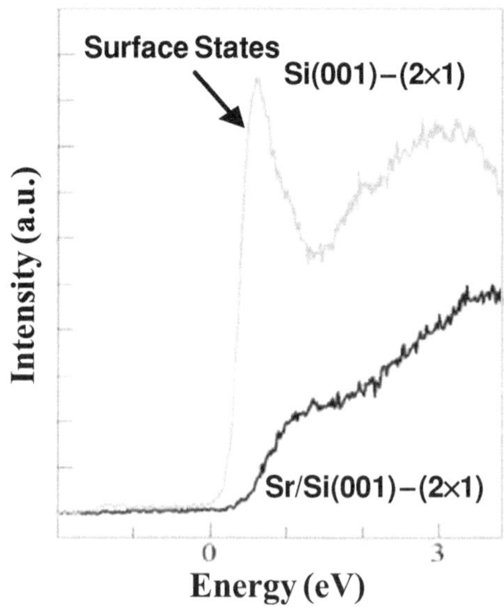

Fig. 8. UPS spectra showing strong surface states near the valance-band edge on clean Si(001) – (2 × 1) and disappearance of such states on Sr/Si(001) – (2 × 1).

energy was observed on p-type Si, indicating more band bending at the Sr/p-Si(001) surface. The combination of these results suggests that the Fermi level pinning position has changed after the Sr adsorption. Since most of the surface states disappear after the Sr deposition, we conclude that the Fermi level pinning at the Sr–Si(001)–(2 × 1) is due to extrinsic residual defects instead of intrinsic surface states.

The oxidation of the Sr covered Si(001) forms a two-dimensional crystalline silicate-like layer (60) instead of an amorphous layer as occurs on the bare Si(001) (61). This result, along with the change of the surface electronic structures upon Sr adsorption, suggests that the presence of Sr on Si(001) serves two purposes during the $SrTiO_3$ growth. First, the Sr adsorption passivates Si by eliminating the chemically reactive states in the Si band gap. Second, the oxidation of the Sr–Si(001) surface results in a silicate-like crystalline layer which not only further stabilizes the surface, but also provides a crystalline template layer for the subsequent $SrTiO_3$ growth. Typically, epitaxial growth of $SrTiO_3$ can only take place in a narrow growth window, indicating that the growth process is kinetically limited. In addition to consideration on diffusion, the template layer provides a local energy minimum which facilitates kinetic processes during $SrTiO_3$ growth. Indeed, XPS results on $SrTiO_3$/Si show that interfacial Si prefers to be oxidized into SiO_x at high temperatures, indicating that $SrTiO_3$/Si interface is thermodynamically unstable. This is discussed further in the next section. The details on the epitaxial growth $SrTiO_3$ on Si are described elsewhere (56–58).

3.2. Band Discontinuity at the $SrTiO_3$/Si Interface

The photoemission determination of the band offset between $SrTiO_3$ and Si, with or without an interlayer between $SrTiO_3$ and Si, was reported by several authors (15, 47, 62). Here we present photoemission measurements of the band offset between Si and $SrTiO_3$, which was grown using the method described in (58). All the measurements were conducted *in situ* without exposure of the specimen to the ambient. Prior to the band-offset measurement, the XPS spectrometer was calibrated against the Au-$4f_{7/2}$ core level and Fermi edge to ensure the accuracy of the energy position. Furthermore, the Sr-3d core level position from a 100 Å thick $SrTiO_3$ film was measured at different biases and with different X-ray intensities. No surface charging was observed.

Figure 9 shows XPS spectra, taken under the normal emission geometry, of core levels and valance bands of the clean n-Si and $SrTiO_3$ films with thickness of 20 and 100 Å, respectively. The separation of the peak positions between Sr-3d and Si-2p from the 20 Å $SrTiO_3$/Si is 34.0 eV. The VBM and core level peak position are 3.1 and 134.2 eV (Sr-3d) for 100 Å thick $SrTiO_3$, and 0.6 and 99.7 eV (Si-2p) for clean Si, respectively. By inserting these values into Eq. (6), one obtains 2.0 ± 0.1 eV for the valance band offset at the $SrTiO_3$/n-Si heterojunction. Using 1.1 and 3.3 eV for Si and $SrTiO_3$ band gaps, one obtains -0.2 ± 0.1 eV for the conduction band offset. These valence band and conduction band offsets agree well with the theoretical values by Robertson and Chen (11), and the value discussed in Section 2.3. These results also agree with the reported experimental values determined by photoemission (47, 62). The significant difference in valance band offset and conduction band offset explains

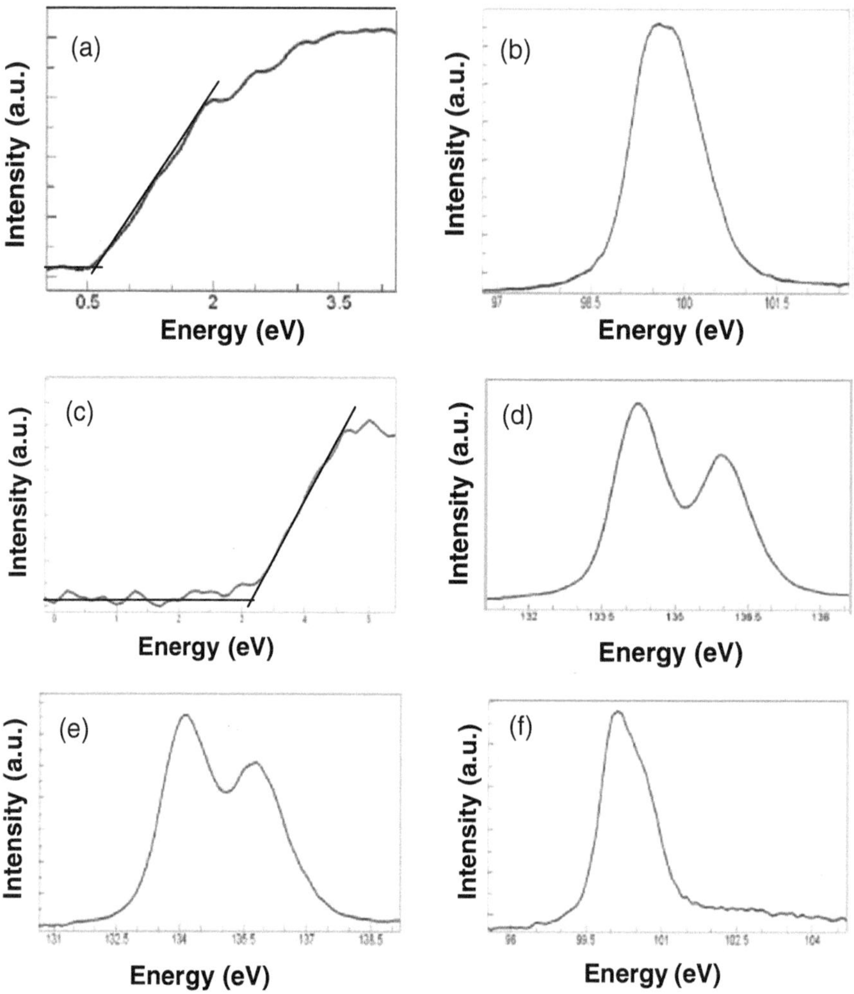

Fig. 9. XPS core level of and VBM spectra at different stages of $SrTiO_3$ growth on n-Si. (a) and (b) are VBM and Si-2p core level of the clean n-Si(001) surface; (c) and (d) are VBM and Sr-3d core level from a 100 Å thick $SrTiO_3$; (e) and (f) are Sr-3d and Si-2p core levels from a 20 Å thick $SrTiO_3$ on n-Si, respectively.

the difference in the leakage current between $SrTiO_3$/Si-based NMOS and PMOS devices (63).

In addition to band offsets, XPS can also provide insight into the band bending at the $SrTiO_3$/Si interface. For example, it was found that the Si-2p core level shifted from 99.7 eV for the clean n-Si(001) to 100.1 eV for the $SrTiO_3$/n-Si heterojunction. Since the Fermi level at the clean Si(001) surface is pinned at approximately 0.4 eV above the VBM, the shift of Si-2p core level to higher energy position for n-Si suggests the original Fermi level pinning on the Si surface is removed, but band remained partially

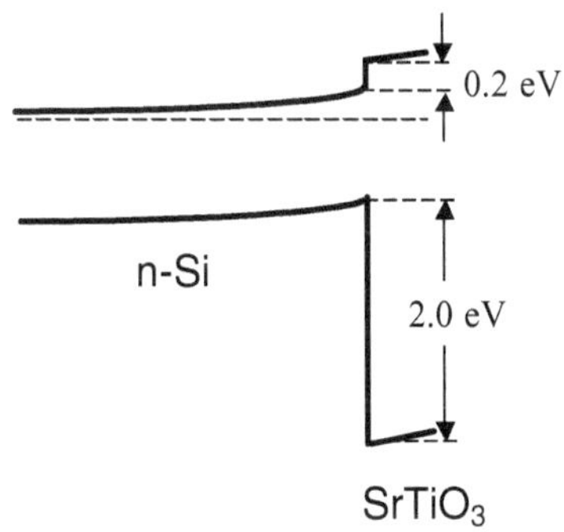

Fig. 10. Band alignments at the $SrTiO_3$/n-Si heterojunction.

bended in Si by approximately 0.3 eV, likely due to defects at the $SrTiO_3$–Si interface and/or difference of the Fermi levels between $SrTiO_3$ film and n-Si substrate. Figure 10 depicts the band offset and band bending at the $SrTiO_3$/n-Si interface.

Chambers et al. also examined the effect of an SiO_2 layer at the Si and $SrTiO_3$ interface on the band offset between Si and $SrTiO_3$(62). The interfacial SiO_2 was introduced by annealing the $SrTiO_3$/Si in an oxygen rich environment. Both the XPS and transmission electron microscopy (TEM) showed that the annealing in oxygen resulted in formation of an SiO_2 layer at the interface between Si and $SrTiO_3$ (57, 64). XPS on the $SrTiO_3$/SiO_2/n-Si showed that the presence of the SiO_2 interlayer did not change the relative band offset between $SrTiO_3$ and Si. This can be understood by the transitivity behavior of the band alignment in a three-layer A–B–C structure as discussed by Kroemer (65):

$$\Delta E_{\mathrm{v}}(\mathrm{A},\mathrm{B}) + \Delta E_{\mathrm{v}}(\mathrm{B},\mathrm{C}) \approx \Delta E_{\mathrm{v}}(\mathrm{A},\mathrm{C}) \tag{9}$$

where $\Delta E_{\mathrm{v}}(\mathrm{A},\mathrm{B})$, $\Delta E_{\mathrm{v}}(\mathrm{B},\mathrm{C})$, and $\Delta E_{\mathrm{v}}(\mathrm{A},\mathrm{C})$ are the valance band offsets between A–B, B–C, and A–C, respectively. We note that the transitivity analysis only focuses on the bulk properties important for band alignment, and does not take the effect of interfacial atomic structure on the band offset into consideration. Such an effect could play an important role in the band alignment as discussed in literatures and in Sections 2 and 4 (15, 17, 37). Additionally, since interfacial pinning is non-commutative, this approach presents further uncertainties in estimating band lineup in a multiplayer structure.

One of the critical issues concerning a gate dielectric is the thermodynamic stability in contact with Si. A high-k dielectric could be viable for CMOS only if it can withstand the various CMOS process conditions. Ion scattering was used to examine the stability of various gate dielectrics including $SrTiO_3$ under different annealing conditions (66–68). It was found that $SrTiO_3$/Si interface was not stable during the high temperature anneal in vacuum and in H_2 (66). In addition to ion scattering, we have used XPS to provide complementary information on interfacial reactions at the dielectric-Si interface. Figure 11 shows the evolution of XPS Si-2p and Ti-2p core levels from $SrTiO_3$/Si annealed at different temperatures in ultrahigh vacuum. Results

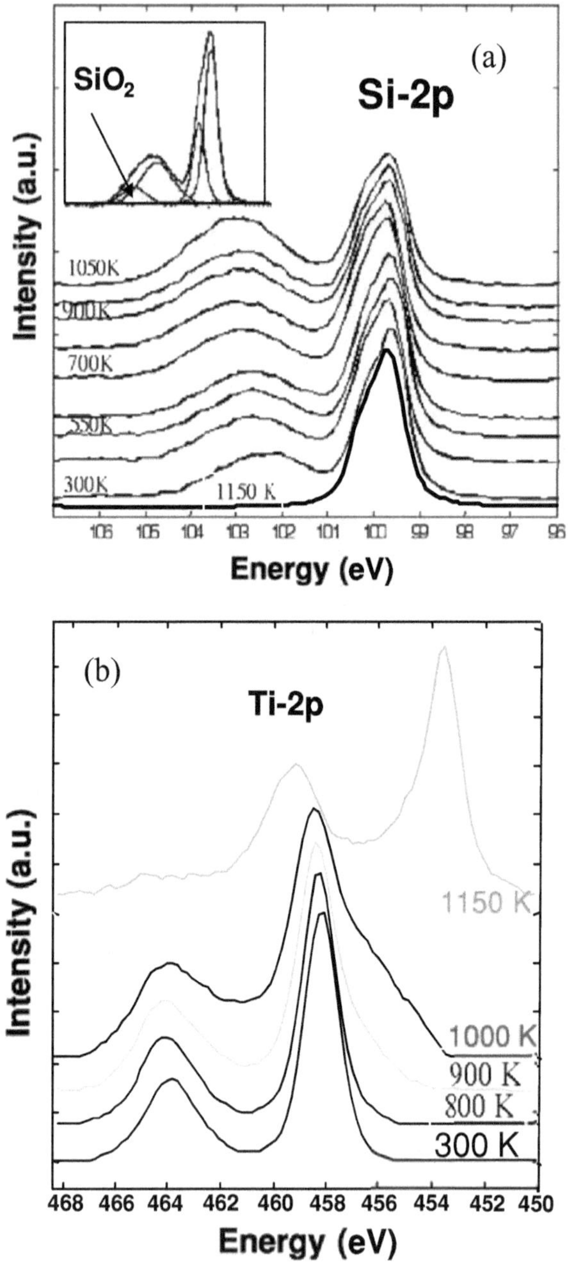

Fig. 11. XPS spectra of Si-2p (a) and Ti-2p (b) core levels showing reactions at a 40 Å $SrTiO_3$/n-Si interface after annealing in UHV at different elevated temperatures. The inset in (a) shows the oxidized Si consisting of silicate and SiO_2, respectively.

show that interface Si reacts with $SrTiO_3$ at elevated temperatures, and the interfacial layer grows as the temperature increased from 300 to 1050 K. This is characterized by the increase of the area of the peak near 101 to 104 eV regime, and by the shift of the peak towards higher binding energy as temperatures increased. Quantitative analysis of the Si-2p core level reveals that the peak in 101 eV to 105 eV regime consists of two components, namely silicate and SiO_2 centered around 102 and 103 eV, as showed in the inset of Fig. 11. XPS spectra also show that $SrTiO_3$ becomes more reduced as the annealing temperature increases. This is evident by the development of the Ti^{3+} shoulder in the XPS Ti-2p core level. The combined results suggest that the interfacial Si becomes oxidized at the expense of titanate reduction. When the annealing temperature reaches 1150 K, virtually all the oxide disappeared on Si, and titanate reduces to metallic titanium, likely in the form of titanium silicide.

3.3. Epitaxial $SrTiO_3$/GaAs System

Crystalline oxides on GaAs could potentially provide an interface with lower defect density and better electronic properties due to structural registry. While the primary motivation for oxide/GaAs has been the development of a gate dielectric for GaAs-based MOSFETs (69), hetero-epitaxy of crystalline oxides on III–Vs also provides new opportunities for integration of various functional oxides with III–V semiconductors.

The epitaxial growth of oxides on GaAs has been investigated for more than a decade. Hong et al. reported epitaxial growth of single domain cubic gadolinium oxide on GaAs (001) using the e-beam deposition (70). Using pulsed laser deposition (PLD), Tarsa et al. showed the growth of highly textured and oriented cube-on-cube MgO films on clean and Sb passivated GaAs substrates, respectively (71). Nshita et al. investigated the growth of NiO on GaAs using the e-beam evaporation (72). Results suggested that epitaxial growth could not be achieved when NiO was grown directly on GaAs. However, by placing an alkali halide layer such as NaCl between NiO and GaAs, single-crystal NiO was grown epitaxially on GaAs.

We have investigated the epitaxial growth of $SrTiO_3$ on GaAs(001) using MBE, and showed that crystalline $SrTiO_3$ could be grown on GaAs and different initial surfaces and growth conditions resulted in different interfacial properties (73, 74). Compared to Si, the GaAs(001) surface is less reactive with oxygen. This allows oxides to be grown on GaAs(001) at higher oxygen pressure. However, the structure and chemistry at the GaAs(001) surface are more complex as the surface can be terminated by Ga or As and involves multiple reconstructions depending upon the surface termination and process conditions. This behavior complicates the initial stage of the oxide growth.

The growth of $SrTiO_3$ was accomplished by the co-deposition of strontium and titanium in the presence of molecular oxygen. A Ti pre-layer was deposited on GaAs(001) prior to the $SrTiO_3$ deposition. XPS and RHEED showed that the Ti prelayer reacted with the surface arsenic and formed disordered TiAs-like species, suggesting that the Ti prelayer was more important for controlling surface chemistry than for structural templating at the early stage of growth (73, 74).

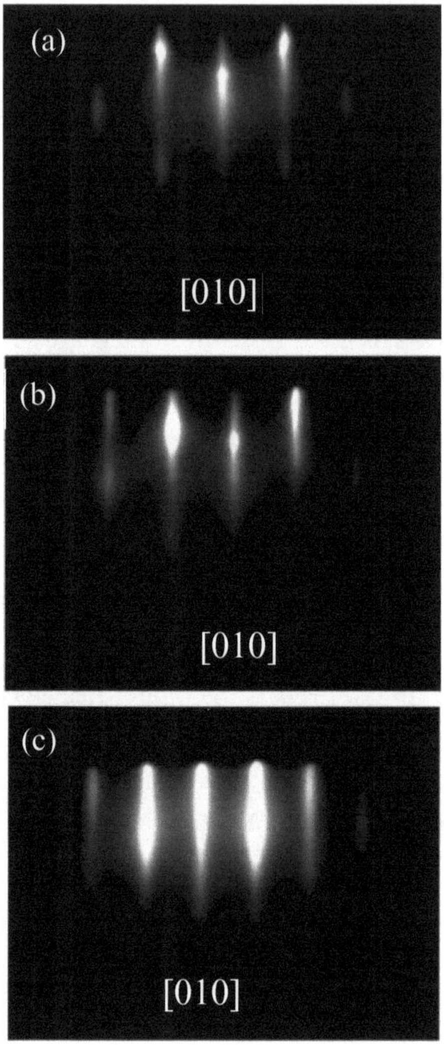

Fig. 12. RHEED images of epitaxial $SrTiO_3$ films on GaAs(001) with film thickness of 8, 16, and 100 Å, respectively. Coherent diffraction is evident on the 8 and 16 Å thick $SrTiO_3$ films.

$SrTiO_3$ can be grown coherently on the GaAs(001) at the early stage of growth. This is characterized by the semicircular diffraction pattern as shown in Fig. 12. The coherent diffraction behavior degraded and $SrTiO_3$ film began to relax when the $SrTiO_3$ thickness exceeded 20 Å due to the ~2% lattice mismatch between $SrTiO_3$ and GaAs. However, the RHEED pattern still displayed low background and streaky diffraction features, suggesting that the smoothness and crystallinity of the film surface remained satisfactory. Indeed the atomic force microscopy images showed that the surface roughness of a 110 Å thick $SrTiO_3$ film was 2.9 Å.

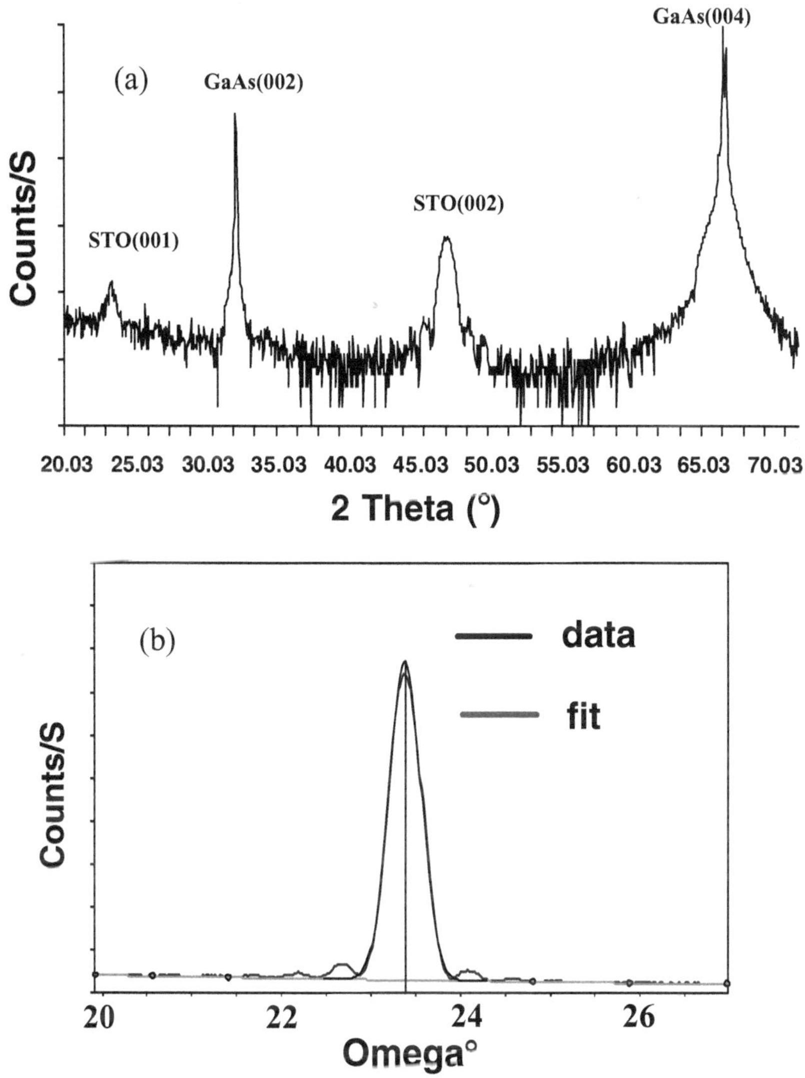

Fig. 13. XRD $\theta - 2\theta$ and rocking curve from a 110 Å $SrTiO_3$ epitaxially grown on GaAs(001).

Figure 13 shows the θ–2θ and rocking curve measurements using X-ray diffraction (XRD) on the 110 Å thick $SrTiO_3$ epitaxially grown on GaAs. The sample was aligned on the GaAs (400) peak prior to collection of the $SrTiO_3$ (200) rocking curve. A Gaussian profile-fit of the rocking curve showed that the full-width-at-half-maximum (FWHM) was 0.42°. As a comparison, the calculated peak width for a 110 Å film using the Scherrer formula is 0.39°. Alternatively, an ideal FWHM value of 0.42° corresponds to a film thickness of 102 Å. These results indicate that the epitaxial $SrTiO_3$

film is of good quality. Additionally, cross-sectional TEM performed on $SrTiO_3$/GaAs showed a strong lattice fringe contrast from $SrTiO_3$ film and an abrupt interface between $SrTiO_3$ and GaAs, suggesting that the $SrTiO_3$ film had good crystallinity and that the interfacial reaction between $SrTiO_3$ and GaAs was under control (74).

3.4. Band Alignment at $SrTiO_3$/GaAs Heterojunction

In order to determine the band alignment at the $SrTiO_3$/GaAs interface, XPS measurements were conducted on $SrTiO_3$/GaAs at different stages of growth. Figure 14 shows valence band edges and core level positions of the clean p-GaAs(001) and $SrTiO_3$ films on the p-GaAs(001) with thickness of 20 and 110 Å, respectively. All the spectra were taken at the normal emission geometry. The VBM and the core level peak position locate at 0.3 and 41.1 eV (As-3$d_{5/2}$) for the clean p-GaAs(001), and 3.0 and 38.2 eV (Ti-3p) for the 110 Å $SrTiO_3$/p-GaAs(001), respectively. The separation between As-3$d_{5/2}$ and Ti-3p in the 20 Å $SrTiO_3$/p-GaAs(001) is 3.1 eV. Based on these values, we obtain a 2.5 ± 0.1 eV valence band offset for $SrTiO_3$/p-GaAs. Using 1.4 and 3.3 eV as the band gaps for $SrTiO_3$ and GaAs, we find that the conduction band offset at the $SrTiO_3$/p-GaAs interface is 0.6 ± 0.1 eV. The positive conduction offset suggests that the $SrTiO_3$/GaAs heterojunction has a type-II band alignment structure, i.e, a structure in which conduction and valance bands of the film are concomitantly lower or higher than that of the substrate. While a type-II structure is not suitable for gate dielectric, it is important for charge separation such as used in solar cells and photocatalysis.

One of the most important issues concerning the oxide/GaAs interface is the Fermi level pinning (75). Extensive efforts have been devoted over the past 40 years in searching for oxides forming the unpinned oxide/GaAs interface. It was reported that Ga_2O_3 films grown on GaAs(001) were free of the Fermi level pinning at the respective interface (76). Recently, Hale et al. showed that the adsorption of Ga_2O on GaAs(001) unpinned the Fermi level at the Ga_2O/GaAs interface (77).

A detailed photoemission study on $SrTiO_3$/GaAs showed that the position of the interfacial Fermi level was sensitive to the growth conditions and the Fermi level could be unpinned through careful control of the initial surface chemistry and growth conditions. Figure 15 shows the As-3d core level from n-GaAs(001) and p-GaAs(001) substrates at different stages of growth. A 0.9 eV separation of the As-3d core levels between the clean, As-terminated n- and p-GaAs is observed on these bare surfaces. Taking in account the doping effect on the measured band bending and the actual Fermi level positions in n- and p-GaAs (54), we conclude that the Fermi level positions at the bare n-GaAs(001) and p-GaAs(001) surfaces are unpinned, and there is a slight band bending with a sum of 0.3 eV at these two surfaces.

With the established starting band position, we can now examine the development of the Fermi level position, or the band bending, at different stages of growth. Results show that upon the deposition of the Ti prelayer on these surfaces, the centroid position of the As-3d core level shifts to approximately 41.5 eV for both n-GaAs and p-GaAs substrates, indicating that the Fermi level pinning takes place at the Ti/GaAs interface. This result is consistent with the reported near mid-gap Fermi level pinning at the

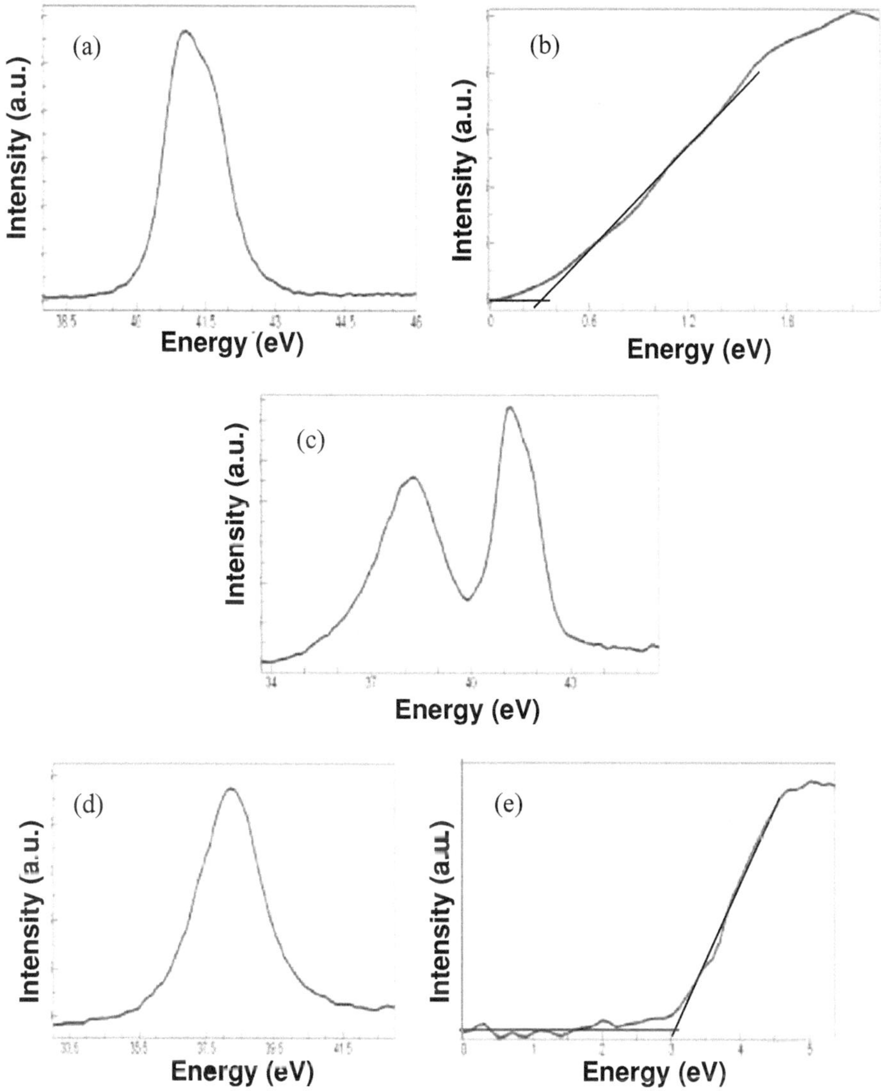

Fig. 14. XPS spectra obtained at different stages of growth of $SrTiO_3$ on p-GaAs. (a) and (b) show As-3d core level and VBM from a clean, As-terminated GaAs(001) surface; (c) shows separation of Ad-3d and Ti-3p core levels from a 20 Å thick $SrTiO_3$/p-GaAs; and (d) and (e) show Ti-3p and VBM of a 110 Å thick $SrTiO_3$ on p-GaAs, respectively.

Ti/GaAs interface (78). The subsequent growth of epitaxial $SrTiO_3$ on GaAs shifts the centroid position of the As-3d core level back close to original positions of the clean GaAs substrates, suggesting that the Fermi level pinning is alleviated upon the epitaxial growth of $SrTiO_3$ on GaAs. The separation of the As-3d core level positions

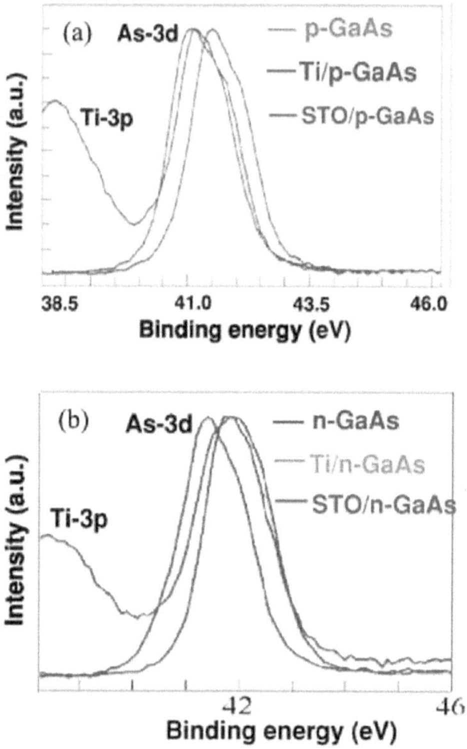

Fig. 15. XPS spectra showing change of As-3d core level position at different stages of $SrTiO_3$ growth on p-GaAs(001) (a) and n-GaAs(001) (b), respectively. The reversal of the As-3d position in Ti/GaAs and $SrTiO_3$/GaAs indicates that Fermi level pinning is alleviated after the $SrTiO_3$ deposition.

between $SrTiO_3$/n-GaAs and $SrTiO_3$/p-GaAs samples is approximately 0.7 eV, less than that of the clean As-terminated n-GaAs and p-GaAs, suggesting an increase of band bending at the $SrTiO_3$/GaAs interface. The amount of band bending in GaAs is approximately 0.2 eV for $SrTiO_3$/n-GaAs and 0.3 eV for $SrTiO_3$/p-GaAs, respectively. The difference in the amount of band bending is likely due to different Fermi level positions in n-GaAs and p-GaAs with respect to that in $SrTiO_3$. Figure 16 shows the band lineup at the $SrTiO_3$/GaAs heterojunction.

4. THEORETICAL INVESTIGATION OF THE BAND OFFSET AT $SrTiO_3$/Si INTERFACE

4.1. Methodology

A seemingly innocent observation of Kleinman (79) that the bulk solid does not have an intrinsic energy scale presents a serious conceptual problem when the relative energies of two materials need to be considered. This is precisely the problem one

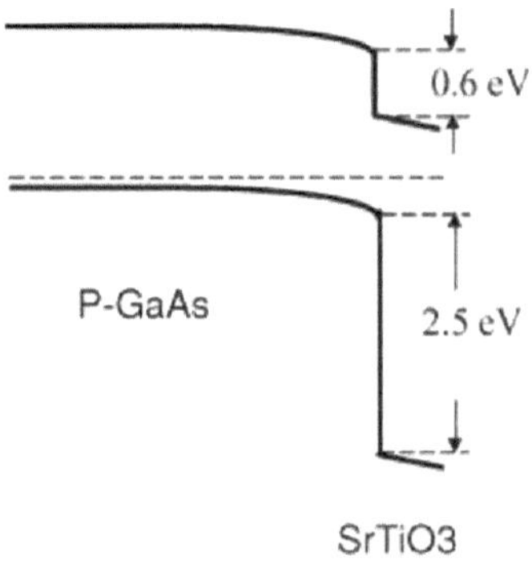

Fig. 16. Band lineup at the $SrTiO_3$/p-GaAs(001) heterojunction.

faces when computing the band offset at a heterojunction. The only known practical solution to this conundrum is to consider a combined system in the same calculation using either slab or super-cell geometry. This is a rather computationally expensive approach, it requires the atomic scale model of the interface often involving a large number of atoms, and results strongly depend on the atomic arrangement. One should also keep in mind that only the valence band discontinuity is reliable, since it is a ground state property, and as such is defined by the self-consistent charge density. Once a self-consistent solution to Kohn–Sham equations is obtained, there are two ways to extract the band offset at the interface.

One way to do the analysis is to consider the density of states of the combined system, and then extract the energies corresponding to different sides of the junction. The density of states analysis is often used in experiment for such an estimate (80). The difficulty obviously lies in the fact that in the total density of states the spatial information about the system is lost. Thus one needs to calculate the so-called local or site-projected density of states in regions far enough from the interface that a bulk-like density of states can be reasonably expected. To calculate the total and partial densities of states, it is convenient to use the Green's function method within the local-orbital formalism. The partial density of states is computed for the atoms in the different layers of the interfacial model and provides the energy-position correlation required for this analysis. The density of states is computed as:

$$N(z) = -\frac{1}{\pi} Tr(\mathrm{Im}(G(z)S^{-1})), \tag{10}$$

where S^{-1} is the inverse of the overlap matrix and $G(z)$ is the matrix element of the resolvent operator $G = (z - H)^{-1}$ in the local orbital basis. This matrix element is given by:

$$G_{\mu v}(z) = \sum_{i,\alpha,\beta,k} \frac{c_{i\alpha}(k)c^*_{i\beta}(k)S_{\mu\alpha}(k)S^+_{\beta v}(k)}{z - \varepsilon_i(k)} \tag{11}$$

where $c_{i\alpha}(k)$ is a local orbital expansion coefficient, and $\varepsilon_i(k)$ is the ith eigenvalue of the Hamiltonian, both depend on the wave vector k. The sum goes over all the basis functions (α, β), over all the eigenvalues of the Hamiltonian i, and over the entire Brillouin zone (the sum over k). It is worth mentioning that this matrix element is not a true representation of the resolvent **G** operator in this space but is related to such by a simple transformation:

$$G = \mathbf{G}S \tag{12}$$

We illustrate how the method works for the Si–SiO_2 structure following the discussion in (81). Figure 17 shows the density of states for a combined Si–SiO_2 structure. To estimate the valence band offset we need to identify the portion of the density of states corresponding to the top of the valence band of Si and of silicon dioxide. We compute the partial densities of states for Si atoms deep in the Si side of the cell, and separately for silicon and oxygen atoms in the stoichiometric oxide layers of the cell away from the interface. Both are shown in Fig. 17 as a dotted and thin solid line, respectively. The valence band tops of Si and oxide correspond to the peaks located at −4.95, and −9.2 eV, respectively. To have a proper representation of the bulk in the partial density of states 8 out of 20 layers of Si were included in the calculation (our method systematically overestimates the band gap of Si due to the minimal sp^3 basis). Only two layers of the oxide are necessary to identify the bulk valence band top. This is in agreement with Kohn's argument that the localization of the wave function is an exponential of the energy band gap. A valence band offset of 4.25 eV is found in close

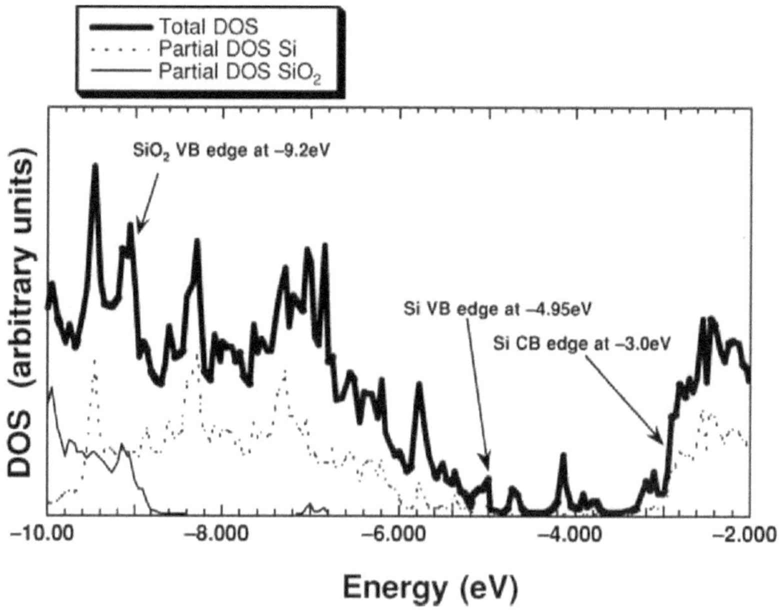

Fig. 17. The density of states computed for the Si–SiO_2 model. The dotted line shows the partial density of states computed for Si atoms in the middle 8 Si layers. The thin solid line shows the contribution from the part of the cell identified as stoichiometric SiO_2. The valence band edges and the Si conduction band edge are indicated with arrows. The valence band offset is estimated to be 4.25 eV. (reproduced with permission from ref. (81)).

agreement with the analysis of ref. (80). Note that all states seen in the middle of the Si gap are coming from the reconstructed bottom Si surface. There are no dangling bonds at the interface in the model, however, there are interface induced states right below the Si conduction band at about -3.2 eV.

Another technique to determine the band discontinuity at the heterojunction is the so-called reference potential method. The method relies on the assumption that a reference level can be introduced on each side of the junction, and then used to line up the band structures. The first theoretical calculation of the band discontinuity based on the idea of a reference potential was done by Frensley and Kroemer who assumed the continuity of the average interstitial potential across the interface (82). A similar idea was also discussed by Harrison (83). The first self-consistent calculation of the interface electronic structure and band discontinuity was performed by Baraff, Appelbaum, and Hamann (BAH) (84). They considered the Ge–GaAs interface and used the self-consistent total potential across the interface to reference the bulk band structure. BAH demonstrated that the deviations of the total potential from its value in the bulk happen only in the interfacial region of the combined slab, and die out within one inter-planar distance. They also traced the reason for such remarkable behavior to insensitivity of the charge density at a fixed point (and thus the total potential) to the spectral changes induced by adding extra layers. Despite the pioneering nature of this work the agreement with experiment was poor. Quantitative agreement with experimental data was achieved by Pickett et al. (85). The modern period starts with the 1986 work of Van de Walle and Martin who used ab initio pseudopotentials to study the Si–Ge interface (14, 86). Demkov and Sankey used the LCAO variation of the method to determine the band discontinuity at the Si–SiO_2 interface (33).

The calculation is done for a large super cell containing the atomic model of the interface. Once the self-consistent charge density is obtained any component of the electron energy can be used as a reference potential. Van de Walle and Martin used the $l = 1$ component of the total electrostatic potential. First, the plane average needs to be calculated:

$$\bar{V}(z) = \left[\frac{1}{\text{area}}\right] \int V(r)\,\mathrm{d}x\,\mathrm{d}y \tag{13}$$

And then the macroscopic average is computed:

$$\tilde{V}(z) = \frac{1}{L_1 L_2} \int_{z-L_1/2}^{z+L_1/2} \int_{z'-L_2/2}^{z'+L_1/2} \bar{V}(z'')\,\mathrm{d}z'\mathrm{d}z'' \tag{14}$$

L_1 and L_2 are the lengths of a single period of the microscopic average on each side of the interface. Away from the interface the value of $\tilde{V}$ is identified with the bulk reference value in the corresponding material. The energy difference between the bulk valence band edge and the reference potential $\Delta V_{\mathrm{RV}} = E_{\mathrm{valce}} - \tilde{V}_{\mathrm{bulk}}$ is determined in two separate calculations. Now the shift in the macroscopic average across the interface $\Delta\tilde{V}_{\mathrm{AB}} = \tilde{V}_{\mathrm{A}} - \tilde{V}_{\mathrm{B}}$ can be used to line up the band structure:

$$\Delta E_{\mathrm{V}}^{\mathrm{AB}} = \Delta E_{\mathrm{RV}}^{\mathrm{A}} - \Delta E_{\mathrm{RV}}^{\mathrm{B}} + \Delta\tilde{V}_{\mathrm{AB}} \tag{15}$$

Within the LDA the calculation is only appropriate for the valence band, which being a ground state entity is properly described. The conduction band offset can then be inferred using the experimental values for the band gaps. Note that Eq. (15) is very similar to Eq. (7) of Section 2.4 with the reference potential $\tilde{V}$ of Eq. (14) playing the role of the core level potential E_{cl}.

4.2. Theoretical Analysis of the Band Alignment at the Si–$SrTiO_3$ Interface

The direct calculation of the band offset requires having an atomic level structural model of the interface. An epitaxial interface such as that between Si and $SrTiO_3$ offers a unique possibility of testing theories of band alignment since the structure can, in principle, be determined precisely. Although presently there still is no consensus concerning this structure, the features essential to understanding the band alignment are available. We now briefly describe the physics driving the formation of the epitaxial interface.

In order to have a high quality film it is preferable to ensure a two dimensional (2D) layer by layer growth of the oxide film on Si. This is achievable if the film wets the substrate. The corresponding growth mode is often referred to as Frank–Van der Merwe growth (87). The thermodynamic condition for the 2D growth is that the surface energy of the substrate is higher than that of the film plus the energy of the interface:

$$\gamma_{\mathrm{substarte}} > \gamma_{\mathrm{film}} + \gamma_{\mathrm{interface}} \tag{16}$$

In the case of $SrTiO_3$ on Si, the lattice mismatch of 1.7% complicates this simple picture, and we are dealing with the so-called Stransky–Krastanov growth mode, where, for films with thicknesses below critical, the growth is two-dimensional but the film is strained, while for thicker films the system reduces the built-in strain either plastically (dislocations) or by switching to a three dimensional island growth mode (87). In our experience the critical thickens of $SrTiO_3$ on Si is about 6–8 nm (10 nm seems large), and for ultra-thin films discussed here layer-by-layer growth is assumed. Zhang et al. have recently reported the theoretical prediction and experimental confirmation that $SrTiO_3$ indeed may wet Si if the growth is done in Sr rich environment (17). As can be seen from Eq. (16), for such a prediction one needs to know the surface energies of the film and substrate and estimate the energy of the interface. The surface energy of Si is about 1710 erg/cm^2 (17). The surface energy of $SrTiO_3$ depends on the termination and chemical environment. In Fig. 18 we plot the surface energy for TiO_2 and SrO terminations as a function of the TiO_2 chemical potential (the bulk reference is set to rutile, and zero chemical potential corresponds to the TiO_2 rich condition). The lower bound of the chemical potential is the formation enthalpy of $SrTiO_3$, and corresponds to SrO rich conditions. The details of the surface energy calculations can be found in references (88, 89). Several important conclusions can be made by analyzing Fig. 18. First, the SrO terminated surface can have a very low surface energy of 800 erg/cm^2 under SrO rich conditions. Second, SrO termination is more stable under a wide range of chemical potential, and thus is a preferred termination in vacuum (hydroxylation of the surface needs to be considered in air). Third, it is difficult to stabilize the TiO_2 termination under SrO rich environment due to a

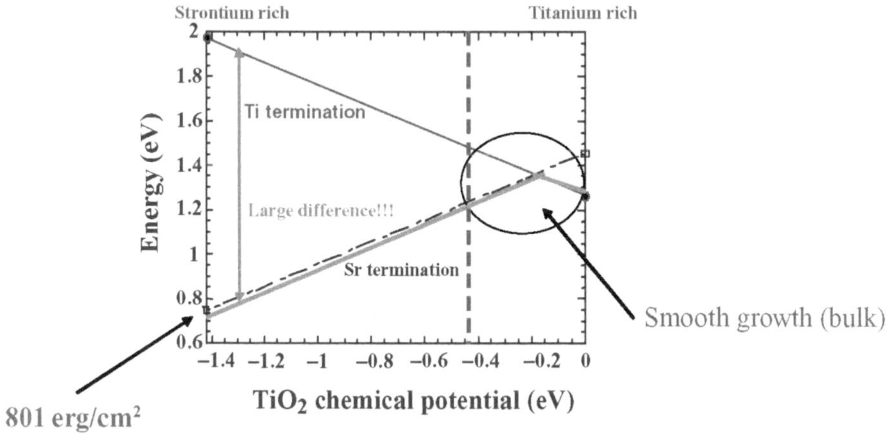

Fig. 18. The energy of SrO and TiO_2 terminated $SrTiO_3(001)$ surface as a function of the TiO_2 chemical potential.

very large difference in the surface energy between the terminations, thus the growth either needs to proceed under the extra Sr coverage (here one relies on Ti diffusion through the topmost Sr rich oxide layer), or under relatively TiO_2 rich conditions (the smooth growth region in Fig. 18).

To estimate the energy of the interface on needs to build an atomic structure and calculate the interface energy in a manner similar to that of the surface energy calculation. Any structure with the interface energy less than 900 erg/cm^2 satisfies the wetting criterion of Eq. (16). Note that finding at least one wetting interface structure means that wetting is thermodynamically possible. In Fig. 19 we show one possible structure (further referred to as 1 ML interface) which is built by connecting the SrO terminated $SrTiO_3$ slab with the unreconstructed Si(001) surface. There is a stoichiometric SrO layer at the interface. The structural relaxation leads to the formation of slightly stretched Si dimers, the structure has 2×1 symmetry. Strontium atoms located above Si dimers are displaced upward by 0.024 nm with respect to those above the troughs. This reflects the fact that the surface area of the Si(001) surface is insufficient to hold 1 ML of SrO[3].

The energy of the stoichiometric 1 ML interface for the system containing different atomic species can be computed in a fashion similar to a surface calculation (17):

$$E = \tfrac{1}{2}(E_{\mathrm{slab}} - N_{\mathrm{Si}}\mu_{\mathrm{Si}} - N_{\mathrm{TiO_2}}\mu_{\mathrm{TiO_2}} - N_{\mathrm{SrO}}\mu_{\mathrm{SrO}}) \qquad (17)$$

Here the energy is given per surface unit cell, and the factor of 1/2 is due to having two interfaces in the super-cell. The chemical potential of Si is set to the bulk Si energy. The chemical potentials of SrO and TiO_2 are related by the equilibrium condition: $\mu_{\mathrm{SrO}} + \mu_{\mathrm{TiO_2}} = \mu_{\mathrm{STO}}$. The $SrTiO_3$ chemical potential is set to its bulk value. Thus the

[3] The covalent and atomic radii of Sr are 1.91 and 2.15 Å, respectively, and the unreconstructed Si (001) surface unit cell lattice vector is only 3.84 Å (90).

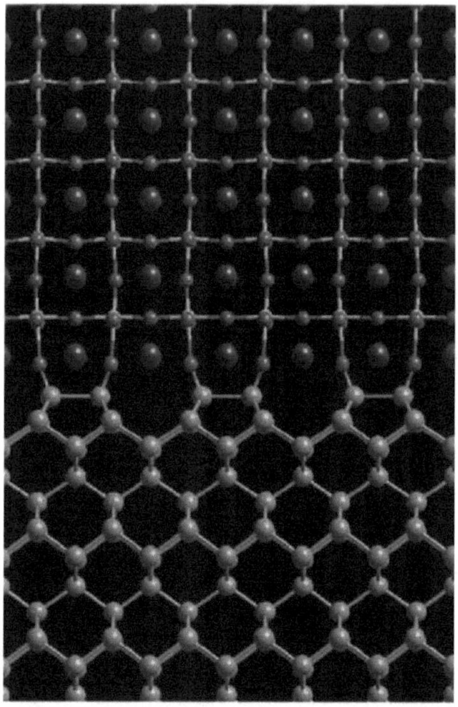

Fig. 19. A (2 × 1) structure of the $Si/SrTiO_3$ interface with 1 ML of Sr at the interface. The Si, O, and Sr are in golden, pink, and red, respectively.

grand thermodynamic potential of the interface is a function of just one variable, and we choose μ_{TiO_2}. In Fig. 20 we show the interface energy of this structure. It can be seen that interface energy can be as low as 574 erg/cm^2 under Sr rich conditions. Luckily, the low interface energy is realized for the same growth conditions as the lowest $SrTiO_3$ surface energy (SrO-termination, Sr rich growth). Therefore the sum of the interface energy and the surface energy of the $SrTiO_3$ film is now only 1433 erg/cm^2, and we conclude that $SrTiO_3$ should wet Si. The structural model discussed here is

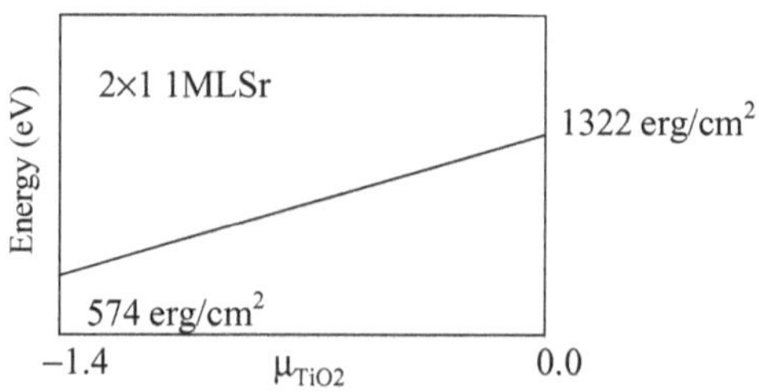

Fig. 20. The energy of the $Si/SrTiO_3$ interface computed for 1 ML structure as a function of the TiO_2 chemical potential.

plausible and consistent with the two-dimensional growth. It does not, however, agree with the experimentally observed (17, 64) +2 oxidation state of Si at the interface (the same is true for models of Först et al. (18)). Work to identify the exact interfacial structure is in progress and results will be published separately. However, the model is close enough to reality to be useful in our discussion of the basics of the barrier formation.

We first use a direct density of states analysis technique of the previous section to compute the valence band offset, and infer the conduction band offset using the experimental band gap values (1.17 and 3.2 eV for Si and $SrTiO_3$, respectively). In addition to the interface described above we also consider one with a half monolayer of Sr in the plane of contact (17). We calculate the total and site projected valence band density of states for a 4.5 nm thick 2 × 1 Si–$SrTiO_3$ slab in vacuum (see Fig. 21). With these considerations we obtain the conduction band offset of 0.87 and 0.23 eV for our models 1/2 and 1 ML, respectively (see Fig. 21(a) and (b)).

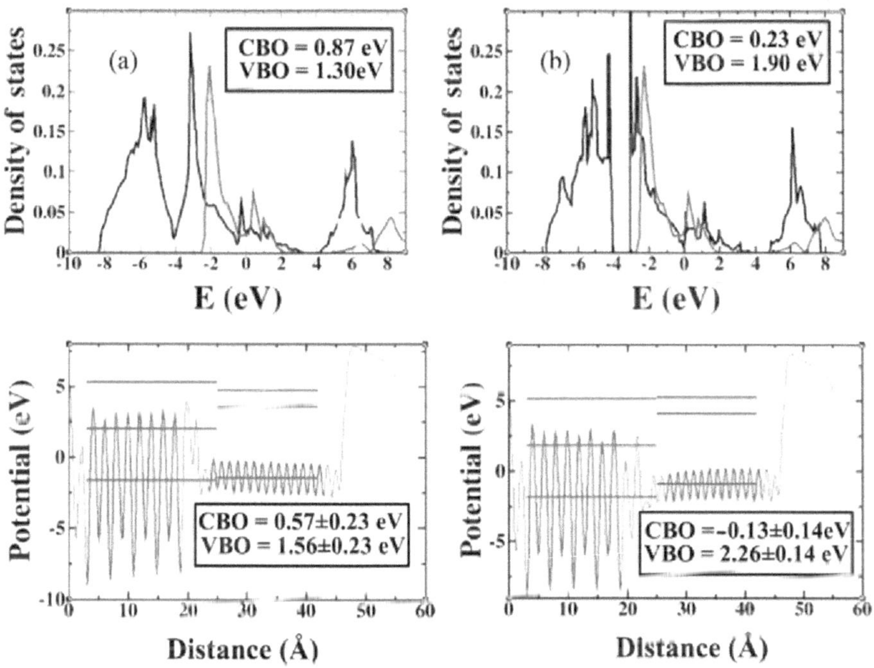

Fig. 21. The band discontinuity at the Si–$SrTiO_3$ interface. (a) For the (2 × 1) structure with 1/2 ML of Sr at the interface we find a sizable conduction band offset in agreement with the strong pinning or the Bardeen limit. The top panel shows the projected density of states analysis (the red curve shows the STO contribution, and the black curve the Si contribution). The bottom panel shows the reference potential calculation. $SrTiO_3$ is on the left side of the simulation cell, and Si is on the right side. The average value of the potential on each side and both bands are indicated with horizontal lines. Using experimental values for the band gaps the conduction band offset is 0.57 eV. (b) For the (2 × 1) structure with a full Sr monolayer at the interface we find a very small conduction band offset in agreement with the unpinned Schottky model. The bottom panel shows the reference potential calculation. $SrTiO_3$ is on the left side of the simulation cell, and Si is on the right side. (reproduced with permission from ref. (17)).

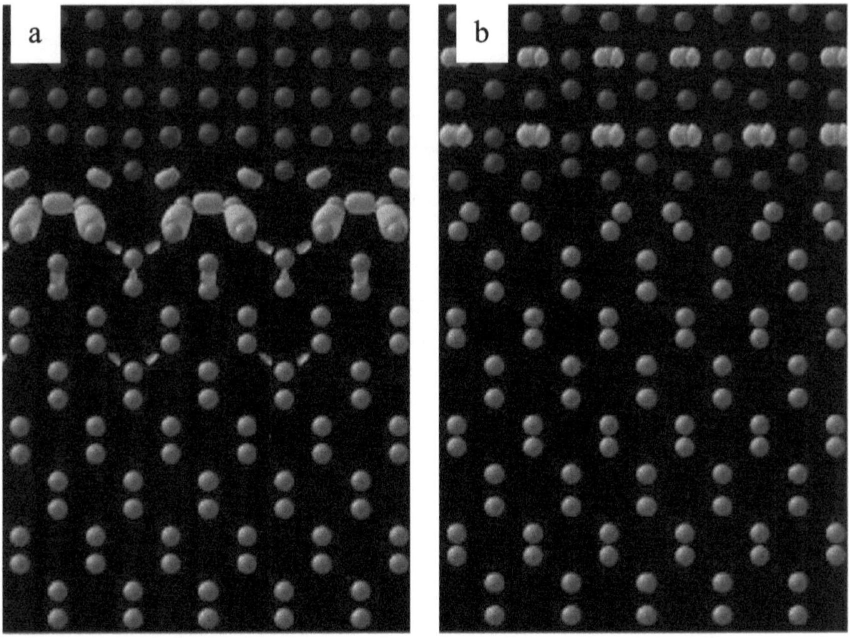

Fig. 22. The electron density obtained by integrating over the states withitn 1 eV window below the Fermi level: (a) 1/2 ML structure and (b) 1 ML structure. The Si, O, and Sr are in blue, red, and green, respectively. (reproduced with permission from ref. (17)).

To verify these results we use the reference potential technique to placing the valence bands with respect to the average electrostatic potential across the slab using two additional bulk calculations as shown in Fig. 21(a) and (b) (lower panels). Two methods agree within the computational error. We now recall that the simple theory of Section 2.3 suggests that the 1/2 ML structure corresponds to the Bardeen limit with the S value ranging between 0.1 and 0.47 (an empirical estimate gives 0.28 (11)), while 1 ML structure corresponds to the Schottky limit. This picture is indeed correct. In Fig. 22(a) and (b) we show the electron density obtained by integrating over the states within a 1 eV window below the Fermi level. In the case of 1/2 ML structure states localized on Si dimers are clearly seen, while no localized charge is observed at the interface for 1 ML structure. The localized states of structure I fall into the $SrTiO_3$ gap. The origin of these states can be explained as follows. Note that the interface layer has the $SrSi_2$ stoichiometry corresponding to a half monolayer of Sr deposited on the Si(001) 2×1 reconstructed surface at the template stage. The top of the valence band for such a template is precisely the dimer localized surface state (the structure is a semiconductor rather than a metal (90)). One can also see the origin of the MIGS model shortcomings, formally equivalent interface models result in almost opposite alignment scenarios depending on the local chemistry in the plane of contact, it is precisely this "interface phase"[4] that is missing in the MIGS concept. In conclusion,

[4] The term was first introduced by Rodney McKee.

we find that a wetting low energy interface structure is characterized by a rather small conduction band offset in agreement with experiment. The challenge is to identify the interfacial structure that has the properly oxidized Si at the interface, is consistent with the high quality STEM images, and has a Schottky type band alignment. The work is under way (91).

5. SUMMARY

We have presented results on the structural and electronic properties at the $SrTiO_3$/semicnductor interfaces from both experimental and theoretical points of view. With the use of the epitaxial oxide/semiconductor system, we have been able to compare and to contrast the experimental results with the theoretical modeling and simulation in a more rigorous fashion. Such comparisons provide us insights into the key parameters that govern the properties at the oxide/semiconductor interface such as band alignment.

ACKNOWLEDGEMENT

The authors are grateful to their colleagues at the Motorola Labs for the many years' collaboration on the fascinating epitaxial oxide/semiconductor system. They are indebted to S. Gan for acquiring the STM and LEED images, and D. McCready and T.C. Eschrich for the XRD measurements. Y.L. is grateful to S. Chambers and D. Baer on many insightful discussions on photoemission measurement on heterostructures and insulators. A.D. is indebt to Rodney McKee and Otto Sankey who helped his understanding of the problem through informal discussions and paper exchange. Part of XRD, LEED and STM experiments were conducted at the Environmental Molecular Sciences Laboratory, a DOE user facility in the Pacific Northwest National Laboratory.

REFERENCES

1. G.D. Wilk, R.M. Wallance, J.M. Anthony, *J. Appl. Phys.* **87**, 484 (2000).
2. R.T. Tung, *Mater. Sci. Eng. R* **35**, 1 (2001).
3. J. Tersoff, *Heterojunction Band Discontinuities: Physics and Device Applications*, eds. F. Capasso, G. Margaritondo (Elsevier Science Publishers B.V., Amsterdam, 1987).
4. W. Schottky, *Z. Physik* **113**, 367 (1939); W. Schottky, *Zeits. f. Physik* **118**, 539 (1942).
5. R.L. Anderson, *Solid-State Electron.* **5**, 341 (1962).
6. J. Bardeen, *Phys. Rev.* **71**, 717 (1947).
7. Ig. Tamm, *Physik. Zeits. Sowjetunion* **1**, 733 (1932).
8. W. Shockley, *Phys. Rev.* **56**, 317 (1939).
9. V. Heine, *Phys. Rev. A* **138**, 1689 (1965).
10. W.R. Frensley, H. Kroemer, *Phys. Rev. B* **16**, 2642 (1977).
11. J. Robertson, C.W. Chen, *Appl. Phys. Lett.* **74**, 1168 (1999); J. Robertson, *J. Vac. Sci. Technol. B* **18**, 1785 (2000).

12. W. Mönch, *Surf. Sci.* **300**, 928 (1994).
13. Y. Yeo, T.J. King, C. Hu, *J. Appl. Phys.* **92**, 7266 (2002).
14. C.G. Van de Walle, *Phys. Rev. B* **39**, 1871 (1989).
15. R.A. McKee, F.J. Walker, M. Buongiorno Nardelli, W.A. Shelton, G.M. Stocks, *Science* **300**, 1726 (2003).
16. R.A. McKee, F.J. Walker, M.F. Chisholm, *Phys. Rev. Lett.* **81**, 3014 (1998); R.A. McKee, F.J. Walker, M.F. Chisholm, *Science* **293**, 461 (2001).
17. X. Zhang, A.A. Demkov, H. Li, X. Hu, Y. Wei, J. Kulik, *Phys. Rev. B* **68**, 125323 (2003).
18. C.J. Först, C.R. Ashman, K. Schwartz, P.E. Blöchl, *Nature* **427**, 53 (2004).
19. P.W. Peacock, J. Robertson, *Appl. Phys. Lett.* **83**, 5497 (2003).
20. J.R. Chelikowsky, M.A. Schlüter, *Phys. Rev. B* **15**, 4020 (1977).
21. S. Pantelides, W.A. Harrison, *Phys. Rev. B* **13**, 2667 (1976).
22. W. Harrison, *J. Vac. Sci. Technol.* **14**, 1016 (1977).
23. J.A. Appelbaum, D.R. Hamann, *Phys. Rev. B* **10**, 4973 (1974).
24. J. Tersoff, *Phys. Rev. B* **30**, 4874 (1984).
25. J. Tersoff, *Phys. Rev. Lett.* **52**, 465 (1984).
26. R.E. Allen, *Phys. Rev. B* **20**, 1454 (1979).
27. W. Kohn, *Phys. Rev.* **115**, 809 (1959); W. Kohn, C. Majumdar, *Phys. Rev.* **138**, A1617 (1965); W. Kohn, J.R. Onffroy, *Phys. Rev. B* **8**, 2485 (1973); J.J. Rehr, W. Kohn, *Phys. Rev. B* **9**, 1981 (1974); J.J. Rehr, W. Kohn, *Phys. Rev. B* **10**, 448 (1974).
28. J. Tersoff, *Phys. Rev. B* **32**, 6968 (1985).
29. T. Boykin, *Phys. Rev. B* **56**, 8107 (1996).
30. J.K. Tomfohr, O.F. Sankey, *Phys. Rev. B* **65**, 245105 (2002).
31. J.K. Tomfohr, O.F. Sankey, *Phys. Stat. Sol. (b)* **233**, 59 (2002).
32. A.A. Demkov, L. Fonseca, J. Tomfohr, O.F. Sankey, *Mater. Res. Soc. Proc.* **786**, E5.6.1 (2004).
33. A.A. Demkov, O.F. Sankey, *Phys. Rev. Lett.* **83**, 2038 (1999).
34. D. Sanchez-Portal, P. Ordejon, E. Artacho, J.M. Soler, *Int. J. Quant. Chem.* **65**, 453 (1999).
35. P. Yu, M. Cardona, *Fundamentals of Semiconductors* (Springer, Berlin, Heidelberg, 1996), p. 325.
36. R.T. Tung, *Phys. Rev. Lett.* **84**, 6078 (2000).
37. N. Dai, L.R. Ram-Mohan, H. Luo, G.L. Yang, F.C. Zhang, M. Dobrowolska, J.K. Furdyna, *Phys. Rev. B* **50**, 18153–18166 (1994).
38. T. Hori, *Gate dielectrics and MOS ULSIs, Principles, Technologies, and Applications* (Springer Verlag, Berlin Heidelberg, 1997).
39. S. Chambers, in *Handbook of Compound Semiconductors*, eds. P.H. Holloway and G.E. McGuire (Noyes Publishing, New York, 1995).
40. C.J. Powell, *Appl. Surf. Sci.* **4**, 492 (1980).
41. T.L. Barr, *Modern ESCA, the Principles and Practice of X-ray Photoelectron Spectroscopy* (CRC Press, Florida, 1994).
42. R.F. Willis, B. Feuerbacher, *Photoemission and the Electronic Properties of Surfaces*, eds. B. Feuerbacher, B. Fitton, R.F. Willis (John Wiley & Sons, Noordwijk, The Netherlands, 1977).
43. R.W. Grant, J.R. Waldrop, E.A. Kraut, *Phys. Rev. Lett.* **40**, 656 (1978).
44. E.A. Kraut, R.W. Grant, J.R. Waldrop, P. Kowalczyk, *Phys. Rev. B* **28**, 1965 (1983).
45. W.X. Ni, J. Knall, G.V. Hansson, *Phys. Rev. B* **36**, 7744 (1987).
46. F. Xu, M. Vos, J.P. Sullivan, Lj. Atanasoska, S.G. Anderson, J.H. Weaver, *Phys. Rev. B* **38**, 7832 (1988).
47. S.A. Chambers, Y. Laing, Z. Yu, R. Droopad, J. Ramdani, K. Eisenbeiser, *Appl. Phys. Lett.* **77**, 1662 (2000).
48. S.A.Chambers, Y. Liang, Y. Gao, *Phys. Rev. B* **61**, 13223 (2000).

49. S. Sayan, E. Garfunkel, S. Suzer, *Appl. Phys. Lett.* **80**, 2135 (2002).
50. K. Hirose, K. Sakano, H. Nohira, T. Hattori, *Phys. Rev. B* **64**, 155352 (2001).
51. S.A. Chambers, T. Droubay, T.C. Kaspar, M. Gutowski, M. van Schilfgaard, *Surf. Sci.* **554**, 81 (2004).
52. C.I. Wu, A. Kahn, *J. Vac. Sci. Technol. B* **16**, 2218 (1998).
53. L.J. Brillson, *Surf. Sci.* **299**, 909 (1994).
54. T. Ogama *J. Appl. Phys.* **64**, 6469 (1988).
55. S. Jeon, F.J. Walker, C.A. Billman, R.A. McKee, H. Hwang, *IEEE Elec. Dev. Letts.* **24**, 218 (2003).
56. Z. Yu, J. Ramdani, J.A. Curless, J.M. Finder, C.D. Overgaard, R. Droopad, K.W. Eisenbeiser, J.A. Jallmark, W.J. Ooms, *J. Vac. Sci. Technol. B* **18**, 1653 (2000).
57. H. Li, X. Hu, Y. Wei, Z. Yu, X. Zhang, R. Droopad, A. Demkov, J. Edwards, K. Moore, W.Ooms, *J. Appl. Phys.* **93**, 4521 (2003).
58. Y. Liang, Y. Wei, X.M. Hu, Z. Yu, R. Droopad, H. Li, K. Moore, *J. Appl. Phys.* **96**, 3413 (2004).
59. Y. Wei, X. Hu, Y. Liang, D.C. Jordan, B. Craigo, R. Droopad, Z. Yu, A. Demkov, J. Edwards, W. Ooms, *J. Vac. Sci. Technol. B* **20**, 1402 (2002).
60. Y. Liang, S. Gan, M. Engelhard, *Appl. Phys. Lett.* **22**, 3591 (2001).
61. T. Engel, *Surf. Sci. Rep.* **18**, 91 (1993).
62. S.A. Chambers, Y. Liang, Z. Yu, Droopad, J. Ramdani, *J. Vac. Sci. Technol. A* **19**, 934 (2001).
63. K. Eisenbeiser, J.M. Finder, Z. Yu, J. Ramdani, J.A. Curless, J.A. Hallmark, R. Droopad, W.J. Ooms, L. Salem, S. Bradshaw, C.D. Overgaard, *Appl. Phys. Lett.* **76**, 1324 (2000).
64. X. Hu, H. Li, Y. Liang, Y. Wei, Z. Yu, D. Marshall, J. Edwards, Jr., R. Droopad, X. Zhang, A. Demkov, K. Moore, J. Kulik, *Appl. Phys. Lett.* **82**, 302 (2002).
65. H. Kroemer, *Proc. NATO Advanced Study Institute on Molecular Beam Epitaxy and Heterostructures*, eds. L.L. Chang, K. Ploog (Martinus Nijhoof, The Hague, 1984).
66. V. Shutthanandan, S. Thevuthasan, Y. Liang, E.M. Adams, Z. Yu, R, Droopad, *Appl. Phys. Lett.* **80**, 1803 (2002).
67. E. Garfunkel, E.P. Gusev, H.C. Lu, T. Gustafsson, M.L. Green, *Fundamental Aspects of Ultrathin Dielectrics on Si-based Devices*, eds. E. Garfunke, E. Gusev, A. Vul, Vol. 47 (NATO Science Series, Dordrecht, The Netherlands, 1997).
68. M. Capel, *Appl. Phys. Lett.* **83**, 3398 (2003).
69. M. Passlack, Z. Yu, R. Droopad, J.K. Abrokwah, D. Dradock, S.I. Yi, M. Hale, J. Sexton, A.C. Kummel, *III–V Semiconductor Heterostructures: Physics, and Devices*, ed. W. Cai (Research Signpost, 2003); J. Sexton, S.I. Yi, M. Hale, P. Kruse, A.A. Demkov, A. Kummel, Displacement of surface arsenic atoms by insertion of oxygen atoms into As–Ga backbonds, *J. Chem. Phys.* **119**, 9191 (2003).
70. M. Hong, J. Kwo, A.R. Kortan, J.P. Mannaerts, A.M. Sergent, *Science* **283**, 1897 (1999).
71. E.J. Tarsa, M. De Gref, D.R. Clarke, A.C. Gossard, J.S. Speck, *J. Appl. Phys.* **7**, 3276 (1993).
72. K. Nishita, A. Koma, K. Saiki, *J. Vac. Sci. Technol. A* **19**, 2282 (2001).
73. Y. Liang, J. Kulik, T.C. Eschrich, R. Droopad, Z. Yu, P. Maniar, *Appl. Phys. Lett.* **85**, 1217 (2004).
74. Y. Liang, J. Kulik, Y. Wei, T. Eschrich, J. Curless, B. Craigo, S. Smith, *Proc. Mat. Res. Soc. Sypm. Proc.* **786**, 312 (2004).
75. C.W. Wilmsen, *Physics and Chemistry of III–V Compound Semiconductor Interfaces* (Plenum Press, New York, 1985).
76. M. Passlack, M. Hong, J.P. Mannaerts, *Appl. Phys. Lett.* **68**, 1099 (1996).
77. M. Hale, S.I. Yi, J.Z. Sexton, A.C. Kummel, *J. Chem. Phys.* **119**, 1062 (2003).
78. L.J. Brillson, *Surf. Sci. Rep.* **2**, 123 (1982).
79. L. Kleinman, *Phys. Rev. B* **24**, 7412 (1981).

80. J.W. Keister, J.E. Rowe, J.J. Kolodziej, H. Niimi, T.E. Madey, G. Lucovsky, *J. Vac. Sci. Technol. B* **17**, 1831 (1999).
81. A.A. Demkov, R. Liu, X. Zhang, H. Loechelt, *J. Vac. Sci. Technol. B* **18**, 2388 (2000).
82. W. Frensley, H. Kroemer, *J. Vac. Sci. Technol.* **13**, 2599 (1976).
83. W. Harrison, *J. Vac. Sci. Technol.* **14**, 1016 (1977).
84. G.A. Baraff, J.A. Appelbaum, D.R. Hamann, *Phys. Rev. Lett.* **38**, 237 (1977).
85. W.E. Pickett, S.G. Louie, M.L. Cohen, *Phys. Rev. B* **17**, 815 (1978).
86. C.G. Van de Walle, R.M. Martin, *Phys. Rev. B* **34**, 5621 (1986); C.G. Van de Walle, R.M. Martin, *Phys. Rev. B* **35**, 8154 (1987).
87. J.Y. Tsao, *Materials Fundamentals of Molecular Beam Epitaxy* (Academic Press, San Diego, CA, 1993).
88. A.A. Demkov, *Phys. Stat. Sol.* (*b*) **226**, 57 (2001).
89. X. Zhang, A.A. Demkov, *J. Vac. Sci. Technol. B* **20**, 1664 (2002).
90. X. Zhang, A.A. Demkov, unpublished.
91. A.A. Demkov, H. Li, D. Müller, X. Hu, Y. Wei, J. Kulik, unpublished.

Chapter 10

FUNCTIONAL STRUCTURES

MATT COPEL

IBM Research Division, T. J. Watson Research Center
P.O. Box 218, Yorktown Heights, NY 10598, USA

1. INTRODUCTION

In this chapter we will discuss contributions that structural characterization techniques can make to selecting alternative dielectrics. Since there is a wide variety of intriguing materials available, with an equally fascinating array of structural properties, we must make some effort to narrow the discussion. So, let us consider what structures will lead to a functional gate dielectric, and use this "wish list" to gain insight on dielectric performance. First, a film needs to be both continuous and insulating if it is going to serve as a gate insulator. So we can start out our list of properties with smooth morphology and a fully oxidized composition, both of which can be measured quite accurately with structural probes. The dielectric needs to have a high permittivity, so that interfacial SiO_2 must be minimized and incorporated silicon must be strictly controlled. Next, a dielectric needs to be insulating not only in the laboratory, but in an actual microelectronic device, which implies that the dielectric must be able to withstand a fair degree of thermal processing without reacting with the silicon substrate and converting into a silicide or otherwise degraded. The extent of the thermal processing for a polysilicon-gated device is usually defined by dopant activation anneals, which can range from 950–1050°C for short times. If a metal is used as a gate material, lower thermal budgets can be expected, but the conventional CMOS process flow must be substantially modified. Oxidative anneals also occur in CMOS processing, and can lead to dielectric degradation. The anticipated difficulty here is that parasitic growth of SiO_2 at the silicon/dielectric interface will lead to a reduction in capacitance. Furthermore, it may be difficult to quantify the degree of oxidation that occurs during CMOS processing. Finally, a dielectric must meet stringent electrical criteria, such as low charge content and low trapping to ensure mobility and threshold stability. Although it is not well understood how the electrical criteria relate to structural properties, they pose some of the most difficult problems in fabricating practical dielectrics.

Much of the data presented below were taken using medium energy ion scattering (MEIS). This is a variant of Rutherford backscattering (RBS) that is very useful for

A.A. Demkov and A. Navrotsky (eds.), Materials Fundamentals of Gate Dielectrics, 349–365.

examining ultra-thin films. As in RBS, we use an energetic beam of light ions such as H^+ or He^+. A medium energy regime is used (100–200 keV) for two reasons. First, the energy is low enough to enable the use of electrostatic energy analysis of backscattered ions. Second, the stopping power, which is the rate at which ions lose energy in the sample, is at a maximum for medium energies. These two factors combine to give sub-nanometer depth resolution. The energy regime is sufficiently high to simplify the ion–solid interaction, so that we can understand the experiment using a screened Rutherford cross-section without complex neutralization effects. (For a more detailed explanation, see (1, 2).)

Many other complementary techniques have had an important role in the topics discussed below. Several results will be illustrated with transmission electron microscopy (TEM), and examples using X-ray scattering and SIMS will be included. Many other techniques have also contributed to our knowledge of alternative gate dielectrics. The sampling of experimental techniques was guided by the familiarity to the author, rather than any judgment of utility.

Now let us turn to the materials that are under consideration. We can group the relevant metal oxides based on their propensity to form silicates; that is, whether the materials readily combine with SiO_2 to form a compound of the form $M_xSi_yO_z$. Lanthana and yttria fall into this category, while zirconia and hafnia tend to remain in a discrete layer that is distinct from the SiO_2. Of course, Zr and Hf silicates do occur, and can be made by codeposition or possibly by interfacial mixing such as observed in some circumstances. Codeposited silicates are, in fact, strong candidates for gate dielectrics (3, 4). But the behavior of deposited HfO_2 and ZrO_2 tends to favor unmixed phases rather than silicate formation, indicating a fundamental difference in the materials properties. Aluminum oxide has also received extensive scrutiny, since it is a wide band-gap insulator. Although the dielectric constant for Al_2O_3 is only 9, there are several favorable points that bring it into consideration. Foremost is the acceptability of aluminum to semiconductor fabrication labs; there is a long history of aluminum interconnect technology, along with extensive research devoted to silicon-on-sapphire. Also, aluminum is chemically "well behaved", forming a highly robust oxide. Below, we will concentrate on these five species, although numerous other candidates, such as $SrTiO_3$, Gd_2O_3 and TiO_2 have been scrutinized. Although this chapter concerns structure rather then electrical characteristics, it is worth noting that the electron–phonon coupling has been predicted to limit mobility in devices with highly polarizable dielectrics (5). From this perspective, silicates and Al_2O_3 are preferred materials.

2. NUCLEATION

Since the ultimate goal is manufacturing of devices, it is not surprising that at an early stage investigators focused on deposition techniques that could be easily implemented conformally on a wafer scale. Thus, chemical vapor deposition (CVD) of metal oxides is a preferred method. A variant of CVD, atomic layer deposition (ALD),

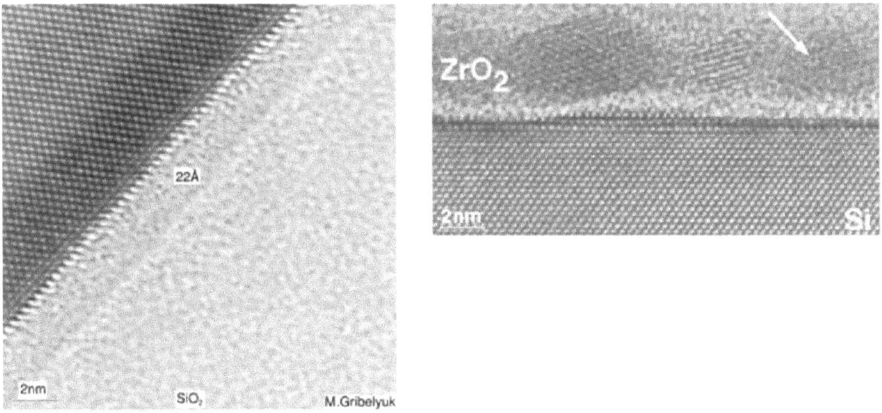

Fig. 1. Transmission electron micrographs of (a) Al_2O_3 and (b) ZrO_2 films grown by atomic layer deposition of HF etched Si(001) substrates (from ref. (7)). While a flat, homogeneous film is observed for Al_2O_3, island growth is observed for ZrO_2 (from ref. (9)).

uses alternating pulses of metal precursor and water. (For a review, see ref. (6).) ALD has the advantage of highly conformal deposition over large wafer areas. One of the disadvantages is that reaction by-products may be incorporated, and it is not clear whether this will limit the use of ALD compared to CVD. A second disadvantage is that ALD films do not always nucleate evenly. The problem is especially acute on hydrogen terminated Si(001). (For CVD, this has not been fully explored yet.) As an example, we will consider nucleation of Al_2O_3 and ZrO_2 on oxidized and HF etched Si(001). In the case of Al_2O_3, which is grown from trimethylaluminum (TMA) and water, nucleation does not appear to pose a significant problem; continuous films as thin as 20–30 Å can be deposited on either oxidized or hydrogen terminated surfaces (Fig. 1) (7). Furthermore, the film thickness does not depend drastically on the substrate preparation. Initial work on these films showed that the interfacial oxide was no more than about 2 Å thick. More recent studies using infrared spectroscopy have shown that some interfacial Si–O bonding does take place during nucleation on hydrogen terminated surfaces (8).

On the other hand, much different results are found for growth of HfO_2 and ZrO_2, where nucleation is inhibited on HF treated substrates (9, 10). A thermally oxidized sample, after ALD growth from $HfCl_4$ and H_2O, exhibits a nearly ideal spectrum with a trapezoidal Hf peak and Si displaced to lower energies, indicating formation of a continuous HfO_2 layer (Fig. 2). In contrast, the same growth sequence on HF etched Si(001) gives rise to much narrower Hf peak with a trailing edge, showing that less Hf has been deposited, with poor morphology. The presence of surface silicon is a certain sign that the film is discontinuous. Quite possibly, the difference in nucleation properties is due to the chlorinated precursor, which may not react with the H–Si bonds of the passivated substrate as readily as with the H–O bonds encountered during normal ALD. Even for oxidized surfaces, there is apparently an incubation

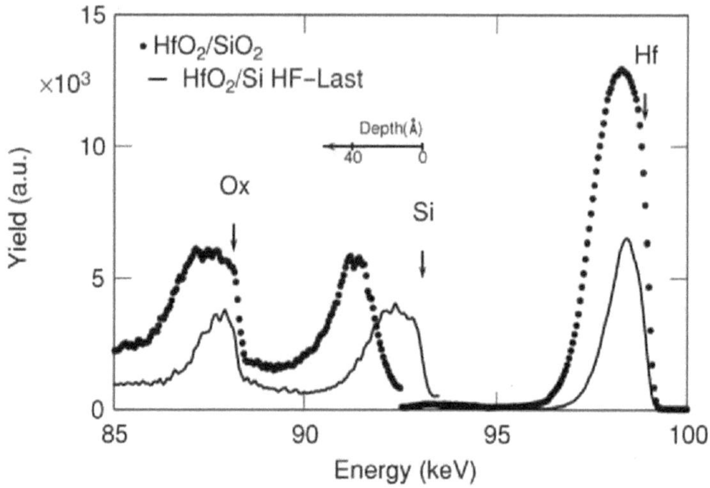

Fig. 2. MEIS spectra showing the effects of substrates preparation on nucleation of atomic layer deposition of HfO_2. When grown on a thermally oxidized substrate, a uniform film nucleates. But growth on hydrogen terminated Si(001) gives islanded growth. The poor morphology can be seen by the shape of the Hf peak, which is smaller and does not have the ideal, trapezoidal lineshape. Also, there is surface Si visible to the ion beam, due to the HfO_2-free regions between islands (from ref. (10)).

time for ALD growth, which indicates uneven nucleation. Investigators have looked at alternative buffer layers such as chemical oxides, which contain numerous surface OH groups, and shown a decreased incubation period (11). Nucleation issues, along with other considerations such as purity and vapor pressure, have contributed to interest in alternative CVD processes, for example growth from Zr *t*-butoxide (12), or Hf tetra-tertbutoxide (13).

3. SILICIDATION

After deposition, a dielectric must withstand the subsequent thermal processing that is required for building a useful device. Thermal cycling may be a blessing in disguise, since post-annealing can lead to performance improvement through densification, impurity desorption, pinhole oxidation or other mechanisms. Silicidation is a major pitfall that must be avoided, since it will render the film conductive. Stability was a high priority early in the selection of materials, and the candidate materials were drawn from a screened list (14). The screening was done on the basis of thermodynamic equilibrium with Si for a limited set of reactions involving oxygen transfer from the metal oxide to form SiO_2. The "approved" oxides are intrinsically stable, and can withstand substantial thermal processing without decomposing *when isolated*. But when there is contact with a silicon substrate, silicidation reactions are

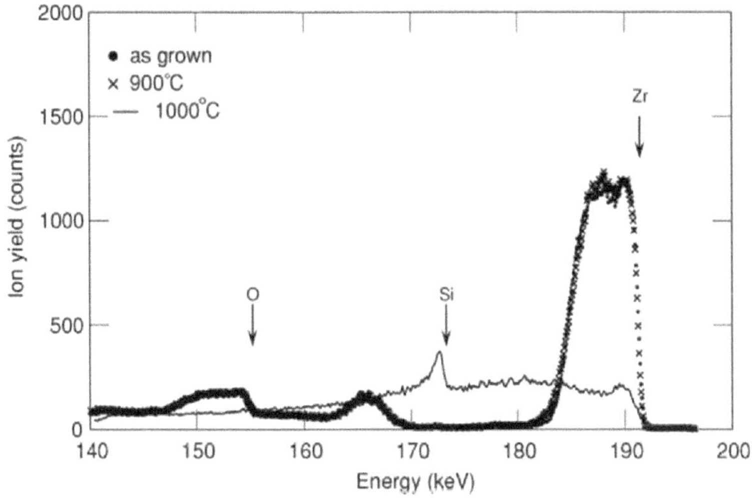

Fig. 3. Effects of vacuum annealing on ZrO_2/SiO_2 films. MEIS spectra show little perceptible change after a 900°C 120 s anneal. But after flashing to 1000°C for 30 s, silicide islands form. As a result, the Zr depth distribution broadens, and the oxygen peak decreases (from ref. (9)).

found. Evidently, substrate/dielectric interactions play a major role in determining film stability.

Two modes of film decomposition have been observed for metal oxides and silicates. The most commonly observed phenomena is the formation of silicides, which generally appear as inhomogeneous islands surrounded by voids. A rarer phenomena is seen for Al_2O_3, where the reaction product desorbs and voids are formed. ZrO_2 silicidation has been studied by numerous groups, and occurs in the range of 900–1000°C. A series of MEIS spectra taken after annealing at increasing temperatures serves as an illustration (Fig. 3). After a 900°C anneal, a well-formed Zr peak is observed, with a narrow trailing edge indicating little intermixing. But after a 1000°C flash, the Zr peak has broadened, and oxygen is no longer present. The film has reacted with the substrate, and the Zr has formed silicide islands, contributing to the width of the peak.

The mechanism for silicidation is still an area of active research, and we may hope that greater understanding will lead to strategies for greater stability. Since the instability is observed in films deposited on silicon, it is reasonable to infer that elemental silicon plays a role in the reaction. One possible reaction sequence involves the creation of volatile SiO, and can be written in unbalanced form as

$$\mathrm{Si} + \mathrm{M}_x\mathrm{O}_y \rightarrow \mathrm{MSi}_n + \mathrm{SiO}\uparrow. \tag{1}$$

This is quite similar to the reduction of SiO_2 by Si, which is believed to take place by the reaction (15, 16)

$$\mathrm{Si} + \mathrm{SiO}_2 \rightarrow 2\mathrm{SiO}\uparrow. \tag{2}$$

Fig. 4. Atomic force micrograph of a yttrium oxide sample after protracted annealing in ultra-high vacuum. The rectangular objects are silicide islands. Material from the surrounding dark regions have been incorporated in the islands. The field of view is 10μ.

In this picture, elemental silicon must be supplied by the substrate for the silicidation reaction to proceed. This could take place either by decomposition of the underlying buffer layer prior to silicidation, or Si outdiffusion through the barrier layer. In either case, modification of the barrier layer could improve robustness against silicidation. Several investigators have observed that silicidation takes place in an inhomogeneous fashion (17, 18). Figure 4 is an atomic force micrograph confirming the observations of ref. (18), but for yttria rather than hafnia. Silicidation is highly localized, causing islands that draw material from nearby areas. Large flat regions of intact metal oxide are co-existent with rectangular silicide islands. The localization could be caused by defects in the buffer layer, allowing contact between the metal oxide and the silicon. The defects may develop during heating, since it is known that SiO_2 desorption is initiated by void formation (16).

An elegant experiment supporting SiO desorption from metal oxide stacks has been described (19). A collector sample was placed in close proximity to a $ZrO_2/SiO_2/Si(001)$ wafer, which was then heated to form a silicide. Afterwards, the collector sample was examined by MEIS, and the desorption product was analyzed. The result was a film of SiO_x, indicating that the volatile product was SiO, rather than metal species.

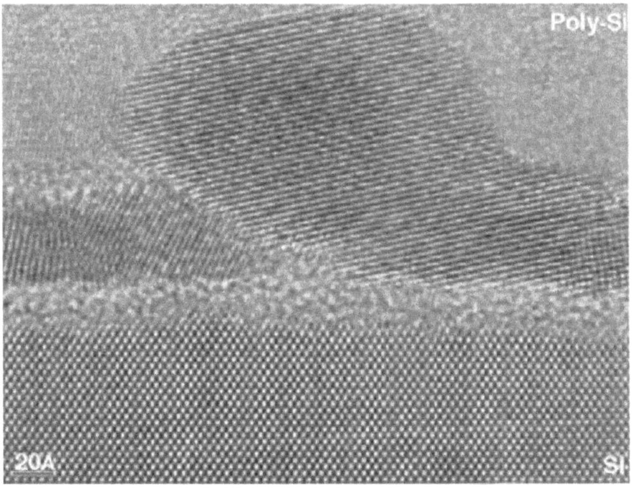

Fig. 5. Transmission electron micrograph of a polysilicon/ZrO_2/SiO_2/Si(001) stack after annealing to 1000°C. A silicide island is centered in the figure (from ref. (17)).

Extensive observations of silicidation vs. oxygen pressure have been used to define the process window for annealing of ZrO_2 and La_2O_3 (20). It was found that at high O_2 pressures where a silicon substrate tends to oxidize, interfacial oxide grows and silicidation is not encountered. At intermediate pressures, neither silicidation nor interface growth occurs. Finally, at low pressures, silicidation occurs. The criterion for metal oxide stability turns out to be quite similar to the criterion for SiO_2 stability on a silicon substrate. As mentioned above, SiO_2/Si(001) decomposition occurs by SiO evolution (15, 16). The pressure temperature dependence has been studied in detail (21). The transition from SiO_2 growth to SiO desorption occurs when the vapor pressure of SiO exceeds the oxidation rate. Thus, for high temperatures (or low oxygen pressures), the rate of oxygen loss exceeds the rate of oxidation and the substrate is etched. For low temperatures (or high oxygen pressures), SiO_2 accumulates more rapidly than SiO desorbs, and oxide accumulates.

Al_2O_3 decomposition is an exceptional case, since it does not form a silicide. Instead, the Al is volatilized, either in the form of Al_2O (22), or in its metallic form, leaving pinholes of roughened silicon (23). After protracted annealing a clean, but scarred, Si(001) surface is left. This has not been observed for other metal oxides, which leave behind silicide islands after decomposition. Of the metals considered here, aluminum is unique in not forming a silicide phase, so it is not surprising that the film decomposition follows a different pathway.

A further complication arises from the use of polysilicon as a gate electrode. Now we have the possibility of metal–oxide reaction with either the substrate or the gate electrode, which does not have a barrier against silicidation (Fig. 5). Indeed, polysilicon–metal oxide interactions have plagued attempts to create ZrO_2 CMOS (17). Efforts to minimize the problem have relied on nitridation schemes.

Fortunately, it appears that HfO_2 based dielectrics offer sufficient stability for fabrication of polysilicon gates (24).

Investigations of silicidation of polysilicon-contacted devices have lead to several alternative proposed mechanisms for decomposition. McIntyre and coworkers have pointed out that the polysilicon deposition temperature plays a key role (25). They have suggested that oxygen vacancies created in ZrO_2 during polysilicon deposition could react with the SiO_2, leading to decomposition. Also, investigators have noted that plasma-deposited polysilicon seems to have a reduced tendency towards silicidation (26). From this evidence, Callegari et al. proposed that the hydrogen evolved from silane pyrolysis leads to ZrO_2 reduction. Another suggested possibility is that SiO may form from decomposition of the buffer layer and interact with the metal oxide to decompose it (27). There is a diverse assortment of mechanisms that have been proposed to explain this important aspect of metal oxide–silicon interactions, and it is likely that this will be a fruitful subject for future work.

4. OXIDATION

If we are to succeed in the goal of producing gate dielectrics with a low equivalent oxide thickness, we must avoid the growth of parasitic SiO_2 and silicates. It turns out that great care must be taken in processing films to avoid interfacial oxide growth. This is not surprising, when we consider that many of materials involved, such as ZrO_2 and Y_2O_3, find common use as catalysts. Because metal oxides are chemically active materials, they may play a non-passive role in SiO_2 growth. Indeed, we should expect that these materials will promote reactions with oxygen containing species.

First, let's look at what happens when we take a $ZrO_2/SiO_2/Si(001)$ film and deliberately expose it to a heavily oxidizing environment. If we take a 38 ± 3 Å layer of ZrO_2 on a 13 ± 3 Å layer of SiO_2 and expose it to an O_2 pressure of 0.1 Torr at 930°C for 2 min, we see several changes in the MEIS spectrum (Fig. 6) (9). The oxygen peak and Si interface peak broaden. Also, a peak due to surface Si appears. Meanwhile, the Zr peak remains nearly unchanged. These results can be explained by the growth of the interfacial SiO_2 layer from 13 to 25 Å, along with the segregation of 3 Å of SiO_2 at the surface. Note that the ZrO_2 has not intermixed with SiO_2; if there was intermixing, the Zr peak would have broadened. Indeed, the presence of a surface spike of SiO_2 confirms that the silicate phase does not readily nucleate. The surface SiO_2 spike is a major inconvenience to analytical techniques such as XPS and Auger spectroscopy, which can easily confuse the segregated SiO_2 with intermixing (silicate formation). Watanabe has taken advantage of surface SiO_2 segregation to create ultra-thin $SiO_2/ZrO_2/SiO_2$ structures, providing buffer layers for both the substrate and polysilicon (28). A detailed analysis of the temperature dependence of interfacial SiO_2 growth has been reported (29).

It is particularly important to consider the pressure dependence of interfacial oxide growth, since this will determine what ambient can be tolerated during post-processing. This question has been studied for Al_2O_3 deposited on HF etched

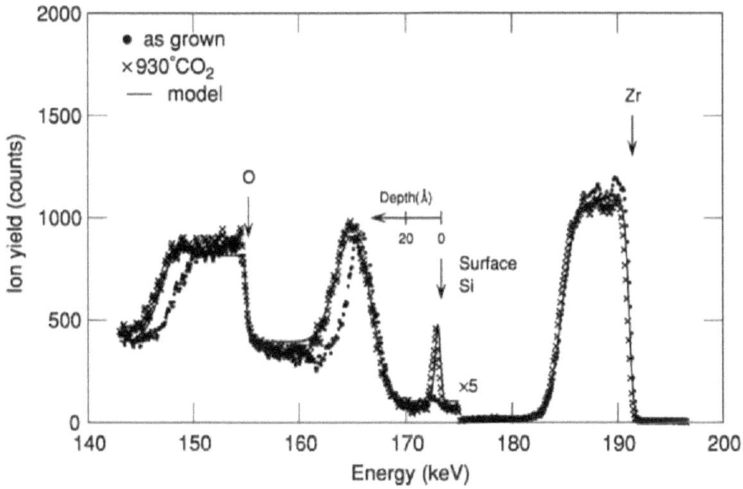

Fig. 6. Oxidation of ZrO_2/SiO_2. After aggressive oxidation, the MEIS shows very little change in the Zr backscatter peak. However, the Si peak has greatly broadened, as well as gaining a surface component. The results indicate growth of SiO_2 at both the interface and the surface (from ref. (9)).

Si(001) (23). The oxidation conditions were chosen to approximate the environment that might be experienced during passivation: 30 min anneals at 600°C with varying O_2 pressures (Fig. 7). For pressures greater than 10^{-6} Torr, 4–6 Å of interfacial SiO_2 grows. This would certainly be a major perturbation on a

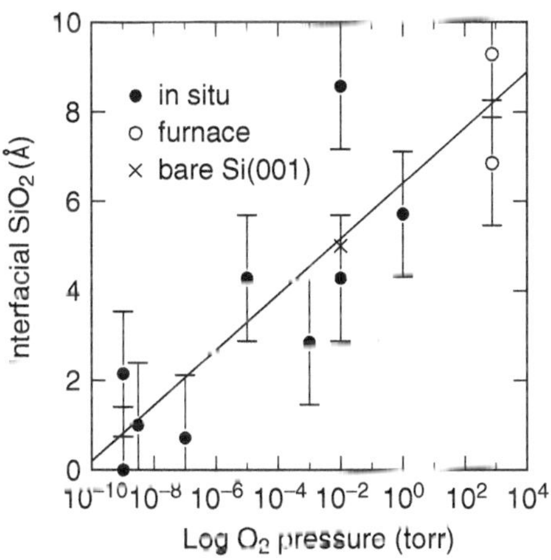

Fig. 7. Pressure dependence of interfacial oxide growth for Al_2O_3/Si(001) films. Substantial quantities of interfacial material are observed after oxidation at low pressures. Large decreases in capacitance can be expected for anneals in uncontrolled ambients for metal oxide/silicon interfaces (from ref. (23)).

sub-nanometer gate dielectric. Furthermore, to achieve partial pressures of oxygen less than 10^{-6} Torr in an atmospheric process, requires ppb purity—a daunting requirement for processing equipment. The pressure threshold for substrate oxidation has also been observed by reflection electron microscopy experiments that have looked at the interfacial step structure (30, 31). The step structure changes at the same pressure that the MEIS experiments see interfacial SiO_2 growth, confirming that very low oxygen pressures can create parasitic oxide.

This section began with an ominous comment about the catalytic properties of metal oxides. With Al_2O_3, you can see this effect in action. Tucked into Fig. 7 is a data point corresponding to oxidation of a bare Si(001) sample in 10^{-2} Torr of O_2, where we have found 5 Å of interfacial oxide. This is roughly the same amount as grows underneath the Al_2O_3 after the same exposure. So the Al_2O_3 layer is completely ineffective at chemically passivating the interface! Normally, we would expect an overlayer to reduce Si oxidation by diffusion-limiting the supply of oxygen. In the case of Al_2O_3, the diffusion limitation must be compensated by its ability to promote the oxidation reaction, probably by converting O_2 to a more reactive species. Since the Al_2O_3 shows no signs of crystallinity, it must be highly defective and it is unlikely that bulk diffusion constants are applicable. Of the various metal oxides discussed in this chapter, it is likely that Al_2O_3 is the least chemically active, hence the least aggressive promoter of silicon oxidation.

So far we have dealt with deliberate oxidation, and used this as a yardstick to judge how much inadvertent substrate oxidation can take place. Now, let's look at advertant oxidation, which can take place from numerous sources. In addition to trace gases that may interact with the sample during post-processing, there are sources that can be difficult to anticipate. Several instances have been reported where the metal oxide incorporates excess oxygen. The excess oxygen can react with the substrate upon heating. La_2O_3 and Y_2O_3 are particularly hygroscopic. Infrared absorption shows that atmospheric exposure causes rapid uptake of OH in Y_2O_3, and reduced uptake for yttrium silicate (32). Even without atmospheric exposure, La_2O_3 films may contain an overabundance of oxygen (33). In a study of Y_2O_3, it was found that a Si capping layer can eliminate excess oxygen, giving stoichiometric material (34). For both yttrium and lanthanum oxides, when there was the extra oxygen, it was observed to combine with the substrate to form interfacial layers. A more insidious source of oxygen is the growth environment itself. Guha et al. showed that for epitaxial $(La_{1-x}Y_x)_2O_3$, the growth of interfacial layers can be reduced simply by eliminating the oxygen flux immediately after growth (35). The lesson from these studies is that careful processing of metal oxide films can help to reduce, if not eliminate, parasitic Si oxidation.

5. INTERACTIONS

In this section we will discuss non-siliciding materials reactions that can take place between a metal oxide and the supporting layers. There are two types of reactions that we will be concerned with. First, the metal oxide can combine with an SiO_2 buffer layer to form a silicate. This may alter both the capacitance of a device and

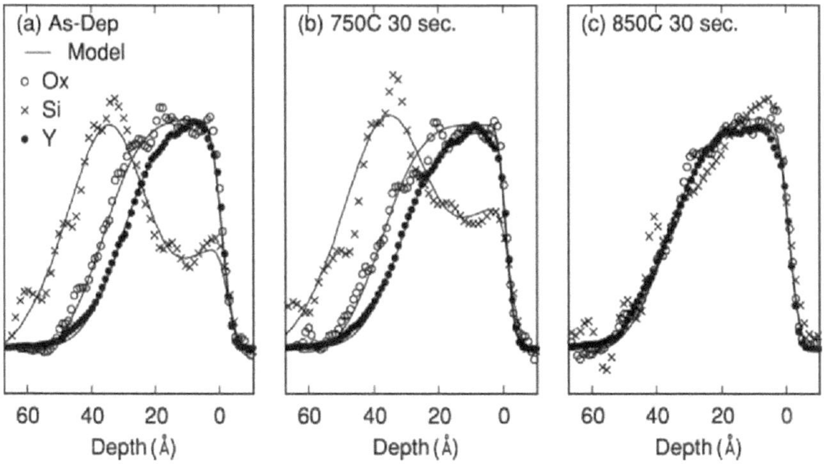

Fig. 8. Depth distributions for yttria/silicon oxynitride measured with MEIS. (a) Initially, the oxygen extends deeper than the yttrium, due to the interfacial layer. (b) After annealing to 750°C, the buffer layer begins to intermix. (c) At 850°C, the layers are fully intermixed and the constituents have identical depth distributions. A component of the Si peak was subtracted to correct for the substrate contribution (from ref. (37)).

the electrical quality of the interface. The former effect could be beneficial, since a silicate may have a higher permittivity than an metal oxide/SiO_2 stack. The latter effect is of greater concern, since we would anticipate that a silicate/Si interface is not as well behaved as a SiO_2/Si interface. A second type of material reaction is metal ion indiffusion. This could possibly disrupt the channel of an FET by Coulombic scattering of electrons from metal ions. In addition, channel doping may alter the threshold voltage of a device, which would be a serious drawback.

In some cases, such as La and Y based dielectrics, silicate formation is easily identified (20, 33, 36). Studies of Y_2O_3 deposited on thin silicon oxynitride layers have shown that high temperature annealing causes intermixing of the two components (37). In fact, the reaction is exothermic for Y_2SiO_5, releasing 53 ± 5 kJ/mol (38). Nanometer scale diffusion is observed after an 850°C RTA, as shown in Fig. 8. In this example, 23 Å of Y_2O_3 was deposited on a 11 Å SiO_xN_y buffer layer. As grown, the oxygen depth distribution goes deeper than the yttrium, because of the oxygen in the buffer layer. After a rapid thermal anneal in UHV to 750°C, there is incomplete intermixing. The silicate fully consumes the buffer layer during an 850°C rapid thermal anneal, indicated by the identical depth distributions of oxygen and yttrium. After the high temperature RTA, the silicon concentration has increased at the surface, confirming that the buffer layer has intermixed with the yttria. The thermal budget for intermixing is well within the limits encountered during CMOS fabrication, so we would expect that silicate formation is unavoidable unless we use a nitridized buffer layer (39).

Silicate formation is much less pronounced for HfO_2 and ZrO_2. We have already shown that oxidation of ZrO_2/SiO_2 structures causes very little change in the Zr depth

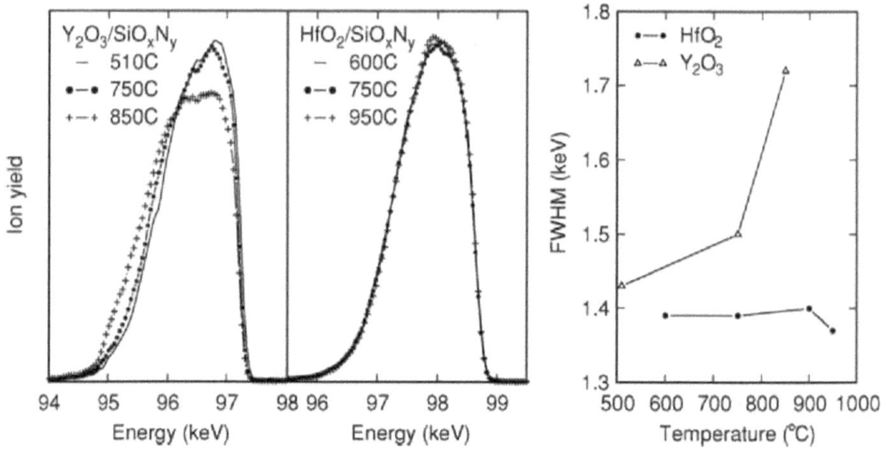

Fig. 9. Effect of annealing yttria and hafnia on silicon oxynitride buffer layers. (a) The yttrium is diluted by intermixing with the buffer layer, broadening the backscatter peak. (b) The hafnium does not intermix, resulting in a highly stable backscatter peak. (c) The peak widths after deconvoluting the detector resolution and energy straggling. The yttrium peak width increases, while the hafnium peak shows only minor changes.

distribution. HfO_2 also shows remarkable stability during annealing, especially when compared to yttrium. For the yttria sample discussed above, silicate formation has the effect of diluting the yttrium concentration with Si, thereby broadening the yttrium backscatter peak (Fig. 9). No such broadening is observed for the Hf peak in a 25 Å HfO_2/10 Å SiO_xN_y sample, even after annealing to temperatures as high as 950°C. So if there is any silicate formation, it must take place during sample growth, rather than post-processing.

Assessing the exact composition of the interface region is a challenging problem, not just because of the demands on our analytical capabilities, but because the degree of interfacial mixing may depend on the deposition technique. Results for energetically deposited films, where there may be significant intermixing, can be expected to differ from CVD films. Analysis of HfO_2 grown by MOCVD has found an interfacial layer of about 9 Å of $HfSiO_4$, decreasing to 5 Å after annealing (13). This may be due to interfacial phase separation, although the simultaneous effects of densification and impurity desorption complicate the analysis. For samples that are made by deposition of Hf or Zr followed by oxidation, a silicide can be formed. Oxidation of a metal silicide should result in a silicate layer, providing phase separation can be avoided (40, 41).

While it is unclear whether silicate formation is a good or a bad thing, it is clear that metal ion penetration into the Si substrate cannot be good for a device. Since the portion of the channel directly underneath the gate dielectric is the main conduit for the electrons, an impurity concentration only needs to penetrate a few nanometers to cause significant charge scattering, reducing mobility and adversely affecting device performance (42). Detecting metal penetration is not a simple task, since we need to

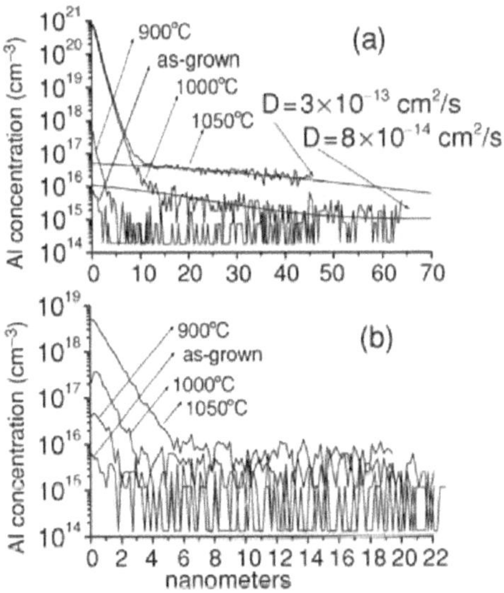

Fig. 10. Al diffusion profiles measured by SIMS for Al_2O_3 deposited on (a) Si(001) and (b) 1 nm silicon oxynitride on Si(001). Rapid thermal anneals were done for 30 s at the temperatures indicated (from ref. (42)).

detect a very low concentration of atoms (10^{16}–10^{19} cm^{-2}) in a region very close to a high concentration, the metal oxide itself. One approach is to chemically etch away the metal oxide and examine the stripped wafer with SIMS. Of course, any residual surface metal can easily be confused indiffused material. A further complication arises from the notoriously effective etch resistance of metal oxides, particularly ones that have been hardened by high temperature treatment. Nonetheless, with careful experimental work one can measure a reliable diffusion profile. For Zr silicate, indiffusion extending 10–20 nm is observed for samples that have been annealed above 1000°C (43, 44). On the other hand, no Hf penetration can be detected from Hf silicate under the same anneal conditions (44, 45).

Studies of diffusion from Al_2O_3 into Si(001) show a more complex profile (Fig. 10) (42). A surface spike is followed by a long tail extending to great depths. The gradient in the long tail matches the diffusivity of Al in Si, and the surface spike corresponds to a supersaturated region. When Al_2O_3 is deposited on a 1 nm thick silicon oxynitride layer, the indiffusion is greatly reduced. Although the intermixing is still enough to cause concern, it demonstrates that nitridized barriers are helpful, and that the SIMS results are not an artifact caused by etch resistance of the annealed metal oxide. In terms of device performance, Guha reported a twofold drop in carrier mobility in annealed Al_2O_3/Si devices due charge scattering from the dissolved Al, which is a well-known dopant. In this application, Al has the distinct disadvantage of being trivalent, unlike Hf and Zr, which are isoelectronic to Si.

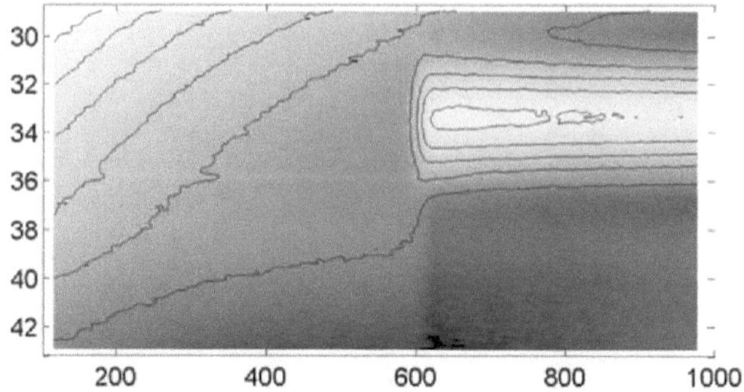

Fig. 11. X-ray diffraction intensity for two-theta scattering from a 5 nm thick HfO_2 film grown by atomic layer deposition and annealed in forming gas. The increase in diffracted intensity at roughly 600°C is due to crystallization of the HfO_2 (from ref. (10)).

6. MICROSTRUCTURE

There are three distinct microstructures that are found in the dielectrics commonly studied; the films may be amorphous, polycrystalline, or epitaxial. Most of the Al_2O_3 samples and the yttrium silicate discussed above were amorphous. ZrO_2 and HfO_2 films are generally polycrystalline after annealing. We have not discussed epitaxial films, but a great deal of effort has gone into epitaxial perovskites such as $SrTiO_3$ and $Ba_xSr_{(1-x)}TiO_3$ (46–48). Epitaxial growth of Y_2O_3 (49) and $(La_{1-x}Y_x)_2O_3$ (50) have also been reported. Since thermal SiO_2 is amorphous, we have at least an existing case of a successful amorphous dielectric. However, it is not clear that a dielectric *must* be amorphous to be successful. Work on HfO_2 has shown that the leakage current does not increase with crystallization (51, 52), bringing question to the notion that grain boundaries are necessarily vulnerable to leakage pathways.

The microstructure of HfO_2 has been studied with X-ray diffraction (XRD), which can probe the sample during heating (10). The intensity of the two-theta diffraction as a function of sample temperature for a ramp rate of 3°C/s shows the microstructure development (Fig. 11). Initially, the film is largely amorphous and no diffraction peak is seen. At about 600°C, the HfO_2 crystallizes, most likely into the monoclinic phase. Not only does XRD graphically illustrate microstructual evolution, it can be used to study the influence of parameters on crystallization. For example, crystallization temperature depends strongly on film thickness (10).

The predilection of materials scientists is to favor amorphous or epitaxial films, since they ought to have reduced susceptibility to indiffusion, as well as the possibility of a defect-free interface. But the optimal microstructure is still far from certain, and is likely to remain uncertain until experiments can be designed that separate the influence of microstructure from the many processing variables that determine microstructure.

7. FUTURE DIRECTIONS

Given what we have discussed in this chapter, what areas are likely to be explored in the future? Some subjects have already been alluded to, such as the optimization of barrier layers both above and beneath the dielectric to eliminate unwanted materials reactions. In addition to nitridized barriers, investigators have already begun to look at composite structures, such as $Al_2O_3/HfO_2/Al_2O_3$ films (53–55). One can also grow alloys of Al_2O_3 and HfO_2 (56). The goal is to combine the nucleation properties of Al_2O_3 with the high permittivity of HfO_2 in an amorphous compound. The chief danger is that the conduction band of HfO_2 may be lower than the conduction band of Al_2O_3, thus creating a high density of electron traps.

In the introduction, we mentioned that calculations of electron–phonon coupling predict that silicates should have a higher mobility than pure metal oxides (5). Hf and Zr silicates are the most successful of this class of materials. But their use gives rise to some interesting structural questions. Because the metal ions are diluted by SiO_2, a lower dielectric constant is expected. Consequently, scaling silicate-based stacks to competitive thicknesses is apt to require careful composition and interface control. Furthermore, the interface thickness may have a crucial role in device mobility. In addition, we know that annealing the silicate separates into metal-rich and metal-poor phases (57). Understanding the science underlying the phase separation and determining the effect of phase separation on electrical properties are subjects of ongoing research (58, 59).

Another subject that is drawing increasing attention is the use of metal gates to replace polysilicon. Although this chapter limited itself to discussion of polysilicon contacts, interactions between gate metals and metal oxides will draw increasing attention as a subject for structural studies. Difficulties in achieving proper threshold voltages for polysilicon devices have been attributed to polysilicon/dielectric interactions (60). This has spurred increased attention to metal gates as well innovative processing techniques.

If it is indeed possible to fabricate low defect metal oxide gate dielectrics on Si, perhaps the technology can be extended to semiconductors with higher mobility. Although Ge lacks a high quality oxide, one could conceivably make an FET with a Ge substrate and a deposited metal oxide gate dielectric. Ideally, the mobility degradation from the metal oxide would be more than compensated by the intrinsic high mobility of the Ge channel. Initial studies have already considered interface formation (61) and transistor performance (62) for metal oxide/Ge devices. But before this can become a viable technology, we need to establish what structures work and why.

Although most of the candidates for metal oxide dielectrics have been examined, there may still be a chance for the introduction of novel materials, providing that their benefits outweigh the advantage of the rapidly growing processing experience with more common materials. Finally, the search for benign, high purity deposition techniques may give improvements in performance that are much needed.

ACKNOWLEDGEMENTS

I thank E. Gusev, S. Guha and E. Cartier for helpful discussions as well as F. M. Ross and C. Cabral for careful reading of this manuscript.

REFERENCES

1. J.F. van der Veen, *Surf. Sci. Rep.* **5**, 199 (1985).
2. M. Copel, *IBM J. Res. Dev.* **44**, 571 (2000).
3. G.D. Wilk, R.M. Wallace, *Appl. Phys. Lett.* **74**, 2854 (1999).
4. G.D. Wilk, R.M. Wallace, J.M. Anthony, *J. Appl. Phys.* **87**, 484 (2000).
5. M.V. Fischetti, D.A. Neumayer, E.A. Cartier, *J. Appl. Phys.* **90**, 4587 (2001).
6. M. Leskala, M. Ritala, Atomic layer deposition, in: *Handbook of Thin Film Materials*, Vol. 1, ed. H.S. Nalwa (Academic, New York, 2002), Chapter 5, p. 103.
7. E.P. Gusev, M. Copel, E. Cartier, I.J.R. Baumvol, C. Krug, M.A. Gribelyuk, *Appl. Phys. Lett.* **78**, 176 (2000).
8. M.M. Frank, Y.J. Chabal, G.D. Wilk, *Appl. Phys. Lett.* **82**, 4758 (2003).
9. M. Copel, M. Gribelyuk, E. Gusev, *Appl. Phys. Lett.* **76**, 436 (2000).
10. E.P. Gusev, C. Cabral Jr., M. Copel, C. D'Emic, M. Gribelyuk, *Microelectr. Eng.* **69**, 145 (2003).
11. M.L. Green, M.-Y. Ho, B. Busch, G.D. Wilk, T. Sorsch, T. Conard, B. Brijs, W. Vander-vorst, P.I. Räisänen, D. Muller, M. Bude, J. Grazul, *J. Appl. Phys.* **92**, 7168 (2002).
12. J.P. Chang, Y.-S. Lin, *Appl. Phys. Lett.* **79**, 3666 (2001).
13. S. Sayan, S. Aravamudhan, B.W. Busch, W.H. Schulte, G.D. Wilk, T. Gustafsson, E. Garfunkel, *J. Vac. Sci. Technol. A.* **20**, 507 (2002).
14. K.J. Hubbard, D.G. Schlom, *J. Mater. Res.* **11**, 2757 (1996).
15. J.J. Lander, J. Morrison, *J. Appl. Phys.* **33**, 2089 (1962).
16. R.M. Tromp, G.W. Rubloff, P. Balk, F.K. LeGoues, E.J. van Loenen, *Phys. Rev. Lett.* **55**, 2332 (1985).
17. M. Gribelyuk, E.P. Gusev, M. Copel, D.A. Buchanan, *J. Appl. Phys.* **92**, 1232 (2002).
18. H. Watanabe, N. Ikarashi, *Appl. Phys. Lett.* **80**, 559 (2002).
19. S. Sayan, E. Garfunkel, T. Nishimura, W.H. Schulte, T. Gustafsson, and G.D. Wilk, Thermal decomposition behavior of the $HfO_2/SiO_2/Si$ system, *J. Appl. Phys.* **94**, 928 (2003).
20. J.P. Maria, D. Wicaksana, A. Kingon, B. Busch, H. Schulte, E. Garfunkel, T. Gustafsson, *J. Appl. Phys.* **90**, 3476 (2001).
21. G. Ghidini, F.W. Smith, *J. Electrochem. Soc.* **129**, 1300 (1982).
22. H.M. Manasevit, W.J. Simpson, *J. Appl. Phys.* **35**, 1349 (1964).
23. M. Copel, E. Cartier, E.P. Gusev, S. Guha, N. Bojarczuk, M. Poppeller, *Appl. Phys. Lett.* **78**, 2670 (2001).
24. E.P. Gusev, et al., *Tech. Digest Int. Electron Devices Meet.* p. 20 (2001).
25. C.M. Perkins, B.B. Triplett, P.C. McIntyre, K.C. Saraswat, E. Shero, *Appl. Phys. Lett.* **81**, 1417 (2002).
26. A. Callegari, E. Gusev, T. Zabel, D. Lacey, M. Gribelyuk, P. Jamison, *Appl. Phys. Lett.* **81**, 4157 (2002).
27. T.S. Jeon, J.M. White, D.L. Kwong, *Appl. Phys. Lett.* **78**, 368 (2001).
28. H. Watanabe, *Appl. Phys. Lett.* **81**, 4221 (2002).
29. B.W. Busch, W.H. Schulte, E. Garfunkel, T. Gustafsson, R.N. W. Qi, J. Lee, *Phys. Rev. B* **62**, R13290 (2000).
30. M. Kundu, N. Miyata, M. Ichikawa, *Appl. Phys. Lett.* **78**, 1517 (2001).
31. M. Kundu, M. Ichikawa, N. Miyata, *J. Appl. Phys.* **91**, 492 (2002).

32. D. Niu, R.W. Ashcraft, G.N. Parsons, *Appl. Phys. Lett.* **77**, 3575 (2002).
33. M. Copel, E. Cartier, F.M. Ross, *Appl. Phys. Lett.* **78**, 1607 (2001).
34. B.W. Busch, J. Kwo, M. Hong, J.P. Mannaerts, B.J. Sapjeta, W.H. Schulte, E. Garfunkel, T. Gustafsson, *Appl. Phys. Lett.* **79**, 2447 (2001).
35. V. Narayanan, S. Guha, M. Copel, N.A. Bojarczuk, P.L. Flaitz, M. Gribelyuk, *Appl. Phys. Lett.* **81**, 4183 (2002).
36. D. Niu, R.W. Ashcroft, Z. Chen, S. Stemmer, G.N. Parsons, *Appl. Phys. Lett.* **81**, 676 (2002).
37. M. Copel, E. Cartier, V. Narayanan, M.C. Reuter, S. Guha, N. Bojarczuk, *Appl. Phys. Lett.* **81**, 4227 (2002).
38. J.-J. Liang, A. Navrostky, T. Ludwig, H.J. Siefert, F. Aldinger, *J. Mater. Res.* **14**, 1181 (1999).
39. J.J. Chambers, G.N. Parsons, *Appl. Phys. Lett.* **77**, 2385 (2000).
40. Y. Hoshino, Y. Kido, K. Yamamoto, S. Hayashi, M. Niwa, *Appl. Phys. Lett.* **81**, 2650 (2002).
41. S. Ramanathan, P.C. McIntyre, *Appl. Phys. Lett.* **80**, 3793 (2002).
42. S. Guha, E.P. Gusev, H. Okorn-Schmidt, M. Copel, L.-Å. Ragnarsson, N.A. Bojarczuk, P. Ronsheim, *Appl. Phys. Lett.* **81**, 2956 (2002).
43. M. Quevedo-Lopez, M. El-Bouanani, S. Addepalli, J.L. Duggan, B.E. Gnade, R.M. Wallace, M.R. Visokay, M. Douglas, M.J. Bevan, L. Colombo, *Appl. Phys. Lett.* **79**, 2958 (2001).
44. M. Quevedo-Lopez, M. El-Bouanani, B.E. Gnade, R.M. Wallace, M.R. Visokay, M. Douglas, M.J. Bevan, L. Colombo, *J. Appt. Phys.* **92**, 3354 (2002).
45. M. Quevedo-Lopez, M. El-Bouanani, S. Addepalli, J.L. Duggan, B.E. Gnade, R.M. Wallace, M.R. Visokay, M. Douglas, M.J. Bevan, L. Colombo, *Appl. Phys. Lett.* **79**, 4192 (2001).
46. R.A. McKee, F.J. Walker, M.F. Chisholm, *Phys. Rev. Lett.* **81**, 3014 (1998).
47. K. Eisenbelser, J.M. Finder, Z. Yu, J. Ramdani, J.A. Curless, J.A. Hallmark, R. Droopad, W.J. Ooms, L. Salem, S. Bredshaw, C.D. Overgaard, *Appl. Phys. Lett.* **76**, 1324 (2000).
48. R.A. McKee, F.J. Walker, M.F. Chisholm, *Science* **293**, 468 (2001).
49. A. Dimoulas, G. Vellianitis, A. Travlos, V. Ioannou-Sougleridis, A.G. Nassiopoulou, *J. Appl. Phys.* **92**, 426 (2002).
50. S. Guha, N.A. Bojarczuk, V. Narayanan, *Appl. Phys. Lett.* **80**, 7066 (2002).
51. H. Kim, P.C. McIntyre, K.C. Saraswat, *Appl. Phys. Lett.* **82**, 106 (2003).
52. H. Kim, A. Marshall, P.C. McIntyre, K.C. Saraswat, *Appl. Phys. Lett.* **84**, (2004)
53. C. Zhao, O. Richard, E. Young, H. Bender, G. Roebben, S. Haukka, S.D. Gendt, M. Houssa, R. Carter, W. Tsai, O.V.D. Biest, M. Heyns, *J. Non-Cryst. Solids* **303**, 149 (2002).
54. C. Zhao, O. Richard, H. Bender, M. Caymax, S.D. Gendt, M. Heyns, E. Young, G. Roebben, O.V.D. Biest, S. Haukka, *Appl. Phys. Lett.* **80**, 2374 (2002).
55. H.S. Chang, S. Jeon, H. Hwang, D.W. Moon, *Appl. Phys. Lett.* **80**, 3387 (2002).
56. M.-Y. Ho, H. Gong, G.D. Wilk, B. Busch, M.L. Green, W.H. Lin, A. See, S.K. Lahiri, M.E. Loomans, P.I. Räsänen, T. Gustafsson, *Appl. Phys. Lett.* **81**, 4218 (2002).
57. D.A. Neumayer, E. Cartier, *J. Appl. Phys.* **90**, 1801 (2000)
58. H. Kim, P.C. McIntyre, *J. Appl. Phys.* **92**, 5094 (2002).
59. S. Stemmer, Z. Chen, C.G. Levi, P.S. Lysaght, B. Foran, J.A. Gisby, J.R. Taylor, *Jpn. J. Appl. Phys.* **42**, 3593 (2003).
60. C. Hobbs, L. Fonseca, V. Dhandapani, S. Samavedam, B. Taylor, J. Grant, L. Dip, D. Triyoso, R. Hegde, D. Gilmer, R. Garcia, D. Roan, L. Lovejoy, R. Rai, L. Hebert, H. Tseng, B. White, P. Tobin, *VLSI Technol. Digest* **9** (2003).
61. C.O. Chui, S. Ramanathan, B.B. Triplett, P.C. McIntyre, K.C. Saraswat, *Electron Device Lett.* **23**, 473–475 (2002).
62. A. Ritenour, S. Yu, M.L. Lee, Z. Lu, W. Bai, A. Pitera, E.A. Fitzgerald, D.L. Kwong, D.A. Antoniadis, *Tech. Digest Int. Electron Devices Meet.* 18.2.1 (2003).

Chapter 11

MECHANISTIC STUDIES OF DIELECTRIC GROWTH ON SILICON

MARTIN M. FRANK[1] AND YVES J. CHABAL[2]

[1]*IBM Thomas J. Watson Research Center, Yorktown Heights, NY 10598, USA*
[2]*Departments of Chemistry and Chemical Biology, Physics and Astronomy, Biomedical Engineering, and Laboratory for Surface Modification, Rutgers University, Piscataway, NJ 08854, USA*

1. INTRODUCTION

Semiconductor surface chemistry has long been recognized as central to the development and optimization of electronic devices. To move towards the required atomic level control of surface reactions, a wide variety of processing techniques have been used, either wet- or gas-phase chemical in nature. Wet chemistry is employed mostly to clean and passivate semiconductor wafers. Gas phase techniques such as chemical vapor deposition (CVD) or molecular beam epitaxy (MBE) play a prominent role in the deposition of films onto such surfaces. Aside from conventional CVD, a novel technique, atomic layer deposition (ALD, described in detail in Section 4), is being developed to grow ultra-thin high-permittivity (high-k) dielectrics on silicon and other semiconductor substrates. Such high-k dielectrics will likely replace the traditional silicon oxide or oxynitride gate insulator in metal-oxide field effect transistors (MOSFETs), to reduce the gate leakage current and hence power dissipation. Dielectrics under consideration include HfO_2, $Hf_xSi_yO_z$, Al_2O_3, and nitrides of these materials. For further information on the MOSFET gate stack and the need for high-k dielectrics, the reader is referred to other chapters in the present book and to a recent review article by Wilk et al. (1).

The characteristics and quality of a gate dielectric depend sensitively on the deposition conditions. For example, atomic details of the Si-dielectric interface determine the band offsets, as discussed in chapter 5 by Robertson and Peacock. In order to devise ways to deposit high-quality gate dielectrics, it is therefore critical to understand the underlying reaction mechanisms that govern surface or film formation. Much has been learned by examining as-grown devices and interfaces *ex situ*. Characterization methods frequently used include imaging techniques such as transmission electron microscopy (TEM), providing spatial resolution down to the atomic level,

A.A. Demkov and A. Navrotsky (eds.), Materials Fundamentals of Gate Dielectrics, 367–401.

and spectroscopic techniques using electrons, ions, or atoms, shedding light on structure and composition of the final surface or film. However, it has been difficult to relate these results unambiguously to processing parameters for lack of *in situ* observations of reaction chemistry under manufacturing conditions. In many cases, for example, surface defects can play an important role in initiating low probability reactions that severely impact the atomic or electronic structure of the final device.

The major stumbling block for the rapid development of advanced devices using new materials or processing conditions is the lack of *fundamental understanding of surface reactions*. While the reactivity on a surface often parallels that of solution or gas-phase molecular chemistry, there are many cases where perturbations, steric interactions, or electronic effects dramatically affect the reaction pathways and the kinetics of surface reactions. It can be misleading to develop growth methods based on gas phase behavior of surface analogs. For instance, a simple scheme that describes the reaction between silane and chlorine gas, such as $SiH_4 + 4Cl_2 \rightarrow SiCl_4 + 4HCl$, may not be a good model of what occurs on a Si surface when attempting to chlorinate it using Cl_2, because of different chemical bonding (Si is bonded to other Si atoms) and steric constraints. Clearly, it may give little insight on what the final surface configuration is. It is therefore important to perform systematic studies to uncover the fundamental mechanisms governing such surface reactions.

A main difficulty in distinguishing various mechanisms results from limitations in accessible parameter space (e.g., pressure, temperature), often due to the constraints imposed by the characterization techniques used. Since the relative importance of various mechanisms may depend on the processing conditions and the surface morphology (e.g., crystal plane, atomic roughness), only by examining each reaction under a variety of conditions can the relevant and dominant mechanisms be identified. Such variations often are problematic in a manufacturing environment. Therefore, fundamental laboratory studies are needed.

In this chapter, we survey the current knowledge of reaction mechanisms relevant to silicon oxidation and nitridation and to the initial growth of high-k dielectrics using ALD. In particular, we highlight the role that *in situ* infrared absorption spectroscopy has played in this field.

Infrared spectroscopy is a versatile and discriminating technique. It is capable of detecting ultra-thin dielectric films and sub-monolayer quantities of the atomic or molecular species that partake in film formation. Applied *ex situ*, infrared spectroscopy thus provides information on film composition, phase, and impurity content (see e.g., ref. (2) for HfO_2 grown by CVD). Furthermore, as most photon spectroscopies, it is applicable in a variety of environments: vacuum, gas phase, and even liquid. Figure 1 shows the two generic geometries for infrared spectroscopy of semiconductor surfaces (3): The straight-through transmission geometry (often performed at the Brewster angle to optimize the throughput, to minimize interferences, and to provide polarization information) is ideal to explore the widest spectral range in vacuum and *in situ* in gas phase environments, while the multiple internal reflection (MIR) geometry is necessary to probe surfaces in liquid environments (via minimization of the liquid ambient absorption for the evanescent field). Infrared spectroscopy

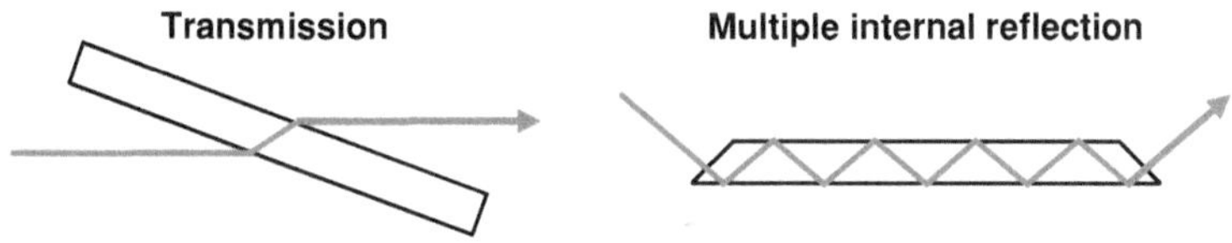

Fig. 1. The two generic geometries for infrared spectroscopy of semiconductor surfaces: transmission and multiple internal reflection.

has thus helped uncover some of the most relevant chemical reactions that govern the formation of important technological interfaces (3, 4). The emphasis of this chapter is on the application of *in situ* infrared spectroscopy to gas phase processing issues.

We first briefly summarize oxidation and nitridation mechanisms on *clean silicon* surfaces in an ultra-high vacuum (UHV) environment (Section 2). We thus illustrate the type of information that can be obtained using infrared spectroscopy and outline the surface chemical concepts that are important for the following sections. We then address in more detail two important applications that both involve processing of the technologically important *H-passivated (H/Si) silicon* surfaces. First, we focus on their *oxidation* in liquid and gas ambients (Section 3), placing this work in context, i.e., providing a concise review of existing literature. Then, we address *growth of high-k dielectrics* on H/Si surfaces using atomic layer deposition, using Al_2O_3 growth as an example (Section 4).

2. INITIAL OXIDATION AND NITRIDATION OF CLEAN SILICON SURFACES

The small oxygen and nitrogen-containing molecules O_2, H_2O, and NH_3 are routinely used in gas phase processing of silicon, in particular for surface oxidation and nitridation. Their adsorption and decomposition are among the most important and fundamental reaction steps in Si surface chemistry. From a scientific point of view, these simple molecules constitute ideal model reactants to uncover mechanistic details fundamental to Si reactivity. For example, contrasting the behavior of O_2 and H_2O oxidation makes it possible to uncover the role of hydrogen in the incorporation and subsequent motion of O atoms in the Si surface region. In turn, the fundamental understanding thus derived is important to control the growth of technologically important ultra-thin oxide and nitride layers on silicon. Early fundamental insight has been obtained on clean Si(100)(2 $\times$ 1) in UHV (Fig. 2). This is the focus of this section.

2.1. O_2 Decomposition

While much experimental and theoretical work has been done over the past two decades, the decomposition pathway for an O_2 molecule on silicon has not been determined unambiguously. For the Si(100)(2 $\times$ 1) surface, for instance, a number of

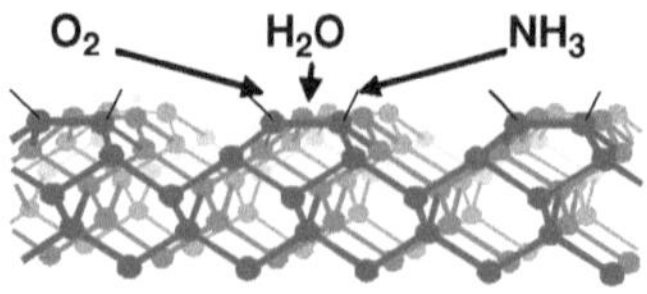

Fig. 2. Schematic representation of a clean Si(100)(2 × 1) surface, with an indication of the process gases regularly used for oxidation and nitridation.

models has been presented, ranging from a peroxide precursor state to direct insertion into the surface Si backbonds (5–9). One of the major difficulties is the highly inhomogeneous nature of the oxidation process, leading to the coexistence of a number of oxidized states and structures (10–14). To fully unravel the decomposition process, it is important to use a technique that can distinguish low concentrations of different species. With a simple transmission geometry to have access to the Si–O vibrations, an *in situ* infrared spectroscopy study of clean Si(100)(2 × 1) surfaces exposed to O_2 was recently performed as a function of sample temperature (15).

The essential results are summarized in Fig. 3. Below room temperature, O_2 adsorbs dissociatively, forming a metastable silanone structure, (O)Si=O, characterized by a distinctive Si=O stretch vibration at 1220 cm^{-1}. Subsequent oxygen insertion from the Si=O configuration into the surface silicon backbond~~to~~ from (O_2)Si— can be either thermally activated (~1 eV barrier) or induced by exposure to atomic hydrogen, forming a spectrally distinctive $(O_2)Si–H_2$ 'dihydride' structure. Importantly, further annealing at temperatures where it is kinetically possible for oxygen to diffuse leads to a highly inhomogeneous surface exposing SiO_2 domains on an otherwise clean Si(100) surface. This is in fact expected: as the oxidation reaction $Si + O_2 \rightarrow SiO_2$ is highly exothermic, oxygen agglomeration should be thermodynamically favorable.

2.2. NH_3 Adsorption

Incorporation of nitrogen is routinely used to stabilize interfaces and prevent oxygen and boron diffusion. Ammonia (NH_3) is often used, but little is understood about the effect of hydrogen on N incorporation. NH_3 adsorption, decomposition, and N

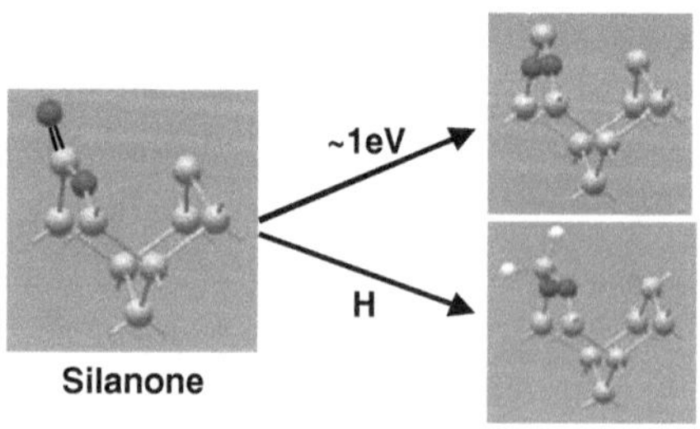

Fig. 3. Schematic representation of a silanone (Si=O) species and of oxygen insertion into a Si backbond via thermal activation (top) or by reaction with two hydrogen atoms (bottom).

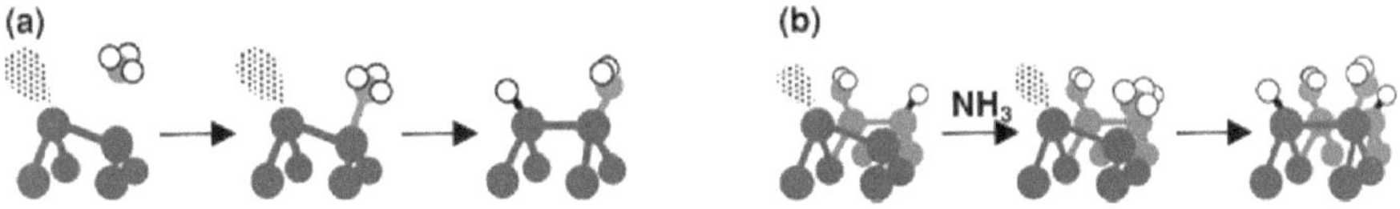

Fig. 4. Schematic representation of (a) NH_3 adsorption on a surface dimer of Si(100)(2 × 1) and of (b) NH_3 adsorption on neighboring dimers.

incorporation were studied on Si(100)(2 × 1) (16). Figure 4(a) summarizes the findings: First, a metastable NH_3 adsorption is identified, followed by dissociation into H and NH_2. Moreover, subtle shifts in N–H vibrational frequencies can be associated with weak interdimer H-interactions, making it possible to infer structural information on the inter-dimer (long-range) structure from a detailed analysis of infrared data. It was thus shown that a dominant zig-zag configuration of the NH_2 groups on each dimer row is formed at saturation coverage (Fig. 4(b)). It arises from the specific dimer buckling induced next to a H–Si–Si–NH_2 dimer. This buckling favors subsequent NH_3 adsorption in a gauche configuration during the adsorption process (16). This study clearly showed that H tends to terminate all dangling bonds, is responsible for the physical arrangement of N atoms at the surface, and prevents N diffusion. As the temperature is raised, the hydrogen and NH_2 in the H–Si–Si–NH_2 structure tend to recombine into NH_3 and desorb. For N to be incorporated into silicon, a partial pressure of NH_3 is required during annealing, resulting in inhomogeneous patches of nitride, separated by clean Si regions. Similar to the case of oxidation, such N aggregation is expected, based on the highly exothermic nature of the nitridation reaction.

2.3 H_2O Adsorption and Decomposition

Water decomposition on Si(100) has been studied extensively, but vibrational spectroscopy (electron energy loss spectroscopy, EELS, and infrared spectroscopy) were needed to clearly show that the adsorption is dissociative into H and OH (17). Furthermore, it was shown that, in contrast to NH_3 adsorption, the long-range arrangement of H–Si–Si–OH dimers is random, with only ~50% in a zig-zag configuration (18). Thermal decomposition of this H–Si–Si–OH surface occurs by O insertion into the Si substrate. In contrast to the case of H–Si–Si–NH_2 discussed above, there is no recombination and desorption. The Si–Si dimer bond is the target for initial oxygen insertion (19). For such studies, the ability to distinguish several different species, such as H–Si–Si–H, H–Si–O–Si–H, H–Si–O–Si(O)H, and so forth, is critical because the surface is highly inhomogeneous with a coexistence of several species at once. Such details of surface structure can be resolved with the high spectral resolution of infrared spectroscopy (18, 20). The concentration of each species derived from quantitative infrared analysis may then be correlated with kinetic Monte Carlo (KMC) simulations of surface reaction and diffusion, yielding important physical insight into the mechanisms for oxygen insertion and agglomeration (21). In particular, KMC offers great insight into the mechanism for oxygen motion (diffusion) in the

Fig. 5. Epoxide structure on a Si(100)(2 × 1) surface.

near surface region and the role that terminal hydrogen plays in blocking a surface hopping channel.

Above hydrogen desorption temperatures (~800 K), the mobility of atomic O in the surface increases and agglomeration into structures containing three to five oxygen atoms is favored. The surface then features ‘epoxide’ configurations, as shown in Fig. 5. Such observations and complementary ab initio calculations have shown that rebonding of the dangling bonds into this epoxide structure is in fact energetically favorable for dimer structures with three or more oxygen atoms. Finally, repeated water exposure/annealing treatments can be used to induce the formation of a thin silicon oxide layer (22, 23), thus establishing a connection with earlier infrared studies of technologically relevant Si/SiO_2 interfaces (24, 25).

In summary, the work described in this section illustrates the methodology that can be used to explore the mechanism of oxidation and nitridation on a variety of surfaces. The next section focuses on the reactivity of H-passivated silicon surfaces.

3. OXIDATION OF H-PASSIVATED SI SURFACES

Despite the tremendous technological importance of H/Si surfaces, much less is known about their initial oxidation mechanisms than for clean Si surfaces in UHV. However, such information would help achieve controlled deposition of oxides in the nanometer regime, e.g., for MOSFETs featuring sub-nm equivalent oxide thickness. The focus in this section therefore is on the initial oxidation regime of the first one or two monolayers. We briefly point out the technological relevance of H/Si surfaces and summarize preparation techniques. We then discuss oxidation by various oxidants in detail. For a review on the kinetics and reaction–diffusion mechanisms of SiO_2 growth on H/Si in the intermediate thickness regime, see, e.g., a review article by Baumvol (26).

H/Si owes its outstanding importance in Si technology to the need for oxide free Si substrates for many growth and deposition processes. Growth on clean Si(100) substrates, as discussed in the previous section, is often not practical when aiming for ultra-thin reliable dielectrics, in particular in CVD-type processes, as surface cleanliness can only be ensured in a UHV environment. In contrast, H-terminated Si is easily prepared in a wet cleaning process and relatively stable against reaction in a clean room environment. Wafers can thus be kept clean and oxide-free until the next

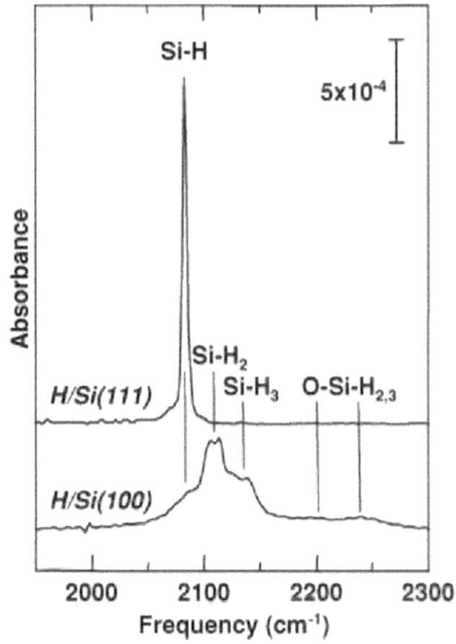

Fig. 6. Transmission infrared spectra of H/Si(100) from an HF(aq) etch and H/Si(111) from an NH_4F(aq) etch.

processing step is performed, making H/Si an ideal substrate. H/Si has been used as a substrate for growth of traditional gate oxides and novel high-k dielectrics, often deposited by CVD or ALD, to optimize the interfacial composition. It has also been used as a substrate prior to evaporation (27) and sputter-deposition (28) of a variety of dielectrics.

The preparation, structure, and properties of H/Si surfaces are well documented (29, 30). HF(aq)-etched and water-rinsed H/Si(100) and H/Si(111) are H-terminated and atomically rough. Atomically flat H/Si(111) can be prepared by a subsequent etch in NH_4F(aq) (31). Transmission infrared spectra of the Si–H stretching region are shown in Fig. 6 and provide details on the surface structure. The atomic perfection of H/Si(111)(1 × 1) is evidenced by a single, sharp Si–H stretching mode at 2084 cm^{-1}, originating from mono-hydrides on flat Si(111) terraces. Atomic roughness of H/Si(100) on the double-layer scale, by contrast, gives rise to a broader, structured Si–H stretching band. This surface features mono- (~2085 cm^{-1}), di- (~2110 cm^{-1}), and tri-hydride (~2135 cm^{-1}) species. Unlike the case of flat H/Si(111), a small concentration of chemical defects is evidenced by a signal at ~2250 cm^{-1} arising from H bonded to oxidized Si atoms (32). This is likely due to oxidation of some step sites during the surface preparation water rinse, probably by dissolved oxygen, as will be discussed below. We note that an HF/ethanol etch instead of the HF/H_2O etch has been reported to result in lower step and defect densities (see e.g., ref. (33)), but so far there have only been few reports on such surfaces as substrates for dielectrics.

As we shall see, a substantial activation barrier is involved in H/Si oxidation. This results in a low reactivity and requiring large reactant exposures for oxidation. A mechanistic picture of H/Si oxidation demands an understanding of the impact of (a) chemical and structural surface defects and of (b) impurities in the reactant, as such imperfections might lead to dramatically higher reactivity. To address these issues experimentally, (a) surfaces with different defect types and densities may be compared, enabling the extraction of site-specific mechanisms, and (b) high purity reactant experiments may be compared with studies under ambient conditions and, ultimately, mixtures of high-purity reactants.

We first summarize results on the interaction of H/Si with liquid water, since a water rinse is an inherent part of H/Si fabrication. Also, comparison of wet and dry reactivity of H/Si will ultimately aid in a full mechanistic understanding of oxidation. We then highlight high-purity gas phase studies involving O_2 and H_2O, which are the most relevant oxidants in high-k gate oxide processes. Kinetics and barriers extracted from these studies can then be invoked to model oxidation and ALD growth. We include a brief review of ambient oxidation studies. Such results are not only important to assess the impact of H/Si contact with clean room environments, but on a fundamental level also help to understand the mechanistic impact of impurities such as H_2O in O_2 vapor.

3.1. Aqueous Chemistry of H-Terminated Si and the Role of Dissolved O_2

The aqueous chemistry of H/Si depends on Si surface orientation, on water pH, and on the concentration of dissolved O_2. We outline the main facts, focusing on cooperative effects involving water and O_2, as this may aid in the understanding of gas-phase oxidation. For more details and references, we refer the reader to a review by Henderson (34).

Ultrapure water, containing only ppb amounts of O_2, etches Si surfaces. Etching likely occurs via successive hydroxylation of Si atoms until they are dissolved in the form of $Si(OH)_4$ (35, 36). OH^- ions probably play the dominant role in the hydroxylation/oxidation process (37), similar to the case of HF(aq) and NH_4F(aq) etching of Si surfaces (38). For water at room temperature and neutral pH, quantum chemical calculations indicate that a H/Si(100) surface site is at least 10^{18} times more likely to react with OH^- than with H_2O (39). Ultimately, etching by boiling water results in atomically flat H/Si(111) with triangular etch pits (likely due to continued oxidation by trace O_2, as discussed below) (35, 36) and in substantially rougher H/Si(100) with pits exposing (111) facets (40). These differences in morphology have been attributed to preferential OH^- attack of the more polarizable Si atoms of $Si–H_2$ and $Si–H_3$ structures at kinks and steps (37, 40).

In view of the proposed etching mechanism, oxidation by OH^- should determine the overall etching rate. In the presence of dissolved O_2, however, the relative reaction rates depend on the relative concentrations of O_2 and of OH^-, and thus on pH (37). At O_2 concentrations in the ppm regime, the oxidation rate exceeds the etch rate, and hence SiO_2 grows (41–43). The actual mechanism of water-induced oxidation is still unknown. However, there is some evidence for a cooperative (e.g., catalytic) effect between the water itself and dissolved O_2. In a study utilizing isotopically marked water it was shown that at least 85% of the O incorporated into a 5–7 Å thick oxide formed at room temperature originates from the H_2O and not from the dissolved O_2 (44). This was taken as indication that O_2 activates the H/Si surface or acts as a catalyst for H_2O oxidation of Si, rather than oxidizing substantial amounts of Si itself. The possibility of isotopic exchange between $H_2^{18}O$ and $^{16}O_2$ or $Si^{16}O_2$ formed from $^{16}O_2$ was not discussed. Rather, the authors proposed that the role of the dissolved OH^-/O_2 couple is to anodically polarize the electrode, driving the H_2O oxidation reaction (44). Water-induced oxidation of phosphorus-doped n^+-Si in fact proceeds dramatically faster than for n- or p^+-Si, supporting a field-assisted oxidation mechanism (41, 42). However, at doping levels of 10^{20} cm^{-3}, chemical effects (e.g., catalysis) due to the dopant atoms may also have to be considered.

A very detailed mechanistic picture of the initial stage of H/Si oxidation by O_2-containing water has been proposed by Cerofolini et al. (45). They suggested that H_2O rapidly attacks a low concentration of F impurities on HF-etched Si, forming OH groups. It is speculated that such sites then bond thermal excitons, i.e., electron–hole pairs, leading to Si–Si back-bond cleavage and thus to the formation of surface Si^- and subsurface Si^+. The subsurface Si^+ then bonds molecular H_2O in the form of a Lewis acid–base adduct, while the surface Si^- ionizes dissolved O_2 and reacts with it, forming $SiOO^-$. Proton transfer from the H_2O adduct to the $SiOO^-$ results in Si–OH (silanol) and Si–OOH (45). In this largely speculative model, O_2 acts as a necessary ingredient for subsurface hydroxylation, and thus ultimately for the formation of Si–O–Si bridges via silanol condensation. However, direct evidence for these reaction steps is not available.

In conclusion, O_2-enhanced oxidation appears to occur through a cooperative effect involving H_2O and/or OH^- and O_2. As discussed below, similar phenomena may be operative in gas-phase oxidation of H/Si. Details of the mechanism are still unclear, however, largely for lack of *in situ* studies.

3.2. O_2 and Air Interaction with H-Terminated Si

3.2.1. Oxidation by nominally pure O_2

The number of experimental studies of H/Si oxidation by molecular oxygen is remarkably small, as compared with observations on clean Si surfaces. In part, this is certainly due to the low reactivity of H-passivated Si. At temperatures below the onset of recombinative H_2 desorption (e.g., above ~600 K on H/Si(100) (46)), substantial oxidation by typical UHV gas exposures in the Langmuir regime does not occur. This was demonstrated for a number of H/Si surfaces prepared in UHV by atomic H exposure of well-defined Si reconstructions, e.g., for mono- and dihydride H/Si(100), both flat and vicinal, and for H/Si(311) surfaces (47, 48). Room temperature O_2 exposures in the 10^3 L regime did not result in observable hydroxyl formation or in the formation of any oxide film exceeding an estimated 0.7 Å in thickness. Dangling bonds appear to be needed for oxidation in this exposure regime. There is no observable reactivity enhancement by steps.

The initial room temperature sticking coefficient of O_2 on HF-etched Si(100) was determined to be 10^{-12}, according to Westermann et al. (49). In line with this, Kawamura et al. (50) found that ca. 7×10^{12} L O_2 were needed for completion of an oxide monolayer on an HF-etched amorphous Si film. Subsequently, growth was accelerated while the H termination was preserved. Even lower sticking coefficients of HF-etched Si(100) emerge from studies by Morita et al. (42), who detected less than 1 Å SiO_2 after one week of 4:1 N_2:O_2 exposure at 1 bar ($\sim 10^{14}$ L O_2); and by Niwano et al. (51), who found that complete surface oxidation in 1 bar O_2 required an exposure of 10^{15} L, as detected by *in situ* infrared spectroscopy. As we shall see, impurities such as H_2O may be responsible for variations in reactivity between different studies.

As is the case for water oxidation, atomic-scale information on the O_2 oxidation mechanism is scarce. Only Cerofolini et al. speculated on an oxidation scheme by O_2 or dry air (45). As discussed above, they predict O_2 dissolved in the water of the HF-last rinse to react with Si^- sites negatively polarized through interaction with H_2O. The same rapid oxidation step may still occur during initial air (or O_2) exposure. On longer time scales, reaction is suggested to occur through O_2 ionization by Si^-, which is created by trapping of a thermally generated exciton. This would explain why the apparent activation energy for oxidation in the initial stage is close to the exciton energy and hence to the Si band gap of ca. 1 eV.

Also photoinduced mechanisms may have to be considered. Photon stimulated H desorption (52) and formation of reactive O radicals in the O_2 gas (53) have been identified as relevant H/Si oxidation channels in the presence of ultraviolet light. Using intense ultrashort laser pulses, even visible or near-infrared light can accelerate oxidation of H/Si(111) (54). This was tentatively attributed to the generation of electron–hole pairs in the Si substrate. Even though photoinduced reaction rates are small, they might turn out to be important in order to understand differences in oxidation rate found in different experiments.

At elevated temperatures relevant for ALD growth of gate oxides, the H/Si reactivity is clearly higher. This was borne out in a comprehensive investigation by *in situ* infrared spectroscopy and XPS (32, 55). Exposure of atomically flat H/Si(100)(2 $\times$1) (prepared by an H_2 anneal of Si(100) (2 $\times$1)) and H/Si(111)(1 $\times$1) (from a NH_4F wet etch) to dry O_2 at 573 K required only $\sim 10^9$ L O_2 for the formation of a near-complete oxide monolayer. On H/Si(111), growth was predominantly lateral in nature, while on H/Si(100) both lateral and vertical growth occurred and the areal density of oxide patches was higher. An *in situ* microscopic study in UHV revealed some direct mechanistic insight into how the initial monolayer oxidation may proceed at elevated temperatures (56). It appears that even a small concentration of isolated dangling bonds is sufficient to facilitate substantial oxidation: at such sites, oxide stripes as long as 15 dimer units were observed after O_2 exposures of only 10 L at 530 K. These stripes probably form via H migration. In these experiments performed at elevated temperature, the H termination is preserved during oxidation, with oxygen insertion into the Si–Si backbonds (32, 55, 56).

A number of density functional calculations have addressed O_2 interaction with H/Si. In the presence of residual dangling bonds, O_2 dissociation at such sites initiates oxidation, which proceeds along dimer rows via H migration, as experimentally observed (56). For ideal H/Si at 0 K, only a physisorbed state was predicted (57). When the reaction does occur, however, O insertion into Si–Si backbonds is energetically favorable over insertion into Si–H bonds (58), consistent with the continued H termination of the surface.

3.2.2. Oxidation by ambient air

Given the large O_2 exposures (10^9–10^{15} L) required to oxidize H/Si surfaces, cleanliness issues due to impurity-mediated reaction may be expected to come into play. Notably, the H_2O content in air or O_2 has been the focus of several studies. Air oxidation of H/Si(100) and H/Si(111) clearly accelerates with increasing humidity (41, 42, 51, 59, 60). Modeling of time-dependent oxidation data only succeeded assuming a two-step oxidation mechanism (59–61): surface oxidation is slow; at or next to sites thus attacked, subsequent faster subsurface oxidation occurs. This may be due to facilitated nucleophilic attack of back bonds polarized by subsurface O and would result in two-dimensional island growth. Scaling of the rate constants with humidity indicated (for H/Si(111)) that H_2O is responsible for initial surface modification, turning those sites hydrophilic which promotes physisorption of additional H_2O. Later, the impact of O_2 seemed higher. Only at high humidity, H loss from the surface occurred during oxidation (59, 60).

Both for air (59–61) and O_2 oxidation (50) at room temperature, the reaction appears to accelerate as soon as a complete monolayer of silicon oxide is formed. This surprising observation might point to a reaction mechanism involving physisorbed precursor molecules that diffuse across the surface until they either desorb into the gas phase or react with the substrate. A model that would result in the observed behavior is based on plausible assumptions regarding barrier heights on pristine and oxidized areas: O_2 or H_2O lateral diffusion barrier: H/Si > H/SiO_2; desorption barrier: H/Si << H/SiO_2; reaction barrier: H/Si > H/SiO_2 (e.g., due to back bond polarization). The vast majority of physisorbed molecules would thus diffuse to H/Si patches where most of them desorb and some react with the substrate; only a negligible number would react on H/SiO_2 patches. As soon as the top Si layer is complete, the fraction that desorbs into the gas phase would drop, resulting in an increased oxidation rate.

Si dopants may influence the H/Si oxidation rate: Oxidation of n^+-Si proceeds much faster than for n- or p^+-Si, again indicating an important impact of substrate electrons on reactivity (41, 42). As for the impact of H/Si step sites, the initial oxidation rate appears to scale linearly with surface roughness (61) (although some authors disagree (51)). This finding was taken as an indication that oxidation starts from steps or defects and continues two-dimensionally (e.g., via Si back bond polarization and nucleophilic attack) until the first monolayer is completed (61). Supporting the notion that defects act as oxide nucleation sites, it was demonstrated that an atomically flat, step-free H/Si surface can remain perfectly oxide-free in humid air for at least 15 min (62).

We note that ambient contaminants other than H_2O, such as organics or radicals, cannot be disregarded. For instance, the presence of organic species has been invoked to account for changes in the electrical surface properties of H/Si (63). These defects may also affect the oxidation rate and are a cause for caution. So far, all air oxidation studies implicitly assumed that other contaminants were either absent or did not influence oxidation. Experiments utilizing high-purity O_2/H_2O mixtures could help remove remaining uncertainties.

3.2.3. Mechanistic *in situ* infrared studies of H/Si oxidation by O_2

In situ infrared studies of O_2 interaction with various H/Si surfaces were performed in order to isolate the main mechanism of H/Si oxidation (64–66). A high O_2 purity was achieved using a purifier, with impurity (e.g., H_2O) levels below 1 ppb. Focus was on comparison of technologically relevant HF-etched Si(100) to structurally well-defined model surfaces. Such substrates can be designed to expose different types of terrace and step sites so that site-specific reaction pathways can be uncovered. Atomically flat H/Si(100)(3 × 1) was prepared by atomic H exposure of clean Si(100) in UHV (67). Flat H/Si(111) and two different types of 9° miscut vicinal surfaces (exposing mono- and dihydride steps, respectively) were prepared by an NH_4F wet etch (29, 30).

These MIR studies take advantage of the fact that the Si–H stretch spectrum is very sensitive to oxygen incorporation and clustering in the first layer of Si–Si bonds. Oxidation kinetics at terrace and step sites can therefore be monitored independently. In addition, Si–O modes can be recorded.

Figure 7(a) shows the Si–H stretch spectrum of a miscut H/Si(111) surface exposing steps terminated by dihydrides ($Si–H_2$). Terrace monohydrides ($\sim 2083\ cm^{-1}$) are clearly distinguished from modes associated with steps ($C_1 = 2095\ cm^{-1}$, $C_2 = 2102\ cm^{-1}$, $C_3 = 2136\ cm^{-1}$: coupled modes originating from step dihydrides and monohydrides below the step) (68). The absorbance per step hydrogen is higher than on the terrace due to their orientation. Upon exposure to O_2 at the typical ALD temperature of 573 K, the intensities of the step-related modes decrease, while the Si–H terrace mode remains nearly unaffected (Fig. 7(b)).

One might speculate that the disappearance of the Si–H modes is due to the removal of surface H or insertion of O into the Si–H bond, resulting in Si–OH formation. Significant hydroxyl formation can be excluded as there is no observable absorption in the range 3658–3750 cm^{-1} expected for isolated hydroxyl groups on silicon or silica surfaces (18, 69) (not shown). Information on the O insertion mechanism into Si backbonds is gathered from examination of the Si–H stretch spectrum in a broader range, extending up to 2250 cm^{-1}. This is demonstrated in Fig. 8 for oxidation of flat H/Si(111): initially, a sharp band due to monohydrides on large flat terraces is observed; during oxidation, bands at ca. 2150, 2200, and 2250 cm^{-1} grow and may be assigned to monohydrides on singly, doubly, and triply oxidized Si sites, respectively (32). In other words, the Si–H band is shifted to higher frequencies by O incorporation into Si–Si backbonds. The integrated Si–H intensity remains nearly constant, suggesting that most H remains on the surface during subsurface oxidation. Most of the oxidized Si–H species have two or three O atoms in their backbonds, with only a low concentration of Si–H with one O atom in their backbonds. This indicates

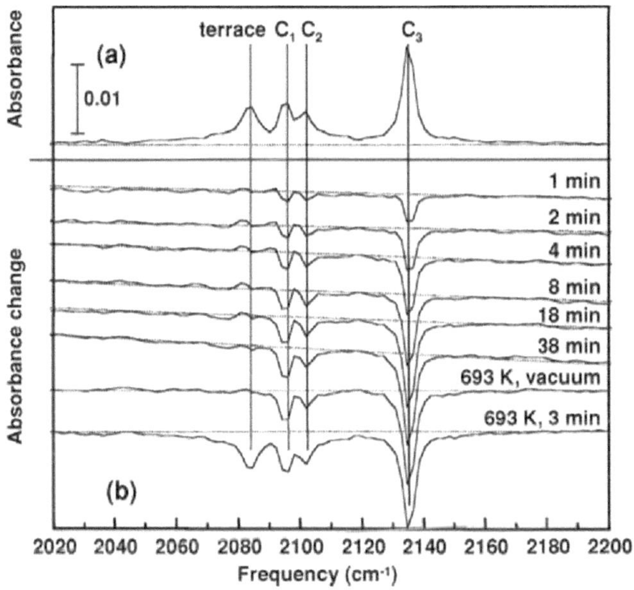

Fig. 7. Multiple internal reflection infrared spectra for H/Si(111) 9° miscut along $<\bar{1}\,\bar{1}\,2>$ (dihydride steps) exposed to 0.02 mbar O_2 at 573 K, subsequently annealed to 693 K, and exposed to O_2 at the same temperature. (a) Absorbance of starting surface; spectral reference: H/Si(111) oxidized by O_2 at 693 K. (b) Difference spectra for the subsequent processing steps; the spectrum of the starting surface is used as a spectral reference, i.e., negative absorbance indicates the loss of species from the surface. C_1, C_2 and C_3 are modes related to step dihydrides. An exposure of 1 min corresponds to a dose of 9×10^5 L O_2 (after (66)).

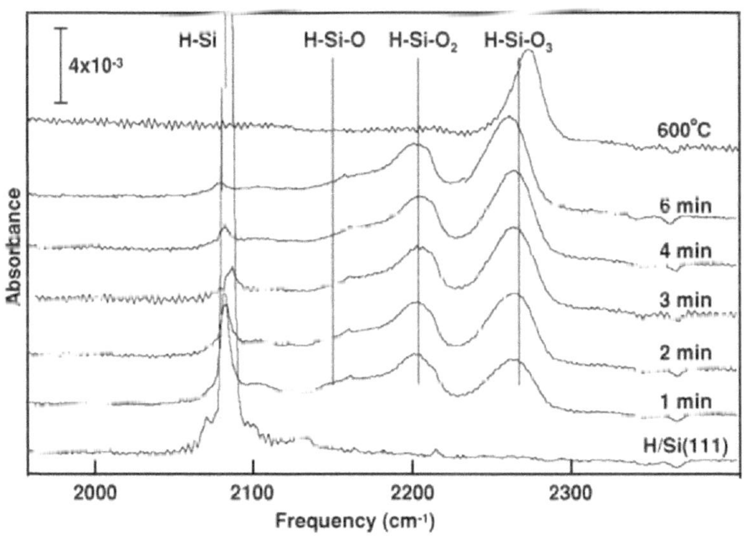

Fig. 8. Multiple internal reflection infrared spectra of flat H/Si(111) exposed to 1.3 mbar O_2 at 643 K. An exposure of 1 min corresponds to a dose of 6×10^7 L O_2 (after (64)).

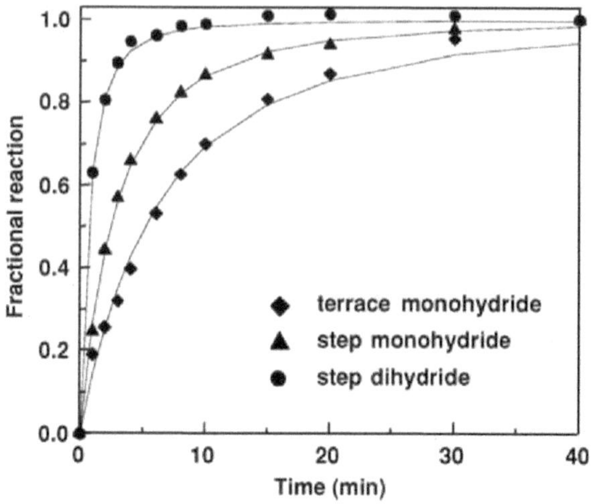

Fig. 9. Kinetics of the Si–H peak decay on flat and stepped H/Si(111) surfaces for terrace hydrogen, step monohydride (on < $11\bar{2}$ > miscut surface), and step dihydride (on < $\bar{1}\bar{1}2$ > miscut surface) in 1.3 mbar O_2 at 573 K. An exposure of 1 min corresponds to a dose of 6×10^7 L O_2 (after (64)).

O agglomeration, possibly due to hopping, and is consistent with findings of other studies (32, 55).

Considering the data from vicinal H/Si(111) in Fig. 7, it is obvious that, at 573 K, O_2 oxidation of dihydride step sites is much more rapid than oxidation of terrace sites. O_2 exposures in the 10^7 L regime are sufficient for complete dihydride step oxidation. In contrast, oxidation of sites below terrace monohydrides is only achieved by comparable O_2 exposures at higher temperatures, e.g., 693 K (Figs. 7(b) and 8).

Similar kinetic experiments were performed for flat H/Si(111) and for H/Si(111) with monohydride steps (64). In Fig. 9, the oxidation kinetics for monohydride on Si(111) terraces are compared to those of mono- and dihydrides at steps. The oxidation rate clearly increases according to: terrace monohydride < step monohydride < step dihydride. In order to obtain a more complete picture of oxidation kinetics at different surface sites, including activation energies, kinetic data for Si–H stretch intensities were recorded at various temperatures and O_2 pressures (64, 65). These data were analyzed using a simple rate equation assuming that the only "reacting" species involved are Si–H and O_2:

$$\frac{\partial[\mathrm{Si-H}]_i}{\partial t} = k_i[\mathrm{Si-H}]_i^{\alpha} P_{\mathrm{O}_2}^{\beta} \tag{1}$$

P_{O_2} denotes the oxygen pressure and $(\mathrm{Si–H})_i$ represents the concentration of Si–H with no oxygen in the Si–Si backbonds, for terrace ($i = 1$), monohydride ($i = 2$), and dihydride ($i = 3$) Si–H species, respectively. By fitting the data in Fig. 9, the order of the reaction in (Si–H) is found to be $\alpha = 1.5 \pm 0.2$ for all hydride species.

To determine the order of the oxygen reactant, O_2 pressure dependent experiments were performed in the range 0.0013–1.3 mbar on flat Si(111) at 643 K, yielding $\beta = 0.7 \pm 0.1$. Finally, from the temperature dependence at constant pressure, the overall apparent activation energies and pre-exponential factors were obtained. For monohydrides on flat H/Si(111), for example, these values are 1.72 ± 0.014 eV and 10^{13}–10^{15} s^{-1}. Assuming similar pre-exponential factors for stepped surfaces, overall apparent activation energies are found lower by 0.04 eV for step monohydrides and by 0.10 eV for step dihydrides. It has to be noted that these numbers are suggestive only, because there is no evidence that the pre-exponential factors should remain unaffected by defects and steps.

Steps not only exhibit enhanced oxidation rates themselves, but also affect the oxidation rate at nearby terrace sites. An understanding of such defect-related phenomena is critical for a complete picture of the reactivity of atomically rough HF-etched Si(100). The impact of steps on terrace Si–H was therefore addressed in detail using H/Si(111) model surfaces (64, 65). The oxidation rate of terrace Si–H was found to increase according to: flat H/Si(111) < monohydride steps < dihydride steps, i.e., steps accelerate the oxidation of adjacent terraces. This phenomenon may be included into the rate Eq. (1) via a step density factor.

The reaction rate constants k at different surface sites on various H/Si surfaces for an O_2 pressure of 0.02 mbar are listed in Table 1. In summary, the following observations were made (comparing rate constants at the same temperature):

- For comparable substrates, k increases according to: terrace monohydride < step monohydride < dihydride < trihydride (by up to one order of magnitude);
- on H/Si(111), k for terrace hydrides increases according to: flat < stepped with monohydrides < stepped with dihydrides.

The experimental data enable conclusions to be drawn concerning the detailed oxidation mechanism (64, 65):

Table 1. Rate constants for the reaction of H-terminated Si sites with 0.02 mbar O_2

Orientation	Structure	Hydride	k (min^{-1})
Si(100)	HF-etched	Mono-	0.068 ± 0.003
	HF-etched	Di-	0.134 ± 0.004
	HF-etched	Tri-	0.27 ± 0.03
	3 × 1	Mono-	0.06 ± 0.01
	3 × 1	Di-	0.69 ± 0.02
Si(111)	Flat	Mono	0.035 ± 0.002
	Flat	*Mono-*	*0.12 ± 0.02*
	Dihydride steps	*Mono- (terrace)*	*0.192 ± 0.007*
	Monohydride steps	*Mono- (step)*	*0.44 ± 0.02*
	Dihydride steps	*Di- (step)*	*1.57 ± 0.04*

Rates are comparable when measured at similar temperatures. 550–553 K in Roman font and 573 K in italics (after (65)).

- Oxidation occurs by direct insertion of O into Si–Si backbonds, i.e., without H removal or hydroxyl formation;
- the activation energies are approximately 1.6–1.7 eV, i.e., lower than typical Si–Si bond energies (e.g., 2.31 eV (70)), indicating that oxidation occurs without Si–Si bond breaking prior to O insertion;
- differences in reaction rate of mono-, di-, and trihydrides are most likely due to different activation energies resulting from variations in steric hindrance;
- comparing the same type of hydride on different substrates, variations in reaction rate are largely consistent with differences in accessibility of the backbonds to O_2; a high reactivity is thus found, e.g., for highly accessible backbonds of strained tilted dihydrides at H/Si(111) steps and for backbonds on rough surfaces in general;
- a minor effect in some cases might be a lowering of the activation energy at strained sites, e.g., for strained mono- and dihydrides on Si(100)(3 × 1), leading to an increased reactivity;
- O hopping immediately after insertion (dissipating the high free energy of oxidation) is a possible cause for the observed enhancement of terrace oxidation by nearby steps/defects and is supported by the failure to detect O_1–Si–H (Fig. 8); such a picture is also in line with O hopping observed in the oxidation of clean Si(100), as discussed in the preceding section;
- a reaction order in P_{O2} of 0.7 shows that interaction of oxygen species simultaneously present at the surface lowers the probability for reaction; this might be due to the presence of a layer of temporarily physisorbed O_2 atop the H/Si surface, whose areal density scales less than linearly with gas phase pressure due to O_2–O_2 collisions.

It remains to be shown whether the oxidation mechanism observed in the low pressure regime applies to atmospheric pressures as well. Therefore, an oxidation experiment was performed in 1 bar O_2 at 433 K (66). Transmission infrared spectra of the H/Si(111) surface were recorded *ex situ* at various oxidation steps (Fig. 10). Si–H stretching modes are shifted to higher frequencies, Si–H bending modes disappear, and Si–O stretching bands typical of TO and LO phonons of SiO_2 are observed, clearly proving oxidation. These data yield a rate constant of 0.052 min^{-1}, fairly close to a value of 0.029 min^{-1} extrapolated from the kinetic data discussed above, obtained at much lower pressures (66). This indicates that the same oxidation mechanism applies across 5 orders of magnitude in pressure and a temperature range of 240 K.

Further extrapolating the kinetics at atmospheric pressure to room temperature, a reactive sticking coefficient of O_2 on H/Si(111) of ca. 10^{-13} is expected (66). This number is identical to room temperature sticking coefficients on H/Si(100) of ca. 10^{-12}–10^{-15} obtained in other studies (42, 49–51), again indicating that the same oxidation mechanism may apply.

3.3. H_2O Vapor Interaction with H-Terminated Si

The reactivity of ultrapure liquid water at H/Si surfaces is dominated by OH^-. In the vapor phase, such (neutral or ionic) species formed by H_2O dissociation clearly cannot

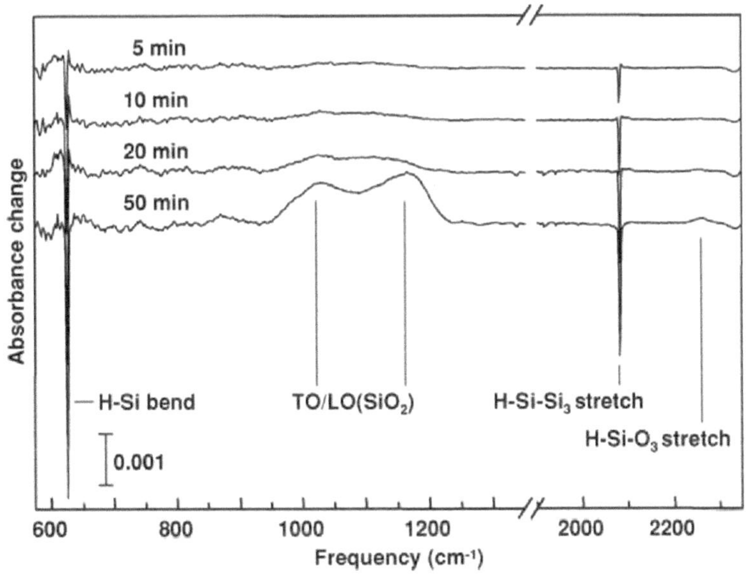

Fig. 10. Transmission infrared spectra of flat H/Si(111) exposed to 1 bar O_2 at 433 K. After each O_2 exposure, spectra were taken *ex situ* in an N_2 purged spectrometer at room temperature. Spectral reference: H/Si(111); negative absorbance indicates the loss of species from the surface. An exposure of 1 min corresponds to a dose of 4.6×10^{10} L O_2 (after (66)).

play a role: assuming that photoinduced fragmentation is insignificant, the fraction of, e.g., OH in the gas phase is ca. 10^{-45} at a temperature of 573 K, as determined by the Boltzmann factor. Molecular oxygen, on the other hand, may be present in water vapor, since the liquid water reservoir contains substantial amounts of dissolved O_2 after contact with atmospheric air. Therefore, special care has to be taken to reduce the oxygen content in the liquid bath by, e.g., bubbling with inert gases or chemical purification (36, 71). Otherwise, the O_2 may be expected to oxidize H/Si, as discussed in the preceding section.

3.3.1. Current knowledge

A low reactivity for water vapor interaction with H/Si was demonstrated by Takagi et al. (72). At room temperature, gas exposures in the Langmuir regime resulted in no detectable water adsorption on H/Si(100)(2 × 1). Water instead physisorbed at 90 K, forming ice clusters, and desorbed below room temperature without reacting with the substrate. 5×10^7 L H_2O had to be supplied to H/Si(111) at room temperature in a later experiment (73) in order to reach an oxide thickness of about 2 Å. The H termination was preserved during formation of thin SiO_2 films, similar to the case of O_2 oxidation discussed above. The authors proposed that generation of activated Si–Si back bonds is the rate-limiting step at comparatively low temperatures. Initial oxidation is extremely slow even at 598 K, while above 723 K more than 3 Å SiO_2 form rapidly. In this high-temperature regime, the oxidation rate is limited by thermal desorption of the H overlayer from the Si substrate (73). This is in line with the findings of Kim et al. (74) who reported that a long (albeit unspecified) H_2O exposure of H/Si(100) in an ALD reactor at up to 573 K results in 1–1.5 Å SiO_2, while at 623 K and above, much thicker oxide films are formed. They interpret their observations as an indication that in fact no reaction should occur below 573 K. In the absence of any direct spectroscopic evidence, oxidation at higher temperatures was tentatively attributed to creation of dangling bonds through H desorption and subsequent OH formation with rapid insertion of the O into Si–Si backbonds. Zaïbi et al. (75), utilizing photoemission yield spectroscopy, also found a reaction threshold at 623 K at which temperature they indirectly inferred a substantial OH concentration on H/Si(111) already upon water doses as low as 50 L.

A number of theoretical studies have addressed the interaction of H_2O with H/Si. The physisorption energies on H/Si(100) and H/Si(111) are estimated at -0.16 eV (76) and -0.13 eV (77), respectively. For hydroxylation via $Si_3{-}Si{-}H + H_2O \rightarrow Si_3{-}O{-}H + H_2$, overall reaction enthalpies of -0.70 eV (39), -0.75 eV (78), and -0.59 to -0.69 eV (79) are reported, with corresponding barrier heights of 2.13, 1.60, and 1.85 – 2.05 eV. There is experimental evidence that trihydride configurations are hydroxylated more easily than dihydride structures (79). For all configurations, however, hydroxylation requires substantial thermal energy to occur and is therefore improbable at room temperature.

3.3.2. Mechanistic *in situ* infrared studies of H_2O interaction with H/Si at elevated temperature

There is clearly a lack of mechanistic *in situ* experiments. The potential influence of O_2 in the water vapor has not been addressed experimentally either. In fact it appears that most authors have been unaware of this potential complication. This situation prompted *in situ* studies of water vapor interaction with H/Si, utilizing transmission infrared spectroscopy, and ensuring a low O_2 concentration by extended bubbling of the water with ultrapure N_2. Motivated by the importance of water-substrate interaction in ALD growth of gate oxides (see next section), most experiments were performed at the typical ALD temperature of 573 K. Surface chemistry was probed *in situ* by transmission infrared spectroscopy in a home-built ALD reactor (80). For such spectroscopic experiments, it is convenient to use D_2O instead of H_2O to separate surface reaction from possible fluctuation due to residual water vapor in the spectrometer, while leaving the oxidation chemistry unaffected (81).

The H–Si stretch bands contrast atomic perfection on H/Si(111) (Fig. 11(a)) and atomic roughness on H/Si(100) (Fig. 11(b)), as discussed above. After D_2O exposure in the 10^8 L regime at 573 K, both H/Si(111) and H/Si(100) remain nearly unchanged chemically (81): the Si–H stretch signals are unaffected, except for a few percent of isotopic H–D exchange (see below).

On water-exposed H/Si(111), there is no evidence of H bonded to oxidized Si atoms (O_n–Si–H, 2130–2300 cm^{-1}) (32), of isolated hydroxyl groups (Si–OD, 2700–2760 cm^{-1}), or of SiO_2 phonon signals (ca. 1000–1200 cm^{-1}) (24). This establishes

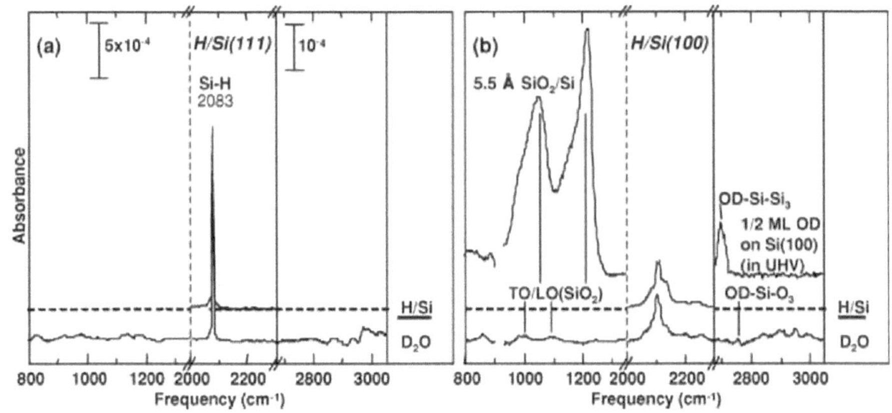

Fig. 11. Transmission infrared spectra of (a) H/Si(111) and (b) H/Si(100) before and after exposure to 1–5 × 10^8 L D_2O (10 mbar) in N_2 carrier gas (1 bar) at a sample temperature of 573 K. Reference spectrum in oxide phonon and O–D stretching regions: H-terminated Si. Note that a signal at ~920 cm^{-1}, due to Si–H bending modes of the H/Si(100) reference surface, has been removed for clarity. Reference spectrum in Si–H stretching region: Si with native oxide. For comparison, we show infrared spectra of well-known oxide and hydroxyl overlayers described in the text (after (81)).

that oxidation and hydroxylation of H/Si(111) are insignificant under the reaction conditions studied.

On water-exposed H/Si(100), there is evidence for incorporated O and isolated OD groups on oxidized Si sites (2760 cm^{-1}) (Fig. 11(b)). The areal density of O and OD species can be quantified by comparison to transmission infrared spectra of well-characterized reference samples (Fig. 11(b)). For SiO_2 phonon intensity calibration, a ~5.5 Å film prepared by wet-chemical oxidation of H/Si(100) in a H_2SO_4:H_2O_2 solution was used (80). For hydroxyl intensity calibration, a well-defined hydroxyl overlayer on Si(100) prepared in ultra-high vacuum was chosen (18, 82). Comparison shows that at most 0.1 Å SiO_2 (5 × 10^{13} O ions per cm^2) are formed, and that the OD areal density amounts to 2%ML (2% of a monolayer; 1.3 × 10^{13} OD per cm^2) (81).

Comparing the reactivity of structurally and chemically defective H/Si(100) to that of completely passivated flat H/Si(111), hydroxylation is found to occur only at certain defects present in low concentration, such as oxide or specific step sites. In conclusion, in the 10^8 L exposure regime, well beyond exposures relevant to ALD processes, O_2-depleted water does not oxidize H/Si(100) and H/Si(111) at 573 K to any significant extent.

On H/Si(111), even higher D_2O exposures of nearly 10^{10} L were studied at 573 K (Fig. 12) (83). On this surface, the high degree of surface order allows to discern infrared features due to different H–Si and D–Si species, originating from isotopic exchange. The spectrum indicates < 1%ML OD; (68 ± 3)%ML Si_3–Si–H; ≤ 3%ML O_3–Si–H; no or very little $O_{1,2}$–Si–H; (24 ± 2)%ML Si_3–Si–D; (10 ± 3)%ML O_3–Si–D; and no or very little $O_{1,2}$–Si–D. Here, 1 ML is defined as the

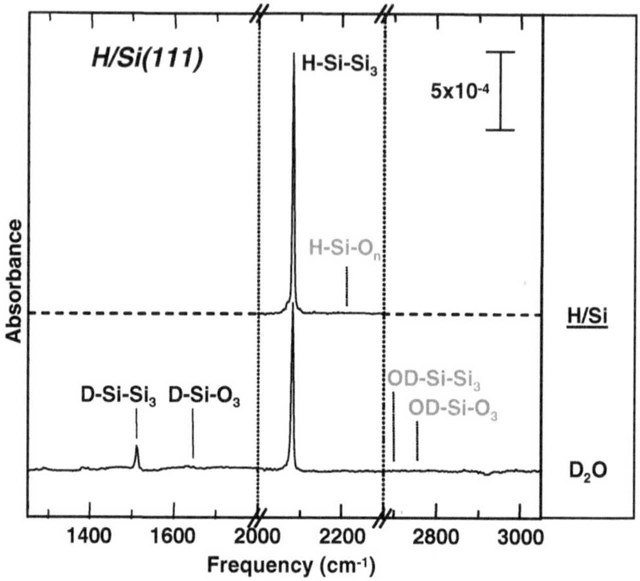

Fig. 12. Transmission infrared spectra of H/Si(111) before and after exposure to 7×10^9 L ultrapure D_2O (10 mbar) in N_2 carrier gas (1 bar). The sample temperature was 573 K (83). Reference spectrum in D–Si and O–D stretching regions: H-terminated Si; reference spectrum in Si–H stretching region: Si with native oxide.

areal density of H atoms on the initial H/Si(111)(1 × 1) surface, i.e., 7.83×10^{14} cm^{-2}. The total areal density of $\sim$30%ML O ions corresponds to ca. 0.6 Å SiO_2 formed in the reaction. In fair agreement with this, a small phonon signal (not shown) indicates the presence of ca. 0.3 Å SiO_2. These observations yield an upper bound for the reactive sticking coefficient of ca. 10^{-10}. While all authors of experimental studies agree that water vapor oxidation of H/Si is a slow process (72–75), the even lower reacitivity found in this work suggests that dissolved oxygen was present in higher quantities in the previous experiments.

The similar areal densities of D and O atoms would be compatible with an initial hydroxylation step, Si_3–Si–H + $D_2O \rightarrow Si_3$–Si–OD + HD, followed by a rapid OD decay with O insertion into a Si–Si backbond. Such hydroxyl decay on pristine Si–Si_3 sites is known to occur below 600 K from studies on Si(100) surfaces in ultra-high vacuum (20). The overall initial oxidation mechanism at elevated temperature therefore may be Si_3–Si–H + $D_2O \rightarrow Si_3$–Si–OD + HD $\rightarrow Si_2$O–Si–D + HD, hydroxylation being the rate-limiting step. A preference for hydroxylation over direct H–D exchange is also supported by a recent density functional study of water reaction pathways with H/Si(100) (78). The activation barrier and the overall enthalpy of reaction for hydroxylation were found to be lower by nearly 0.7 eV than for proton exchange. However, as discussed in detail for O_2 oxidation of H/Si, pathways and barriers may be different for different Si surface orientations. Therefore, comparison of H/Si(100) and H/Si(111) may be difficult. More experimental and theoretical studies are needed.

It is noteworthy that, for oxidation at 673 K, appreciable quantities of stable OD groups were observed once the first SiO_2 monolayer was completed (not shown) (83). Such OD groups are bonded in the local arrangement O_3–Si–OD and are thus likely stabilized against decay by the backbonded O. We note that hydroxyl groups on pristine Si sites at room temperature are kinetically stable (20), which may imply that the oxidation mechanism is fundamentally different at 300 K than at elevated temperatures.

In conclusion, *in situ* studies have shown that the reactivity of H/Si with respect to water vapor at 573 K is lower than with respect to O_2. Weighing all available evidence regarding the reactivity of H/Si between room temperature and 573 K, it appears to increase in the order: $H_2O < O_2 <$ humid air. This surprising finding may indicate that there is a cooperative effect of H_2O and O_2, such as a two-step oxidation mechanism in which one species opens up the surface while the other is needed for substantial oxidation. Alternatively, the reactivity of ambient air might be due to impurity species not yet considered, for example hydrocarbons, ozone, or radicals. In order to come to a conclusive answer, high purity gas experiments utilizing O_2/H_2O mixtures and impurity addition to O_2 or H_2O will be required, similarly to the careful studies that have been performed in the liquid phase. Such experiments have to be coupled with corresponding quantum chemical calculations.

4. ATOMIC LAYER DEPOSITION OF GATE DIELECTRICS

ALD was first introduced as a growth technique in the 1980s and has since been used in various thin film applications, most importantly in electroluminescent display technology and in microelectronics. This technique, also known as atomic layer chemical vapor deposition (ALCVD), provides a higher degree of film uniformity, conformality, and/or thickness control than other chemical growth techniques, such as regular metal–organic chemical vapor deposition (MOCVD). Ritala and Leskelä have thoroughly reviewed the ALD technique, precursors, and applications (84). In the present context, we will limit ourselves to a short description of the principles and implementation of ALD. Then, we will address fundamental reaction mechanisms, using Al_2O_3 growth as a typical example.

ALD is based on self-saturating surface reactions. Molecular precursors are brought to the sample surface in alternating pulses, separated by an inert gas purge. Each reactant undergoes a self-terminating surface reaction, depositing a monolayer of material or less. In commercial reactors, the precursors are usually carried by an inert gas (usually N_2) at sub-atmospheric pressure, while in research reactors pure precursor pulses have been utilized as well. The surface is thus exposed to a pulse series according to A–N_2–B–N_2–A–. . . . For example, the most frequently employed Al_2O_3 growth process uses A = $Al(CH_3)_3$ (trimethylaluminum, TMA) and B = H_2O. Figure 13 schematically illustrates the initial Al_2O_3 growth phase on a hydroxylated

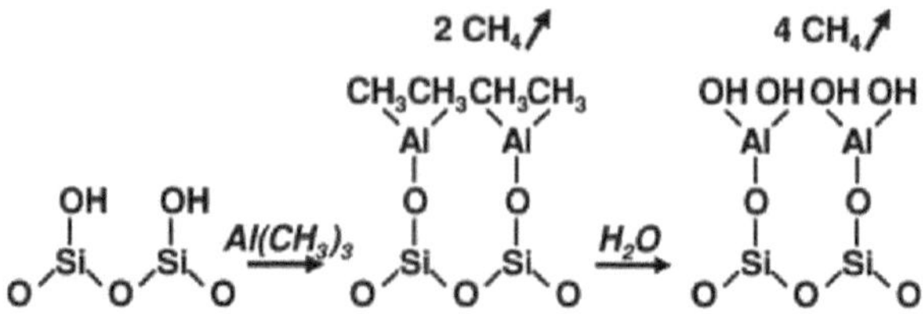

Fig. 13. Idealized scheme of the first Al_2O_3 ALD cycle on hydroxylated SiO_2 using $Al(CH_3)_3$ and water.

SiO_2 surface (disregarding chemical crosslinking as discussed below). The first $Al(CH_3)_3$ pulse reacts with the surface hydroxyl groups, releasing CH_4 (methane), until all reactive sites have been consumed and the reaction stops. A water pulse then replaces the surface $-CH_3$ (methyl) groups by hydroxyl groups, again releasing CH_4. In this way, the chemical ingredients to form a monolayer of Al_2O_3 are deposited, exposing reactive sites available for the second pulse cycle. According to this scheme, it should is possible to deposit a compound (or an element (84)) on a substrate in a highly conformal fashion, with a maximum growth rate of one monolayer per pulse and with thickness and uniformity control at the angstrom level.

However, this idealized picture of ALD clearly neglects a number of potentially important phenomena, such as reaction of the organometallic precursor with more than one hydroxyl group[1,2] (i.e., $(\text{–OH}^*)_n + Al(CH_3)_3 \rightarrow (\text{–O–})_n Al(CH_3)^*_{3-n} + n\ CH_4$) incomplete hydroxyl consumption, steric hindrance, or crosslinking of metal ions through oxygen bridges. In fact, such crosslinking is clearly necessary to form a closed oxide film. In practice, severe deviations from ideality take place. Notably, growth rates are usually lower than one monolayer per pulse. In cases involving unfavorable substrates, even non-constant growth rates have been observed, resulting in film thicknesses that increase non-linearly with the number of cycles. For example, when attempting to grow alternative dielectric materials for MOSFET gate stack applications directly on H/Si, the initial growth rate usually is very low exhibiting a long lag period before linear growth occurs. Al_2O_3 growth from $Al(CH_3)_3$ and water (85–87) and HfO_2 growth from $HfCl_4$ and water (88, 89) are two prominent cases that exhibit such an încubation period.' The surfaces of such films are rough (88–90) and unwanted interfacial SiO_2 is often formed (85, 86, 91).

4.1. Current Knowledge on ALD Mechanisms

Widely varying film qualities raise questions as to the underlying atomistic nucleation and growth mechanisms. There is a critical need to understand these mechanisms in order to achieve better control over device structures and properties. Two approaches have been employed to uncover reaction mechanisms: (a) indirect interpretation of

[1] Surface hydroxyl groups are not the only sites available for $Al(CH_3)_3$ reaction (see below). This does not affect the self-saturating nature of the process.
[2] In this section, *'denotes surface species.

kinetic growth data and of properties of as-deposited films (e.g., thickness, roughness, crystallinity, density, and stoichiometry); and (b) direct *in situ* observation of surface reactions and dielectric film growth.

Based on indirect methods, differences in the initial ALD growth rate on different substrates have been correlated to variations in areal density of nucleation sites. For example, for HfO_2 growth from $HfCl_4$ and water on Si the observed increase in initial growth rate (H/Si $<<$ thermally grown SiO_2/Si $<$ wet-chemically grown SiO_2/Si) correlates with hydroxyl group density on the substrates (88). This indicates that such hydroxyl groups are the reaction sites for incoming $HfCl_4$. Inhomogeneous HfO_2 nucleation due to the low concentration of OH groups on H/Si substrates also rationalizes the observed high film roughness. Finally, steady-state growth rates lower than one monolayer per pulse are usually attributed to steric hindrance due to ligands of the chemisorbed species (88, 92). Recently, such fundamental phenomena have been explicitly incorporated into kinetic models of ALD in order to quantitatively predict growth kinetics and film quality (93, 94). Conversely, theoretical modeling of experimental kinetics may enable the extraction of fundamental reaction parameters.

However, air exposure of as-deposited films may lead to long-term changes in the dielectric material (95) as well as at the high-k/Si interface (86). Also, we shall see that a large number of different elementary reaction steps and adsorption geometries are conceivable. Direct *in situ* observation of ALD surface reactions therefore is indispensable to achieve a full understanding of film formation and impurity incorporation. We exemplify the status of the field by focusing on ALD growth of Al_2O_3 from $Al(CH_3)_3$ and water. It is crucial to understand these reactions, since Al_2O_3, Al-containing ternary oxides (e.g., Hf aluminate), and Al_2O_3-containing multilayer systems (e.g., Al_2O_3–HfO_2 multilayers) are under consideration as gate dielectrics (90, 96) or as tunnel barriers in nonvolatile floating-gate field-effect-transistor (Flash) memories (97).

Important insight into initial $Al(CH_3)_3$ interaction with SiO_2 substrates has been obtained by infrared spectroscopy. Early studies were performed on porous (98–102) and mesoporous (103) high-area SiO_2 substrates. They showed that $Al(CH_3)_3$ reacts with surface siloxane bridges (Si–O–Si) through dissociation and with silanol groups (Si–OH) through a ligand-exchange reaction. Isolated silanol groups are completely eliminated, while a substantial concentration of vicinal silanol groups remain unreacted (99). These reactions result in the formation of 30–40% Si-bonded $(CH_3)_n$ ($n = 1$–3) and 60–70% O-bonded $Al(CH_3)_n$ ($n = 1, 2$) (101, 103). The saturation Al coverage is determined by steric hindrance due to the presence of the methyl groups (101). However, surface chemistry during subsequent cycles of the $Al(CH_3)_3$–water process was not investigated with infrared spectroscopy in these early studies, and growth on H/Si substrates was not addressed.

For flat Si substrates, *In situ* studies of Al_2O_3 growth have mostly been based on the detection of released reaction products by quadrupole mass spectrometry (QMS) (104, 105) and measurement of deposited areal mass density using a quartz crystal microbalance (105, 106). It was thus shown that throughout film growth between 1 and 2 hydroxyl groups are consumed per $Al(CH_3)_3$ adsorbed (105), as is the case for

$Al(CH_3)_3$ reaction with SiO_2 films (101). The temperature has no marked effect on the growth mechanism in the range 150–400°C. However, the growth rate is highest at 250°C, being limited by kinetics at low temperatures and by the concentration of surface OH groups at high temperatures (104, 105).

Experimental observations of $Al(CH_3)_3$ interaction with hydroxylated SiO_2 surfaces were recently complemented by density functional theory (DFT) calculations of reaction pathways and energetics (107). It was shown that the reaction of surface –OH* with $Al(CH_3)_3$ is initiated by Al–O charge transfer, resulting in a weakly chemisorbed state (0.46 eV exothermic). The saddle point energy for precursor decomposition is 1.04 eV higher. Most $Al(CH_3)_3$ therefore desorbs back into the gas phase which explains the low reactive sticking coefficient (high required gas exposures) commonly observed. For the case in which the reaction runs to completion and CH_4 is released, it is 1.91 eV exothermic.

Also the cyclic Al_2O_3 growth reactions have been addressed by DFT (108). Similarly, the reactions of Al–OH* with $Al(CH_3)_3$ and of $Al(CH_3)_n$ ($n = 1, 2$) with H_2O both are initiated by a Lewis acid–base interaction between O lone pair electrons and the empty Al p orbital. A metastable complex is formed in both cases, from which the reaction runs to completion and CH_4 is released. The overall activation barriers are low (<0.17 eV) and all reactions are exothermic (by 1.3–1.7 eV).

4.2. Mechanistic Infrared Studies of Al_2O_3 ALD on H/Si

The lack of experimental data concerning adsorbed species during individual ALD reaction cycles prompted an *in situ* infrared absorption study of Al_2O_3 growth on H-passivated and on oxidized Si using sequential $Al(CH_3)_3$ and water exposures at 300°C (80, 81). The aim was to detect the characteristic absorption bands of atomic and molecular surface species (–H, $-CH_3$, hydroxyls, etc.), as well as the phonon bands of thin films formed (Al_2O_3, SiO_2).

Commercial ALD machines rarely provide the possibility to attach spectroscopic tools. Therefore, a dedicated model reactor has been constructed (Fig. 14) (80). It is small, making it possible to fit it inside the sample compartment of a commercial infrared spectrometer. The rectangular Si sample is held in a vertical geometry by Ta clips for resistive heating and can be rotated to enable angle-dependent infrared measurements. Reactant gases carried in purified N_2 gas are introduced into the chamber via separate gas lines. In such growth studies, it is very important to prevent dielectric growth on the chamber windows. This was ensured by window shutters that are closed during precursor pulses, with a constant N_2 purge of the space between the shutters and the infrared windows.

The power of this spectroscopic approach is illustrated by infrared spectra of the O–H, O–D, and C–H stretching regions for various Si surfaces before and after $Al(CH_3)_3$ exposure (Fig. 15). Firstly, surface hydroxyl groups are clearly detected on the starting substrates, as exemplified by the spectrum of a wet thermal oxide grown *in situ* on H-terminated Si using D_2O: A sharp signal at 2760 cm^{-1} arises from isolated OD groups bonded to oxidized Si sites (109, 110). A weak broad band extending down by several hundred cm^{-1} is due to groups of H-bonded hydroxyls (69). The integrated band intensities can be used as a measure of hydroxyl area

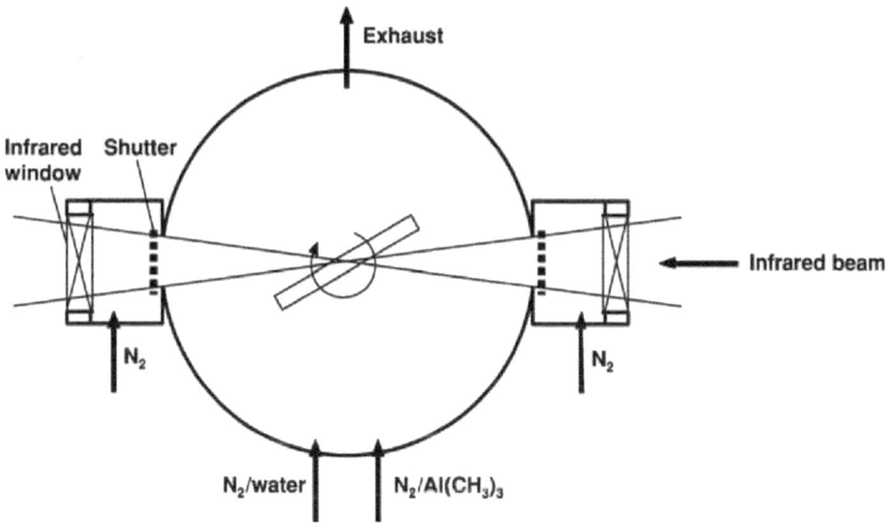

Fig. 14. Schematic drawing (top view) of an *in situ* infrared transmission cell used for mechanistic ALD studies (not to scale) (from (80), reproduced with permission).

densities. Calibration was achieved by comparison to well-characterized OD + D and OH + H monolayers on Si(100) prepared in ultra-high vacuum (Fig. 11). These surfaces contain 1/2 ML of hydroxyl groups bonded to (non-oxidized) Si surface atoms (2698 and 3658 cm^{-1}, respectively), corresponding to a number density of $3.39\ nm^{-2} = 3.39 \times 10^{14}\ cm^{-2}$ (18, 82). For isolated OD groups on the wet thermal oxide (2760 cm^{-1}), a number density of 1.8 nm^{-2} was thus obtained. Quantification of the broad signal from H-bonded hydroxyls in the monolayer regime is not as accurate due to baseline instabilities.

When initiating Al_2O_3 growth by exposing SiO_2 substrates to $Al(CH_3)_3$, a consumption of surface hydroxyl groups is expected. As Fig. 15 illustrates, spectral changes were indeed observed in the region of hydroxyl stretch and C–H stretch vibrations. In these difference spectra, obtained by plotting the TMA-induced absorbance change, spectral regions with reduced absorbance indicate the loss of molecular surface species due to surface chemical reactions. For instance, the sharp 'anti-absorption' features are due to the consumption of isolated OH groups (3741–3746 cm^{-1}) from the chemical and dry thermal oxides and of isolated OD groups (2760 cm^{-1}) from the wet thermal oxide grown from D_2O. Their low-frequency shoulders and the weak broad bands originate from the loss of H-bonded hydroxyl groups which participate in the bonding with their O atoms only, or with both their O and H/D atoms, respectively (69).

Interestingly, the OH areal density lost from the chemical oxides (3–4 nm^{-2} of isolated and weakly H-bonded hydroxyls) is close to the areal density of Al ions deposited on ozone oxide, which was found to be 4.2 nm^{-2} by Rutherford backscattering spectrometry (111). However, we note that the hydroxyl–Al ratio is not necessarily unity: on the one hand, $Al(CH_3)_3$ may react with more than one hydroxyl group

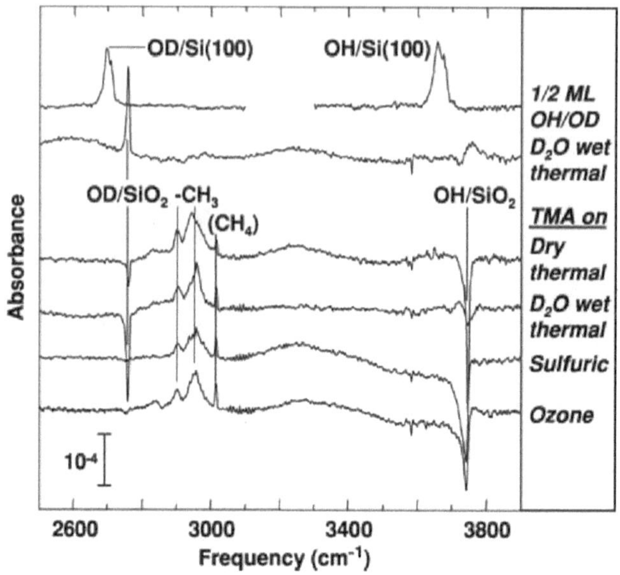

Fig. 15. *In situ* transmission infrared spectra of various chemical and thermal SiO_2/Si(100) surfaces (80) after $Al(CH_3)_3$ exposure in the C–H and O–D stretching regions (reference: as-prepared SiO_2/Si(100) surfaces annealed to 300°C; i.e., difference spectra are plotted). For comparison, spectra taken from an *in situ* grown wet thermal oxide before $Al(CH_3)_3$ exposure (reference: H-terminated Si) and calibration spectra from 1/2 ML OH and OD groups on Si(100), prepared in ultra-high vacuum (18, 82) (reference: clean Si(100)), are shown (from (80), reproduced with permission).

or with oxygen bridges (101); on the other hand, steric hindrance between methyl groups may limit the maximum areal density of deposited Al species, independent of the exact nature of the substrate (101), which might result in unreacted H-bonded hydroxyl groups (69). In line with self-saturation through steric hindrance, the infrared spectra (Fig. 15) indeed indicate similar $Al(CH_3)_3$-induced methyl areal densities on all oxides, as judged from the characteristic C–H stretch signals (~2945–2960 and 2902 cm^{-1}).

We have seen that $Al(CH_3)_3$-induced nucleation on SiO_2 substrates is fairly well understood. This is not the case for H/Si substrates. Usually, an incubation period of slower growth is found, resulting in films that are thinner by an equivalent of 4 to ~15 cycles than what is expected for ideal linear growth (85–87, 111). It is unclear which factors determine the extent of this incubation period. Also, it is desirable to devise ways to induce linear ALD growth right from the start, as this would lead to dielectric films with much lower roughness. It has been attempted to activate the H/Si surface through hydroxylation using 20 H_2O pulses prior to the first $Al(CH_3)_3$ exposure, in order to induce homogeneous nucleation and linear growth (87). This approach had to be unsuccessful, since the H/Si surface is virtually inert towards H_2O even at 300°C, as discussed earlier in this chapter.

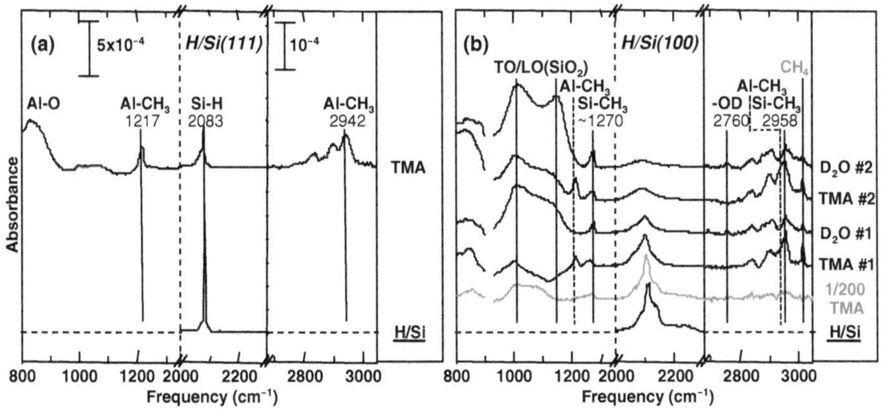

Fig. 16. *In situ* transmission infrared spectra of H/Si before and after precursor exposure in N_2 carrier gas (1 bar) at 573 K: (a) H/Si(111) exposed to 5×10^7 L $Al(CH_3)_3$ (1 mbar) and (b) H/Si(100) exposed to 2×10^5 L $Al(CH_3)_3$ (1 mbar), followed by sequential pulses of 5×10^7 L $Al(CH_3)_3$ (1 mbar) and 5×10^8 L D_2O (10 mbar). Reference spectra in oxide phonon, CH_3 bending, C–H, and O–D stretching regions: H-terminated Si. Note that a signal at $\sim$920 cm^{-1}, due to Si–H bending modes of the H/Si(100) reference surface, has been removed for clarity. Reference spectra in Si–H stretching region: Si with native oxide. The symbols TO and LO mark transverse- and longitudinal-optical phonon modes of SiO_2 (after (81)).

Infrared spectroscopy was employed to provide insight into the mechanisms of Al_2O_3 nucleation and initial growth. Spectra for the $Al(CH_3)_3$–H_2O process on H/Si were recorded for both (111) and (100) surfaces, to establish whether the difference in defect density affects the nucleation mechanism (Fig. 16) (81). Extended $Al(CH_3)_3$ pulses (5×10^7 L $Al(CH_3)_3$; exposure ca. 10^4 times higher than in standard commercial processes) react with both surface orientations (Fig. 16(a) and (b)). This is in sharp contrast to water exposure, as we saw earlier. Focusing first on atomically flat H/Si(111) (Fig. 16(a)), $Al(CH_3)_3$ reaction is evidenced by a drop of the hydrogen concentration (i.e., the Si–H stretch area) and the rise of modes due to Al–CH_3 (and not Si–CH_3) bonding, at 1217 and 2942 cm^{-1} (112–116). On atomically rough H/Si(100) (Fig. 16(b)), by contrast, low $Al(CH_3)_3$ exposures give rise to Si–CH_3 bonding. Clearly, methyl transfer to Si occurs at defect sites, for example at step edges or oxidized Si atoms. Only at higher exposures is Al–$(CH_3)_n$ formed. On both surface orientations, oxide phonons below 1100 cm^{-1} (24, 117) show that oxygen insertion results in Si–O–Al bond formation. The oxygen appears to originate from trace impurities in the gas pulse, as supported by DFT calculations (118) and discussed in more detail below. Non-oxygen-containing Si–Al–CH_3 arrangements probably are also formed (81).

During the second half of the first ALD cycle, when water is exposed to $Al(CH_3)_3$-functionalized H/Si(100), rapid sub-surface oxidation of Si occurs (Fig. 16(b)): 1.2 Å SiO_2 is formed. Aluminum clearly catalyzes H/Si oxidation, as has been observed previously for Si oxidation (119). In addition, Al-bonded CH_3 is replaced by OD, as expected. Additional Si–CH_3 forms via transfer of methyl groups from Al sites,

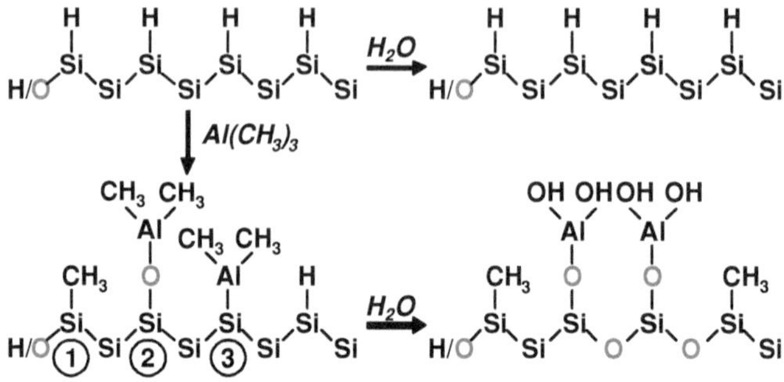

Fig. 17. Simplified scheme of water and $Al(CH_3)_3$ reactions with H/Si. H/O'symbolizes H or O atoms bonded to a Si structural or chemical defect site. The numbers 1–3 indicate the sequence of methyl formation. On H/Si(100), we have observed species 1 and 2, while on step- and defect-free H/Si(111) species 2 and 3 are formed. A low degree of hydroxylation of SiO_2 sites (~0.02 ML) has been omitted (from (81), reproduced with permission).

most likely due to the presence of O–Si sites favoring CH_3 accommodation. Figure 17 summarizes the results discussed so far.

Data from the second ALD cycle confirm the accepted ALD steady-state growth mechanism (120): The second $Al(CH_3)_3$ pulse reacts with surface OD groups, forming Al–CH_3; the second water pulse then forms new Al–OD groups (Fig. 16(b)). During subsequent cycles, amorphous Al_2O_3 (954 cm^{-1}) grows and most (but not all) Si–CH_3 decomposes, as expected (Fig. 18: 2–16 cycles). Interface formation is not complete after the first cycle as commonly assumed. Instead, SiO_2 continues to form for several cycles. The SiO_2 thickness ultimately formed is controlled by the water exposure per pulse (81). In the model study employing 5×10^8 L water per pulse, a thickness of 4 Å is reached, while smaller values are found for films grown in commercial reactors that utilize typical exposures of e.g., 10^4 L water per pulse (81).

The mechanistic surface chemical insight gained in this model study suggests a route to achieving more linear ALD growth kinetics and, concomitantly, lower film roughness on H/Si. Increasing the first $Al(CH_3)_3$ exposure should promote initial growth. This was demonstrated for Al_2O_3 growth from $Al(CH_3)_3$ and water and for HfO_2 growth from $HfCl_4$ and water in a commercial reactor, otherwise using standard precursor exposures (111). Linear ALD kinetics were confirmed by narrow nuclear resonance profiling (NRP) and RBS. Organic functionalization schemes thus hold promise for controlling nucleation of a wide range of dielectric materials.

Returning to atomic scale insight, the infrared observations motivated detailed DFT calculations of reaction energetics and pathways relevant in ALD precursor interaction with H/Si surfaces (78, 118). H/Si(100)(2 × 1) was chosen as a model substrate. The barrier for $Al(CH_3)_3$ reaction forming Si–$Al(CH_3)_2^*$ was found to be 0.3 eV lower than for water-induced hydroxylation (78). This supports the experimental observation that $Al(CH_3)_3$ and not H_2O reacts with the H/Si surface. In addition, the finding of substantial Si–O–Al bonding and O incorporation into the substrate upon $Al(CH_3)_3$ exposure was taken into account by comparing the reactivities of the

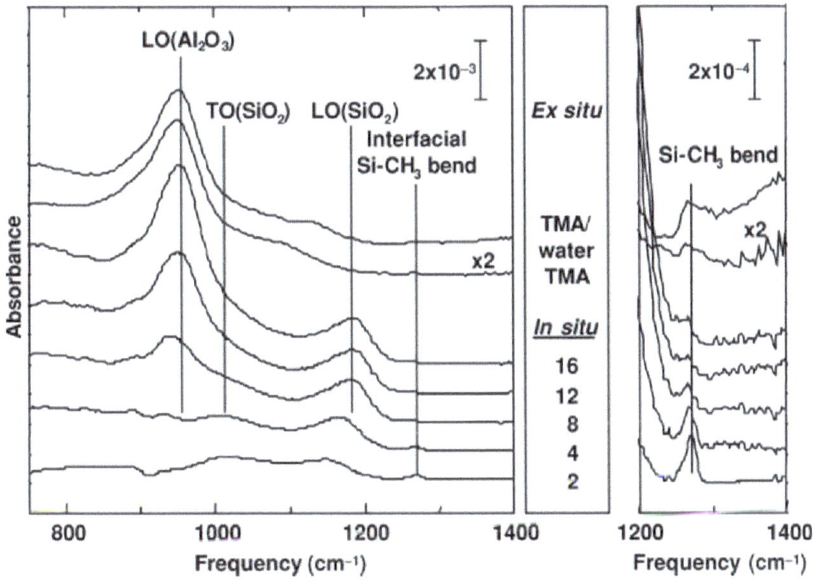

Fig. 18. Transmission infrared spectra of H/Si(100) after exposure to $Al(CH_3)_3$–water cycles. Reference spectrum: H/Si. Bottom spectra taken *in situ* after the number of $Al(CH_3)_3$–D_2O cycles indicated (growth on both wafer surfaces). Top spectra taken *ex situ* from films grown from 34 $Al(CH_3)_3$–H_2O cycles in a commercial ALD reactor; all pulses are short, except for the first cycle: long $Al(CH_3)_3$/short H_2O labeled 'TMA', long $Al(CH_3)_3$/long H_2O labeled 'TMA/water' (growth on one wafer surface only; spectra multiplied by a factor of two) (after (81)).

pristine precursor molecules with that of an important impurity, dimethylaluminumhydroxide ($Al(CH_3)_2OH$) (118). This species may originate from an exothermic and barrierless gas phase side reaction of $Al(CH_3)_3$ with H_2O. Two reaction pathways of $Al(CH_3)_2OH$ were shown to be particularly significant for interface formation: (a) the reaction forming $–O–Al(CH_3)_2^*$, which is the most exothermic of all reactions of the three molecules studied; and (b) surface hydroxylation, whose activation energy is 1.1 eV lower for $Al(CH_3)_2OH$ than for H_2O.

These results show that precursor purity is of prime concern when studying reaction mechanisms relevant to nucleation and Si/oxide interface formation in ALD. This might explain differences observed in the length of incubation period in different studies, with other experimental parameters being similar (85–87). Clearly, a deeper understanding of the impact of impurities on film structure will open opportunities for controlled addition of trace reactants to optimize the final film quality.

5. CONCLUSIONS AND DIRECTIONS FOR FUTURE RESEARCH

Dielectric growth on Si substrates constitutes a cornerstone of microelectronic fabrication and thus has tremendous commercial impact. Constant improvements in surface and growth chemistry are needed to fulfill ever more stringent requirements on atomic structure and composition of dielectric films arising from continued scaling of electronic devices. We have shown that true understanding of the interdependencies between processing parameters and film quality can only be gained through *in situ* characterization of the surface reactions involved. Commercial considerations aside, Si surface chemistry is of prime academic interest in itself as Si constitutes *the* model substrate for all of semiconductor surface chemistry.

Whether under UHV conditions or in processing environments, *optical spectroscopy* is a powerful technique to characterize surface reactions as they proceed. Infrared spectroscopy is particularly attractive as it enables detection both of molecular or atomic adsorbates and of dielectric films formed. However, it requires effort to avoid window contamination and to achieve sufficient spectral sensitivity.

We have seen that the initial *oxidation and nitridation of clean Si surfaces* under UHV conditions is quite well understood. Adsorption and decomposition pathways of O_2, H_2O, and NH_3 on Si(100)(2 $\times$ 1) have been characterized thoroughly. In the case of H-containing molecules they are very similar, involving dissociation on surface dimers. Thermally activated reactions differ, however, and can lead to NH_3 desorption on the one hand and O insertion on the other hand. Infrared spectroscopy has played a crucial role in elucidating such reaction steps as well as details of O migration and agglomeration in near-surface regions. Such processes determine the quality (e.g., the uniformity) of the final oxide or nitride films, and are therefore of highest importance.

In-depth characterization of the *reaction mechanisms on H-passivated Si surfaces* is in many ways more demanding than is the case for clean surfaces. While UHV conditions are not required in order to prepare and study such substrates, tremendous effort is necessary to avoid reactions by minority species (impurities) that may dominate the reaction at such high pressures. This issue has so far prevented a full mechanistic understanding of H/Si oxidation, even though a substantial body of work exists. Careful studies utilizing high purity gases can yield approximate kinetic parameters for some basic reactions of simple molecules, including their dependence of the type of surface site involved. In broad terms, H/Si reactivity increases according to terrace site $<$ step site'and $H_2O < O_2 <$ humid air.' The surprising finding of enhanced oxidation by humid air may arise from a cooperative effect of H_2O and O_2 or from impurity species not yet considered, such as O_3 or radicals. High purity gas experiments utilizing O_2/H_2O mixtures and/or added impurities, in the dark or under well-characterized illumination to control photochemical reactions, and coupled with corresponding quantum chemical calculations, will be required to resolve this issue.

A comprehensive understanding of Si surface reactions is particularly crucial for complex growth processes such as *atomic layer deposition of metal oxides*. Using Al_2O_3 growth as an example, we have seen that a number of surface phenomena give rise to deviations from simple layer-by-layer growth. For example, metal-catalyzed subsurface oxidation continues until a sufficiently thick metal oxide layer protects the Si substrate; and relatively stable Si-bonded methyl groups are created initially and are not fully decomposed in the course of film growth. We envision that differences in reaction kinetics and energetics for different desirable and undesirable species may be exploited via careful tuning of gas exposures and growth temperatures. Film quality may thus be optimized, e.g., minimizing interfacial layer thickness and impurity concentration. Rational design of such processes will require determination of thermodynamic quantities not yet known in many cases. Also, the impact of impurities on nucleation and growth is only starting to be appreciated fully. This is still a largely open field of research.

While *Si surface termination other than by H* is beyond the scope of the present chapter, it is important to point out that alternative functionalization schemes may provide added flexibility to dielectric deposition onto Si, e.g., to optimize film nucleation and substrate passivation. Organic species appear particularly attractive due to the wide range of terminal functional groups conceivable, and due to the stability of the Si–C bond. The incredibly rich field of organic Si surface chemistry has only recently started to be explored in depth (see, e.g., reviews by Buriak et al. (121) and Bent et al. (122, 123)). Renewed and increased interest may be expected from researchers in the fields of organic electronics and biosensors. Most likely, H/Si surfaces will remain the preferred starting substrates for most future functionalization processes, underscoring the importance of fundamental work on these surfaces.

Si surface chemistry for electronics is a thriving field. Optimization of growth processes and development of novel reaction schemes drives the implementation of new materials into traditional devices, and possibly the fabrication of entirely new structures. *In situ* characterization of reaction mechanisms is indispensable, providing atomic scale insight into interface and film formation.

ACKNOWLEDGEMENTS

The experimental work performed at Rutgers University was supported by International SEMATECH (Contract number 306106). A substantial part of the work presented was performed at Agere Systems at Lucent Technologies'Bell Laboratories (Murray Hill, New Jersey). M.M.F. was supported, in part, by a fellowship within the Postdoc Program of the German Academic Exchange Service (DAAD). Among many others, the authors wish to thank Edward E. Chaban, Stan B. Christman, Eric Garfunkel, Martin L. Green, Mathew D. Halls, Krishnan Raghavachari, Glen D. Wilk, and Xiang Zhang for fruitful collaboration and stimulating discussions.

REFERENCES

1. G.D. Wilk, R.M. Wallace, J.M. Anthony, *J. Appl. Phys.* **89**, 5243 (2001).
2. M.M. Frank, S. Sayan, S. Dörmann, T.J. Emge, L.S. Wielunski, E. Garfunkel, Y.J. Chabal, *Mater. Sci. Eng. B* **109**, 6 (2004).
3. Y.J. Chabal, *Surf. Sci. Rep.* **8**, 211 (1988).
4. Y.J. Chabal, *Chemistry and Physics of Solid Surfaces VII*, eds. R. Vanselow, R.F. Howe (Springer, Berlin, 1988), pp. 108–150.
5. G.F. Cerofolini, G. LaBruna, L. Meda, *Appl. Surf. Sci.* **93**, 255 (1996).

6. K. Kato, T. Uda, K. Terakura, *Phys. Rev. Lett.* **80**, 2000 (1998).
7. A.A. Demkov, O.F. Sankey, *Phys. Rev. Lett.* **83**, 2038 (1999).
8. Y. Widjaja, C.B. Musgrave, *J. Chem. Phys.* **116**, 5774 (2002).
9. T. Hoshino, M. Tsuda, S. Oikawa, I. Ohdomari, *Phys. Rev. B* **50**, 14999 (1994).
10. R. Ludeke, A. Koma, *Phys. Rev. Lett.* **34**, 1170 (1975).
11. H. Ibach, J.E. Rowe, *Phys. Rev. B* **10**, 710 (1974).
12. P. Morgen, U. Höfer, W. Wurth, E. Umbach, *Phys. Rev. B* **39**, 3720 (1989).
13. U. Höfer, P. Morgen, W. Wurth, E. Umbach, *Phys. Rev. B* **40**, 1130 (1989).
14. I.W. Lyo, P. Avouris, B. Schubert, R. Hoffmann, *J. Phys. Chem.* **94**, 4400 (1990).
15. Y.J. Chabal, K. Raghavachari, X. Zhang, E. Garfunkel, *Phys. Rev. B* **66**, 161315 (2002).
16. K.T. Queeney, Y.J. Chabal, K. Raghavachari, *Phys. Rev. Lett.* **86**, 1046 (2001).
17. Y.J. Chabal, S.B. Christman, *Phys. Rev. B* **29**, 6974 (1984).
18. A.B. Gurevich, B.B. Stefanov, M.K. Weldon, Y.J. Chabal, K. Raghavachari, *Phys. Rev. B* **58**, R13434 (1998).
19. M.K. Weldon, B.B. Stefanov, K. Raghavachari, Y.J. Chabal, *Phys. Rev. Lett.* **79**, 2851 (1997).
20. M.K. Weldon, K.T. Queeney, A.B. Gurevich, B.B. Stefanov, Y.J. Chabal, K. Raghavachari, *J. Chem. Phys.* **113**, 2440 (2000).
21. A. Estève, Y.J. Chabal, K. Raghavachari, M.K. Weldon, K.T. Queeney, M.D. Rouhani, *J. Appl. Phys.* **90**, 6000 (2001).
22. K.T. Queeney, Y.J. Chabal, M.K. Weldon, K. Raghavachari, *Phys. Stat. Sol. A* **175**, 77 (1999).
23. K.T. Queeney, M.K. Weldon, Y.J. Chabal, K. Raghavachari, *J. Chem. Phys.* **119**, 2307 (2003).
24. K.T. Queeney, M.K. Weldon, J.P. Chang, Y.J. Chabal, A.B. Gurevich, J. Sapjeta, R.L. Opila, *J. Appl. Phys.* **87**, 1322 (2000).
25. Y.J. Chabal, M.K. Weldon, K.T. Queeney, A. Estève, *Fundamental Aspects of Silicon Oxidation*, Vol. 46, ed. Y.J. Chabal (Springer, Berlin, Heidelberg, New York, 2001), pp. 143–159.
26. I.J. R. Baumvol, *Surf. Sci. Rep.* **36**, 1 (1999).
27. H. Harris, K. Choi, N. Mehta, A. Chandolu, N. Biswas, G. Kipshidze, S. Nikishin, S. Gangopadhyay, H. Temkin, *Appl. Phys. Lett.* **81**, 1065 (2002).
28. K. Onishi, C.S. Kang, R. Choi, H.J. Cho, S. Gopalan, R.E. Nieh, S.A. Krishnan, J.C. Lee, *IEEE Trans. Electron Devices* **50**, 384 (2003).
29. W. Kern, *Handbook of Semiconductor Wafer Cleaning Technology: Science, Technology, and Applications* (Noyes Publications, Park Ridge, NJ, USA, 1993).
30. Y.J. Chabal, G.S. Higashi, K. Raghavachari, V.A. Burrows, *J. Vac. Sci. Technol. A* **7**, 2104 (1989).
31. G.S. Higashi, Y.J. Chabal, G.W. Trucks, K. Raghavachari, *Appl. Phys. Lett.* **56**, 656 (1990).
32. T. Hattori, T. Aiba, E. Iijima, Y. Okube, H. Nohira, N. Tate, M. Katayama, *Appl. Surf. Sci.* **104/105**, 323 (1996).
33. K. Choi, H. Harris, S. Gangopadhyay, H. Temkin, *J. Vac. Sci. Technol. A* **21**, 718 (2003).
34. M.A. Henderson, *Surf. Sci. Rep.* **46**, 5 (2002).
35. S. Watanabe, Y. Sugita, *Surf. Sci.* **327**, 1 (1995).
36. S. Watanabe, Y. Sugita, *Appl. Surf. Sci.* **107**, 90 (1996).
37. S.P. Garcia, H.L. Bao, M.A. Hines, *Surf. Sci.* **541**, 252 (2003).
38. P. Jakob, Y.J. Chabal, K. Raghavachari, R.S. Becker, A.J. Becker, *Surf. Sci.* **275**, 407 (1992).
39. T. Tada, R. Yoshimura, *Phys. Lett. A* **220**, 224 (1996).
40. T. Imai, Y. Kurioka, N. Nagataki, M. Okuyama, Y. Hamakawa, *Appl. Surf. Sci.* **113/114**, 398 (1997).

41. M. Morita, T. Ohmi, E. Hasegawa, M. Kawakami, K. Suma, *Appl. Phys. Lett.* **55**, 562 (1989).
42. M. Morita, T. Ohmi, E. Hasegawa, M. Kawakami, M. Ohwada, *J. Appl. Phys.* **68**, 1272 (1990).
43. H. Ogawa, K. Ishikawa, C. Inomata, S. Fujimura, *J. Appl. Phys.* **79**, 472 (1996).
44. M.L. W. Vanderzwan, J.A. Bardwell, G.I. Sproule, M.J. Graham, *Appl. Phys. Lett.* **64**, 446 (1994).
45. G.F. Cerofolini, L. Meda, *J. Non-Cryst. Solids* **216**, 140 (1997).
46. G. Hess, M. Russell, B. Gong, P. Parkinson, J.G. Ekerdt, *J. Vac. Sci. Technol. B* **15**, 1129 (1997).
47. K. Yamamoto, M. Hasegawa, *J. Vac. Sci. Technol. B* **12**, 2493 (1994).
48. H. Ikeda, T. Yamada, K. Hotta, S. Zaima, Y. Yasuda, *Appl. Surf. Sci.* **101**, 431 (1996).
49. J. Westermann, H. Nienhaus, W. Mönch, *Surf. Sci.* **311**, 101 (1994).
50. K. Kawamura, S. Ishizuka, H. Sakaue, Y. Horiike, *Jpn. J. Appl. Phys. Pt. 1* **30**, 3215 (1991).
51. M. Niwano, J. Kageyama, K. Kurita, K. Kinashi, I. Takahashi, N. Miyamoto, *J. Appl. Phys.* **76**, 2157 (1994).
52. M.P. Stewart, J.M. Buriak, *J. Am. Chem. Soc.* **123**, 7821 (2001).
53. S.A. Mitchell, *J. Phys. Chem. B* **107**, 9388 (2003).
54. S. Ye, T. Saito, S. Nihonyanagi, K. Uosaki, P.B. Miranda, D. Kim, Y.-R. Shen, *Surf. Sci.* **476**, 121 (2001).
55. T. Hattori, H. Nohira, *Fundamental Aspects of Silicon Oxidation*, ed. Y.J. Chabal (Springer, Berlin, 2001), pp. 61–88.
56. K. Kato, H. Kajiyama, S. Heike, T. Hashizume, T. Uda, *Phys. Rev. Lett.* **86**, 2842 (2001).
57. A. Estève, M.D. Rouhani, D. Estève, *Thin Solid Films* **344**, 350 (1999).
58. K. Sakata, A. Tachibana, S. Zaima, Y. Yasuda, *Jpn. J. Appl. Phys. Pt. 1* **37**, 4962 (1998).
59. T. Miura, M. Niwano, D. Shoji, N. Miyamoto, *J. Appl. Phys.* **79**, 4373 (1996).
60. T. Miura, M. Niwano, D. Shoji, N. Miyamoto, *Appl. Surf. Sci.* **101**, 454 (1996).
61. W. Henrion, M. Rebien, H. Angermann, A. Röseler, *Appl. Surf. Sci.* **202**, 199 (2002).
62. M.C. Hersam, N.P. Guisinger, J.W. Lyding, D.S. Thompson, J.S. Moore, *Appl. Phys. Lett.* **78**, 886 (2001).
63. Y.J. Liu, D.M. Waugh, H.Z. Yu, *Appl. Phys. Lett.* **81**, 4967 (2002).
64. X. Zhang, Y.J. Chabal, S.B. Christman, E.E. Chaban, E. Garfunkel, *J. Vac. Sci. Technol. A* **19**, 1725 (2001).
65. X. Zhang, E. Garfunkel, Y.J. Chabal, S.B. Christman, E.E. Chaban, *Appl. Phys. Lett.* **79**, 4051 (2001).
66. X. Zhang, Ph.D. Thesis, Rutgers University (2002).
67. Y.J. Chabal, K. Raghavachari, *Phys. Rev. Lett.* **54**, 1055 (1985).
68. K. Raghavachari, P. Jakob, Y.J. Chabal, *Chem. Phys. Lett.* **206**, 156 (1993).
69. B.A. Morrow, A.J. McFarlan, *J. Non-Cryst. Solids* **120**, 61 (1990).
70. R.T. Sanderson, *Chemical Bonds and Bond Energy* (Academic Press, New York, 1976).
71. Y. Yagi, T. Imaoka, Y. Ksama, T. Ohmi, *IEEE Trans. Semicond. Manufact.* **5**, 121 (1992).
72. N. Takagi, N. Minami, T. Furukawa, M. Nishijima, *Surf. Sci.* **297**, L43 (1993).
73. S. Takami, Y. Egashira, H. Komiyama, *Jpn. J. Appl. Phys. Pt. 1* **36**, 2288 (1997).
74. Y.B. Kim, M. Tuominen, I. Raaijmakers, R. de Blank, R. Wilhelm, S. Haukka, *Electrochem. Solid State Lett.* **3**, 346 (2000).
75. M.A. Zabi, C.A. S´ebenne, J.P. Lacharme, *Surf. Rev. Lett.* **8**, 25 (2001).
76. K. Akagi, M. Tsukada, *Thin Solid Films* **344**, 397 (1999).
77. S. Picaud, C. Girardet, *Surf. Sci.* **258**, 210 (1991).
78. M.D. Halls, K. Raghavachari, *J. Chem. Phys.* **118**, 10221 (2003).
79. R.Q. Zhang, W.C. Lu, S.T. Lee, *Appl. Phys. Lett.* **80**, 4223 (2002).
80. M.M. Frank, Y.J. Chabal, G.D. Wilk, *Mater. Res. Soc. Symp. Proc.* **745**, 41 (2003).

81. M.M. Frank, Y.J. Chabal, G.D. Wilk, *Appl. Phys. Lett.* **82**, 4758 (2003).
82. A.B. Gurevich, B.B. Stefanov, M.K. Weldon, Y.J. Chabal (unpublished).
83. M.M. Frank, Y.J. Chabal (unpublished).
84. M. Ritala, M. Leskelä, *Handbook of Thin Film Materials*, Vol. 234, ed. H.S. Nalwa (Academic Press, 2002), pp. 183–198.
85. H. Bender, T. Conard, H. Nohira, J. Petry, O. Richard, C. Zhao, B. Brijs, W. Besling, C. Detavernier, W. Vandervorst, M. Caymax, S.D. Gendt, J. Chen, J. Kluth, W. Tsai, J.W. Maes, *Extended Abstracts of International Workshop on Gate Insulator*, IWGI 2001 (IEEE Cat. No. 01EX537), 86.
86. V. Cosnier, H. Bender, A. Caymax, J. Chen, T. Conard, H. Nohira, O. Richard, W. Tsai, W. Vandervorst, E. Young, C. Zhao, S.D. Gendt, A. Heyns, J.W. H. Maes, M. Tuominen, N. Rochat, M. Olivier, A. Chabli, *Extended Abstracts of International Workshop on Gate Insulator*, IWGI 2001 (IEEE Cat. No. 01EX537), 226 (2001).
87. L.G. Gosset, J.F. Damlencourt, O. Renault, D. Rouchon, P. Holliger, A. Ermolieff, I. Trimaille, J.J. Ganem, F. Martin, M.N. Semeria, *J. Non-Cryst. Solids* **303**, 17 (2002).
88. M.L. Green, M.-Y. Ho, B. Busch, G.D. Wilk, T. Sorsch, T. Conard, B. Brijs, W. Vandervorst, P.I. Räisänen, D. Muller, M. Bude, J. Grazul, *J. Appl. Phys.* **92**, 7168 (2002).
89. E.P. Gusev, J.C. Cabral, M. Copel, C. DEmic, M. Gribelyuk, *Microelec. Eng.* **69**, 145 (2003).
90. K. Kukli, M. Ritala, M. Leskelä, T. Sajavaara, J. Keinonen, *J. Mater. Sci.-Mater. Electron.* **14**, 361 (2003).
91. B.W. Busch, O. Pluchery, Y.J. Chabal, D.A. Muller, R.L. Opila, J.R. Kwo, E. Garfunkel, *MRS Bull.* **27**, 206 (2002).
92. M. Ylilammi, *Thin Solid Films* **279**, 124 (1996).
93. M.A. Alam, M.L. Green, *J. Appl. Phys.* **94**, 3403 (2003).
94. R.L. Puurunen, *Chem. Vap. Deposition* **10**, 159 (2004).
95. T. Gougousi, D. Niu, R.W. Ashcraft, G.N. Parsons, *Appl. Phys. Lett.* **83**, 3543 (2003).
96. M.-Y. Ho, H. Gong, G.D. Wilk, B.W. Busch, M.L. Green, W.H. Lin, A. See, S.K. Lahiri, M.E. Loomans, P.I. Räisänen, T. Gustafsson, *Appl. Phys. Lett.* **81**, 4218 (2002).
97. J.D. Casperson, L.D. Bell, H.A. Atwater, *J. Appl. Phys.* **92**, 261 (2002).
98. M.E. Bartram, T.A. Michalske, J.W. R. Jr., *J. Phys. Chem. B* **95**, 4453 (1991).
99. B.A. Morrow, A.J. McFarlan, *The Colloid Chemistry of Silica*, ed. H.E. Bergna (American Chemical Society, Washington, DC, 1994), Vol. 234, pp. 183–198.
100. E.L. Lakomaa, A. Root, T. Suntola, *Appl. Surf. Sci.* **107**, 107 (1996).
101. R.L. Puurunen, A. Root, S. Haukka, E.I. Iiskola, M. Lindblad, A.O. I. Krause, *J. Phys. Chem. B* **104**, 6599 (2000).
102. R.L. Puurunen, A. Root, P. Sarv, S. Haukka, E.I. Iiskola, M. Lindblad, A.O. I. Krause, *Appl. Surf. Sci.* **165**, 193 (2000).
103. R. Anwander, C. Palm, O. Groeger, G. Engelhardt, *Organometallics* **17**, 2027 (1998).
104. M. Juppo, A. Rahtu, M. Ritala, M. Leskelä, *Langmuir* **16**, 4034 (2000).
105. A. Rahtu, T. Alaranta, M. Ritala, *Langmuir* **17**, 6506 (2001).
106. J.W. Elam, M.D. Groner, S.M. George, *Rev. Sci. Instrum.* **73**, 2981 (2002).
107. L. Jeloaica, A. Estève, M.D. Rouhani, D. Estève, *Appl. Phys. Lett.* **83**, 542 (2003).
108. Y. Widjaja, C.B. Musgrave, *Appl. Phys. Lett.* **80**, 3304 (2002).
109. F.H. V. Cauwelaert, P.A. Jacobs, J.B. Uytterhoeven, *J. Phys. Chem.* **76**, 1434 (1972).
110. H.E. Bergna, *The Colloid Chemistry of Silica*, Vol. 234, ed. H.E. Bergna (American Chemical Society, Washington, DC, 1994), pp. 1–47.
111. M.M. Frank, Y.J. Chabal, M.L. Green, A. Delabie, B. Brijs, G.D. Wilk, M.-Y. Ho, E.B. O. da Rosa, I.J. R. Baumvol, F.C. Stedile, *Appl. Phys. Lett.* **83**, 740 (2003).
112. Y. Imaizumi, Y. Zhang, Y. Tsusaka, T. Urisu, S. Sato, *J. Mol. Struct.* **352/353**, 447 (1995).
113. R.L. Puurunen, M. Lindblad, A. Root, A.O. I. Krause, *Phys. Chem. Chem. Phys.* **3**, 1093 (2001).

114. C.J. Pouchert, *The Aldrich Library of FT-IR Spectra*, 1st Edition (Aldrich Chemical Company, Milwaukee, WI, 1985).
115. J.A. Glass, E.A. Wovchko, J.T. Yates, *Surf. Sci.* **338**, 125 (1995).
116. A. Fidélis, F. Ozanam, J.N. Chazalviel, *Surf. Sci.* **444**, L7 (2000).
117. W. Mozgawa, M. Sitarz, M. Rokita, *J. Mol. Struct.* **512**, 251 (1999).
118. M.D. Halls, K. Raghavachari, M.M. Frank, Y.J. Chabal, *Phys. Rev. B* **68**, 161302(R) (2003).
119. S.W. Lim, F. Machuca, H.M. Liao, R.P. Chiarello, R.C. Helms, *J. Electrochem. Soc.* **147**, 1136 (2000).
120. A.W. Ott, J.W. Klaus, J.M. Johnson, S.M. George, *Thin Solid Films* **292**, 135 (1997).
121. J.M. Buriak, *Chem. Rev.* **102**, 1271 (2002).
122. S.F. Bent, *Surf. Sci.* **500**, 879 (2002).
123. S.F. Bent, *J. Phys. Chem. B* **106**, 2830 (2002).

Chapter 12

METHODOLOGY FOR DEVELOPMENT OF HIGH-κ STACKED GATE DIELECTRICS ON III–V SEMICONDUCTORS

MATTHIAS PASSLACK

Freescale Semiconductor, Tempe, AZ 85284, USA

ABSTRACT

A comprehensive methodology for the development of gate dielectrics on III–V semiconductors is presented. This methodology has been motivated by the tremendous difficulties encountered during the development of gate dielectrics on GaAs. The understanding that modern gate dielectrics are typically layered structures with the immediate dielectric/semiconductor interface having substantially different (and often mutually exclusive) requirements compared to the bulk of the dielectric film in terms of materials, manufacturing, and suitable characterization techniques, is at the core of the proposed methodology. While capacitor-based characterization methods such as capacitance voltage measurements which require to maintain quasi-equilibrium in the semiconductor remain an essential component, non-equilibrium techniques such as photoluminescence intensity have become a necessary ingredient. The application of the proposed methodology has led to high-κ stacked gate oxides on GaAs displaying a broad minimum of interface state density $D_{it} \leq 2 \times 10^{11}$ cm^{-2} eV^{-1} on n-type GaAs suggesting a U-shaped D_{it} distribution, an oxide relative dielectric constant of 20.8 ± 1, a breakdown field exceeding 4 MV/cm, and leakage currents of $\cong 2 \times 10^{-8}$ A/cm^2 at an electric field of 1 MV/cm (SiO_2 equivalent field = 5.3 MV/cm). Potential extensions of the proposed methodology to high-κ gate dielectric development on elemental semiconductors such as Si and Ge and wide bandgap semiconductors such as GaN are further discussed.

1. INTRODUCTION

The quest for III–V gate oxides has been fueled by scientific curiosity and commercial opportunity for almost four decades. More recently, the demand for III–V semiconductors in high volume applications such as wireless and fiber optic communications

A.A. Demkov and A. Navrotsky (eds.), Materials Fundamentals of Gate Dielectrics, 403–467.

has sparked even stronger interest in III–V based metal–oxide–semiconductor (MOS) field effect transistors (FET); the past and present lack of III–V MOSFET devices has limited functionality, scalability, performance, and broader market acceptance of III–V technologies. Excellent reviews of early gate insulator research on III–V semiconductors were published by Croydon and Parker [1] and Wilmsen [2] in 1981 and 1985, respectively. The conclusions of almost 20 years of effort in the field were not encouraging: attempts at adopting the wildly successful thermal oxidation technique of silicon had failed and insulator deposition techniques did not look more promising [1–10]. Consequently, the field had been nearly abandoned. However, the pioneering work by Spicer [11], Pianetta [12] and others had greatly advanced the understanding of III–V semiconductor surface reactivity and would later support the foundation of modern III–V MOS technology concerning the requirement of surface preparation under ultra-high vacuum conditions.

In 1988, the first deposition of gallium oxide thin films on GaAs by a reactive oxidation technique for gate dielectric applications was reported by Callegari et al. [13]. Later, Ga_2O_3/GaAs interfaces with low interface state density D_{it} were manufactured by an *in-situ* molecular beam epitaxy (MBE) technique using electron beam evaporation from a high purity, single crystal gadolinium gallium garnet ($Gd_3Ga_5O_{12}$); inversion (under light illumination) and accumulation were reported in n- and p-type MOS capacitors. [14–16]. Secondary ion mass spectroscopy (SIMS) and Rutherford backscattering spectroscopy (RBS) revealed that the deposited $Gd_xGa_{0.4-x}O_{0.6}$ layer was essentially free of Gd at the oxide/GaAs interface and that the Gd concentration was enhanced towards the oxide surface. It was suggested as early as 1995 that the low D_{it} property of the oxide–GaAs interface is solely provided by Ga_2O_3 and that the oxide resistivity is related to the depth profile of the Gd concentration [14, 15]. More recently, interfaces formed by a high purity, amorphous Ga_2O_3 bulk layer on a GaAs (001) surface manufactured by evaporation of a polycrystalline Ga_2O_3 source from a high-temperature effusion cell by MBE demonstrated low D_{it} [17–19], and a well performing, self-aligned enhancement mode GaAs-based metal–oxide–semiconductor heterostructure field effect transistors using Ga_2O_3 as gate oxide was demonstrated [20]. However, the large leakage currents inherent to Ga_2O_3 films [19, 21] limit their usefulness as a gate oxide for field effect transistor applications. Although $Gd_xGa_{0.4-x}O_{0.6}$ films showed dramatically reduced leakage current density [21, 22], large frequency dispersion and significant $C–V$ stretch-out was found in capacitance–voltage ($C–V$) measurements when $Gd_xGa_{0.4-x}O_{0.6}$ was directly grown on GaAs [21].

The purposes of this chapter are to (1) introduce the reader to the methodology applied to the development of high-κ $Gd_xGa_{0.4-x}O_{0.6}/Ga_2O_3$ dielectric stacks on GaAs, (2) clarify the distinctively different roles of Gd mole fraction and Ga_2O_3 template thickness in amorphous high-κ $Gd_xGa_{0.4-x}O_{0.6}/Ga_2O_3$ dielectric stacks on GaAs (see also [23]), (3) present state-of-the-art data of high-κ $Gd_xGa_{0.4-x}O_{0.6}/Ga_2O_3$ dielectric stacks on GaAs and put them in context to previously published results, and (4) enable the reader to understand how certain aspects of the proposed methodology may be applicable to the development of MOS technologies on other wide bandgap semiconductors such as nitrides and on elemental semiconductors such as Si and Ge.

2. EXPERIMENT

2.1. Wafer Fabrication

The GaAs epitaxial layer structure employed in this work has been optimized for *C–V* and photoluminescence intensity (PL-I) characterization and comprises a 1.4 μm, $2 - 2.5 \times 10^{16}$ cm^{-3} Si doped, active top layer and a 0.2 μm, 2×10^{18} cm^{-3} Si doped blocking layer grown on 2 in. (100) GaAs n^+ substrates by MBE. A sharp and streaky As-stabilized GaAs(001)–(2 × 4) reflection high energy electron diffraction (RHEED) pattern is obtained after epi-layer growth, indicating an atomically ordered and flat GaAs(001) surface. Subsequent to epitaxial layer growth, the wafer was either transferred *in-situ* under ultra-high vacuum (UHV) to an oxide growth chamber or the GaAs surface was covered by a protective arsenic cap layer [24]. In the latter case, the As cap layer was desorbed after loading the wafer into an UHV oxide growth chamber and before Ga_2O_3 growth commenced using effusive evaporation of a polycrystalline Ga_2O_3 source material from a high temperature effusion cell. Effusive evaporation allows highly reproducible Ga_2O_3 growth rates as low as 0.01 Å/s and produces stoichiometric gallium oxide films (Ga_2O_3); further details about Ga_2O_3 growth can be found in [17]. Following Ga_2O_3 growth, a $Gd_xGa_{0.4-x}O_{0.6}$ layer was grown using additional molecular beams of Gd (provided by a second high temperature effusion cell) and oxygen supplied by a leak valve or a plasma source. All oxide films have been found to be amorphous as shown by RHEED. Further, two baseline structures, one with an AlGaAs window layer and the other without oxide have been grown using the identical epitaxial layer structure consisting of active and blocking layers as described above.

Structures from three different oxide growth scenarios on GaAs are considered in this chapter: (1) Ga_2O_3 bulk films (as-deposited and postdeposition hydrogen plasma passivated), (2) $Gd_xGa_{0.4-x}O_{0.6}$ layers, and (3) optimized $Gd_xGa_{0.4-x}O_{0.6}/Ga_2O_3$ dielectric stacks where Ga_2O_3 is referred to as Ga_2O_3 template in the following. The fabricated oxide films have been characterized by atomic force microscopy (AFM), ellipsometry, cross sectional high resolution transmission electron microscopy (TEM), and RBS. The Ga_2O_3 template thickness has been typically determined based on the template growth time and the Ga_2O_3 growth rate of Ga_2O_3 reference wafers. High resolution cross sectional TEM has been used in a few selected cases to verify the Ga_2O_3 template thickness (see Fig. 1); good agreement with the above growth rate approach has been observed. Optimized $Gd_xGa_{0.4-x}O_{0.6}/Ga_2O_3$ dielectric stacks discussed in this chapter have a Ga_2O_3 template thickness in between 9 and 13 Å. The Gd, Ga, and oxygen mole fractions of the $Gd_xGa_{0.4-x}O_{0.6}$ films have been measured by RBS. The average oxygen mole percent as determined by RBS is 61 ± 5 at.% Therefore, all oxide stacks are assumed to be stoichiometric in respect to oxygen and an oxygen mole percent of 60 at.% used throughout this chapter.

Postdeposition hydrogen exposure of as-deposited Ga_2O_3/GaAs structures with different oxide thickness has been conducted in a Tegal 6000 etching tool where the atomic hydrogen is provided by an RF plasma discharge device with a frequency of 13.56 MHz. The electrical interface data have been correlated to the presence or absence of hydrogen at the Ga_2O_3/GaAs interface as determined by SIMS.

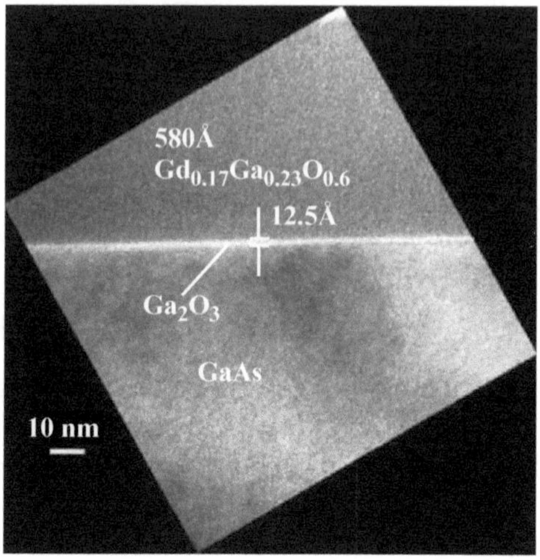

Fig. 1. *Cross sectional high resolution TEM image* of a $Gd_xGa_{0.4-x}O_{0.6}/Ga_2O_3$ dielectric stack on GaAs. The Ga_2O_3 template thickness is 12.5 Å.

Conversion of SIMS ion counts to concentrations has been accomplished by using relative sensitivity factors derived from ion implanted Ga_2O_3 standards which were analyzed along with our test samples.

2.2. Interface Characterization Techniques

Interface characterization of the proposed gate oxide structures rests on three foundations: (1) Atomic level measurements of Ga_2O chemisorption on GaAs including scanning tunneling microscopy (STM), scanning tunneling spectroscopy (STS), and atomic level modeling using density-functional theory (DFT); (2) Ga_2O_3/GaAs interface characterization of Ga_2O_3 templates grown on GaAs based on a PL-intensity (PL-I) technique; and (3) Metal/$Gd_xGa_{0.4-x}O_{0.6}/Ga_2O_3$/GaAs capacitor characterization using C–V measurements, where the latter two techniques are combined with a fully numerical semiconductor drift-diffusion model for data analysis and interpretation (see Fig. 2). Note that each of the three material systems is best characterized by a different set of techniques. It should be further emphasized that only the combination of all three sets of techniques has fundamentally enabled the success of gate dielectric development on GaAs at Motorola. This chapter is limited to the second and third set of techniques; the reader is referred to ref. [25] for details on atomic level measurement instrumentation and techniques. Consequently, this paragraph focuses on the instrumentation and sample preparation side of interface characterization by PL-I and C–V measurements.

In PL-I measurements [26, 27] the dependence of the measured GaAs photoluminescence intensity $\Im_{PL}$ on the light intensity entering the semiconductor structure I_0' (cm^{-2} s^{-1}) is used to determine the interface recombination velocity S and

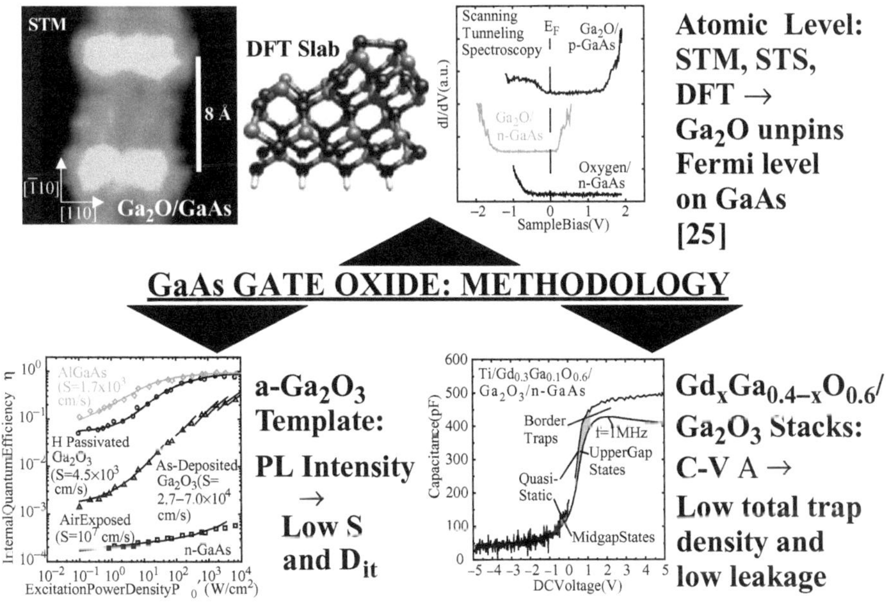

Fig. 2. Interface characterization of the proposed gate oxide structures rests on three foundations: (1) Atomic level studies of Ga_2O chemisorption on GaAs, (2) Ga_2O_3/GaAs interface characterization of Ga_2O_3 templates grown on GaAs based on a recombination technique (PL-intensity), and (3) Metal/$Gd_xGa_{0.4-x}O_{0.6}$/Ga_2O_3/GaAs capacitor characterization using C–V measurements.

the interface state density D_{it}. Here, $I_0' = T_{ex}I_0$ where I_0 and T_{ex} are the incident light intensity on the sample surface and the optical transmissivity of the sample surface at the excitation wavelength, respectively. In order to unequivocally map the measured characteristics onto the calculated curves, a range of $P_0' = I_0' h\nu$ of five orders of magnitude including very high injection ($p \cong n >> N_D^+$, $P_0' \geq 10^3$ W/cm^2), high injection ($p \cong n \cong N_D^+$, 10^3 W/cm^2 $> P_0' \geq 10$ W/cm^2), and low injection ($p < n$, N_D^+, $P_0' < 10$ W/cm^2) is required (n-type semiconductor). Both baseline structures with low D_{it} ($\eta \cong 1$, AlGaAs–GaAs interface) and very high D_{it} ($\eta \cong 0$, bare GaAs surface) must be provided. Here, P_0', h, ν, p, n, N_D^+, and η are the excitation power density entering the semiconductor structure, Plancks' constant, the photon frequency at the excitation wavelength, the semiconductor hole and electron concentrations, the density of ionized shallow donors, and the internal quantum efficiency, respectively. One of the distinct advantages of the PL-I technique is its ability to contactless characterize interfaces and that it does not require dedicated test structures such as capacitors which significantly reduces the cycle time. Other advantages of the PL-I technique include the exclusion of effects caused by bulk properties of the insulating film and the capability to characterize semiconductor surfaces, semiconductor–semiconductor interfaces such as AlGaAs–GaAs, and semiconductor–insulator interfaces with extremely thin insulating films or with

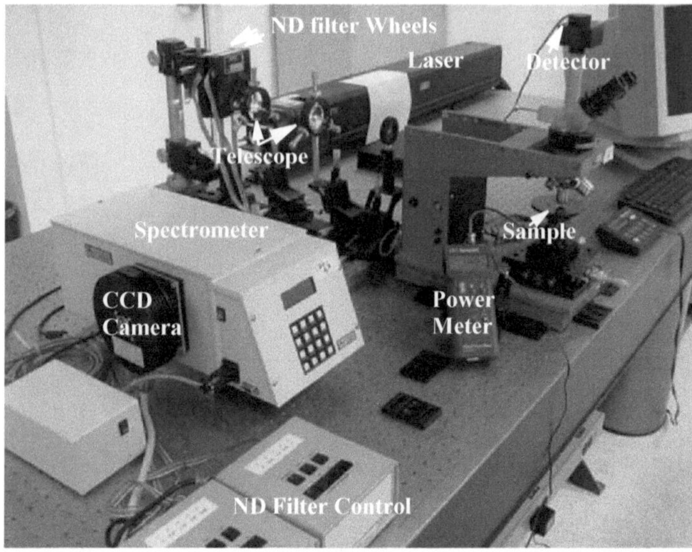

Fig. 3. Optical measurement system employed for PL-I measurements.

insulators having excessive leakage currents. Under conditions discussed later and the provision of appropriate standards, PL-I allows the characterization of the immediate oxide/III–V semiconductor interface free of the ambiguities which often plague the interpretation of C–V measurements on non-ideal MOS capacitors. Further, traditional capacitance–voltage methods such as the quasi-static/high frequency technique [28] and Terman's method [29] are not suitable for interface state density measurements in the Ga_2O_3/GaAs system; quasistatic data are unavailable due to leakage currents and the Terman method fails since midgap D_{it} observed in the Ga_2O_3/GaAs system is below its detection limit. Further advantages and limitations of the PL-I technique are detailed in Section 3. The PL-I measurement setup is briefly described in the following, the reader is referred to ref. [26] for further details.

Figure 3 shows a photograph where essential components of the optical measurement system are labeled. Excitation is provided by an argon ion laser with a nominal maximum power output of 5 W and emitting at $\lambda_o = 514.5$ nm. A first Oriel motorized filter wheel system equipped with one set of Oriel absorptive neutral density filters with optical densities of 1, 2, 3, and 4, and a second Oriel motorized filter wheel system equipped with another set of Oriel absorptive neutral density filters with optical densities of 0.3, 0.5, and 0.8 provide a maximum combined attenuation of 6.9×10^4. A telescope comprising two lenses with a focal length of 120 and 80 mm, respectively, is used to adjust the focal plane of the incoming laser beam which enters through the microscope illumination port at the microscope's backside. The microscope is equipped with a customized stage comprising a Newport 406 dual axis translation stage having two DM-13 differential micrometers with 0.07 μm resolution, and a 488 Newport rotary platform with 10 μm vertical resolution. The collimated luminescence beam emitted from a test sample positioned on the stage of the microscope exits on the microscopes' top where a 514.5 nm laser line filter is used to filter out laser light reflected by the test samples' surface. The collimated luminescence beam is attenuated as required using Oriel glass metallic neutral density filters with optical densities of 1, 2, 3 and 4 providing a maximum attenuation of 3.97×10^3. Subsequently, the PL signal is coupled into a spectrometer manufactured by Acton Research Corporation using various mirrors and a lens with a focal length of 100 mm (not shown). The PL spectra are detected by a model ST-6 CCD camera, manufactured by the Santa Barbara Instruments Group, Inc. When very short cycle time is required, the integrated luminescence signal is measured directly on top of the microscope using a Newport 835 optical power meter with a Newport 818 SL low power detector in conjunction with two Newport cut-on 830 nm filters and one Newport 950 nm short pass filter 10SWF-950. This configuration is shown in Fig. 3. The latter method, however, is less sensitive and does not allow the examination of luminescence spectra. The laser spot size is measured by translating the cleaved edge of a test sample in x- or y-direction across the laser beam using a differential micrometer (see Fig. 4) while the reflected laser power is measured on top of the microscope. The FWHM of the laser spot is typically adjusted to $\cong 35\ \mu m$.

Capacitor measurements have been done in a light tight enclosure and using a Keithley Model 82 C–V analyzer. The system was calibrated using a standard set of calibration capacitors provided by the manufacturer. The standard calibration does not include parasitic effects originating from manipulators, probe arms, and

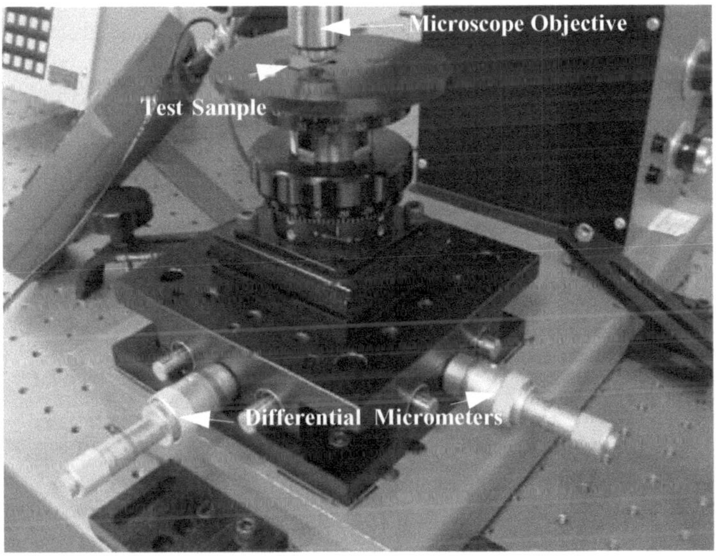

Fig. 4. Detail of the optical measurement system showing the measurement stage with the differential micrometers. The laser spot size is measured by translating the cleaved edge of a test sample in x- or y-direction across the laser beam using the differential micrometers.

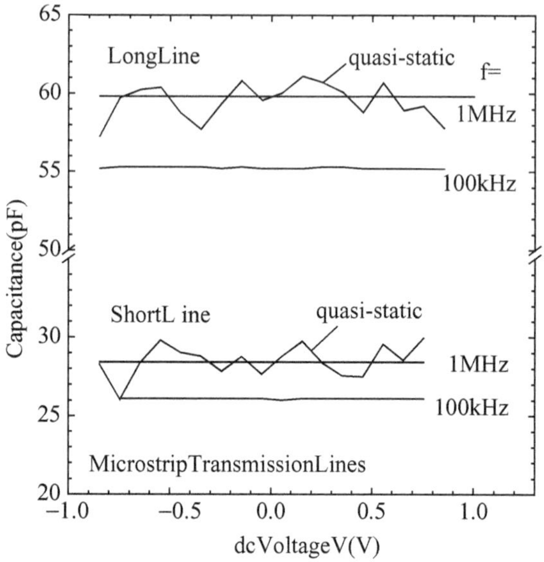

Fig. 5. Measured quasi-static, 100 kHz, and 1 MHz C–V curves obtained from microstrip transmission lines on dispersion free substrate.

the chuck, and a systematic capacitance error has been identified. This is illustrated in Fig. 5 which shows measured quasi-static, 100 kHz, and 1 MHz C–V curves obtained from microstrip transmission lines on dispersion free substrate (951 Green Tape manufactured by Dupont [30]). When using the average quasi-static capacitance as a reference, the measured 100 kHz and 1 MHz capacitances are typically 6.9% lower and 1.5% higher, respectively. Although this difference is typically neglected (it is inconsequential for the derivation of interface state density), it is considered when 100 kHz and 1 MHz C–V curves are scrutinized for frequency dispersion in MOS capacitors. Such capacitors with areas of 1.96×10^{-3} and 4.9×10^{-4} cm^2 have been manufactured by depositing Ti/Au circular dots onto the oxide surface using a shadow mask and a blanket Ge/Ni/Au layer on the backside of the substrates. The polarity of the dc voltage is used with respect to the metal dot on the oxide surface. Utmost care has been taken to assure that all requirements for correct C–V data interpretation are met including (1) quasi-equilibrium conditions during C–V sweep, (2) actual high frequency C–V data, (3) accurate oxide dielectric constant, and (4) provision of quasi-static C–V data. dc voltage sweep rates have been varied by two orders of magnitude and range from a fast sweep rate of 5.5 V/min (step $\Delta V = 50$ mV, delay $\tau_d = 0.5$ s), over a medium sweep rate of 0.55 V/min (step $\Delta V = 20$ mV, delay $\tau_d = 2$ s), to a slow sweep rate of 0.055 V/min (step $\Delta V = 10$ mV, delay $\tau_d = 10$ s), and ac capacitance measurements are done at 100 kHz and 1 MHz. Note that all of the above requirements comprise significant challenges for the material system investigated. C–V and PL-I data interpretation is also based on the provision of GaAs ionized dopant concentration by an independent electrochemical depth profiling technique (Polaron).

3. MODEL DESCRIPTION

Numerical models based on the classical semiconductor equations (drift-diffusion model) have found widespread use to predict and calculate the performance of semiconductor devices including heterostructure devices [31]. We earlier extended the use of such a fully numerical model to the interpretation of and parameter extraction from PL-I and C–V data taking into account the dual role of interfacial defects in creating both nonradiative interface recombination and interface charge. Details of the model are described in refs. [32, 33]; the model is briefly revisited here and improvements and revisions are discussed further below. The system of equations is solved for a one-dimensional domain (see Fig. 6) using a standard finite difference technique. The one-dimensional model comprises Poisson's equation and the steady-state continuity equations as outlined in the following. Poisson's equation reads

$$\frac{\mathrm{d}}{\mathrm{d}x}\left(\varepsilon\frac{\mathrm{d}\varphi}{\mathrm{d}x}\right) = -q(p - n + N_{\mathrm{D}}^{+} - N_{\mathrm{A}}^{-} + p_{\mathrm{t}}^{+} - n_{\mathrm{t}}^{-}), \tag{1}$$

where $\varphi, x, q, \varepsilon, N_{\mathrm{D}}^{+}, N_{\mathrm{A}}^{-}, p_{\mathrm{t}}^{+}, n_{\mathrm{t}}^{-}$ are the electrostatic potential, the spatial coordinate, the electronic charge, the dielectric constant, the density of ionized shallow donors and acceptors, and the volume density of charged donor and acceptor type traps,

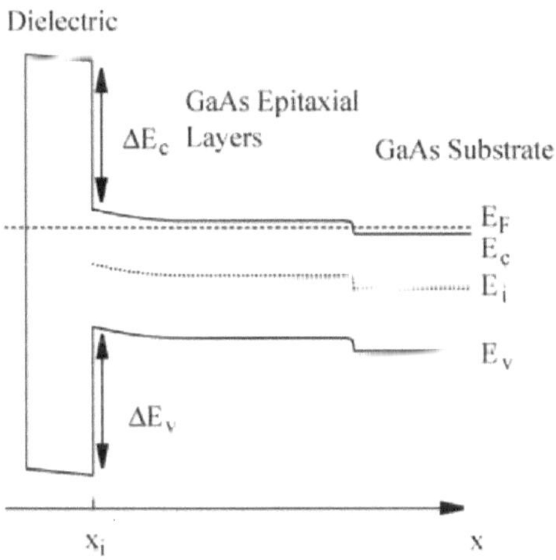

Fig. 6. The one-dimensional domain for which the system of equations is solved. Here, E_{c}, E_{v}, E_{i}, E_{F}, ΔE_{c}, and ΔE_{v} are the conduction and valence band edge, the intrinsic energy level, the Fermi energy, and the conduction and valence band offset, respectively.

respectively. The steady-state continuity equations for electrons and holes read

$$\frac{1}{q}\cdot\frac{\mathrm{d}J_\mathrm{n}}{\mathrm{d}x} = R_\mathrm{SRH} + R_\mathrm{rad} - G'_\mathrm{n}, \tag{2}$$

$$-\frac{1}{q}\cdot\frac{\mathrm{d}J_\mathrm{p}}{\mathrm{d}x} = R_\mathrm{SRH} + R_\mathrm{rad} - G'_\mathrm{p}, \tag{3}$$

respectively. Here, J_n, J_p, R_SRH, R_rad, and G'_n and G'_p are the electron and hole current density, the Shockley–Read–Hall recombination rate, the radiative band-to-band recombination rate, and the optical generation rate for electrons and holes, respectively. We have used the standard expressions for $R_\mathrm{rad} = B[np - n_\mathrm{i}^2]$, $G'_\mathrm{n} = G'_\mathrm{p} = \alpha I'_0 e^{\alpha(x-x_\mathrm{i})}$ where B, n_i, α are the radiative band-to-band recombination coefficient, the intrinsic carrier density, and the optical absorption coefficient, respectively. Note that the calculated photoluminescence signal I_PL ($\mathrm{cm}^{-2}\,\mathrm{s}^{-1}$) is obtained by integration of R_rad over x and that the calculated internal quantum efficiency $\eta = I_\mathrm{PL}/I'_0$.

3.1. Interface and Surface States

The dual role of interfacial defects in creating both nonradiative interface recombination and interface charge are implemented in the model as described below. Nonradiative recombination due to interface/surface states is presented first, interface charge second and finally, the extraction of interface parameters such as recombination velocity, capture cross section, and interface state density is discussed in the context of both interface recombination and charge. Throughout the chapter, the following conventions are used: trap state energy E_t (eV), state density per unit volume (cm^{-3}) N_t, state density per unit volume and unit energy ($\mathrm{cm}^{-3}\,\mathrm{eV}^{-1}$) D_t, state density per unit area (cm^{-2}) $N_\mathrm{it} = N_\mathrm{t}\delta$, and state density per unit area and unit energy (cm^{-2} eV^{-1}) $D_\mathrm{it} = D_\mathrm{t}\delta$ where δ is the depth of the surface/interface region.

In general, R_SRH can be written for states of density D_t and energy E_t within the bandgap of a semiconductor E_G as follows:

$$R_\mathrm{SRH}\,(\mathrm{cm}^{-3}\,\mathrm{s}^{-1}) = \left(np - n_\mathrm{i}^2\right)\int_{E_\mathrm{v}}^{E_\mathrm{c}} \frac{D_\mathrm{t}\{E_\mathrm{t}\}}{\frac{1}{\bar{v}_\mathrm{th}\sigma_\mathrm{p}}(n+n_1) + \frac{1}{\bar{v}_\mathrm{th}\sigma_\mathrm{n}}(p+p_1)}\mathrm{d}E_\mathrm{t}, \tag{4}$$

where $n_1 = n_\mathrm{i}e^{\frac{E_\mathrm{t}-E_\mathrm{i}}{kT}}$, $p_1 = n_\mathrm{i}e^{-\frac{E_\mathrm{t}-E_\mathrm{i}}{kT}}$. Here, $\sigma_\mathrm{n,p}$, $\bar{v}_\mathrm{th}$, k, and T are the electron and hole capture cross sections, the average thermal velocity, the Boltzmann constant, and the temperature in Kelvin, respectively. To understand the implementation and application of [4] in experimental PL-I data fitting and parameter extraction, the specific cases of discrete and distributed surface/interface states are discussed in the following. For interface/surface states located at a discrete energy E_t with density N_t and with spatial extension δ, [4] reads

$$R_\mathrm{SRH}\,(\mathrm{cm}^{-2}\,\mathrm{s}^{-1}) = \frac{np - n_\mathrm{i}^2}{\frac{\tau_\mathrm{p}}{\delta}(n+n_1) + \frac{\tau_\mathrm{n}}{\delta}(p+p_1)}, \tag{5}$$

where $\tau_p = 1/(\sigma_p\ \bar{v}_{th}\ N_t)$, $\tau_n = 1/(\sigma_n\ \bar{v}_{th}\ N_t)$ are the nonradiative electron and hole lifetime, respectively. For surface/interface states, the expressions for the surface recombination velocities $S_n = \delta/\tau_n$ and $S_p = \delta/\tau_p$ are commonly used. A fit to measured PL-I curves is typically obtained by adjusting $S_{n,p}$ with the parameters $\sigma_{n,p}$ and N_t not explicitly known. Note that this model is used for interfaces with a pinned Fermi level only.

For surface/interface states distributed throughout the bandgap, a fit to measured PL-I data based on [4] can be in general accomplished by varying the capture cross sections $\sigma_{n,p}$ if $D_t\{E_t\}$ is know. However, since fitting measured PL-I curves with both $\sigma_{n,p}$ and $D_t\{E_t\}$ as variables is not practical, the following approach is proposed. Assuming a constant D_t throughout the bandgap, [4] can be re-written as

$$R_{SRH}\,(\mathrm{cm}^{-2}\,\mathrm{s}^{-1}) = \left(np - n_i^2\right) \int\limits_{E_v}^{E_c} \frac{\mathrm{d}E_t/E_G}{\dfrac{\tau_p}{\delta}(n+n_1) + \dfrac{\tau_n}{\delta}(p+p_1)}, \tag{6}$$

where the product of D_t and the semiconductor bandgap E_G replaces N_t ($N_t = D_t\ E_G$) in the above definitions for $\tau_{n,p}$. Again, a fit to measured PL-I curves can be done by adjusting $\tau_{n,p}$ ($S_{n,p}$) without the parameters $\sigma_{n,p}$ and D_t explicitly known. As will be shown in Section 4, excellent fits to measured PL-I curves are accomplished under the assumption of constant D_t throughout the GaAs bandgap in the case of hydrogen passivated Ga_2O_3/GaAs interfaces. However, PL-I data obtained from as-deposited Ga_2O_3/GaAs interfaces cannot be fitted under the assumption of constant D_t. In the latter case, $D_t\{E_t\}$ is piecewise approximated by fitting each segment j of a measured PL-I curve with an appropriate nonradiative carrier lifetime $\tau_j = \tau_p = \tau_n$ (or $S_j = S_p = S_n$; the equivalence of $\tau_p = \tau_n$ is discussed later in this paragraph) according to [6] and using the following formalism. Note that this formalism is not implemented in the source code but is merely used to construct a step-like $D_t\{E_t\}$ dependence based on τ_j and average energy position of the Fermi level E_{Fj} obtained from the simulator for each fitted segment j of the PL-I curve. For dielectric/n-type semiconductor interfaces with low D_t ($n \geq p$ on semiconductor surface for any given light intensity I_0'), [4] reads

$$R_{SRH}\,(\mathrm{cm}^{-2}\,\mathrm{s}^{-1}) \cong \left(np - n_i^2\right) \int\limits_{E_{Fn}^*}^{E_{Fn}} \frac{D_t\{E_t\}}{\dfrac{n}{\delta\,\bar{v}_{th}\sigma_p} + \dfrac{p}{\delta\,\bar{v}_{th}\sigma_n}} \mathrm{d}E_t, \tag{7}$$

where E_{Fn} is the quasi-Fermi level for electrons and E_{Fn}^* is the mirror image of E_{Fn} relative to the intrinsic energy level E_i. For constant capture cross sections $\sigma_{n,p}$ throughout the semiconductor bandgap, symmetrical D_t distribution with respect to the intrinsic energy level E_i, approximation of $D_t\{E_t\}$ by a step-like function with steps j of constant D_{tj} and average quasi-Fermi level energy for electrons E_{Fnj}, [7] simplifies to

$$R_{\mathrm{SRH}}\,(\mathrm{cm}^{-2}\,\mathrm{s}^{-1}) = \frac{np - n_{\mathrm{i}}^2}{\dfrac{n}{\delta\,\bar{v}_{\mathrm{th}}\sigma_{\mathrm{p}}} + \dfrac{p}{\delta\,\bar{v}_{\mathrm{th}}\sigma_{\mathrm{n}}}} \left(D_{\mathrm{t1}}(E_{\mathrm{Fn1}} - E^*_{\mathrm{Fn1}}) + \sum_{j=2}^{n} 2D_{\mathrm{t}j}(E_{\mathrm{Fn}j} - E_{\mathrm{Fn}(j-1)}) \right) \tag{8}$$

with

$$D_{\mathrm{t}j} = \frac{\hat{D}_{\mathrm{t}j}(E_{\mathrm{Fn}j} - E^*_{\mathrm{Fn}j}) - \hat{D}_{\mathrm{t}(j-1)}(E_{\mathrm{Fn}(j-1)} - E^*_{\mathrm{Fn}(j-1)})}{2(E_{\mathrm{Fn}j} - E_{\mathrm{Fn}(j-1)})} \quad (j \geq 2), \tag{9}$$

where $\hat{D}_{\mathrm{t1}} = D_{\mathrm{t1}}$ and $\hat{D}_{\mathrm{t}j} = \hat{D}_{\mathrm{t}(j-1)}\frac{\tau_{j-1}}{\tau_j}$ $(j \geq 2)$. This approach will become more clear in Section 4 when applied to as-deposited Ga_2O_3/GaAs interfaces. Note that neither D_{t} nor D_{t1} can be explicitly determined from [6] and [8], respectively.

The interface trap charge n_{t} originating from a a state with discrete energy E_{t} and of density N_{t} reads [34]

$$n_{\mathrm{t}}^- = N_{\mathrm{tA}} \frac{\sigma_{\mathrm{n}} n + \sigma_{\mathrm{p}} p_1}{\sigma_{\mathrm{n}}(n + n_1) + \sigma_{\mathrm{p}}(p + p_1)} \tag{10}$$

and

$$p_{\mathrm{t}}^+ = N_{\mathrm{tD}} \frac{\sigma_{\mathrm{p}} p + \sigma_{\mathrm{n}} n_1}{\sigma_{\mathrm{n}}(n + n_1) + \sigma_{\mathrm{p}}(p + p_1)}, \tag{11}$$

for acceptor type traps N_{tA} and donor type traps N_{tD}, respectively. Again, this model is used for interfaces with a pinned Fermi level only. Analogous, the charge for surface/interface states distributed throughout the bandgap with density D_{t} reads

$$n_{\mathrm{t}}^- = \int_{E_{\mathrm{v}}}^{E_{\mathrm{c}}} \frac{\sigma_{\mathrm{n}} n + \sigma_{\mathrm{p}} p_1}{\sigma_{\mathrm{n}}(n + n_1) + \sigma_{\mathrm{p}}(p + p_1)} D_{\mathrm{tA}} \mathrm{d}E_{\mathrm{t}} \tag{12}$$

and

$$p_{\mathrm{t}}^+ = \int_{E_{\mathrm{v}}}^{E_{\mathrm{c}}} \frac{\sigma_{\mathrm{p}} p + \sigma_{\mathrm{n}} n_1}{\sigma_{\mathrm{n}}(n + n_1) + \sigma_{\mathrm{p}}(p + p_1)} D_{\mathrm{tD}} \mathrm{d}E_{\mathrm{t}}, \tag{13}$$

for acceptor type traps D_{tA} and donor type traps D_{tD}, respectively. The total interface charge $n_{\mathrm{it}} = \delta(p_{\mathrm{t}}^+ - n_{\mathrm{t}}^-)$.

Hitherto, the implementation of interface/surface charge and recombination has been discussed. Before an approach for interface parameter extraction is proposed in Subsection 3.2, simplified representations of interface/surface charge and recombination are discussed in the following for the specific cases of low injection and high injection to further facilitate the readers' understanding. Figure 7 illustrates the case of (a) low injection ($N_{\mathrm{D}}^+ = n > p$) and (b) high injection ($n \cong p > N_{\mathrm{D}}^+$) for an

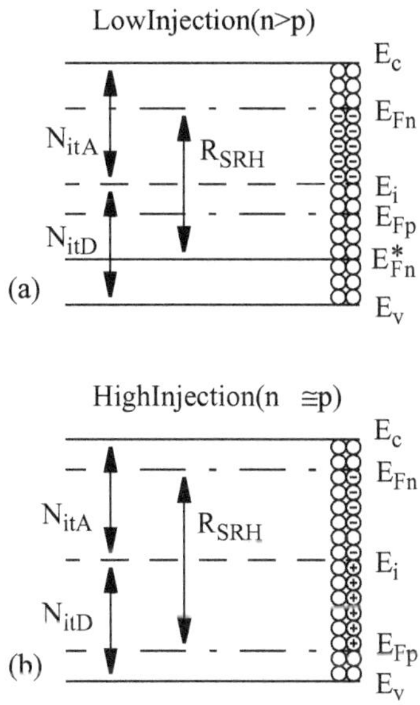

Fig. 7. Simplified energy band diagrams illustrating the case of (a) low injection ($n > p$) and (b) high injection ($n \cong p$) for an n-type semiconductor.

n-type semiconductor. Acceptor type states are assumed above the intrinsic level E_i and donor type interface states are located below the intrinsic level E_i, e.g., the charge neutrality level is located at the intrinsic energy level. For low injection (Fig. 7(a)), all states in between E_{Fn} and the mirror image of E_{Fn} (E^*_{Fn}) contribute to nonradiative recombination (see Eq. (7)), but only the contributing acceptor states are charged, all contributing donor states are neutral (see Eqs. (12) and (13)). Thus, the net interface charge in low injection on n-type semiconductors is negative. In high injection (Fig. 7(b)), interface states situated in between E_{Fn} and E_{Fp} ($E_{Fn} = E^*_{Fn}$) contribute to nonradiative recombination (see Eq. (4)), and 50% of contributing acceptor and contributing donor states are charged (see Eqs. (12) and (13)). Obviously, interface charge is diminished in high injection, a property which is taken advantage of in interface parameter extraction as discussed further below. In the following discussion of nonradiative lifetimes τ and interface/surface recombination velocity S in high injection, only the states N_{it} which contribute to nonradiative recombination are considered. The interface recombination velocities for holes S_p and electrons S_n read

$$S_p = \bar{v}_{th}\left(\sigma_p^- N_{itA}^- + \sigma_p^x N_{itD}^x\right), \qquad S_n = \bar{v}_{th}\left(\sigma_n^+ N_{itD}^+ + \sigma_n^x N_{itA}^x\right), \tag{14}$$

where the indices A and D identify acceptor and donor type states, and the $+$, $-$, and x superscripts denote positively charged, negatively charged, and neutral states, respectively. It is typically assumed that capture cross sections of charged states are significantly larger than capture cross sections of neutral states ($\sigma_p^- \gg \sigma_p^x$, $\sigma_n^+ \gg \sigma_n^x$). With the charge neutrality level at midgap ($N_{itA}^- = N_{itD}^n = N_{itD}^+ = N_{itA}^n = N_{it}$), surface

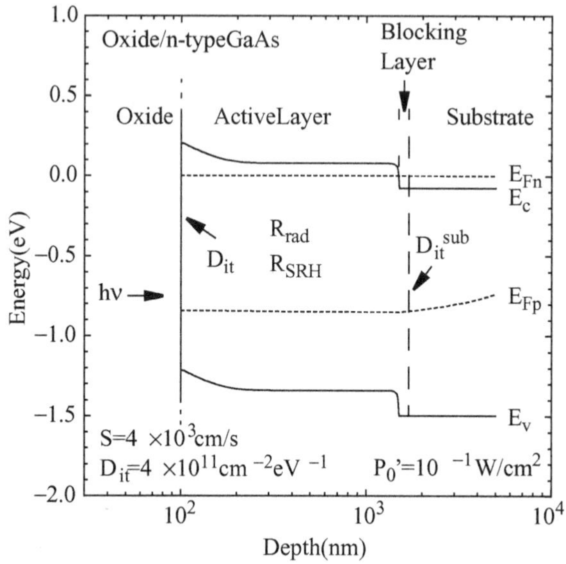

Fig. 8. GaAs energy band diagram in low injection ($P_0' = 10^{-1}$ W/cm^2) based on the n-type epitaxial layer structure optimized for PL-I measurements. The structure comprises two interfaces, an oxide/GaAs active layer interface having an interface state density D_{it} and a blocking layer/substrate interface with D_{it}^{sub} with both interfaces being sources of charge and nonradiative recombination.

recombination velocities can be approximated by $S_p \cong v_{th} N_{it} \sigma_p^-$ and $S_n \cong v_{th} N_{it}\, \sigma_n^+$. It is further reasonable to assume that positively and negatively charged states have similar cross section ($\sigma_p^- \cong \sigma_n^+$) which implies $S_n \cong S_p \cong S\,(\tau_p \cong \tau_n \cong \tau)$. Note that the equality of S_p and S_n (τ_p and τ_n) is irrelevant for low injection where S_p dominates (n-type semiconductor). In the following, the charge neutrality level is always assumed to be located at the intrinsic energy level.

3.2. Photoluminescence Intensity (PL-I)

In this subsection, the approach for interface parameter extraction is discussed. Figure 8 shows a GaAs energy band diagram in low injection ($P_0' = 10^{-1}$ W/cm^2) based on the n-type epitaxial layer structure optimized for PL-I measurements. The structure comprises two interfaces, an oxide/GaAs active layer interface having an interface state density D_{it} and a blocking layer/substrate interface with D_{it}^{sub} with both interfaces being sources of charge and nonradiative recombination. Further, nonradiative bulk recombination is present in the epi-layers (active and blocking layer) and the substrate. Consequently, a total of six parameters related to nonradiative recombination and interface charge are present in the structure: the oxide/GaAs interface charge n_{it} and recombination velocity S, the epi/substrate interface charge n_{it}^{sub} and recombination velocity S_{sub}, and the nonradiative bulk lifetime of the GaAs epilayers τ_{epi} and of the substrate τ_{sub}. It is demonstrated in the following how the design of the

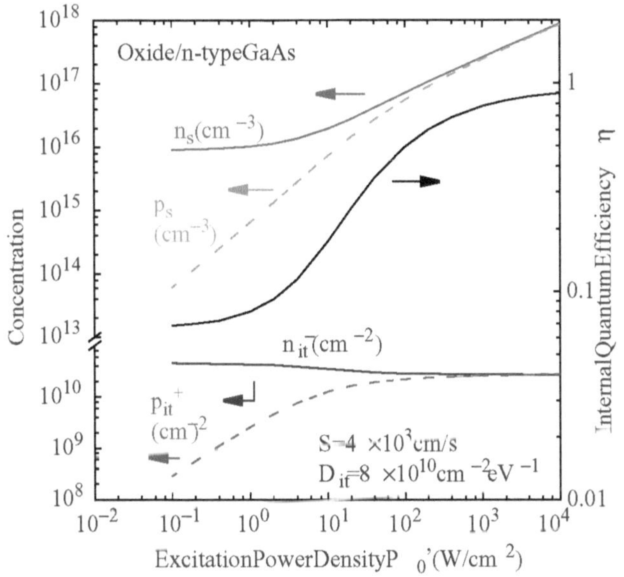

Fig. 9. Electron and hole concentrations n_s and p_s, negative and positive charge n_{it}^- and p_{it}^+ on the GaAs surface, as well as the internal quantum efficiency η as a function of excitation density P_0' of the structure shown in Fig. 8 for a typical low D_{it} oxide/GaAs interface ($N_D^+ = 2 \times 10^{16}$ cm^{-3}).

structure facilitates the parameter extraction such that S and D_{it} at the oxide/GaAs interface can be determined.

Figure 9 depicts the electron and hole concentrations n_s and p_s, the negative and positive surface charge n_{it}^- and p_{it}^+ on the GaAs surface, as well as the internal quantum efficiency η as is a function of excitation density P_0' of the structure shown in Fig. 8 for a typical low D_{it} oxide/GaAs interface. Clearly visible are the modes of low injection ($p_s << n_s \cong N_D^+$), high injection ($p_s \cong n_s \cong N_D^+$), and very high injection ($p_s \cong n_s >> N_D^+$). Figure 9 illustrates that interface charge is dominated by negatively charged acceptors in low injection and diminishes in very high injection ($n_{it} \cong 0$). The effect of interface charge on internal quantum efficiency is further elucidated in Fig. 10 for different injection modes. As can be seen, both negative and positive interface charge is inconsequential for the determination of η in very high injection ($P_0' = 10^4$ W/cm^2). This is a consequence of the absence of band bending in high injection even when interface charge is present. On the other hand, negative and positive interface charge have a dramatic and opposite effect on η in low injection ($P_0' -$ 10^{-1} W/cm^2). When negative interface charge is present, band bending is enhanced and minority carriers (holes) are increasingly driven into the interface recombination sink reducing η, while positive interface charge lowers surface band bending thus discouraging minority carriers from getting to the interface and increasing η in turn. These distinct effects of interface charge on η are later applied to interface parameter extraction in Section 4.

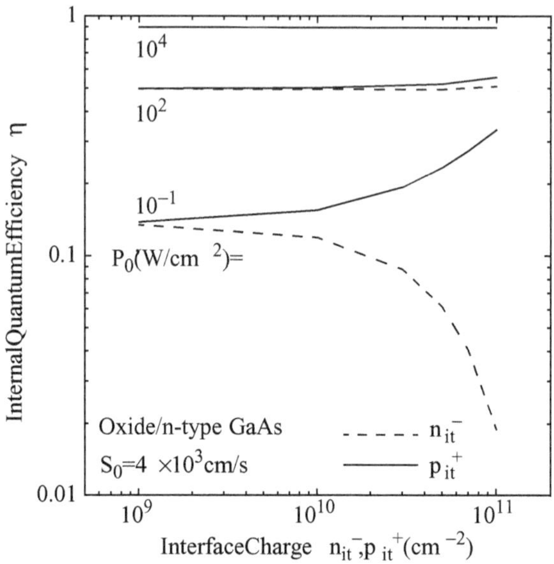

Fig. 10. Internal quantum efficiency as a function of negative and positive charge n_{it}^- and p_{it}^+ on the GaAs surface for different injection modes.

Figure 11 illustrates the interplay between oxide/GaAs interface recombination velocity S and nonradiative bulk lifetime τ_{epi} for (a) $n_{it} = 0$ and (b) $D_{it} = 8 \times 10^{10}$ cm^{-2} eV^{-1}. Since the diffusion length of holes $L_p > 9$ μm in good quality epitaxial GaAs ($\tau_{epi} \geq 100$ ns [35], hole mobility $\cong 300$ cm^2/Vs) is significantly longer than the thickness of the active GaAs layer, nonradiative recombination in the bulk active layer τ_{epi} and oxide/GaAs interface recombination S have very similar effects on η. When band bending is absent ($n_{it} = 0$, Fig. 11(a)), the effects of S and τ_{epi} are interchangeable, i.e., η is unaffected for $S + (\Delta x/\tau_{epi})$ = constant where Δx is the thickness of the active layer (1.4 μm). When band bending is present ($D_{it} = 8 \times 10^{10}$ cm^{-2} eV^{-1}, Fig. 11(b)), surface recombination S has a somewhat stronger effect on η in low injection as expected. In the following, an essentially infinite τ_{epi} (1 ms) is always used in the model for simplicity. Note that this assumption provides an upper estimate of S and only affects the inferred oxide/GaAs interface recombination velocity when S is very low; a lifetime τ_{epi} of 100 ns results in an equivalent S of 1.4×10^3 cm/s.

Figure 12 depicts the effects of epi/substrate interface states and nonradiative substrate recombination including interface recombination velocity S_{sub} and substrate carrier lifetime τ_{sub} on η. As apparent in Fig. 12(a), variations of τ_{sub} and S_{sub} over two orders of magnitude affect η only slightly over the entire range of excitation power densities investigated. Figure 12(b) reveals the reason for this behavior: the n^-/n^+ junction formed by the active layer and the blocking layer acts as an effective barrier against minority carriers (holes) and effectively confines the radiative recombination to the active layer. This is illustrated in Fig. 12(b) using the examples of very small

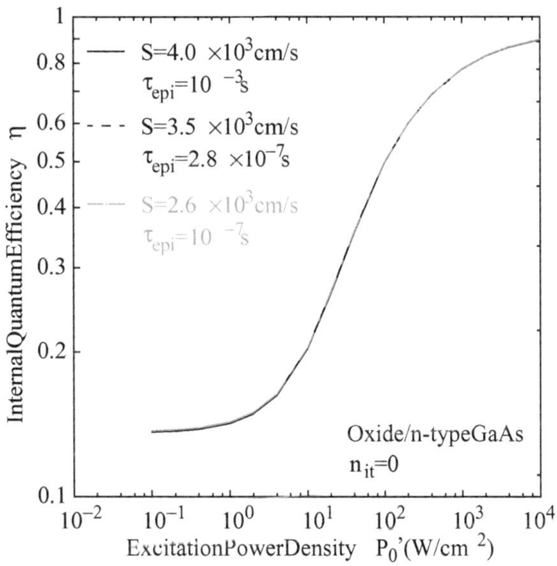

(a)

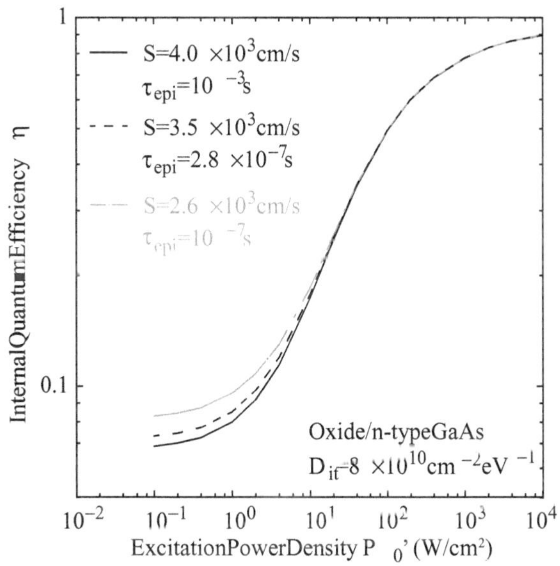

(b)

Fig. 11. Interplay between oxide/GaAs interface recombination velocity S and nonradiative bulk lifetime τ_{epi} epi for (a) $n_{it} = 0$ and (b) $D_{it} = 8 \times 10^{10}$ cm^{-2} eV^{-1}.

epi/substrate interface recombination ($S_{sub} = 10^3$ cm/s) and high S_{sub} of 10^5 cm/s: η is reduce by only 19 and 5% when S_{sub} is increased by two orders of magnitude in very high injection ($P_0' = 10^4$ W/cm^2) and low injection ($P_0' = 10^{-1}$ W/cm^2), respectively. Further, interface charge at the epi/substrate interface is neglected since an estimated

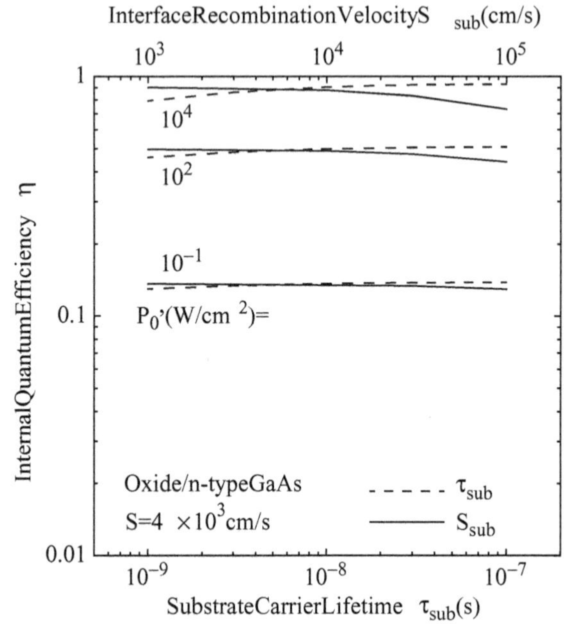

(a)

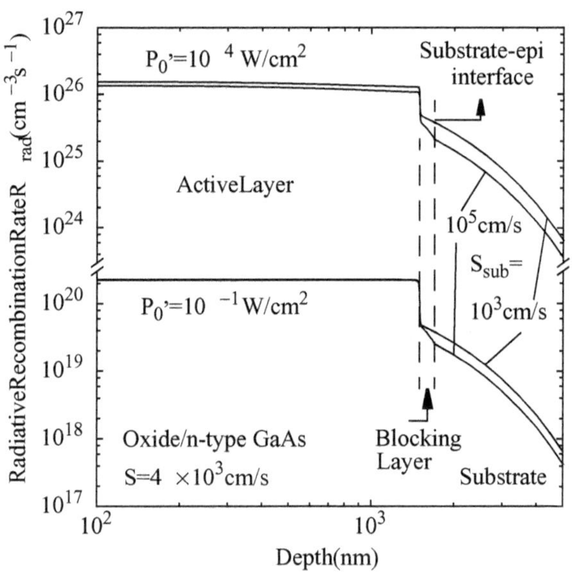

(b)

Fig. 12. Effects of epi/substrate interface states and nonradiative substrate recombination including (a) interface recombination velocity S_{sub} and substrate carrier lifetime τ_{sub} on η, and (b) interface recombination velocity S_{sub} on the radiative recombination depth profile for selected cases in (a).

depletion layer width of 5 nm is much shorter than the thickness of the blocking layer of 200 nm ($N_D^+ \cong 2 \times 10^{18}$ cm^{-3} on either side of the interface and assuming an already very high interface charge of 10^{12} cm^{-2} for a GaAs homointerface). This is confirmed in electrochemical depth profiling which shows an essentially undisturbed ionized donor concentration in the blocking layer. Thus, any interface charge located at the epi/substrate interface does not alter the energy band diagram in the active layer. In summary, the n^-/n^+ junction formed by the two epitaxial layers (active layer/blocking layer) effectively screens nonradiative recombination sources located at the epi/substrate interface and in the substrate itself such that the effect of substrate parameters including S_{sub}, n_{it}^{sub}, and τ_{sub} on η is diminished.

Based on the above findings, the following approach for extraction of interface parameters at the oxide/GaAs interface is proposed. First, the (very) high injection branch of a measured PL-I curve is fitted by adjusting $\tau = \tau_p, \tau_n$ (or S) in Eq. (6) with interface charge n_{it} (Eqs. (12) and (13)) being inconsequential. Second, the low injection branch of the measured PL-I curve is fitted using the above determined τ (or S) and by adjusting $D_t = D_{tA} = D_{tD}$ in Eqs. (12) and (13) with $\sigma = \sigma_p^- = \sigma_n^+$. Third, the capture cross section σ is obtained from $\sigma = 1/(\tau \bar{v}_{th} D_t E_G)$. For the other remaining four parameters the following is used: $\tau_{epi} = 1$ ms, $n_{it}^{sub} = 0$, $S_{sub} = 0$, and $\tau_{sub} = 10$ ns [35]. *Note that the outstanding features of this approach are the ability to characterize the quality of an interface using only one parameter* (τ) *determined in (very) high injection and the absence of ambiguities which often plague the interpretation of C–V measurements on non-ideal MOS capacitors.* Further, the capability to use different injection modes to extract interface properties sets this approach clearly apart from time-resolved PL measurements which are typically restricted to low injection.

3.3 Capacitance–Voltage (C–V)

The interpretation of C–V curves on non-ideal MOS systems is often problematic. This problem is compounded by the fact that C–V measurements are done exactly with the objective to characterize the non-ideality of a MOS system. For materials with a bandgap wider than that of Si and/or MOS systems with significantly higher leakage currents than that of thick thermal oxide on Si, a condition called steady-state deep depletion can significantly complicate the interpretation of C–V curves [32]. Since steady-state deep depletion involves nonequilibrium conditions in the semiconductor (finite slope of F_F and $F_{Fn} \neq E_{Fp}$), it requires the inclusion of carrier transport which is typically described by current continuity equations for electrons and holes (Eqs. (2) and (3)). This is in sharp contrast to the classical treatment of the MOS system which is based on the assumption of quasi-equilibrium and electrostatics only (Eq. (1)). Further, steady-state depletion is distinctively different in its causes and effects compared to the classical dynamic condition of deep depletion which occurs when the surface potential is altered on a time scale shorter than the minority carrier response time [28]. Steady-state deep depletion is not a dynamic condition and typically occurs when the leakage current of the dielectric exceeds a certain limit. This limit is a function of various variables including the semiconductor bandgap and

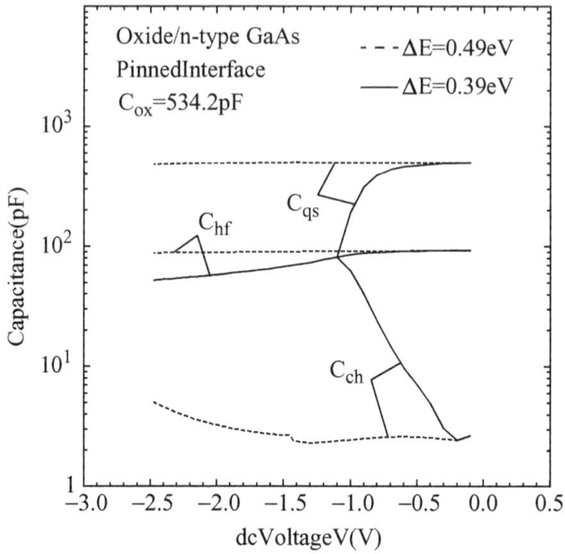

Fig. 13. Calculated high frequency, quasi-static, and equivalent channel capacitances C_{hf}, C_{qs}, and C_{ch}, respectively, as a function of dc voltage V for two hypothetical oxide/n-GaAs MOS systems with a pinned Fermi level ($N_{\mathrm{D}}^{+} = 2 \times 10^{16}$ cm^{-3}, area $= 1.96 \times 10^{-3}$ cm^{2}).

the surface potential. Since the condition of steady-state deep depletion continues to confuse the interpretation of C–V data and is further a typical occurrence in the systems considered in this chapter, its underlying principles are revisited and discussed in more detail in the following for two specific, frequently observed cases.

3.3.1. Fermi level pinned or unpinned?

Figure 13 shows calculated high frequency, quasistatic, and equivalent channel capacitances C_{hf}, C_{qs}, and C_{ch}, respectively, as a function of dc voltage V for two hypothetical oxide/n-GaAs MOS systems with a pinned Fermi level. The Fermi level has been virtually pinned by placing a donor and acceptor level with a density $\mathrm{N}_{\mathrm{it}} = 5 \times 10^{12}$ cm^{-2} each, 50 meV below and above the GaAs intrinsic energy level E_{i}, respectively. Whereas C_{hf} and C_{qs} have their usual meanings well known from C–V measurements, C_{ch} is defined as a dc capacitance with $C_{\mathrm{ch}} = \mathrm{d}n/\mathrm{d}V$ and can be considered as a measure analogous to a MOSFETs' dc transconductance g_{m}. The band offsets are determined by the difference between an assumed oxide bandgap and the GaAs bandgap using a 50/50 splitting ratio ($\Delta E = \Delta E_{\mathrm{c}} = \Delta E_{\mathrm{v}}$). As clearly seen in Fig. 13, the character of the C–V curves fundamentally changes when ΔE is only slightly lowered from 0.49 eV (dashed lines) to 0.39 eV (solid lines). The typical behavior of C_{hf}, C_{qs}, and $C_{\mathrm{ch}} \cong 0$ which one would expect for a pinned interface is observed for $\Delta E = 0.49$ eV, however, all three capacitances show a distinctively different behavior for $\Delta E = 0.39$ eV, in particular for $V \leq -1$ V. The latter condition is called steady-state deep depletion and described in detail in ref. [32]. In short, the

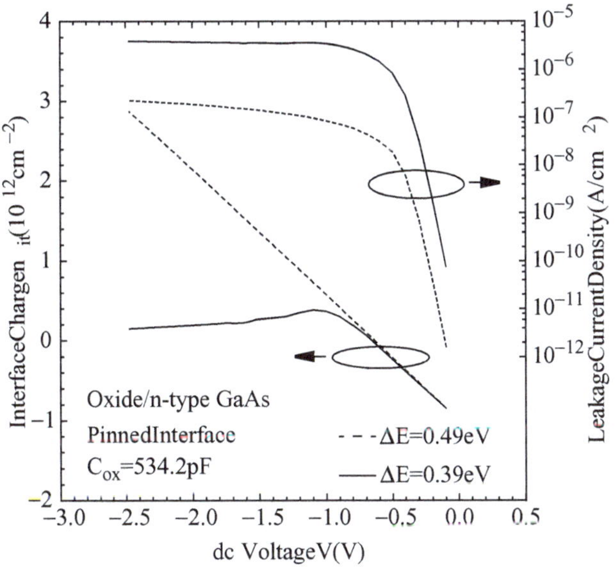

Fig. 14. Calculated interface charge n_{it} and leakage current density of the structures discussed in Figure 13 ($N_D^+ = 2 \times 10^{16}$ cm^{-3}, area $= 1.96 \times 10^{-3}$ cm^2).

requirement of current continuity throughout the structure breaches quasi-equilibrium and the movement of the quasi-Fermi levels at the interface with respect to the band edges is diminished when the quasi-Fermi levels fall below E_i (see also Figs. 17 and 18). The criterion of $\Delta E_c < 0.39$ eV was derived in ref. [32] for the occurrence of steady-state deep depletion in GaAs with a pinned interfacial Fermi level.

As illustrated in Fig. 14, which shows the interface charge n_{it} of the structures discussed in Fig. 13, a pinned interface virtually approaches the behavior of an unpinned interface in steady state deep depletion ($V < -1$V): n_{it} is very small and independent of dc voltage V (solid line, $V < -1$V). Further shown in Fig. 14 is the leakage current density which provides another criterion for the occurrence of steady-state deep depletion. Although the criterion for ΔE_c derived in ref. [32] is correct, it is considered to be impractical: The leakage current at room temperature in real systems with non-ideal dielectrics is often not determined by the band offsets at the interface but by the properties of traps inside the bandgap of the dielectric, i.e., the leakage current is dominated by trap assisted tunneling in the dielectric layer (the model considered here only includes charge transport within the bands). Therefore, the leakage current is used to demonstrate the criterion for the occurrence of steady-state deep depletion: leakage currents exceeding approximately 0.1–1 μA/cm^2 trigger steady state deep depletion in the system considered here. Note that the specific quantification of gate voltage at the onset of steady-state deep depletion and leakage current density required to trigger steady-state deep depletion depend on a number of parameters, in particular the energy at which the Fermi level is pinned on the GaAs interface and the degree of pinning (density and distribution of interface states). Coincidentally, a number of depletion mode GaAs based MOSFETs can be found in the recent literature which show the above signature of steady-state deep depletion in a GaAs based MOS system with a Fermi level more or less strongly pinned at the interface: g_m drops more or less abruptly when the gate voltage exceeds about -1 V [36–38] (compare also to C_{ch} shown in Fig. 13 for $\Delta E = 0.39$ eV, solid line). A further analysis of the above referenced MOSFET data using standard two-dimensional device simulation also revealed that the point ($\cong -1$ V) where g_m starts to drop with increasing gate bias roughly coincides with the midgap position of the Fermi level on the GaAs surface at the source side of the gate. In summary, (a) D_{it} analysis using standard techniques such as the quasi-static/high frequency technique and Terman's method are not applicable for capacitors when operating in steady-state deep depletion, (b) steady-state deep depletion can be triggered by leakage current densities of 0.1–1 μA/cm^2 on GaAs based MOS capacitors with a pinned Fermi level, and (c) depletion mode devices can work well up to a certain gate bias with a pinned interface Fermi level when operated in steady-state deep depletion.

3.3.2. Fermi level pinning or surface inversion?

Figure 15 shows the calculated high frequency C_{hf} and quasi-static capacitance C_{qs} as a function of dc voltage V for two hypothetical oxide/n-GaAs MOS systems with $D_{it} = 0$ and ΔE determined as explained above. While the C–V curves show typical inversion behavior for $\Delta E = 1.14$ eV, steady-state deep depletion is observed for $\Delta E = 0.94$ eV. The ΔE_v required for inversion to occur on n-type GaAs was

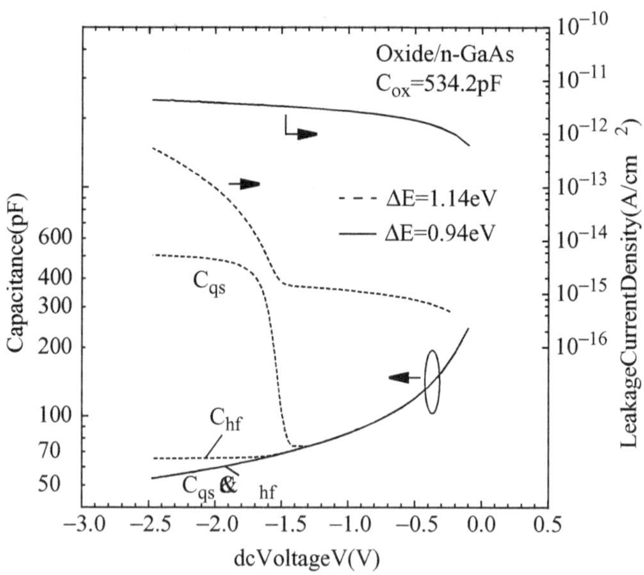

Fig. 15. Calculated high frequency C_{hf} and quasi-static capacitance C_{qs} as a function of dc voltage V for two hypothetical oxide/n-GaAs MOS systems with $D_{it} = 0$ ($N_D^+ = 2 \times 10^{16}$ cm^{-3}, area $= 1.96 \times 10^{-3}$ cm^2).

estimated to $\Delta E_v > 1.14$ eV in ref. [32], a criterion which does not appear to be unrealistic. However, for reasons explained above, this criterion, although correct, is impractical because the leakage current at room temperature is often dominated by trap assisted tunneling in the dielectric layer in non-ideal MOS systems. Therefore, the leakage currents for both cases of inversion and steady-state deep depletion are also depicted in Fig. 15. As apparent from Fig. 15, a leakage current density well below 1 pA/cm^2 and possibly closer to 10 fA/cm^2 is required for inversion to occur in a GaAs based MOS system. Since the lowest leakage current densities reported for GaAs based MOS systems are around 1 nA/cm^2, it must be concluded that it is impossible to observe inversion in C–V measurements on GaAs based MOS systems when measured in the dark. Note that the leakage current density required for inversion to occur is roughly proportional to $1/n_i$ [32] which largely relaxes the requirement for smaller bandgap semiconductor such as Si and Ge but makes the observation of inversion on wider bandgap semiconductors all but impossible in the dark. It should be emphasized that the above discussion is based on a steady-state solution: in steady-state deep depletion, the leakage current supported by the dielectric (far) exceeds the flow of minority carriers provided by the semiconductor, i.e., even if one waited for an infinite amount of time, the surface will never invert. Only if the above derived steady-state criterion for inversion is met, the classical inversion carrier response time [28] is applicable to the dynamics of inversion layer formation. Consequently, MOS systems with wider bandgap semiconductors show no frequency dispersion in C–V measurements in the dark under negative dc voltage (n-type) when the Fermi level is unpinned, i.e., $C_{qs} = C_{hf}$. Any frequency dispersion observed in this bias range is due to interface states and indicates a high D_{it} (see also dashed lines in Fig. 13 for $\Delta E = 0.49$ eV).

3.4. C–V vs. PL-Intensity: A Comparison

While capacitor based characterization methods such as C–V measurements which require to maintain quasi-equilibrium in the semiconductor remain an essential component of interface characterization, non-equilibrium techniques such as PL-I have become, according to our experience, a enabler and necessary ingredient for gate dielectric development on wider bandgap semiconductors. Besides the many ambiguities inherent to the interpretation of measured C–V characteristics on non-ideal MOS systems, the portion of the bandgap which is accessible to quasi-equilibrium techniques is reduced for semiconductors with bandgaps larger or smaller than that of Si. The bandgap range accessible to ac and quasi-static C–V methods is illustrated in Fig. 16 for a temperature of (a) 300 K and (b) 600 K. The solid lines are obtained using

$$\tau_{C/E} = \frac{1}{v_{th}\sigma n_i} \exp\left(\frac{-|E_t - E_i|}{kT}\right), \quad (15)$$

where $\tau_{C/E}$ is the trap capture/emission time constant. The boundaries of the shaded areas are calculated using $\omega\tau_{C/E} = 1$ where ω is the angular frequency ($\omega = 2\pi f$) for 50 Hz $\leq f \leq$ 10 MHz (quasi-static and ac) and $\tau_{C/E} = 10$ s (quasi-static) where 10 s is considered the practical limit for the delay time τ_d in quasi-static C–V

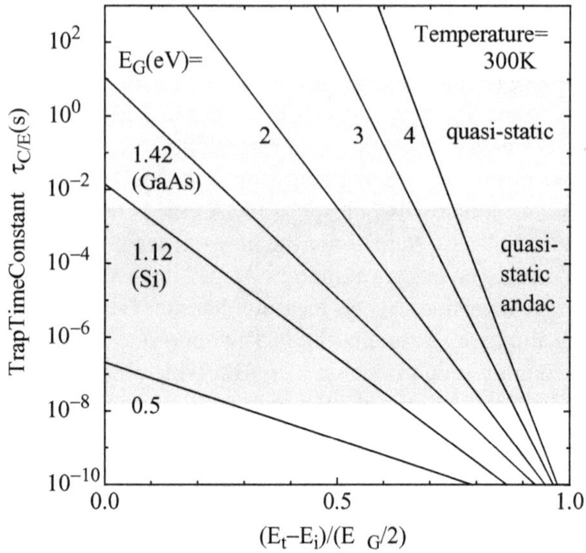

(a)

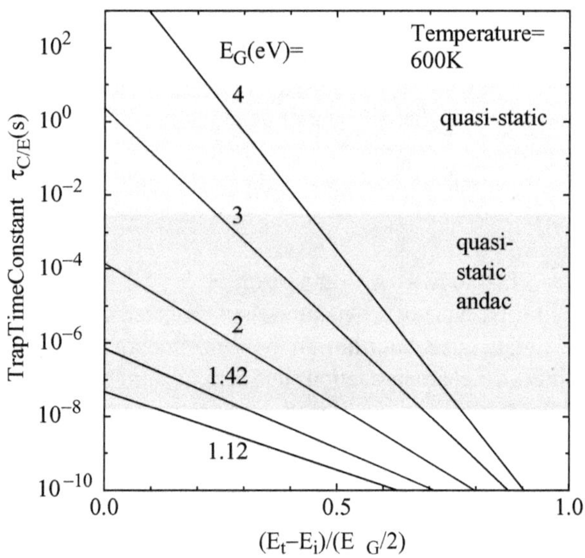

(b)

Fig. 16. Bandgap range accessible to ac and quasi-static *C–V* methods for a temperature of (a) 300 K and (b) 600 K.

measurements. While ac methods (50 Hz $\leq f \leq$ 10 MHz) and quasi-static methods (delay time $\tau_d \leq 10$ s) cover about 45 and 75% of the GaAs bandgap at 300 K, the bandgap range drops to 20 and 35%, respectively, for semiconductors such as GaN with a bandgap slightly higher than 3 eV. Most of the interface states are simply not seen by *C–V* techniques and *C–V* curves look artificially good. Note that the shaded areas in Fig. 16 provide guidelines only; the accessibly bandgap energy range depends on the actual capture cross sections and may be slightly different in specific cases. One approach to address the issue of limited bandgap accessibility is to take measurements at elevated temperatures (see Fig. 16(b)), however, high temperature measurements are often problematic, in particular for high-κ dielectrics. Therefore, a non-equilibrium technique such as PL-I should be considered an essential and enabling tool for interface development on wide bandgap semiconductors such as GaN. Instead of relying on trap capture/emission time constants which are an exponential function of trap energy and semiconductor bandgap, non-equilibrium techniques rest on carrier lifetimes which are typically of the order of nano- or microseconds and independent of the semiconductor bandgap. Since PL-I is a relatively new method, many questions about the technique have been raised in the past. The following comparison of C–V and PL-I attempts to give first answers.

Table 1 compares equipment and technical requirements, means of excitation, capabilities, and data interpretation aspects of PL-I to C–V. Some basic properties are discussed here, other features are covered throughout this chapter. The particular strength of PL-I is at the front end where it provides one definite measure of interface quality for dielectric films of any insulating properties and thickness (bulk film to monolayer as long as the interface is stable in the respective environment), any bandgap, with fast turnaround, and without the need to manufacture specific test structures such as capacitors. The PL-I technique is a perfect tool to make go-no go decisions for a particular material system and it is therefore used as a screening tool prior to C–V measurements which are more time consuming and where data interpretation is complex and prone to error for non-ideal systems. Without any doubt, PL-I has become our most robust interface characterization method during the development of high-κ gate dielectrics on GaAs. Ultimately, C–V measurements provide the total trap concentration in the entire stacked gate dielectric applicable to the prediction of MOSFET performance including interface traps, border traps, traps located at oxide/oxide interfaces (if applicable), mobile ionic charge, fixed charge, etc. C–V measurements further furnish the distribution of interface states in energy space over a larger portion of the bandgap, however, with serious limitations for wide and very small bandgap semiconductors. It is correct to state that the development of stacked gate dielectrics on GaAs at Motorola has built on the synergy and complementary character of C–V and PL-I. Our effort would most likely have failed if (1) one of the techniques of C–V or PL-I had not been available and, (2) the data provided by both techniques had not been evaluated in their entirety. The implementation of this approach is now discussed in Section 4.

4. RESULTS AND DISCUSSION

After the experimental and theoretical foundations have been introduced in Section 2 and Section 3, the application of the proposed methodology for the development of high-κ $Gd_xGa_{0.4-x}O_{0.6}/Ga_2O_3$ dielectric stacks on GaAs is now discussed and

Table 1. Comparison of C–V and PL-I equipment, technical requirements, means of excitation, capabilities, and data interpretation

	C–V	PL-I
Equipment		
Instrumentation	**Standard, commercially available**	Non-standard, customized[(a)]
Cost	\$00–\$00k (prober and C–V system)[(b)]	\$00–300k [(b)]
Technical requirements		
Standards	**Not required**	Required
Epitaxial structure	Not required, but many material systems use epitaxial layers	Required
Sample processing	Yes (capacitor)	**No**
Semiconductor bandgap	**Indirect and direct**	Direct
Quasi-equilibrium	Required	**Not relevant**
Dielectric	Electrically insulating, bulk film	**monolayer to bulk film**
Means of excitation	Voltage, electric field	Photon flux
Capabilities		
Accessible bandgap range	Function of bandgap	**Not a function of bandgap**
Energy resolution	**Yes**	Limited to vicinity of band edges
Type of states probed	Immediate interface, border traps, internal dielectric/dielectric interfaces	Immediate interface
Interface parameters provided	D_{it} (E_t)	S, (D_{it}, σ)
Data interpretation		
Required parameters	Ionized doping concentration	Ionized doping concentration (epilayer nonradiative lifetime)
Prone to ambiguities	Yes, on non-ideal systems	**Not in very high injection**

[a]Could be implemented in existing commercial tools such as the photoluminescence mappers manufactured by Accent Optical Technologies.
[b]Dependent on level of automation.

state-of-the-art data are presented, when possible, in context with previously published data. The stacked dielectric structure is virtually built step by step in this chapter while the feasibility of materials, manufacturing techniques, and characterization methods is investigated for each step and in correlation with previous steps. While the first step, the study of chemisorption of Ga_2O molecules on GaAs, is not a subject of this chapter, it is an integral part of the methodology. The ability to investigate the structural and electrical interface properties under (sub)-monolayer coverage is a powerful tool and in particular suitable when oxide molecules are deposited by nonreactive methods. The reader is referred to ref. [25] for details of oxide monolayer formation when Ga_2O molecules are chemisorbed on GaAs. The finding that chemisorption of Ga_2O molecules creates a charge balanced (2×2) surface order on GaAs(001) that is electronically unpinned [25] has provided the first step onto which all subsequent work carried out with the ultimate objective to provide a stacked gate dielectric with low D_{it} and adequate dielectric properties had to be built on. Analogous efforts are now also underway for high-κ dielectrics on silicon, see for example, the work by Tao et al. [39] based on valence-mended semiconductor surfaces [40], and chapters by Yong and Demkov as well as McKee in this book.

4.1. Ga_2O_3/GaAs

After a monolayer of Ga_2O chemisorbs on the GaAs surface, growth proceeds via formation of an amorphous Ga_2O_3 film [17], the second step in our proposed methodology. The presence of a bulk oxide film typically facilitates the use of quasi-equilibrium interface characterization methods such as C–V. Here, C–V data of metal/AlGaAs/GaAs capacitors are discussed first and the findings are later extended to Ga_2O_3/GaAs interfaces. This approach is chosen because the AlGaAs/GaAs system is well established and all material parameters are known, i.e., the AlGaAs system is considered to be a perfect vehicle to illustrate the difficulties encountered with C–V techniques. It is subsequently demonstrated that PL-I overcomes the limitations of the C–V technique and allows to extract the interface parameters in the material systems discussed here.

Figure 17 shows the measured quasi-static and 100 kHz C–V curves (solid lines) as well as an ideal ($D_{it} = 0$) calculated C–V curve (dashed line) of a $Ti/Al_{0.58}Ga_{0.42}As/GaAs$ capacitor. Note that AlGaAs/GaAs has been grown as a

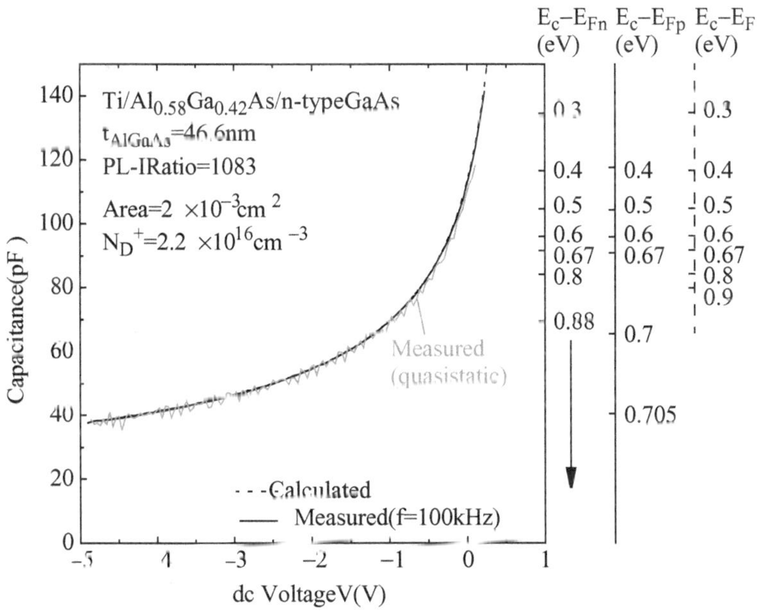

Fig. 17. Measured quasi-static and 100 kHz C–V curves (solid lines) as well as an ideal ($D_{it} = 0$) calculated C–V curve (dashed line) of a $Ti/Al_{0.58}Ga_{0.42}As/GaAs$ capacitor.

baseline structure (see Section 2). The y-axes on the right-hand side show the quasi-Fermi level position with reference to the conduction band edge E_c ($E_c - E_{Fn}$, $E_c - E_{Fp}$) and a hypothetical Fermi level position $E_c - E_F$ which is calculated under the assumption that quasi-equilibrium prevails in GaAs. Note that Fig. 17 also shows a quantity termed PL-I ratio. This parameter is obtained from PL-I measurements and an indicator of interface quality; it is further discussed below in connection with the application of the PL-I technique to the interfaces discussed here. In the following, a number of characteristic features is pointed out such as (a) the occurrence of steady-state deep depletion for $E_F < E_i$ ($E_i = E_c - 0.67$ eV), (b) $C_{qs} \cong C_{100\,kHz}$, and (c) the breakdown of C_{qs} and $C_{100\,kHz}$. Apparently, the movement of the quasi-Fermi levels with respect to the band edges is diminished for $E_{Fn}, E_{Fp} < E_i$, the hallmark of steady-state deep depletion. To further elucidate this point, an energy band diagram for steady-state deep depletion is shown in Fig. 18 ($V = -1$ V). As can be seen from Fig. 18, both E_{Fp} and E_{Fn} bend in the GaAs and reside around midgap at the interface in steady-state deep depletion. Note that trap occupancy is determined by E_{Fp} when $E_{Fn}, E_{Fp} < E_i$ (Eqs. (12) and (13)) and that $E_{Fp} - E_c$ is virtually constant for $E_{Fp} < E_i$; i.e., trap occupancy remains constant and C_{qs} and $C_{100\,kHz}$ do not contain any information relevant for D_{it} analysis. Above E_i, quasi-equilibrium prevails ($E_{Fn} \cong E_{Fp} \cong E_F$), however, D_{it} is apparently too small to be either detected by the quasi-static/high frequency method ($C_{qs} \cong C_{100\,kHz}$) [28] and the Terman method (slopes of $C_{100\,kHz}$ and calculated ideal C–V are virtually identical) [29]. The resolution limits of the quasi-static/high frequency technique and Terman's method

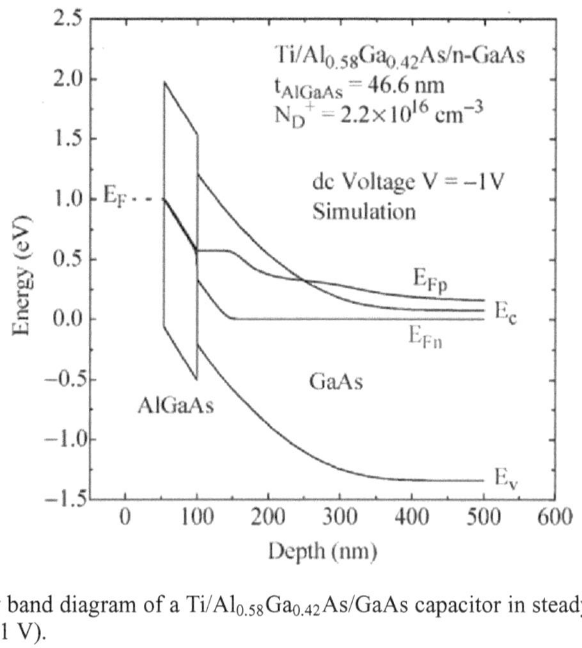

Fig. 18. Energy band diagram of a Ti/$Al_{0.58}Ga_{0.42}As$/GaAs capacitor in steady-state deep depletion ($V = -1$ V).

are $\cong 10^{10}$ cm^{-2} eV^{-1} and mid 10^{11} cm^{-2} eV^{-1}, respectively. Although such systems may have a significant higher D_{it} in the vicinity of the band edges, this range is not accessible by C–V since C_{qs} and C_{hf} break down at $E_c - E_F \cong 0.4$ and 0.3 eV, respectively [41]. Consequently, the quasi-static/high frequency method and the Terman method could be applied only in between $E_i < E_t < 0.4$ eV and $E_i < E_t < 0.3$ eV, respectively, but no information in regard to D_{it} is obtained within these energy ranges (see above).

Figure 19 shows measured 100 kHz C–V curves (solid lines) and ideal ($D_{it} = 0$) calculated C–V curves (dashed lines) of (a) a Pt/Ga_2O_3/GaAs capacitor with an as-deposited Ga_2O_3/GaAs interface, (b) a Ti/Ga_2O_3/GaAs capacitor with a post-deposition hydrogen annealed Ga_2O_3/GaAs interface, and (c) a Ti/GaAs Schottky diode. The latter is manufactured from a GaAs baseline wafer grown without oxide (see Section 2). Note that quasi-static curves could not be acquired from any of the structures discussed in Fig. 19. Otherwise, the C–V curves in Fig. 19 resemble those shown in Fig. 17, the only difference is the energy at which the 100 kHz C–V measurement breaks down with $E_c - E_F \cong 0.35 - 0.25$ eV (comparable to AlGaAs/GaAs) and $E_c - E_F \cong E_i = 0.67$ eV for Ga_2O_3/GaAs capacitors and Ti/GaAs Schottky diodes, respectively. Note that the ionized donor concentration is independently determined by electrochemical depth profiling (see Fig. 20) and used

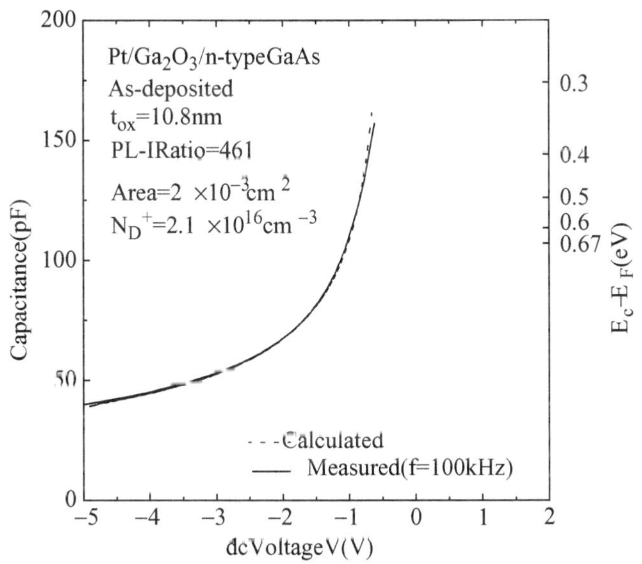

(a)

Fig. 19. Measured 100 kHz C–V curves (solid lines) and ideal ($D_{it} = 0$) calculated C–V curves (dashed lines) of (a) a Pt/Ga_2O_3/GaAs capacitor with an as-deposited Ga_2O_3/GaAs interface, (b) a Ti/Ga_2O_3/GaAs capacitor with a postdeposition hydrogen annealed Ga_2O_3/GaAs interface, and (c) a Ti/GaAs Schottky diode. The ionized donor concentration is independently determined by electrochemical depth profiling.

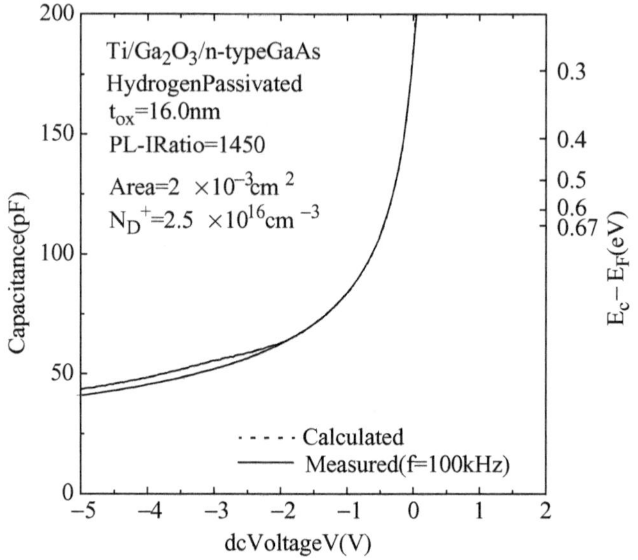

(b)

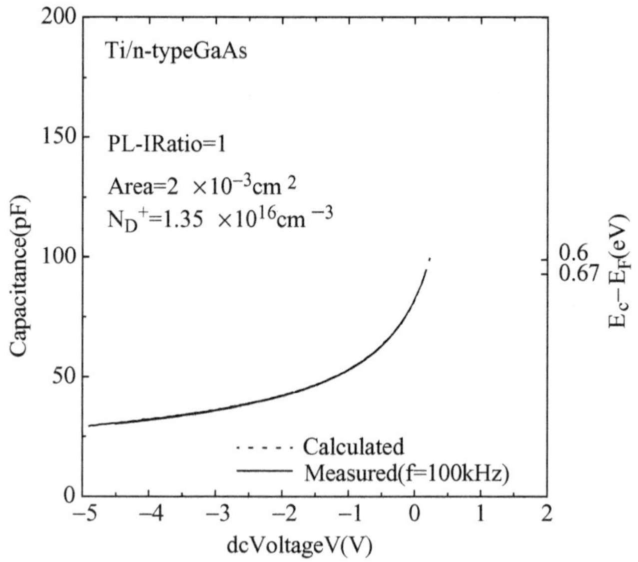

(c)

Fig. 19. (*continued*)

as an input parameter for the calculation of C–V curves. Figure 21 shows measured 100 kHz (solid lines) and 1 MHz (dashed lines) C–V curves for all four systems discussed above, Ti/$Al_{0.58}Ga_{0.42}As$/GaAs, as-deposited Pt/Ga_2O_3/GaAs, and hydrogen passivated Ti/Ga_2O_3/GaAs capacitors as well as Ti/GaAs Schottky diodes. None of the systems shows any frequency dispersion. Remarkably, systems which were

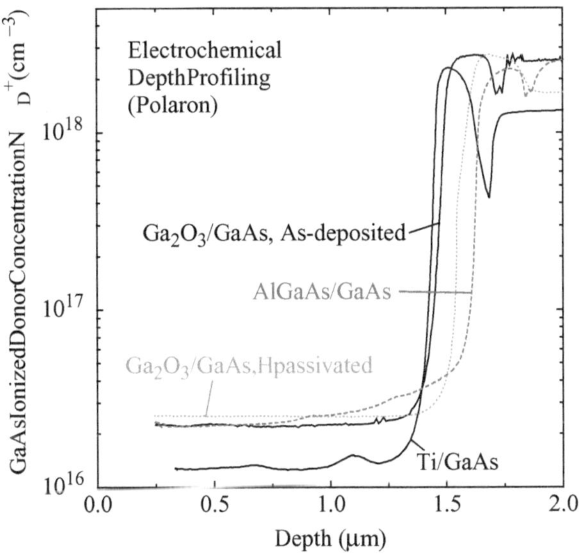

Fig. 20. GaAs ionized donor concentration as a function of depth. The following average doping concentrations are measured for a depth ≤0.8 μm (maximum depletion layer depth in C–V measurements). As-deposited and H passivated Ga_2O_3/GaAs samples, Ti/GaAs sample, and AlGaAs/GaAs sample: 2.2×10^{16} cm^{-3}, 2.5×10^{16} cm^{-3}, 1.3×10^{16} cm^{-3}, and 2.2×10^{16} cm^{-3}, respectively.

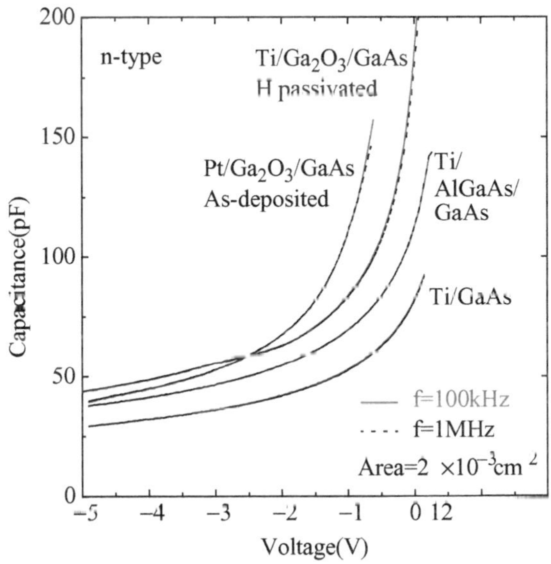

Fig. 21. Measured 100 kHz (solid lines) and 1 MHz (dashed lines) C–V curves for Ti/$Al_{0.58}$ $Ga_{0.42}$As/GaAs, as-deposited Pt/Ga_2O_3/GaAs, and hydrogen passivated Ti/Ga_2O_3/GaAs capacitors as well as Ti/GaAs Schottky diodes.

extensively investigated in the past and are well known to be set apart by approximately four orders of magnitude in surface/interface recombination velocity (AlGaAs/GaAs $\cong 10^3$ cm/s, Ti/GaAs Schottky diode $> 10^7$ cm/s, see also the PL-I ratios provided for each structure in Figs. 17 and 19), produce identical features in C–V measurements and further, such C–V measurements entirely fail to provide any information about the investigated interfaces. It is now tempting to discard the Ga_2O_3/GaAs system because satisfactory C–V data are not obtained and some have argued that this system is of no use for MOSFET applications. Instead, we have chosen to develop an appropriate technique (PL-I) which allowed to fully characterize the electrical properties of the Ga_2O_3/GaAs interface. This approach has enabled Motorola to eventually proceed to the third step of the proposed methodology, the development of stacked gate dielectrics on GaAs which provide both low D_{it} and adequate dielectric properties as discussed in the following.

Figure 22 shows measured normalized PL spectra for selected test structures at various representative excitation densities. Starting at the highest curve for each test structure, $P_0' = 6.6 \times 10^3, 7.1 \times 10^1, 8.0 \times 10^{-1}$ W/cm^2 for postdeposition hydrogen passivated Ga_2O_3/GaAs, 6.9×10^3, 7.4×10^1, 8.4×10^{-1} W/cm^2 for as-deposited Ga_2O_3/GaAs, and 6.5×10^3, 8.0×10^{-1} W/cm^2 for the air exposed GaAs surface (baseline wafer), respectively. The PL spectra are acquired over an intensity range of more than eight orders of magnitude. The spectra shown in Fig. 22 are normalized

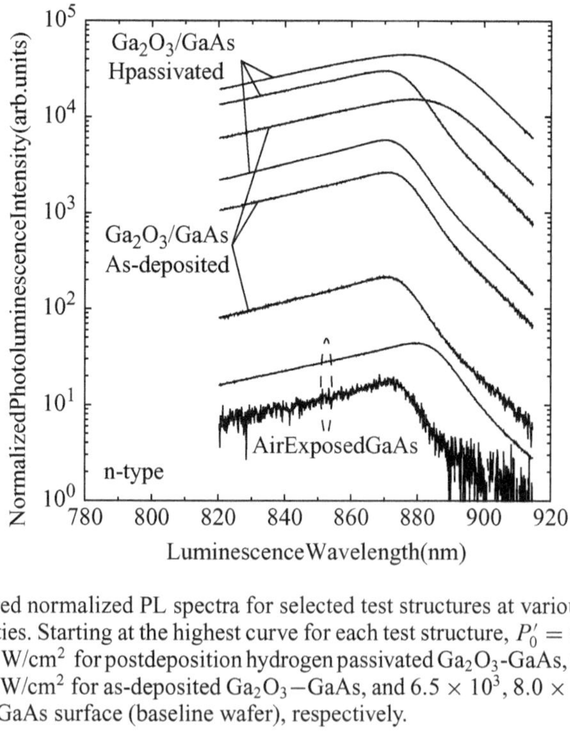

Fig. 22. Measured normalized PL spectra for selected test structures at various representative excitation densities. Starting at the highest curve for each test structure, $P_0' = 6.6 \times 10^3, 7.1 \times 10^1, 8.0 \times 10^{-1}$ W/cm^2 for postdeposition hydrogen passivated Ga_2O_3-GaAs, $6.9 \times 10^3, 7.4 \times 10^1, 8.4 \times 10^{-1}$ W/cm^2 for as-deposited Ga_2O_3–GaAs, and $6.5 \times 10^3, 8.0 \times 10^{-1}$ W/cm^2 for the air exposed GaAs surface (baseline wafer), respectively.

to the excitation density P_0' entering the semiconductor while the integration time, the attenuation of the PL signal by neutral density filters (if applicable), and the optical transmissivity of the sample surface at the peak position of the PL at 870 nm are taken into account. Some features of the spectra such as broadening and shift of PL peak position for higher P_0' are discussed in more detail in ref. [26]. The measured photoluminescence intensity $\Im_{PL}$ (arbitrary units) is obtained by integrating the normalized spectra shown in Fig. 22 over the wavelength range from 820 to 910 nm. The PL-I ratio which is frequently used as a simple but reliable measure of interface quality is defined as the ratio between the intensity $\Im_{PL}$ measured for a sample under investigation and the $\Im_{PL}$ intensity of the air-exposed, baseline epitaxial wafer, corrected for any differences in optical transmissivity of the samples surfaces. The PL-I ratio is further defined at the highest excitation density used ($P_0 = 10$ kW/cm^2). The calculated maximum PL-I ratio is in between 1400 and 1500 for the epitaxial layer structure employed in this study (see Fig. 23).

Figure 24 illustrates the two-step procedure as proposed in Section 3 and now employed to determine (a) the interface recombination velocity S, and (b) the interface state density D_{it} and capture cross section σ for a typical postdeposition hydrogen annealed Ga_2O_3/GaAs interface. Figure 24(a) illustrates the shift of the measured $\Im_{PL}$ intensity curve (diamonds) along the y-axis conducted with the objective to match a calculated η-P_0' dependence (solid lines) in (very) high injection. Note that there is only one model parameter (τ or S in Eq. (6)) with the interface charge (Eqs. (12) and (13)) being inconsequential in (very) high injection. The measured $\Im_{PL}$

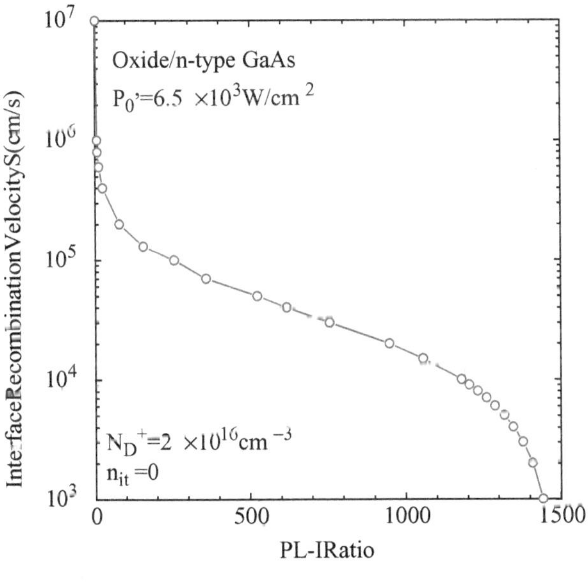

Fig. 23. Calculated dependence between interface recombination velocy S and PL-I ratio for the optimized test structure ($n_{it} = 0$). The calculated maximum PL-I ratio is in between 1400 and 1500.

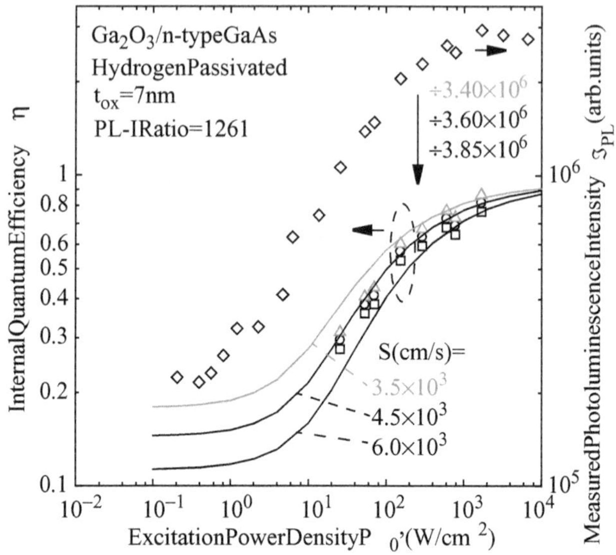

(a)

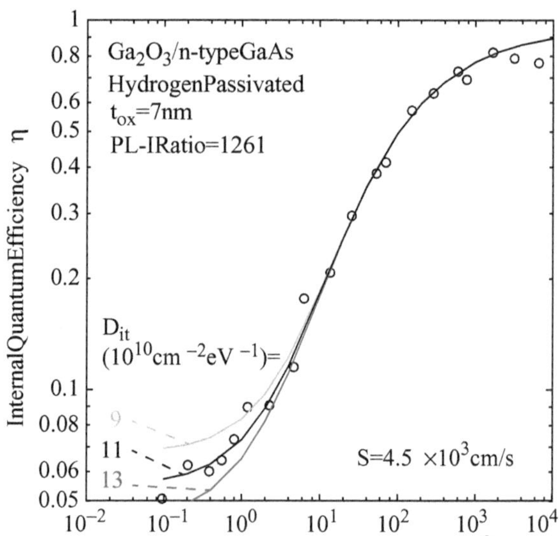

(b)

Fig. 24. Extraction of (a) the interface recombination velocity S with n_{it} neglected, and (b) the interface state density D_{it} and capture cross section σ for a typical postdeposition annealed Ga_2O_3–GaAs interface (see text for further explanation). The Ga_2O_3/GaAs structure with a 7 nm thick Ga_2O_3 film was passivated in the Tegal 6000 using the following parameter set: RF power 100 W, pressure 40 mTorr, hydrogen flow 10 sccm, and exposure time 90 s. Note that virtually identical $\Im_{PL}$ vs. P_0' curves are measured after successful hydrogen postdeposition annealing irrespective of Ga_2O_3 film thickness ($7.1 \leq t_{ox} \leq 37.9$ nm in this chapter).

intensity curve is divided by slightly different factors (3.4×10^6 (triangles), 3.6×10^6 (circles), and 3.85×10^6 (squares) in this case) when attempting to fit to a calculated η-P_0' dependence ($S = 3.5 \times 10^3, 4.5 \times 10^3, 6 \times 10^3$ cm/s, respectively). Apparently, only the fit with $S = 4.5 \times 10^3$ (factor 3.6×10^6, circles) is satisfactory, the other calculated curves provide a slope either to small ($S = 3.5 \times 10^3$) or too large ($S = 6 \times 10^3$). Consequently, S can be reliably inferred with an accuracy of approximately ±25%Subsequently, D_{it} is determined in low injection by adjusting $D_t = D_{tA} = D_{tD}$ in Eqs. (12) and (13) and placing the charge neutrality level at E_i (see Fig. 24 (b)). In summary, the best fit to the measured $\Im_{PL}$ intensity curve of a postdeposition annealed Ga_2O_3/GaAs interface is obtained for $S = 4.5 \times 10^3$ cm/s and $D_{it} = 1.1 \times 10^{11}$ cm^{-2} eV^{-1} which results in a capture cross section of 2.9×10^{-15} cm^2. Note that the fit has been accomplished assuming a constant D_{it} throughout the GaAs bandgap. As outlined above, D_{it} quantification is dependent on the position of the charge neutrality level in the model. Charge neutrality is expected to occur close to midgap and the charge neutrality level has been consequently situated at the intrinsic energy in this chapter. This energy position is more realistic than the valence band maximum E_v used in earlier reports [26, 33].

Figure 25 illustrates the interface parameter extraction for a typical as-deposited Ga_2O_3/GaAs interface including the fit (a) in (very) high injection to obtain S, (b) in low injection to obtain D_{it} and σ, and (c) a summary of the fits depicted in (a) and (b). Note that virtually identical $\Im_{PL}$ vs. P_0' curves are measured for

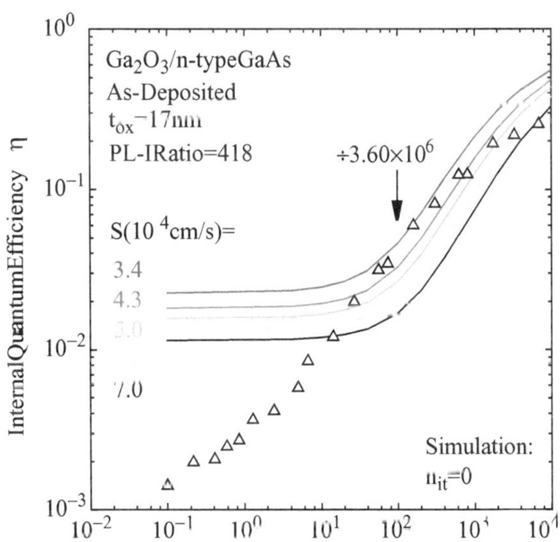

(a)

Fig. 25. Interface parameter extraction for a typical as-deposited Ga_2O_3/GaAs interface including the fit (a) in (very) high injection to obtain S with n_{it} neglected, (b) in low injection to obtain D_{it} and σ, and (c) a summary of the fits depicted in (a) and (b).

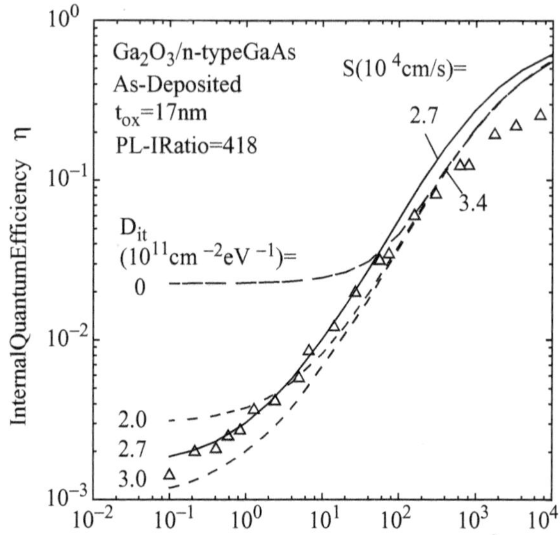

(b)

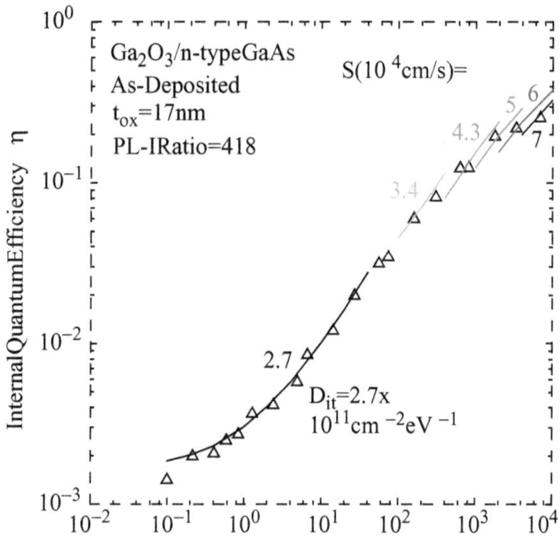

(c)

Fig. 25. (*continued*)

all manufactured Ga_2O_3 film thicknesses ($7.1 \leq t_{ox} \leq 105.1$ nm). As apparent in Fig. 25(a), the measured $\Im_{PL}$ intensity curve (divided by the same factor as the measured curve for the postdeposition annealed Ga_2O_3/GaAs interface in Fig. 24) intersects various calculated η-P_0' dependencies in (very) high injection, i.e., a fit with constant D_{it} throughout the GaAs bandgap is not feasible in this case. Using $S = 3.4 \times 10^4$ cm/s which provides a reasonable fit to the data points with the lowest P_0' in high injection, a fit in low injection is attempted (dashed lines) as shown in Fig. 25(b). However, satisfactory D_{it} fits cannot be obtained for $S = 3.4 \times 10^4$ cm/s, but a fit to $S = 2.7 \times 10^4$ cm/s and $D_{it} = 2.7 \times 10^{11}$ cm^{-2} eV^{-1} is eventually successful (solid line). This also indicates that the effects of S and D_{it} on η in low injection are not interchangeable and unique solutions for S and D_{it} may be obtained even when D_{it} is not constant throughout the bandgap. In summary, the best fit to the measured $\Im_{PL}$ intensity curve of a typical as-deposited Ga_2O_3/GaAs interface is obtained for 2.7×10^4 cm/s $\leq S \leq 7.0 \times 10^4$ cm/s and a midgap $D_{it} = 2.7 \times 10^{11}$ cm^{-2} eV^{-1} which results in a capture cross section of 6.9×10^{-15} cm^2. The dependence of S on injection level as demonstrated in Fig. 25 is further used to approximate a D_{it} vs. E_t relationship in the GaAs bandgap using Eqs. (8) and (9). As apparent in Fig. 26, D_{it} increases to as high as 8.5×10^{12} cm^{-2} eV^{-1} in the vicinity of the band edges for as deposited Ga_2O_3/GaAs interfaces.

Figure 27 summarizes the measured (symbols) and calculated (lines) internal quantum efficiencies η as a function of excitation power density P_0' for typical as-deposited Ga_2O_3/GaAs structures (triangles), typical postdeposition hydrogen annealed Ga_2O_3/GaAs structures (circles), a baseline air exposed GaAs epitaxial wafer (squares), and a baseline GaAs epitaxial wafer with an AlGaAs window layer (diamonds). Note that the measured $\Im_{PL}$ intensity curves of all investigated structures are divided by virtually identical factors. The parameter extraction for the AlGaAs/GaAs

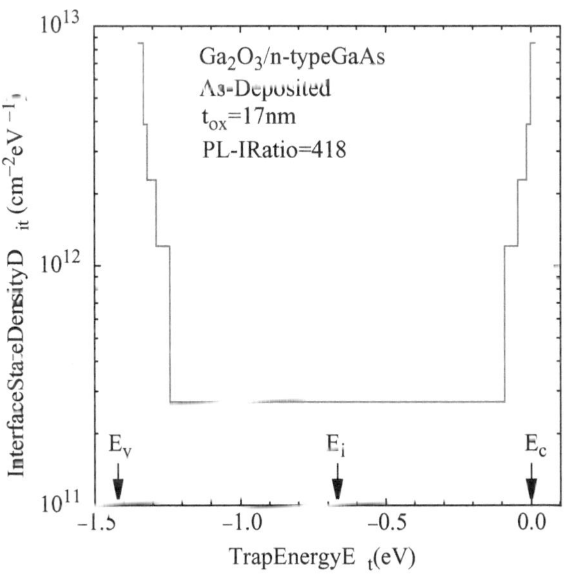

Fig. 26. Estimated D_{it} distribution in the GaAs bandgap for a Ga_2O_3/GaAs structure. D_{it} increases to as high as 8.5×10^{12} cm^{-2} eV^{-1} in the vicinity of the band edges.

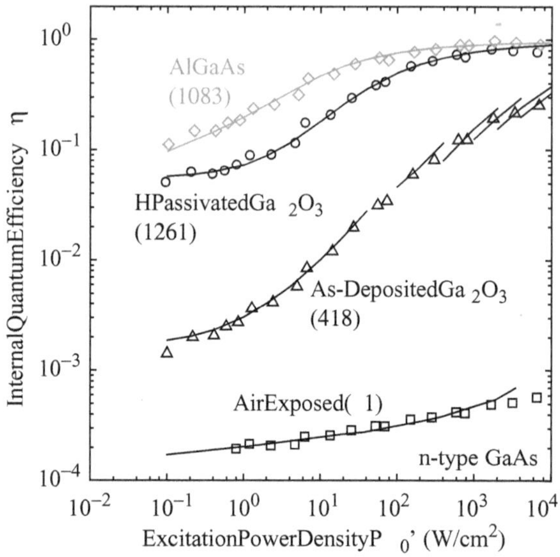

Fig. 27. Measured (symbols) and calculated (lines) internal quantum efficiencies η as a function of excitation power density P_0' for typical as-deposited Ga_2O_3–GaAs structures (triangles), typical postdeposition hydrogen annealed Ga_2O_3–GaAs structures (circles), a baseline air exposed GaAs epitaxial wafer (squares), and a baseline GaAs epitaxial wafer with an AlGaAs window layer (diamonds). The corresponding PL-I ratios are shown in parentheses. The $\Im_{PL}$ intensities of the measured $\Im_{PL}$ vs. P_0' curves were divided by 2.7×10^6 (AlGaAs/GaAs), 3.6×10^6 (as-deposited and H passivated Ga_2O_3/n–GaAs), and 4.0×10^6 (air exposed n-GaAs).

and the air exposed epitaxial baseline wafers follows procedures similar to those outlined above for the Ga_2O_3/GaAs interface and is not further discussed here. When comparing Fig. 27 with Figs. 17, 19, and 21, the advantage of the PL-I technique over C–V techniques is evident for the material systems discussed so far. Table 2 summarizes the extracted interface parameters.

As discussed earlier in Section 3, the internal quantum efficiency is a particular suitable screening tool in (very) high injection where band bending is diminished and the quality of the interface is assessed using only one model parameter, the interface recombination velocity S. As apparent from Fig. 27, η of low D_{it} interfaces (AlGaAs–, Ga_2O_3–GaAs) differs by more than two orders of magnitude from that of pinned interfaces (air exposed GaAs); low D_{it} and pinned interfaces on GaAs typically fall into two groups as illustrated in Fig. 28 (see also [42]). Clearly, Al_2O_3–, SiO_2–, and MgO–GaAs interfaces do not warrant any further consideration. PL spectra of an intensity comparable to the lower group were also measured for Mo, Zr, Ti, Ta, and Gd oxides on GaAs.

It should be noted that although evidence of low D_{it} at as-deposited amorphous Ga_2O_3/GaAs interfaces was provided as early as 1996 [14], this chapter represents the first report on postdeposition hydrogen passivation of the Ga_2O_3/GaAs system. The beneficial effects of hydrogen as a passivant in SiO_2/Si MOS systems have been recognized for decades; annealing in forming gas (a mixture of N_2 and typically

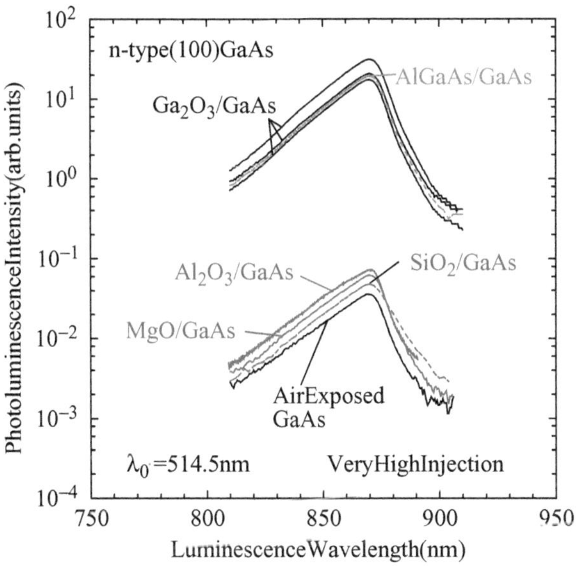

Fig. 28. Low D_{it} and pinned interfaces on GaAs typically fall into two groups: [1] AlGaAs– and Ga_2O_3–GaAs, and [2] other oxides (Al_2O_3–, SiO_2–, MgO–GaAs, and air exposed GaAs in this figure).

5–8% H_2) is a standard procedure in silicon technology. This has been generally ascribed to the passivation reaction of the silicon dangling bond structure with atomic hydrogen [43–45]. More recently, reductions in interface state density D_{it} have been reported for other Si based MOS systems such as SiO_2/SiC [46], SiO_2/SiGe [47], and $SrTiO_3$/Si [48] under molecular and/or atomic hydrogen flux. For GaAs, reports were restricted to GaAs surface treatments where both beneficial (see, e.g., [49]) and detrimental (see, e.g., [27]) effects due to hydrogen surface exposure were observed. However, if surface passivation was accomplished, surface properties degraded rapidly over time and the deposition of dielectric materials on top of the passivated surface inevitably destroyed the surface passivation.

For the MBE grown Ga_2O_3/GaAs structures reported in this chapter, the midgap interface state density D_{it} has been reduced by a factor of 2.5 after hydrogen plasma exposure and the increase of D_{it} towards the band edges observed for as-deposited Ga_2O_3/GaAs interfaces is not found after hydrogen plasma exposure (see Table 2). The latter is important for enhancement-mode MOSFETs where mobility and carrier density can be significantly affected by a high interface state density in the vicinity of the band edge. Further, the observed postdeposition hydrogen passivation of Ga_2O_3/GaAs interfaces is absolutely stable over time; identical results are measured after years of exposure in air. In order to verify that the observed passivation is due to hydrogen, the electrical interface data derived from PL-I measurements have been correlated to the presence or absence of hydrogen at the Ga_2O_3/GaAs interface as determined by SIMS. Figure 29 shows typical hydrogen concentration depth profiles of Ga_2O_3/GaAs structures where (a) Ga_2O_3/GaAs interface passivation is observed in PL-I

Table 2. Ga_2O_3–GaAs, AlGaAs–GaAs interface and air exposed GaAs surface properties as determined by PL-I (n-type GaAs)

Structure	Interface recombination velocity S (cm/s)	Interface state density D_{it} ($cm^{-2}\ eV^{-1}$)	Capture cross section $\sigma (cm^2)$	PL-I ratio	Integrated PL divided by	Interface hydrogen concentration (cm^{-3})
Ga_2O_3–GaAs As-Deposited	2.7×10^4 (low injection) 7×10^4 (very high injection)	2.7×10^{11} (midgap) 8.5×10^{12} (E_c)	6.9×10^{-15}	418	3.6×10^6	At or below detection limit ($\cong 5 \times 10^{18}\ cm^{-3}$)
Ga_2O_3–GaAs Hydrogen plasma	4.5×10^3	1.1×10^{11}	2.9×10^{-15}	1261	3.6×10^6	$>10^{20}$
$Al_{0.58}Ga_{0.42}As$–GaAs[(a)]	1.7×10^3	–[(b)]	–[(b)]	1083	2.7×10^6	Not applicable
Air exposed surface	10^7[(c)]	$N_{itA} = N_{itD} = 5 \times 10^{12}\ cm^{-2}$[(c)]	–[(b)]	1	4.0×10^6	Not applicable

(a) AlGaAs surface parameters are as follows: $N_{itA} = N_{itD} = 10^{12}\ cm^{-2}$, $S = 10^5$ cm/s; τ(AlGaAs) = 1 ns.
(b) Cannot be reliably determined.
(c) Standard GaAs surface parameters assumed, no fit to experimental data performed.

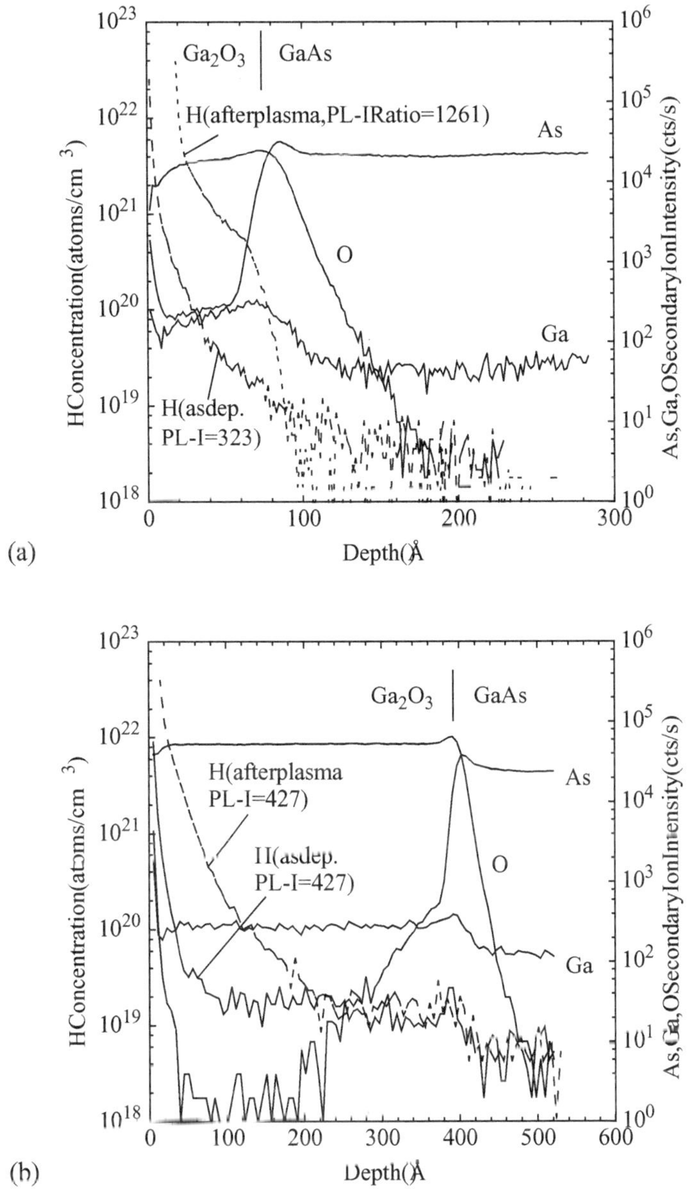

Fig. 29 Hydrogen concentration and secondary ion intensity (As, Ga, O) depth profiles of Ga_2O_3–GaAs structures as measured by SIMS with (a) a 7.4 nm thin Ga_2O_3 layer and (b) a 39.3 nm thick Ga_2O_3 layer where in (a) Ga_2O_3/GaAs interface passivation is observed after plasma exposure, and (b) Ga_2O_3/GaAs interface passivation is not observed in PL-I. The passivation was done in the Tegal 6000 using the following parameter set: RF power 100 W, pressure 40 mTorr, hydrogen flow 10 sccm, and exposure time 90 s. The elevated surface hydrogen concentration is due to surface hydrocarbons on all samples.

measurements after plasma exposure, and (b) Ga_2O_3/GaAs interface passivation is not observed in PL-I. The hydrogen noise floor in our SIMS measurements is apparent from Fig. 29(b) and is estimated to $1 - 2 \times 10^{19}$ cm^{-3} and $5 \times 10^{18} - 10^{19}$ cm^{-3} in Ga_2O_3 and GaAs, respectively. Whenever H is not detected at the Ga_2O_3/GaAs interface by SIMS (as-deposited case in Figure 29(a) and both "as-deposited" and "after H plasma" curves shown in Fig. 29(b)), the PL-I measured is represented by the as-deposited Ga_2O_3 curve (triangles) shown in Fig. 27. The typical η vs. P_0' curve of H passivated Ga_2O_3 in Fig. 27 (circles) is obtained only when interfacial hydrogen is detectable by SIMS, e.g., when an interfacial hydrogen concentration above 10^{20} cm^{-3} is measured after postdeposition hydrogen plasma exposure (see Fig. 29(a)). Note that no interface passivation by hydrogen could be observed when the plasma source was turned off. Further, S, F, C, N, O, He, Ar, Kr, Cl, and Br based plasmas did not passivate the Ga_2O_3/GaAs interface.

The second step of the methodology is now complete and the stage is set to proceed to the third, final step: the realization of high-κ gate oxide stacks using a Ga_2O_3 template which provides the unique low D_{it} property on GaAs as established during the first two steps of the methodology. Before that, a brief digression into $Gd_xGa_{0.4-x}O_{0.6}$/GaAs structures further strengthens the case for the proposed methodology.

4.2. $Gd_xGa_{0.4-x}O_{0.6}$/GaAs

Figure 30 shows (a) measured quasi-static, 100 kHz, and 1 MHz C–V curves for a slow sweep rate of 0.055 V/min (solid lines), and (b) measured 1 MHz curves for sweep rates of 5.5 V/min (fast) and 0.055 V/min (slow) in comparison with quasi-equilibrium data calculated for a D_{it} of zero (dashed line) for a Ti/$Gd_{0.19}Ga_{0.21}O_{0.6}$/GaAs MOS capacitor. Note the absence of a Ga_2O_3 template and the low PL-I ratio of 96. Large frequency dispersion is observed for both positive and negative dc bias (Fig. 30(a)). Since frequency dispersion persists up to the highest measurement frequency of 1 MHz, a high frequency C–V curve is not obtained and both the quasi-static/high frequency and the Terman methods are not applicable for D_{it} analysis; both techniques require the provision of a high frequency C–V curve. Figure 30(b) further emphasizes the requirement of very slow sweep rates when wider bandgap systems are investigated. A fast sweep rate of 5.5 V/min produces a 1 MHz C–V curve far too optimistic and there is no assurance that the slow sweep rate of 0.055 V/min provides quasi-equilibrium conditions. Further, C–V measurements quickly become technically not feasible and impractical beyond the above conditions. For example, it takes 3 hours to acquire one C–V curve and the quasi-static instrumentation is pushed to and above its limitations (further discussed below) when the slow sweep rate of 0.055 V/min is used; ac C–V measurements above 1 MHz represent significant technical challenges which are not met with standard C–V instrumentation. Consequently, a lower limit of $\cong 2 \times 10^{13}$ cm^{-2} eV^{-1} can be estimated at best for midgap D_{it} based on the data depicted in Fig. 30 (see Fig. 44).

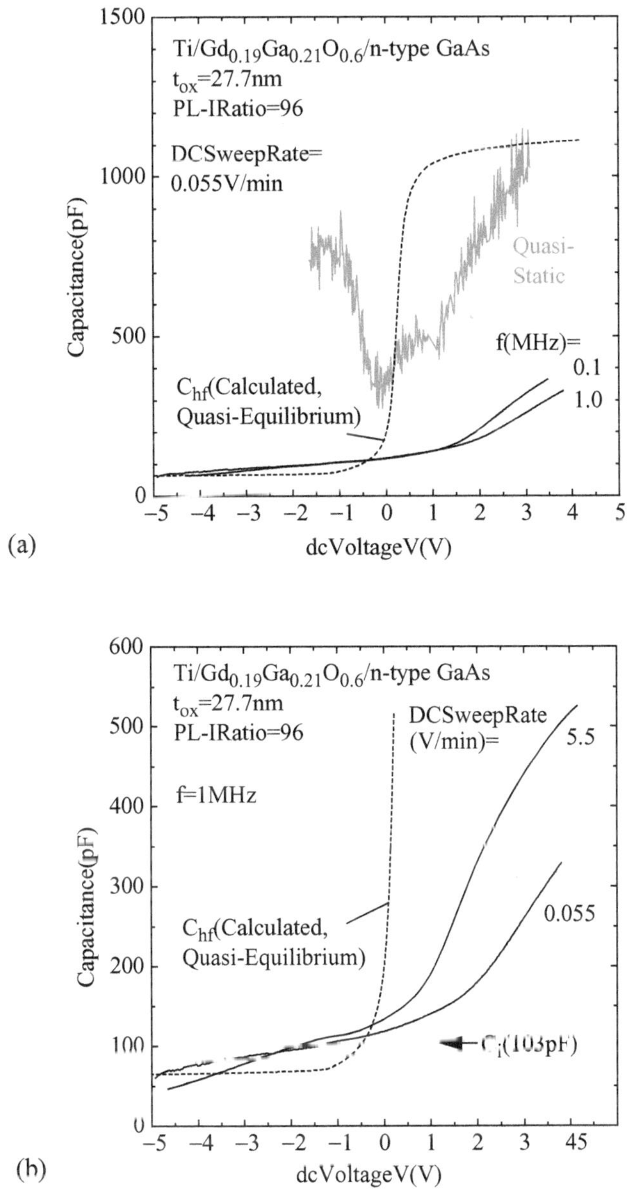

Fig. 30. (a) Measured quasi-static, 100 kHz, and 1 MHz C–V curves for a slow sweep rate of 0.055 V/min (solid lines), and (b) measured 1 MHz curves for sweep rates of 5.5 V/min (fast) and 0.055 V/min (slow) in comparison with quasi-equilibrium data calculated for a D_{it} of zero (dashed line) for a $Ti/Gd_{0.19}Ga_{0.21}O_{0.6}/GaAs$ MOS capacitor (area $= 1.96 \times 10^{-3}$ cm^2, for calculated curve: $N_D^+ = 2 \times 10^{16}$ cm^{-3}, $k = 20.8$).

The operational modes of the Ti/$Gd_{0.19}Ga_{0.21}O_{0.6}$/GaAs MOS capacitor can also be easily identified in Fig. 30(b). A plateau for negative dc voltage is observed with $C \cong C_i$, where C_i is the MOS capacitance when $E_F = E_i$ on the GaAs surface. Consequently, this plateau is clearly due to midgap interface states and not due to inversion carriers. Also note that the leakage current density required for inversion to occur in GaAs based MOS capacitors is below 0.1 pA/cm^2 (see Fig. 15) and that the leakage current of the structure depicted in Fig. 30 ($\cong$10 nA/cm^2) is more than five orders of magnitude higher. C–V data with similar large frequency dispersion and pronounced capacitance plateau under negative dc voltage were reported earlier for $Gd_xGa_yO_z$/GaAs MOS capacitors [21, 50]. However, in refs. [21, 50], the frequency dispersion, which was observed for f as high as 10 kHz under negative bias and for leakage currents >1 nA/cm^2, was erroneously ascribed to surface inversion: the leakage current criterion for surface inversion was violated by at least four orders of magnitude, and the classical inversion carrier response time τ_{inv} which is in excess of 1.5×10^4 s for the reported doping concentration [28], would only have allowed minority carriers to respond for $f \leq 10^{-5}$ Hz, a discrepancy of 9 orders of magnitude. It should be reiterated here that inversion carriers cannot be observed under any practical measurement conditions in the dark on GaAs and wider bandgap semiconductors (see Section 3); only light illumination allows the observation of surface inversion when D_{it} is sufficiently low [14–16]. The measurement of C–V characteristics under illumination was later abandoned because of the additional level of complexity introduced into the data interpretation when extracting D_{it} [32].

If growth parameters are not optimized, Gd easily accumulates at the oxide/GaAs interface. Figure 31 shows experimental and simulated RBS data of an oxide/GaAs

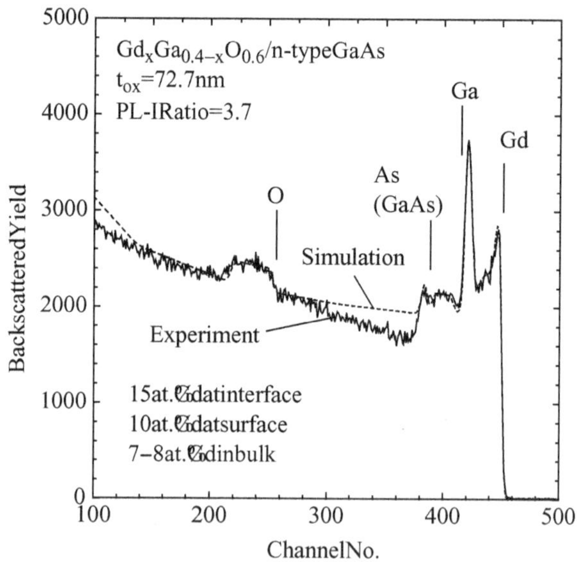

Fig. 31. Experimental (solid line) and simulated (dashed line) RBS data of a $Gd_xGa_{0.4-x}O_{0.6}$/GaAs structure with an enhanced interfacial Gd concentration.

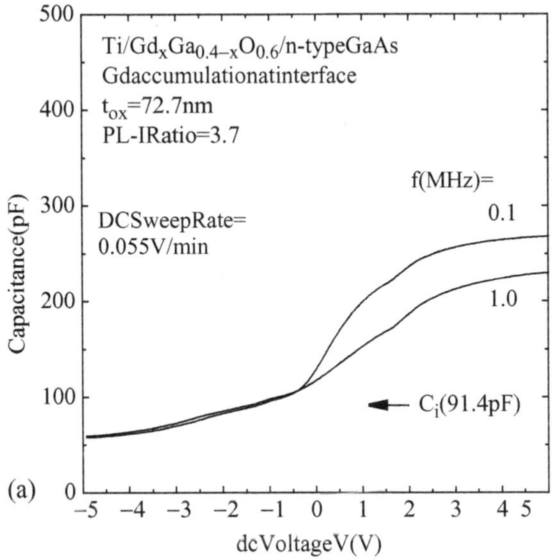

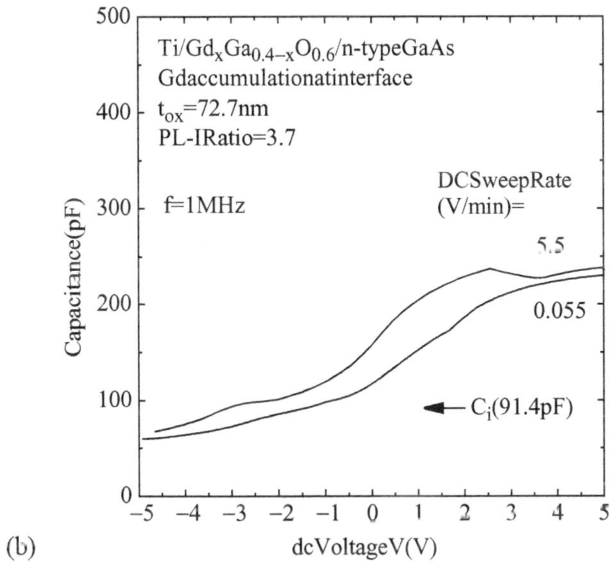

Fig. 32. (a) Measured 100 kHz and 1 MHz C–V curves for a slow sweep rate of 0.055 V/min, and (b) measured 1 MHz curves for sweep rates of 5.5 V/min (fast) and 0.055 V/min (slow). (area = 1.96×10^{-3} cm^2, for C$_i$ calculation: $N_D^+ = 2 \times 10^{16}$ cm^{-3}, k = 20.8.)

structure with an enhanced interfacial Gd concentration. The electrical properties of such an interface are degraded further as illustrated in Fig. 32 which shows (a) measured 100 kHz and 1 MHz C–V curves for a slow sweep rate of 0.055 V/min, and (b) measured 1 MHz curves for sweep rates of 5.5 V/min (fast) and 0.055 V/min (slow). Even more pronounced frequency and dc sweep dispersions are observed and the PL-I ratio of 3.7 indicates electrical interface properties virtually identical to an air exposed surface, i.e., a native oxide/GaAs interface.

4.3. $Gd_xGa_{0.4-x}O_{0.6}/Ga_2O_3/GaAs$

The final and third step, the development of high-κ $Gd_xGa_{0.4-x}O_{0.6}/Ga_2O_3$ dielectric stacks on GaAs is now discussed and state-of-the-art data are presented, when possible, in context with previously published data. To begin with, the distinctively different roles of Gd mole fraction and Ga_2O_3 template thickness in high-κ $Gd_xGa_{0.4-x}O_{0.6}/Ga_2O_3$ dielectric stacks on GaAs are clarified. In a first experiment, the effects of both Ga_2O_3 template thickness and Gd mole fraction in $Gd_xGa_{0.4-x}O_{0.6}/Ga_2O_3$ stacks have been studied. Figure 33(a) shows the high frequency (100 kHz) C–V traces of various wafers numbered 00–06 through 00–45. Throughout this wafer sequence, the Gd mole percent was increased from 8.8 to 17.3 and the Ga_2O_3 template thickness was simultaneously reduced from 73 to 12 Å. The final wafer 00–45 has a $Gd_xGa_{0.4-x}O_{0.6}$ film with 17.3 at.%Gd and a Ga_2O_3 template with a thickness of 12 Å. As it is apparent from Fig. 33(a), both a Ga_2O_3 template thickness exceeding 12 Å and a Gd mole percent below 17 at.% causes degradation of the C–V curves. This degradation is clearly visible in the form of kinks and reduced capacitance under positive bias. Further, the inset of Fig. 33(a) reveals a constant PL-I ratio throughout all studied wafers which indicates that the oxide/GaAs interface properties remained virtually unchanged. Thus, the observed degradation of the C–V curves must be ascribed to charge trapping at defect sites in the oxide; the C–V curves only adopt a qualitatively correct behavior after the number of available oxide defects has been reduced by thinning the Ga_2O_3 template and increasing the Gd mole fraction in the $Gd_xGa_{0.4-x}O_{0.6}$ layer. This experiment also demonstrates the unique synergy between the two techniques of PL-I (which only characterizes the immediate interface) and C–V, which reflects charge trapping in the entire structure. But how much thinner can the Ga_2O_3 template be made without interface degradation? This issue is addressed in a second experiment described in the next paragraph.

Figure 33(b) shows the high frequency (100 kHz) C–V traces of GaAs wafers with a Ga_2O_3 template thickness $\leq$11 Å and a Gd mole percent of $\cong$19. As evident from the C–V curves, the reduction of the Ga_2O_3 template thickness to below 9–11 Å results in significant C–V stretch-out and simultaneously, the PL-I ratio falls (see inset of Fig. 33(b)). Since *both* degradation of C–V curves and PL-I ratio is observed, it must be concluded that oxide/GaAs interface properties are degrading when the template thickness falls below 10–12 Å. The observed gradual degradation of oxide/GaAs interface properties with decreasing Ga_2O_3 thickness is tentatively attributed to two factors. First, the Ga_2O_3 surface roughness as determined by AFM on a Ga_2O_3 template of approximate 10–12 Å thickness is 5.8 Å (root-mean square). Second, diffusion of Gd through the Ga_2O_3 template towards the oxide/GaAs interface is possible. Both roughness of Ga_2O_3 template surface and diffusion of Gd through the template will increase the amount of Gd at the oxide/GaAs interface causing

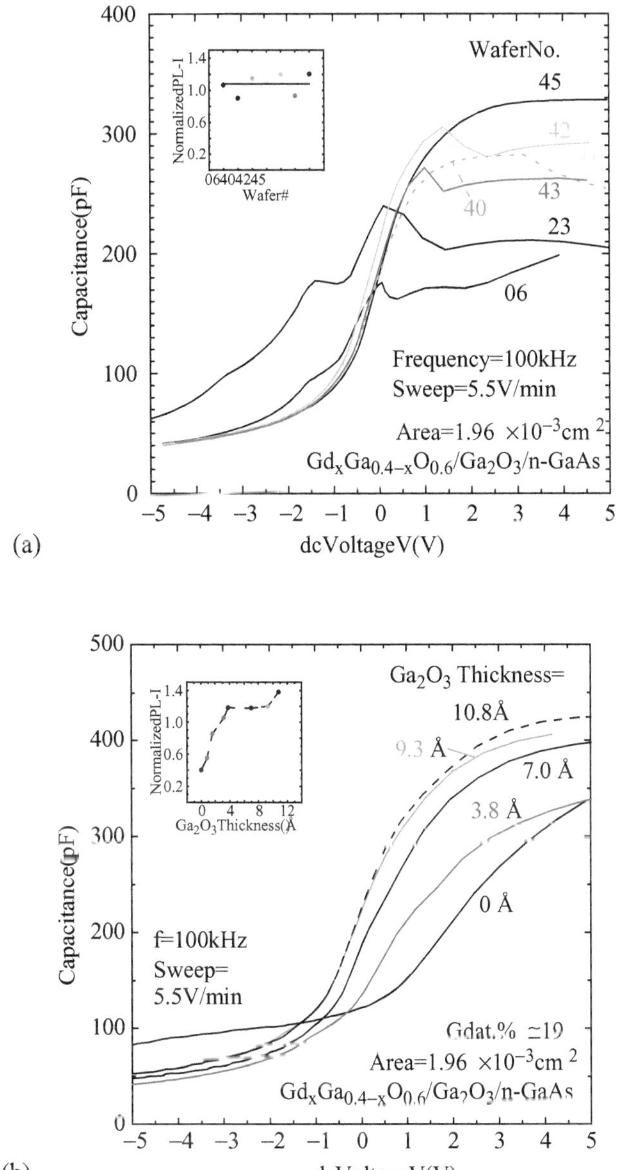

Fig. 33. (a) High frequency (100 kHz) C–V traces of the wafers 00–06 through 00–45. All C–V curves are normalized to the thickness $t_{ox} = 607$ Å of wafer 00–45 to facilitate comparison between the different wafers. (b) High frequency (100 kHz) C–V traces of wafers with Ga_2O_3 template thickness as a parameter (fast sweep rate of 5.5 V/min). Again, all C–V curves are normalized to the thickness $t_{ox} = 607$ Å of wafer 00–45 to facilitate comparison between the wafers.

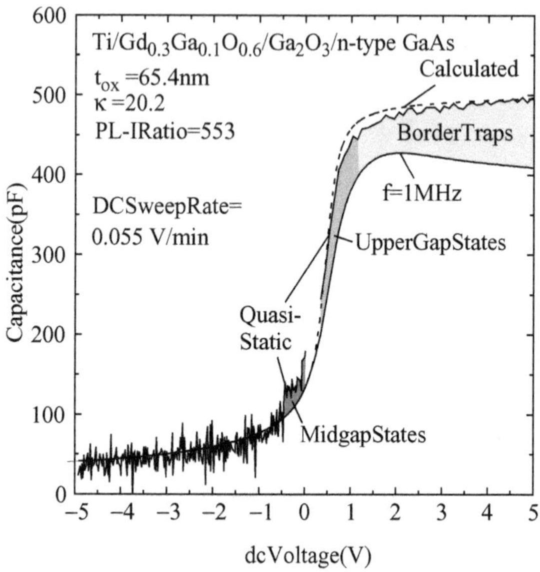

Fig. 34. Measured quasi-static and 1 MHz C–V traces of an optimized $Gd_{0.3}Ga_{0.1}O_{0.6}/Ga_2O_3$ dielectric stack on GaAs with a Ga_2O_3 template thickness of 13 Å (solid lines) in comparison with data calculated for a D_{it} of zero and a metal workfunction of 4.64 eV (dashed line). The dc sweep rate is 0.055 V/min except for the quasi-static data under positive dc voltage (5.5 V/min). (area $= 1.96 \times 10^{-3}$ cm^2, $N_D^+ = 2.5 \times 10^{16}$ cm^{-3}, oxide capacitance $C_{ox} =$ 536 pF, $C_f = 324$ pF).

gradual degradation of interface properties with decreasing Ga_2O_3 template thickness. Consequently, optimum gate oxide stack and oxide/GaAs interface properties have been obtained with a Ga_2O_3 template thickness of 9–12 Å and a minimum Gd mole percent of 15–17 at.%, respectively.

The typical modes of operation of an optimized, low D_{it} $Gd_xGa_{0.4-x}O_{0.6}/Ga_2O_3$ dielectric stack on n-type GaAs shall be now discussed based on typical C–V curves. Figure 34 shows measured quasi-static and 1 MHz C–V traces of an optimized $Gd_{0.3}Ga_{0.1}O_{0.6}/Ga_2O_3$ dielectric stack on GaAs with a Ga_2O_3 template thickness of 13 Å (solid lines) in comparison with data calculated for a D_{it} of zero and a metal workfunction of 4.64 eV (dashed line). The shaded areas illustrate the different charge trapping events either due to midgap states, upper gap states, or border traps. The typical modes of operation are as follows: (a) steady-state deep depletion with $C_{1\,\mathrm{MHz}} \cong C_{qs}$ for negative dc voltage, (b) depletion and the observance of midgap states in the quasi-static C–V curve around a dc voltage of 0 V, (c) depletion and the observance of upper gap states as indicated by $C_{qs} > C_{1\,\mathrm{MHz}}$ and the slope of the measured 1 MHz C–V being lower as the slope of the calculated C–V for slightly positive dc voltage, and (d) accumulation for $C_{1\,\mathrm{MHz}} > C_f = 324$ pF and the dominance of border traps for positive dc voltage where C_f is the flatband capacitance. Figure 35 shows experimental and simulated RBS data of the $Gd_{0.3}Ga_{0.1}O_{0.6}/Ga_2O_3$

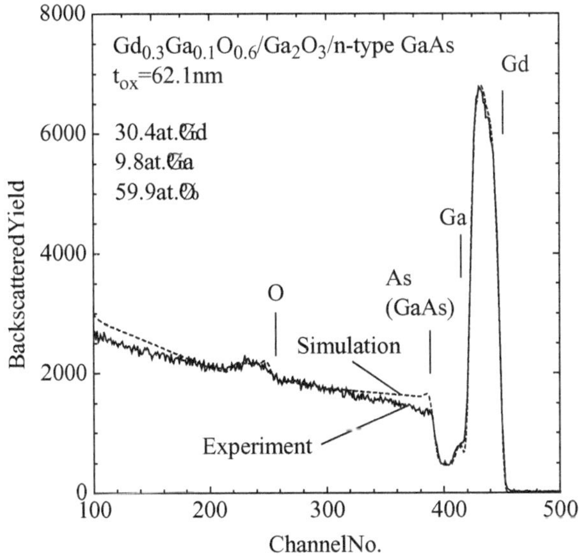

Fig. 35. Experimental (solid line) and simulated (dashed line) RBS data of a $Gd_{0.3}Ga_{0.1}O_{0.6}/Ga_2O_3$ dielectric stack on GaAs. The oxide atomic percentages of Gd, Ga, and O are 30.4, 9.8, and 59.9, respectively.

dielectric stack on GaAs, the atomic percentages of Gd, Ga, and O are 30.4, 9.8, and 59.9, respectively; Gd accumulation is not observed at the interface. It should be emphasized here that the $Gd_xGa_{0.4-x}O_{0.6}/Ga_2O_3$ system appears to be unique in its properties, attempts to use other bulk materials on Ga_2O_3 templates have failed either due to diffusion related issues and subsequent destruction of the low D_{it} property of the Ga_2O_3/GaAs interface (SiO_x, Mo_xO_y, MgO, Zr_xO_y, Al_xO_y, TiO_y, Ta_xO_y, Gd_xO_y, Sr, Ba) or the creation of a second interface between the template and the overlying bulk material (SiN). In the following, the extraction of the oxide relative dielectric constant k is discussed and the measured C–V curves are scrutinized in detail for frequency and dc sweep dispersion.

Figure 36(a) shows measured quasi-static C–V curves of an optimized $Gd_{0.3}Ga_{0.1}O_{0.6}/Ga_2O_3$ dielectric stack on n-type GaAs (solid lines) for three different sweep rates: 5.5 V/min (fast), 0.55 V/min (medium), and 0.055 V/min (slow) in comparison with data calculated for a D_{it} of zero (dashed line). The Ga_2O_3 template thickness is 10 Å. First, some specifics of quasi-static capacitance measurements are outlined to further illustrate the apparent difficulties seen in Figure 36(a) related to quasi-static measurements on MOS capacitors with finite leakage currents. The quasi-static capacitance is obtained from $Q_{dis}/\Delta V$ with the displacement charge $Q_{dis} = Q_{total} - Q_{dc}$ (this is simplified, for details see the Keithley Model 595 Instruction Manual [51]), where Q_{total} and Q_{dc} are the total charge measured subsequent to a voltage step ΔV, and the charge due to dc leakage current, respectively. Further, the maximum charge Q_{max} accepted by the quasi-static meter is a function of the

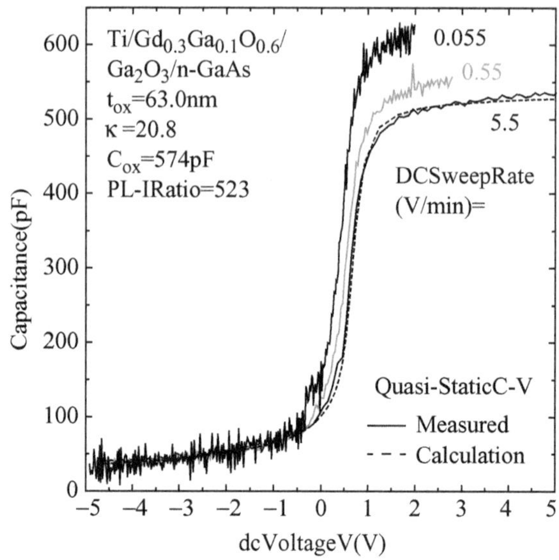

(a)

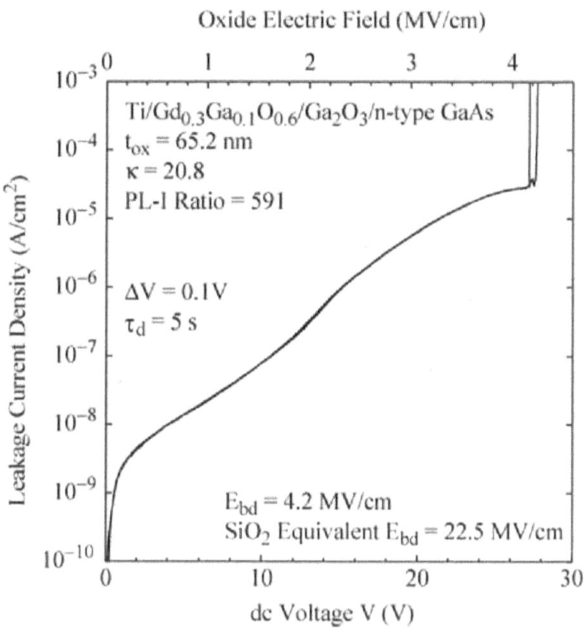

(b)

Fig. 36. Measured (a) quasi-static C–V curves of an optimized $Gd_{0.3}Ga_{0.1}O_{0.6}/Ga_2O_3$ dielectric stack on n-type GaAs (solid lines) for three different sweep rates: 5.5 V/min (fast), 0.55 V/min (medium), and 0.055 V/min (slow) in comparison with data calculated for a D_{it} of zero (dashed line). The measurement range is 20 and 2 nF for a sweep rate of 0.055 V/min and for the other sweep rates, respectively. (b) Measured leakage current as a function of dc voltage. The leakage current density is $\cong$10 nA/cm^2 at 3V. E_{bd} is the dielectric oxide breakdown field. For (a) area = 1.96× 10^{-3} cm^2, $N_D^+ = 2 \times 10^{16}$ cm^{-3}, C_{ox} = 574 pF, C_f = 322 pF, and (b) area = 4.9×10^{-4} cm^2.

Table 3. Displacement and leakage current in quasi-static measurements

						Q_{max}(As)	
Sweep	τ_d (s)	ΔV (V)	Q_{dis}(As)	Q_{dc}(As)	Q_{dc}/Q_{dis}	2 nF	20 nF
Fast	0.5	0.05	2.5×10^{-11}	9.8×10^{-12}	0.39	1×10^{-10}	1×10^{-9}
Medium	2	0.02	1.0×10^{-11}	3.9×10^{-11}	3.9	4×10^{-11}	4×10^{-10}
Slow	10	0.01	5.0×10^{-12}	2.0×10^{-10}	39	2×10^{-11}	2×10^{-10}

instruments' capacitance measurement range (2, 20 nF). Table 3 shows estimated Q_{dis}, Q_{dc}, and Q_{max} for all three sweep rates using a capacitance of 500 pF, a leakage current density of 10 nA/cm^2 (see Fig. 36(b)), and a capacitor area of 1.96×10^{-3} cm^2. It is apparent from Table 3 that Q_{dc} increases linearly with the delay time τ_d and that the ratio of Q_{dc}/Q_{dis} is inversely proportional to the sweep rate; further, Q_{max} places limits on the feasible capacitance measurement range. This has two implications: (1) The quasi-static measurement breaks down when the total charge collected exceeds the measurement limit of the C–V meter (Q_{max}) and consequently, a larger capacitance measurement range needs to be selected for slow sweep rate, and (2) $Q_{dc}/Q_{dis} \gg 1$ for slow sweep rate. Both cases may affect the accuracy of the quasi-static measurement, in particular at very slow sweep rates.

Revisiting Fig. 36(a), the breakdown of the quasi-static capacitance with decreasing sweep rate can be clearly observed: whereas a quasi-static capacitance is measured over the entire dc voltage sweep up to +5 V at a fast sweep rate of 5.5 V/min, the quasi-static capacitance measurement breaks down (instrument overload) at $\cong$+3 and $\cong$+2 V for sweep rates of 0.55 and 0.055 V/min, respectively. In the latter case, a measurement range of 20 nF had to be selected. The oxide dielectric constant is obtained by fitting the calculated accumulation branch of the C–V to the corresponding measured quasi-static curve; a k value of 20.8 is inferred when an ideal MOS model (dashed line) is fitted to the measured quasi-static C–V curve for a fast sweep rate of 5.5 V/min. Since the majority of slow border traps does not respond during fast sweep, an ideal MOS model (see Fig. 37(b)) is considered to be a reasonable assumption. However, for a slow sweep rate of 0.055 V/min, the majority of border traps does respond to the dc voltage sweep and the measured quasi-static capacitance in accumulation should approximate the oxide capacitance (see Fig. 37(c)). This is confirmed by the measured quasi-static curve for slow sweep rate (0.055 V/min) in Fig. 36(a). Note that border traps have been identified as the major component of the interface state capacitance C_{it} shown in Fig. 37 in accumulation (see also Fig. 45).

Consistent oxide dielectric constants have been measured on all film thicknesses t_{ox} investigated ($32.8 \leq t_{ox} \leq 65.5$ nm). Figure 38 shows a detail of the measured 100 kHz C–V for slow sweep rate (solid line) in comparison with data calculated for a D_{it} of zero, a metal workfunction of 4.73 eV, and an ionized donor concentration of 2×10^{16} cm^{-3} (dashed line). Although the ionized donor concentration could be easily obtained by fitting the ideal C–V data to the measured C–V in steady-state deep depletion, the ionized donor concentration is again treated as an input parameter for

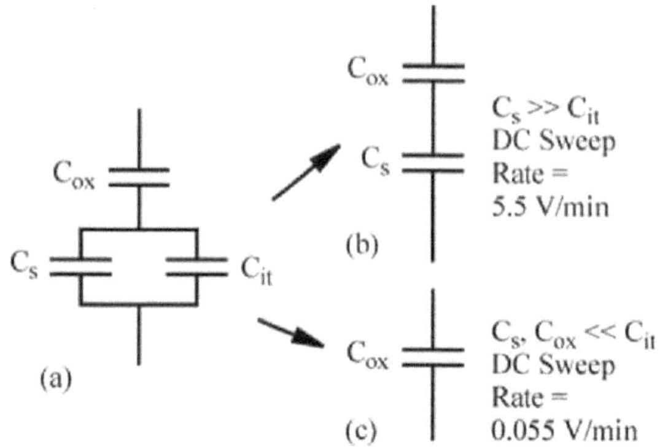

Fig. 37. Equivalent circuits of (a) a MOS capacitor, (b) simplification for $C_s \gg C_{it}$ (ideal MOS capacitor), and (c) simplification for C_s, $C_{ox} \ll C_{it}$ in accumulation. C_s and C_{it} are the semiconductor capacitance (depletion, accumulation capacitance) and the capacitance due to interface states, respectively.

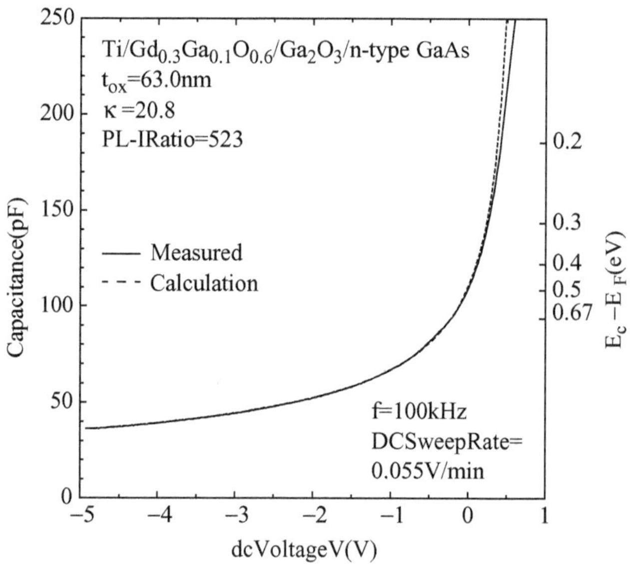

Fig. 38. Detail of the measured 100 kHz C–V for slow sweep rate (solid line) in comparison with data calculated for a D_{it} of zero and a metal workfunction of 4.73 eV (dashed line). (area = 1.96×10^{-3} cm^2, $N_D^+ = 2 \times 10^{16}$ cm^{-3}, $C_{ox} = 574$ pF, $C_f = 322$ pF).

C–V data interpretation. Consequently, the GaAs ionized donor concentrations are obtained independently by electrochemical depth profiling and measure 2–2.5 $\times 10^{16}$ cm^{-3} for the $Gd_xGa_{0.4-x}O_{0.6}/Ga_2O_3/GaAs$ structures discussed in this section.

Figure 39 shows measured quasi-static and 100 kHz C–V curves of the high-κ $Gd_xGa_{0.4-x}O_{0.6}/Ga_2O_3$ dielectric stack of Fig. 36 for the sweep rates of

(a) 5.5 V/min, (b) 0.55 V/min, and (c) 0.055 V/min. It is apparent from Fig. 39 that the slow sweep rate of 0.055 V/min ($\tau_d = 10$ s) is required to detect states in the vicinity of E_i, if faster sweep rates are used $C_{qs} \cong C_{100\,kHz}$ and artificially low D_{it} numbers are obtained by the quasi-static/high frequency technique as discussed

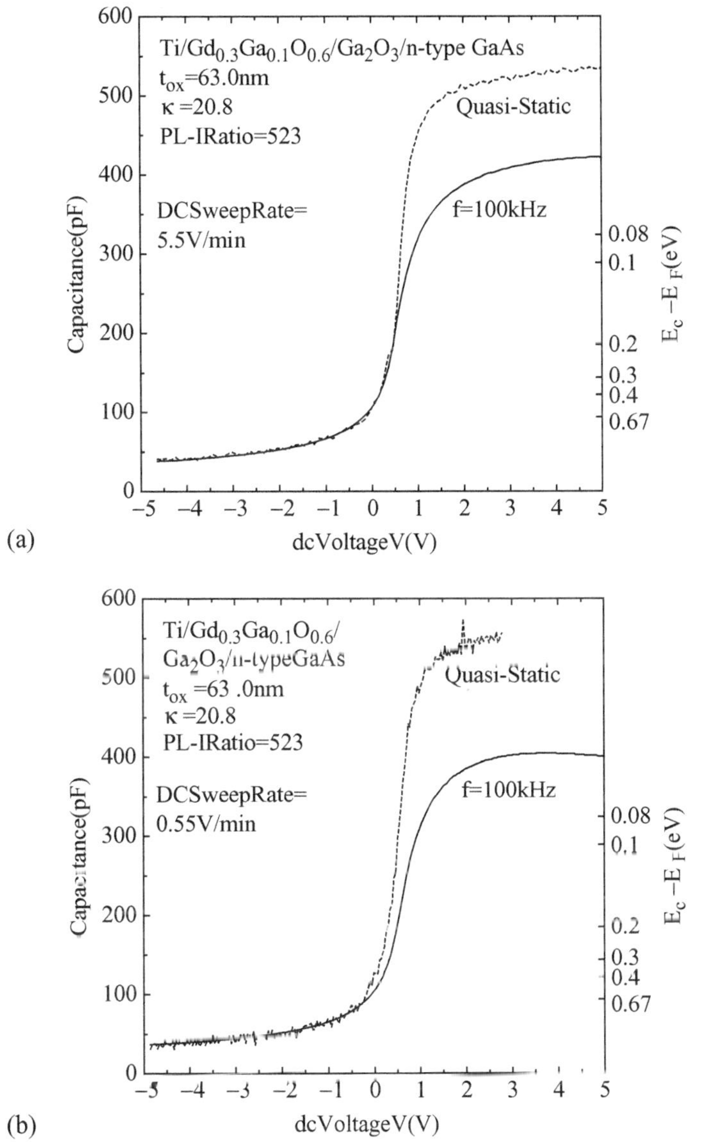

Fig. 39. Measured quasi-static and 100 kHz C–V curves of the high-k $Gd_{0.3}Ga_{0.1}O_{0.6}/Ga_2O_3$ dielectric stack of Fig. 36 for the sweep rates of (a) 5.5 V/min, (b) 0.55 V/min, and (c) 0.055 V/min. The metal workfunction of the calculated curve in (c) is 4.73 eV.

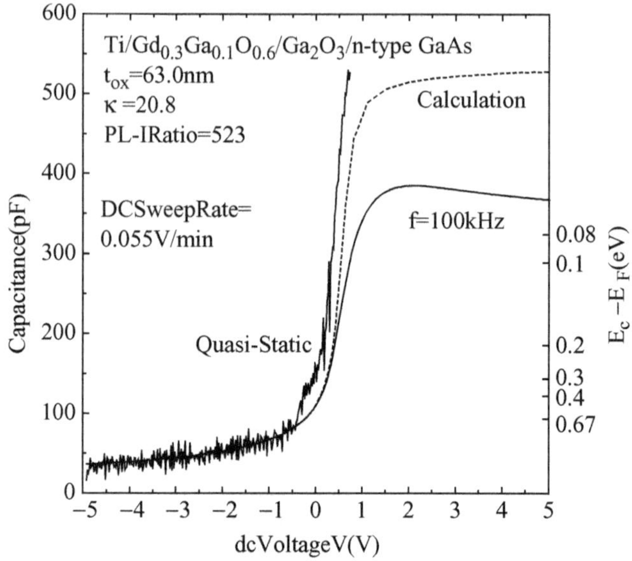

(c)

Fig. 39. *(continued)*

further below. Note that the capture cross sections derived by PL-I for Ga_2O_3/GaAs interfaces provide emission/capture time constants for traps located at E_i in between 8 and 19 s. It is therefore reasonable to assume that the majority of states located at and in the vicinity of E_i are reflected in the quasi-static capacitance at the slow sweep rate with $\tau_d = 10$ s. Figure 40 shows a direct comparison of measured 100 kHz C–V curves of the high-κ $Gd_xGa_{0.4-x}O_{0.6}/Ga_2O_3$ dielectric stack of Fig. 36 for slow (solid line) and fast (dashed line) sweep rate. Since virtually no sweep rate dispersion is observed for $C_{100\,kHz} < C_f$ and $C_{100\,kHz} \cong C_f$, it is concluded that quasi-equilibrium prevails in GaAs in depletion and around the flatband condition. The sweep dispersion observed in accumulation is attributed to border traps as discussed further below in more detail. Figure 41 shows a direct comparison of C–V curves measured at slow sweep rate of the high-κ $Gd_xGa_{0.4-x}O_{0.6}/Ga_2O_3$ dielectric stack of Fig. 36 for a frequency of 100 kHz (solid line) and 1 MHz (dashed line). The complete absence of frequency dispersion indicates that a high frequency C–V curve where only the GaAs majority carriers respond to the ac signal is indeed obtained over the entire dc voltage range. Consequently, all requirements for correct C–V data interpretation and D_{it} extraction are met including (1) quasi-equilibrium conditions during C–V sweep, (2) provision of actual high frequency C–V data and an accurate oxide dielectric constant. The final step of C–V data interpretation, D_{it} extraction, is discussed in the following.

Figure 42 shows the D_{it} distribution in the upper part of the GaAs bandgap as obtained from optimized high-κ $Gd_xGa_{0.4-x}O_{0.6}/Ga_2O_3$ dielectric stacks on n-type GaAs using the quasi-static/high frequency and the Terman techniques for the sweep

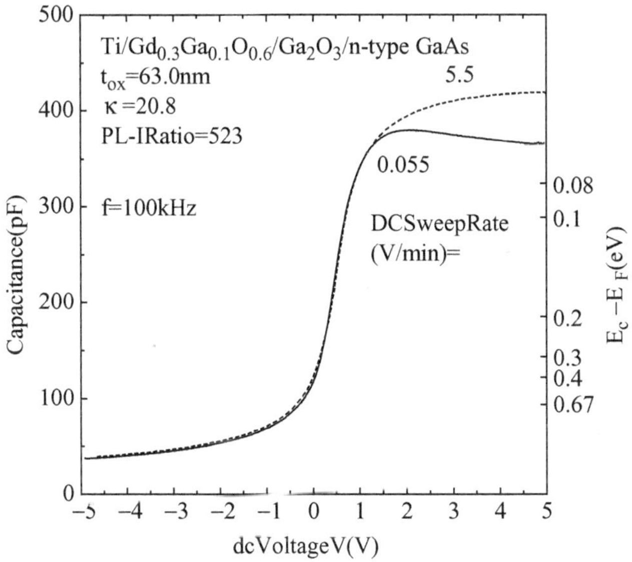

Fig. 40. Comparison of measured 100 kHz *C–V* curves of the high-κ $Gd_{0.3}Ga_{0.1}O_{0.6}/Ga_2O_3$ dielectric stack of Fig. 36 for slow (solid line) and fast (dashed line) sweep rate.

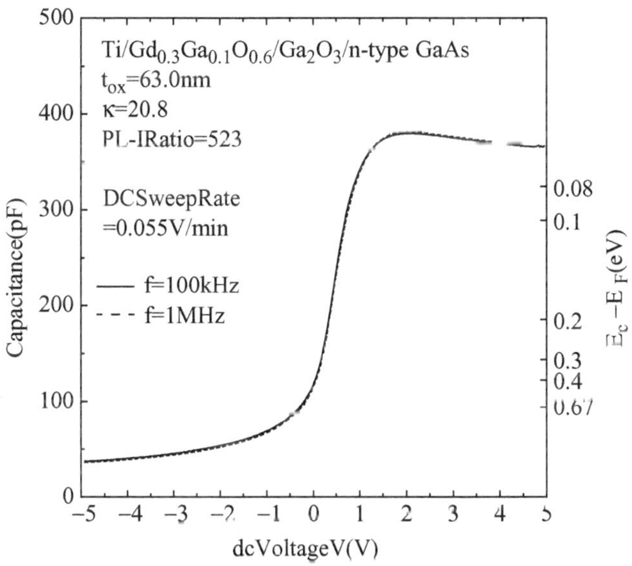

Fig. 41. Comparison of *C–V* curves measured at slow sweep rate of the high-κ $Gd_{0.3}Ga_{0.1}O_{0.6}/Ga_2O_3$ dielectric stack of Fig. 36 for a frequency of 100 kHz (solid line) and 1 MHz (dashed line).

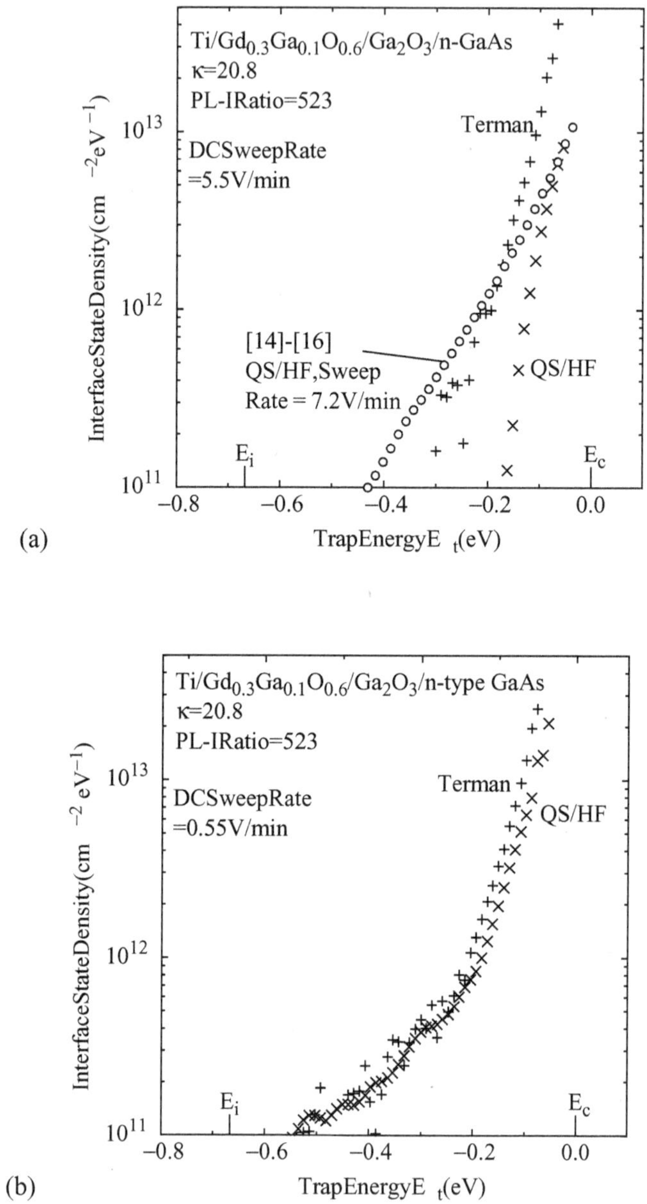

Fig. 42. D_{it} distribution in the upper part of the GaAs bandgap as obtained from optimized high-κ $Gd_{0.3}Ga_{0.1}O_{0.6}/Ga_2O_3$ dielectric stacks on n-type GaAs using the quasi-static/high frequency and the Terman techniques for the sweep rates of (a) 5.5 V/min, (b) 0.55 V/min, and (c) 0.055 V/min.

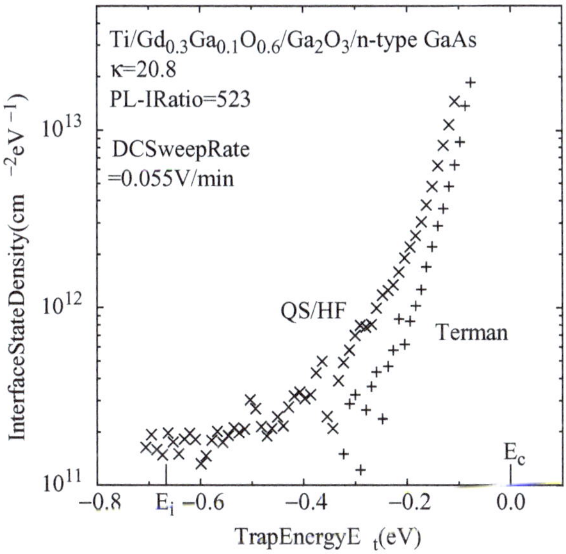

(c)

Fig. 42. *(continued)*

rates of (a) 5.5 V/min, (b) 0.55 V/min, and (c) 0.055 V/min. When compared to D_{it} derived using Terman's method, the D_{it} obtained by the quasi-static/high frequency method is less, comparable, and larger for fast, medium, and slow sweep rate, respectively. It is also obvious that interface states in the vicinity of E_i can only be detected by the quasi-static/high frequency method and at very slow sweep rate (0.055 V/min). The midgap interface state density thus obtained is $\simeq 1.5 \times 10^{11}$ cm^{-2} eV^{-1}. Note that Fig. 42(a) also shows a D_{it} distribution obtained from re-evaluating earlier published data [14] by applying the above discussed C–V analysis methodology. A k value of 19.1 is obtained which is comparable to the oxide dielectric constants reported in this chapter but higher than 14.2 which was originally derived [14]. Further, the sweep rate used in [14] was very fast with 7.2 V/min and the D_{it} obtained from [14] and depicted in Fig. 42(a) is significantly higher than the D_{it} of the high-κ $Gd_xGa_{0.4-x}O_{0.6}/Ga_2O_3$ dielectric stacks reported in this chapter based on an identical extraction method (quasi-static/high frequency) and a comparable sweep rate.

Figure 43 shows a comparison of the D_{it} distributions in the upper part of the GaAs bandgap as obtained from optimized high-κ $Gd_xGa_{0.4-x}O_{0.6}/Ga_2O_3$ dielectric stacks on n-type GaAs using different sweep rates obtained from (a) the quasi-static/high frequency technique, and (b) the Terman method. As expected, more interface states are detected by the quasi-static/high frequency technique when the sweep rate is lowered, this applies in particular to states located at E_i and in its vicinity. In contrast, the D_{it} distribution obtained by the Terman method is largely independent of sweep rate because detectable states are confined to the vicinity of the band edge where trap time constants are short. Also nicely visible in Fig. 43(b) is the detection limit

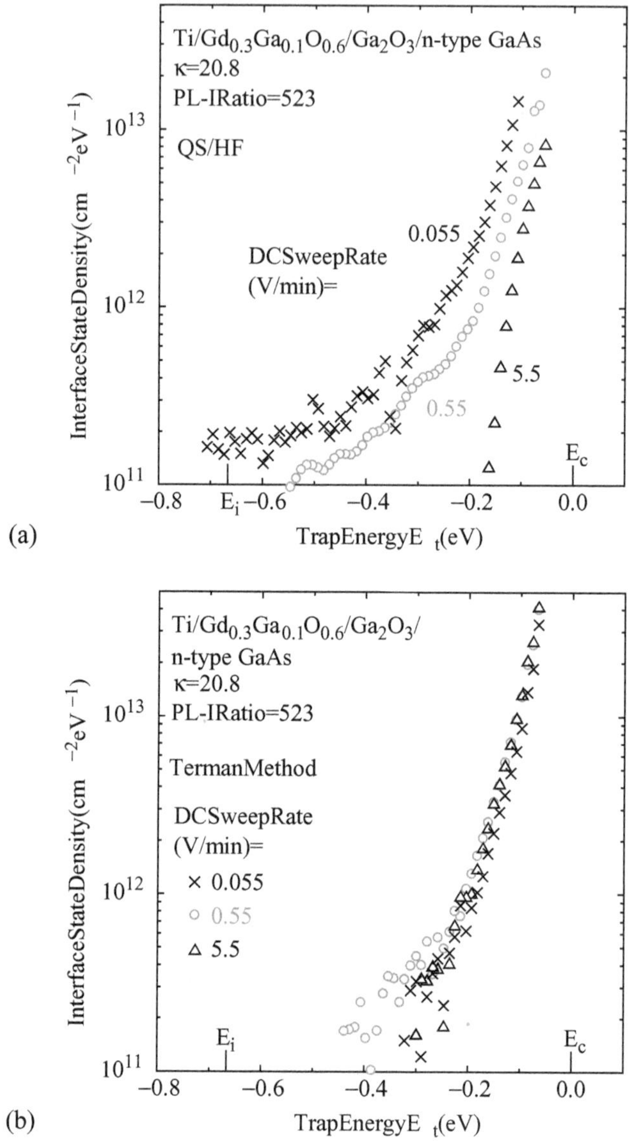

Fig. 43. D_{it} distributions in the upper part of the GaAs bandgap as obtained from optimized high-κ $Gd_{0.3}Ga_{0.1}O_{0.6}/Ga_2O_3$ dielectric stacks on n-type GaAs using different sweep rates obtained from (a) the quasi-static/high frequency technique, and (b) the Terman method.

of the Terman method of mid 10^{11} cm^{-2} eV^{-1}. Summarizing the data depicted in Figs. 42 and 43, the most realistic D_{it} distribution in the upper half of the GaAs bandgap probably comprises the D_{it} midgap plateau of $\cong 1.5 \times 10^{11}$ cm^{-2} eV^{-1} (quasi-static/high frequency method, sweep rate of 0.055 V/min) and the D_{it}

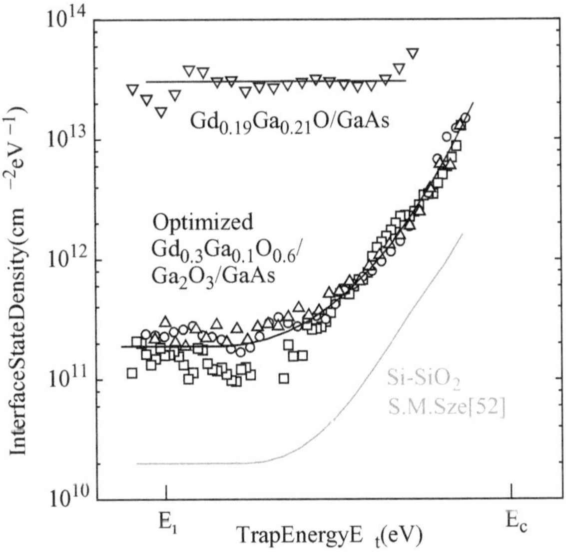

Fig. 44. D_{it} distributions of optimized high-κ $Gd_{0.3}Ga_{0.1}O_{0.6}/Ga_2O_3$ dielectric stacks on n-type GaAs, $Gd_{0.19}Ga_{0.21}O_{0.6}$/n-GaAs structures, and of a generic SiO_2/Si interface in the upper half of the semiconductor bandgap.

distribution obtained by the Terman method above 5×10^{11} cm^{-2} eV^{-1} at any sweep rate. Note that the D_{it} distributions obtained by Terman's method (Fig. 43(b)) are comparable to the D_{it} distribution of the quasi-static/high frequency technique for medium sweep rate (Fig. 42(b)); the quasi-static/high frequency technique apparently underestimates D_{it} for fast sweep (Fig. 42(a)) and overestimates D_{it} for slow sweep (Fig. 42(c)) above 5×10^{11} cm^{-2} eV^{-1}. The latter may be due to the imminent breakdown of the quasi-static capacitance at low sweep rate (see Fig. 39(c)). Finally, Fig. 44 summarizes D_{it} distributions of optimized high-κ $Gd_{0.3}Ga_{0.1}O_{0.6}/Ga_2O_3$ dielectric stacks on n-type GaAs, $Gd_{0.19}Ga_{0.21}O_{0.6}$/n-GaAs structures, and of a generic SiO_2/Si interface in the upper half of the semiconductor bandgap. Note that the curve obtained for $Gd_{0.19}Ga_{0.21}O_{0.6}$/n-GaAs structures constitutes a lower limit since real high frequency C–V data could not be obtained for this structure.

Eventually, the issue of border traps shall be discussed. Border traps become visible mainly above the flatband condition of $E_c - E_F = 79$ meV ($N_D^+ - 2 \times 10^{16}$ cm^{-3}) and consequently, these traps are only partially reflected by the D_{it} distributions depicted earlier in this chapter. These traps can be assessed by hysteresis measurements of the ac capacitance as illustrated in Fig. 45 where the dc voltage is swept up and subsequently swept down and the 100 kHz capacitance is monitored. Typically, a clockwise hysteresis indicative of trap densities above 10^{12} cm^{-2} is obtained when the dc voltage is swept to +5 V, however, a small counterclockwise hysteresis (average hysteresis = 0.11 V, equivalent to a charge of 2×10^{11} cm^{-2}) is observed when the dc voltage is only swept up to flatband at a low sweep rate of

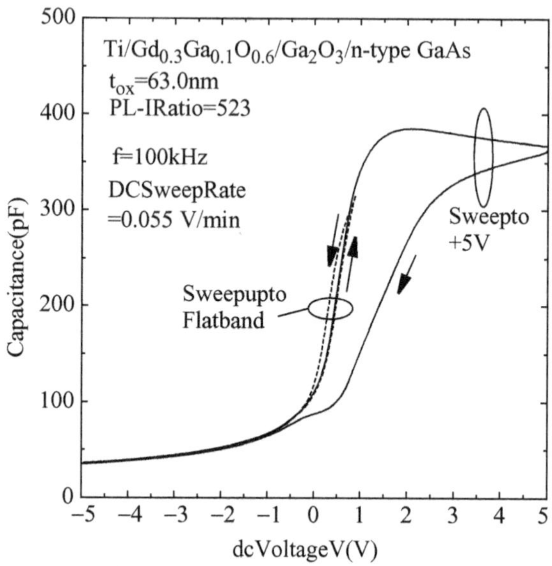

Fig. 45. The dc voltage is swept up and subsequently swept down and the 100 kHz capacitance is monitored. Typically, a clockwise hysteresis indicative of trap densities above 10^{12} cm^{-2} is obtained when the dc voltage is swept to +5 V, however, a small counterclockwise hysteresis (average hysteresis = 0.11 V, equivalent to a charge of 2×10^{11} cm^{-2}) is observed when the dc voltage is only swept up to flatband at a low sweep rate of 0.055 V/min.

0.055 V/min. Note that all interface states below the Fermi level, irrespective of their energy, must be occupied when flatband is reached because capture time constants of these interface states are of the order of nanoseconds for $E_F \cong E_c$. When the dc voltage is swept back down, trapped electrons will cause a clockwise hysteresis since the trap emission time constants depend exponentially on the traps'energy positions in the bandgap. Because such a clockwise hysteresis is not observed when the dc voltage is swept up to flatband and back down, charge trapping due to interface states is small and not the dominant mechanism below flatband and other sources producing a counterclockwise hysteresis such as mobile ions need to be considered. It is further understood that capture and emission time constants of interface traps are of the order of nanoseconds or less above flatband and thus, are not the origin of the observed hysteresis when the dc voltage is swept to +5 V. Considering that the dc sweep above 0 V takes about 3 hours when the dc voltage is swept up to +5 V, the observed traps above flatband must have time constants of the order of minutes or even hours. Such time constants are easily observed for border traps with a separation as small as 1 nm from the interface. Note that border traps have very small effective capture cross section and therefore, are not observed by PL-I. The issue of border traps should be either addressed by further improvements of the material system or device designs which alleviate the effect of border traps.

5. CONCLUSIONS

A methodology for the development of gate dielectrics on III–V semiconductors has been proposed. This methodology involves three levels: (1) atomic scale engineering using oxide molecules with (sub)monolayer coverage of the semiconductor surface, (2) engineering of the electrical properties of the immediate oxide/semiconductor interface using oxide templates, and (3) engineering of the bulk properties of the high-κ gate dielectric in conjunction with the oxide template.

This methodology has allowed for the separate and independent control of (a) the oxide/semiconductor interface properties and (b) the bulk properties of the high-κ dielectric bulk film while (c) maintaining the requirements common to all necessary components. This approach has further become necessary due to largely different and sometimes mutually exclusive requirements for each of the components in high-κ dielectric stacks. A summary and a first attempt to extend the proposed methodology beyond GaAs is presented in Table 4. Note that the intent here is to point out possibilities and not to provide a complete picture. The application of the proposed methodology has led to a manufacturable process for high-κ stacked gate oxides on

Table 4. Proposed methodology for development of high-κ gate dielectrics stacks

	GaAs (this work)	Elemental semiconductor Ge, Si etc.	Other compound semiconductors P, N, Sb based etc.
Atomic Level			
Material	Ga_2O	Oxide molecules manufactured by non-reactive deposition technique	Ga_2O (similar surface chemistry), other?
Manufacturing	MBE	UHV, MBE	UHV, MBE
Characterization	STM, STS, LEED[(a)], RHEED	STM, STS, XPS[(b)], LEED, RHEED	STM, STS, XPS, LEED, RHEED
Template			
Material	Ga_2O_3	SiO_x, HfO, other	Ga_2O_3, other
Manufacturing	MBE	ALD[(c)], CVD[(d)], MBE etc.	UHV, MBE, ALD, CVD
Characterization	PL-I, C–V	Applicable recombination based techniques (e.g. [53], [54]), XPS	PL-I, C–V, XPS
Bulk			
Material	$Gd_xGa_{0.4-x}O_{0.6}$	Compatible with template (diffusion, second interface etc.)	Compatible with template (diffusion, second interface etc.)
Manufacturing	MBE	ALD, CVD, MBE etc.	UHV, MBE, ALD, CVD
Characterization	C–V, PL-I	C–V, MOSFET	C–V, PL-I, MOSFET

[(a)] low energy electron diffraction
[(b)] x-ray photoelectron spectroscopy
[(c)] atomic layer deposition
[(d)] chemical vapor deposition

GaAs with a broad minimum of interface state density $D_{it} \leq 2 \times 10^{11}$ cm^{-2} eV^{-1} on n-type GaAs suggesting a U-shaped D_{it} distribution, an oxide relative dielectric constant of 20.8 ± 1, and leakage currents of ≅2 × 10^{-8} A/cm^2 at an electric field of 1 MV/cm (SiO_2 equivalent field = 5.3 MV/cm).

ACKNOWLEDGEMENTS

This work would not have been possible without numerous contributions, inspirations, and encouragement by others. The gate oxide and epitaxial layer growth by D.W. Braddock and his team at Osemi, Inc. as well as by Z. Yu, R. Droopad, and C. Overgaard of Motorola should be specifically acknowledged. Prof. A.C. Kummel and his team (M.J. Hale, S.-I. Yi, J.Z. Sexton, T.J. Grassman, D.L. Winn) at the University of San Diego have provided one of the pillars of the proposed methodology: the investigation of the oxide/GaAs interface on the atomic level. A. Kummel further contributed to the development and understanding of hydrogen passivation of Ga_2O_3/GaAs interfaces. N. Medendorp (formerly Motorola) made major contributions to the development of oxide evaporation techniques and gate oxide processing; R. Gregory of Motorola has provided the RBS data. I would like to acknowledge the contributions made by many engineers and technicians of Motorola's Microelectronics and Physical Sciences Laboratories and Process and Materials Characterization Lab. I also would like to thank T. Büyüklimanli of Evans East for the significant effort he undertook to provide SIMS data for this new material system. T.D. Harris and G. Zydzik, both formerly of AT&T Bell Laboratories, who were instrumental in planting the early seeds for interface characterization via photoluminescence and for Ga_2O_3 deposition on GaAs, respectively, are especially acknowledged. Finally, I would like to thank K. Johnson and J.K. Abrokwah for their firm support. This chapter would not have been written without P. Maniar's constant encouragement and support. This work was supported in part by the Department of Defense (Grant No. MDA904-93-C-L042).

REFERENCES

1. W.F. Croydon, E.H.C. Parker, *Dielectric Films on Gallium Arsenide* (Gordon and Breach Scientific Publishers: New York, 1981).
2. *Physics and Chemistry of III–V Compound Semiconductor Interfaces*. ed. C.W. Wilmsen (Plenum Press: New York, 1985).
3. H. Becke, R. Hall, J. White, Gallium arsenide MOS transistors, *Solid-State Electron* **8**, 813–823 (1965).
4. T. Ito, Y. Sakai, The GaAs inversion-type MIS transistors, *Solid-State Electron* **17**(7), 751–759 (1974).
5. T. Mimura, K. Odani, N. Yokoyama, Y. Nakayama, M. Fukuta, GaAs microwave MOSFET's, *IEEE Trans. Electron Devices* **25**(6), 573–579 (1978).

6. G.G. Fountain, R.A. Rudder, S.V. Hattangady, R.J. Markunas, J.A. Hutchby, Demonstration of an n-Channel Inversion Mode GaAs MISFET, in: *IEDM Tech. Dig.* Dec. 1989, pp. 887–889.
7. A. Colquhoun, E. Kohn, H.L. Hartnagel, Improved enhancement/depletion GaAs MOSFET using anodic oxide as the gate insulator, *IEEE Trans. Electron Devices* **25**(3), 375–376 (1978).
8. T. Mimura, K. Odani, N. Yokoyama, M. Fukuta, New structure of enhancement-mode GaAs microwave M.O.S.F.F.E.T., *Electron. Lett.* **14**(16), 500–502 (1978).
9. K. Kamimura Y. Sakai, The properties of GaAs-Al_2O_3 and InP-Al_2O_3 interfaces and the fabrication of MIS field-effect transistors, *Thin Solid Films* **56**, 215–223 (1979).
10. B. Bayraktaroglu, E. Kohn, H.L. Hartnagel, First anodic-oxide GaAs M.O.S.F.E.T. S based on easy technological processes, *Electron. Lett.* **12**(2), 53–54 (1976).
11. W.E. Spicer, I. Lindau, P. Skeath, C.Y. Su, P. Chye, Unified defect model and beyond, *J. Vac. Sci. Technol.* **17**, 1019–1027 (1980).
12. P. Pianetta, I. Lindau, C.M. Garner, W.E Spicer, Photoemission studies of the initial stages of oxidation of GaSb and InP, *Surf. Sci.* **88**, 439–460 (1979).
13. A. Callegari, P.D. Hoh, D.A. Buchanan, D. Lacey, Unpinned gallium oxide/GaAs interface by hydrogen and nitrogen surface plasma treatment, *Appl. Phys. Lett.* **54**(4), 332–334 (1989).
14. M. Passlack, M. Hong, J.P. Mannaerts, Quasi-static and high frequency capacitance-voltage characterization of Ga_2O_3–GaAs structures fabricated by in-situ molecular-beam epitaxy, *Appl. Phys. Lett.* **68**(8), 1099–1101 (1996).
15. M. Passlack, M. Hong, J.P. Mannaerts, S.N.G. Chu, R.L. Opila, N. Moriya, *In-situ* Ga_2O_3 process for GaAs inversion/accumulation device and surface passivation applications, in: *IEDM Tech. Dig.* Dec. 1995, pp. 383–386.
16. M. Passlack, M. Hong, J.P. Mannaerts, R.L. Opila, S.N.G. Chu, N. Moriya, F. Ren, J.R. Kwo Low D_{it}, thermodynamically stable Ga_2O_3–GaAs interfaces: Fabrication, characterization, and modeling, *IEEE Trans. Electron. Dev.* **ED-44**(2), 214–225 (1997).
17. Z. Yu, R. Droopad, C. Overgaard, M. Passlack, J.K. Abrokwah, Growth and physical properties of Ga_2O_3 thin films on GaAs(001) substrate by molecular beam epitaxy, *Appl. Phys. Lett.* **82**(18), 2978–2980 (2003).
18. M. Passlack, J.K. Abrokwah, Z. Yu, R. Droopad, C. Overgaard, H. Kawayoshi, Thermally induced oxide crystallinity and interface destruction in Ga_2O_3–GaAs structures, *Appl. Phys. Lett.* **82**(11), 1691–1693 (2003).
19. M. Passlack, Z. Yu, R. Droopad, J.K. Abrokwah, D. Braddock, S.-I. Yi, M. Hale, J. Sexton, A.C. Kummel, *III–V Semiconductor Heterojunctions: Physics and Devices* ed. W.Z. Cai, (Transworld Research Publisher: Kerala, India, 2003), pp. 327–355.
20. M. Passlack, J.K. Abrokwah, R. Droopad, Z. Yu, C. Overgaard, S.-I. Yi, M. Hale, J. Sexton, A.C. Kummel, Self-aligned GaAs p-channel heterostructure field effect transistor, *IEEE Electron Device Lett.* **23**(9), 508–510 (2002).
21. J. Kwo, D.W. Murphy, M. Hong, J. P. Mannaerts, R.L. Opila, R.L. Masaitis, A.M. Sergent, Passivation of GaAs using gallium–gadolinium oxides, *J. Vac. Sci. Technol.* **B17**(3), 1294–1297 (1999).
22. M. Passlack, N. Medendorp, R. Gregory, S. Zollner, D. Braddock, Optical and electrical properties of amorphous $Gd_xGa_{0.4-x}O_{0.6}$ films in $Gd_xGa_{0.4-x}O_{0.6}/Ga_2O_3$ gate dielectric stacks on GaAs, submitted for publication.
23. M. Passlack, N. Medendorp, R. Gregory, D. Braddock, The role of Ga_2O_3 template thickness and gadolinium mole fraction in $Gd_xGa_{0.4-x}O_{0.6}/Ga_2O_3$ gate dielectric stacks on GaAs, *Appl. Phys. Lett.* **83**(25), 5262–5264 (2003).
24. M. Passlack, R. Droopad, Z. Yu, C. Overgaard, B. Bowers, J. Abrokwah, Nonradiative recombination at GaAs homointerfaces fabricated using an As cap deposition/removal process, *Appl. Phys. Lett.* **72**(24), 3163–3165 (1998).

25. M. Hale, S.I. Yi, J.Z. Sexton, A.C. Kummel, M. Passlack, Scanning tunneling microscopy and spectroscopy of gallium oxide deposition and oxidation on GaAs(001)-c(2 × 8)/(2 × 4), *J. Chemical Physics* **119**(13), 6719–6728 (2003).
26. M. Passlack, R.N. Legge, D. Convey, Z. Yu, J.K. Abrokwah, Optical measurement system for characterizing compound semiconductor interface and surface states, *IEEE Trans. Instrum. Meas.* **47**(5), 1362–1366 (1998).
27. T. Sawada, K. Numata, S. Tohdoh, T. Saitoh, H. Hasegawa, *In-situ* characterization of compound semiconductor surfaces by novel photoluminescence surface state spectroscopy, *Jpn. J. Appl. Phys.* **32**, Part 1, 511–517 (1993).
28. E.H. Nicollian J.R. Brews, *MOS Physics and Technology* (Wiley: New York, 1982).
29. L.M. Terman, An investigation of surface states at a silicon/silicon oxide interface employing metal-oxide-silicon diodes, *Solid-State Electron.* **5**(5), 285–299 (1962).
30. D. Amey, S.J. Horowitz, High frequency characterization of advanced ceramic materials, *Proc. Ann. IEPS Conf.* 573–582 (1996).
31. See, for example, M.S. Lundstrom R.J. Schuelke, Numerical analysis of heterostructure semiconductor devices, *IEEE Trans. Electron. Dev.* **ED-30**(9), 1151–1159 (1983).
32. M. Passlack, M. Hong, E.F. Schubert, G.J. Zydzik, J.P. Mannaerts, W.S. Hobson, T.D. Harris, Advancing metal–oxide–semiconductor theory: Steady-state nonequilibrium conditions, *J. Appl. Phys.* **81**, 7647–7661 (1997).
33. M. Passlack, Z. Yu, R. Droopad, B. Bowers, C. Overgaard, J. Abrokwah, A.C. Kummel, Interface charge and nonradiative carrier recombination in Ga_2O_3–GaAs interface structures, *J. Vac. Sci. Technol.* **B17**(1), 49–52, (Jan./Feb. 1999).
34. See, for example, A.S. Grove, *Physics and Technology of Semiconductor Devices* (John Wiley and Sons: New York, 1967).
35. R.K. Ahrenkiel, in *Minority Carriers in III–V Semiconductors: Physics and Applications* ed. R.K. Ahrenkiel and M.S. Lundstrom, *Semiconductors and Semimetals,* Vol. **39**, (Academic: Boston, MA, 1993), pp. 119–141.
36. P.D. Ye, G.D. Wilk, B. Yang, J. Kwo, S.N.G. Chu, S. Nakahara, H.-J.L. Gossmann, J.P. Mannaerts, M. Hong, K.K. Ng, J. Bude, GaAs metal–oxide–semiconductor field-effect transistor with nanometer thin dielectric grown by atomic layer deposition, *Appl. Phys. Lett.* **83**(1), 180–182 (2003).
37. P.D. Ye, G.D. Wilk, J. Kwo, B. Yang, H.-J.L. Gossmann, M. Frei, S.N.G. Chu, J.P. Mannaerts, M. Sergent, M. Hong, K.K. Ng, J. Bude, GaAs MOSFET with oxide gate dielectric grown by atomic layer deposition, *IEEE Electron Device Lett.* **24**(4), 209–211 (2003).
38. B. Yang, P.D. Ye, J. Kwo, M.R. Frei, H.-J.L. Gossmann, J.P. Mannaerts, M. Sergent, M. Hong, K.K. Ng, J. Bude, DC and RF characteristics of depletion-mode GaAs MOSFET employing a thin $Ga_2O_3(Gd_2O_3)$ gate dielectric layer, *Proc. IEEE GaAs IC Symp.* 139 (2002).
39. M. Tao, D. Udeshi, N. Basit, E. Maldonado, W.P. Kirk, Removal of dangling bonds and surface states on silicon (001) with a monolayer of selenium, *Appl. Phys. Lett.* **82**(10), 1559–1561 (2003).
40. E. Kaxiras, Semiconductor-surface restoration by valence-mending adsorbates: Application to Si(100):S and Si(100):Se, *Phys. Rev. B* **43**(8), 6824–6827 (1991).
41. M. Passlack, M. Hong, J.P. Mannaerts, T.H. Chiu, C.A. Mendonca, J.C. Centanni, Capacitance-voltage and current-voltage characterization of $Al_xGa_{1-x}As$-GaAs structures, *J. Appl. Phys.* **78**, 7091–7098 (1995).
42. M. Passlack, M. Hong, J.P. Mannaerts, J.R. Kwo, L.W. Tu, Recombination velocity at oxide-GaAs interfaces fabricated by in situ molecular beam epitaxy, *Appl. Phys. Lett.* **68**(25), 3605–3607 (1996).
43. J.H. Stathis, E. Cartier, The role of atomic hydrogen in degradation and breakdown of SiO_2 films, *Extended Abstracts 1996 International Conference on Solid State Devices and Materials* pp. 791–793, (1996).

44. M.C. Hersam, N.P. Guisinger, J. Lee, K. Cheng, J.W. Lyding, Variable temperature study of the passivation of dangling bonds at Si(100)-2 $\times$ 1 reconstructed surfaces with H and D, *Appl. Phys. Lett.* **80**(2), 201–203 (2002).
45. J.R. Chavez, R.A.B. Devine, W.M. Shedd, Radiation sensitivity reduction in deuterium annealed Si–SiO_2 structures, *Appl. Phys. Lett.* **80**(2), 213–215 (2002).
46. J. Senzaki, K. Kojima, S. Harada, R. Kosugi, S. Suzuki, T. Suzuki, K. Fukuda, Excellent effects of hydrogen postoxidation annealing on inversion channel mobility of 4H-SiC MOSFET fabricated on (1120) face, *IEEE Electron Device Lett.* **23**(1), 13–15 (2002).
47. T. Ngai, X. Chen, J. Chen, S.K. Banerjee, Improving SiO_2/SiGe interface of SiGe p-metal–oxide–silicon field-effect transistors using water vapor annealing, *Appl. Phys. Lett.* **80**(10), 1773–1775 (2002).
48. R.J. Browne, E.A. Ogryzlo, K. Eisenbeiser, Z. Yu, R. Droopad, C.D. Overgaard, Passivation of defects at the $SrTiO_3$/Si interface with H and H_2, *Appl. Phys. Lett.* **80**(15), 2699–2700 (2002).
49. E.S. Aydil, K.P. Giapis, R.A. Gottscho, V.M. Donnelly, E. Yoon, Ammonia plasma passivation of GaAs in downstream microwave and radio-frequency parallel plate plasma reactors, *J. Vac. Sci. Technol.* **B11**(2), 195–205 (1993).
50. J. Kwo, M. Hong, A.R. Kortan, D.W. Murphy, J. P. Mannaerts, A.M. Sergent, Y.C. Wang, K.C. Hsieh, The $(Ga_2O_3)_{1-x}(Gd_2O_3)_x$ Oxides with $x = 0$–1.0 for GaAs Passivation, *Mater. Res. Soc. Symp.* **573**, 57–67 (1999).
51. Model 595 Quasistatic CV Meter, *Instruction Manual* (Keithley Instruments Cleveland, OH, 1986), p. 3-3.
52. S.M. Sze, *Physics of Semiconductor Devices* (Wiley: New York, 1981), p. 385.
53. E. Yablonovitch, D.L. Atlara, C.C. Chang, T. Gmitter, T.B. Bright, Unusually low surface-recombination velocity on silicon and germanium surfaces, *Phys. Rev. Lett.* **57**(2), 249–252 (1986).
54. M.L. Green, J.-U. Sachse, G. Higashi, L.C. Feldman, T. Boone, D. Brasen, B.E. Weir, Rapid technique for determination of Si/SiO_2 and Si/Si-O-N interface state densities based on measurement of recombination lifetimes, *The Physics and Chemistry of SiO_2 and the Si-SiO_2 Interface* Vol. 96-1 (The Electrochemical Society: Pennington, NJ, 1996), pp. 555–567.

INDEX

Kolibri 360 GmbH
Ferdinand-Jühlke-Straße 7
99095 Erfurt, Deutschland
produktsicherheit@kolibri360.de